Introduction to Biomass Energy Conversions

Introduction to Biomass Energy Conversions explores biomass energy conversions and characterization using practical examples and real-world scenarios. It begins with biomass resource estimation and extends to commercialization pathways for economical biomass conversion into high-value materials, chemicals, and fuels.

With extended discussions of new sustainability issues in biofuels production, such as carbon capture and sequestration, the second edition has been updated with carbon footprint work, life cycle analysis, the growing circular economy, and newer research directions of biomass resources, such as graphene production from biochar. This book covers thermo-chemical conversion processes, including torrefaction, pyrolysis, gasification and advanced gasification, biomass liquefaction, and combustion.

This book is intended for senior undergraduate students taking Renewable Energy Conversions, Bio Energy, Biomass Energy, Introduction to Biofuels, and Sustainability Engineering courses.

This book also features end-of-chapter problems, exercises, and case studies with a Solutions Manual available for instructors.

Introduction to Biomass Energy Conversions

Second Edition

Sergio C. Capareda

CRC Press
Taylor & Francis Group
Boca Raton London New York

CRC Press is an imprint of the
Taylor & Francis Group, an **informa** business

Second edition published 2023
by CRC Press
6000 Broken Sound Parkway NW, Suite 300, Boca Raton, FL 33487-2742

and by CRC Press
4 Park Square, Milton Park, Abingdon, Oxon, OX14 4RN

CRC Press is an imprint of Taylor & Francis Group, LLC

First edition published by CRC Press 2013

ISBN: 978-1-032-27833-9 (hbk)
ISBN: 978-1-032-27837-7 (pbk)
ISBN: 978-1-003-29430-6 (ebk)
ISBN: 978-1-032-27855-1 (eBook+)

DOI: 10.1201/9781003294306

Typeset in Times
by codeMantra

Access the Instructor and Student Resources: https://routledgetextbooks.com/textbooks/instructor_downloads/

I dedicate this book to my loving family—my wife Evamarie, daughter Alexa Jean, and son Karlo Anton—for providing endless inspiration and support.

Contents

List of Figures xxiii
List of Tables xxix
Foreword xxxiii
Preface xxxiv
Acknowledgments xxxvi
Author xxxvii

Chapter 1 Biomass as Energy Source 1

Learning Objectives 1
1.1 Introduction 1
1.2 US Energy Use and Needs 3
1.3 Biomass Energy Usage and Overall Energy Needs in Other Countries 4
1.4 Future of Agriculture 4
1.5 Advantages and Disadvantages in Use of Biomass as Energy Source 6
1.5.1 Advantages 7
1.5.2 Disadvantages 7
1.6 Sources of Biomass Available for Energy Use 8
1.6.1 Crop Residues 8
1.6.2 Fuelwood 10
1.6.3 Aquatic Biomass 11
1.6.4 Sugar Crops 12
1.6.5 Oil Crops 13
1.6.6 Animal Manure 17
1.6.7 Municipal Solid Wastes 19
1.6.8 Energy Farming 20
1.7 Units and Conversions 27
1.8 Conclusion 28
1.9 Problems and Discussion Issues 29
1.9.1 Simple Definitions 29
1.9.2 Energy Units 29
1.9.3 Net Energy Trends in United States 30
1.9.4 Contribution of Energy Sources to the US Economy 30
1.9.5 Moisture Removal in Biomass 30
1.9.6 Energy from Imported Crude Oil 30
1.9.7 Pure Sugar to Ethanol Conversion 30
1.9.8 Biogas Utilization in Households 30
1.9.9 Acreage Needed to Build Power Plants 31
1.9.10 Air-to-Fuel Ratios for the Combustion of Fuel 31
1.9.11 Ideal Biomass Feedstock for Methane Production 31
1.9.12 Ideal Biomass Feedstock for Bioethanol Production 31

1.9.13 Ideal Biomass Feedstock for Biodiesel Production with the Highest Yield 31
1.9.14 Biomass Feedstock Production and Delivery Project 31
1.9.15 Circular Bioeconomy Strategies for a Dairy Farm 31
References 32
Assess Your Knowledge 33
Additional Learning Resources 35

Chapter 2 Biomass Conversion Processes 36

Learning Objectives 36
2.1 Introduction 36
2.2 Overview of Biomass Conversion Processes 37
2.3 Chemical Conversion Processes 39
2.4 Biological Conversion Processes 41
2.4.1 Ethanol Fermentation 42
2.4.2 Biogas Production 43
2.5 Thermal Conversion Processes 45
2.5.1 Torrefaction 46
2.5.2 Pyrolysis 47
2.5.3 Gasification 50
2.5.4 Combustion 54
2.6 Hybrid Conversion Processes 55
2.6.1 Biological and Thermal Conversions 55
2.6.2 Thermal and Biological Conversions 56
2.7 Applications of Biomass Conversion Products 57
2.7.1 Heat Energy 58
2.7.2 Electrical Energy 58
2.7.3 Combined Heat and Power 60
2.7.4 Mechanical Energy 60
2.7.5 Liquid Biofuels or Bio-Oil Production 62
2.7.6 Syngas Production and Use 63
2.7.7 Biochar Production 64
2.8 Conclusion 65
2.9 Problems and Discussion Issues 66
2.9.1 Heat Energy Conversion Efficiency 66
2.9.2 Energy of Feedstock and Biofuel Product 66
2.9.3 Ethanol Yield per Unit of Area 66
2.9.4 Number of Animals Required for Power Generation 66
2.9.5 Pyrolysis Conversion Efficiency 66
2.9.6 Gasification Carbon Efficiency 67
2.9.7 Area Required for a Power Plant 67
2.9.8 Energy Required to Raise Water Temperature 67
2.9.9 Biogas Conversion Efficiency 67
2.9.10 Steam Cycle Conversion Efficiency 68

2.9.11 Major Gases in Gasification Systems ... 68
2.9.12 Major Gases in Anaerobic Digestion Process ... 68
2.9.13 Thermal Conversion Systems ... 68
2.9.14 Advantages of Thermal Conversion Systems versus Biological Conversions ... 68
2.9.15 Topping versus Bottoming Cycle ... 68
References ... 68
Assess Your Knowledge ... 70
Additional Learning Resources ... 72

Chapter 3 Biomass Properties for Thermal Conversion ... 73

Learning Objectives ... 73
3.1 Introduction ... 73
3.2 Physical Properties of Biomass ... 75
3.2.1 Particle Density ... 75
3.2.2 Bulk Density ... 75
3.2.3 Particle Size and Particle Size Distribution ... 76
3.2.4 Angle of Repose ... 77
3.2.5 Angle of Friction ... 78
3.2.6 Heat Capacity and Thermal Conductivity of Biomass ... 78
3.3 Important Thermal Related Properties ... 78
3.3.1 Proximate Analysis ... 79
3.3.2 Ultimate Analysis ... 83
3.3.3 Heating Value Analysis ... 86
3.4 Other Standard Methods for Biomass Analysis ... 91
3.5 Conclusion ... 93
3.6 Problems and Discussion Issues ... 94
3.6.1 Proximate Analysis ... 94
3.6.2 Reporting Heating Values of Biomass ... 94
3.6.3 Approximate Chemical Formula of Biomass ... 94
3.6.4 Stoichiometric Combustion and A/F Calculations for Biomass ... 94
3.6.5 Sustainability of Biomass Use ... 95
3.6.6 Proximate Analysis Calculations ... 95
3.6.7 Air-to-Fuel Ratio Calculations ... 95
3.6.8 Heating Value Equivalents ... 95
3.6.9 Biomass Requirements for Power Generation ... 95
3.6.10 Heating Value Calculations ... 95
3.6.11 Slagging and Fouling Factors ... 96
3.6.12 Biomass Feedstock with the Highest Carbon Content ... 96
3.6.13 Biomass with Lowest Ash Content ... 96

3.6.14 Biomass with Highest Energy Content 96
3.6.15 Cotton Gin Trash for Electrical Power 96
3.6.16 Dairy Manure for Electrical Power 96
References 96
Assess Your Knowledge 97
Additional Learning Resources 100

Chapter 4 Biomass Properties for Biological Conversion 101

Learning Objectives 101
4.1 Introduction 101
4.2 Properties of Biomass Important for Biological Conversion 103
4.3 Standard Methods for Analysis and Examples 106
4.3.1 Preparation of Samples for Compositional Analysis (LAP No. 42620) 107
4.3.2 Determination of Percent Solids in Biomass 108
4.3.3 Determination of Extractives in Biomass 109
4.3.4 Percent Total Solids of Extractives-Free Biomass 110
4.3.5 Percent Protein of Extractives-Free, ODW Biomass 111
4.3.6 ODW of Extractives-Free Sample 112
4.3.7 Percent Acid-Insoluble Residue 113
4.3.8 Percent AIL (Extractives Free) 114
4.3.9 Percent ASL (Extractives Free) 115
4.3.10 Carbohydrates in Biomass 119
4.3.11 Total Composition of *Nannochloropsisoculata* 122
4.4 Summary of ASTM Procedures for Compositional Analysis 122
4.5 Biomass Pretreatment 123
4.6 Conclusion 124
4.7 Problems and Discussion Issues 125
4.7.1 Calculation of Percent Total Solids 125
4.7.2 Calculation of Percent Ash Content (Dry Basis) 125
4.7.3 Calculation of Percent Extractives 126
4.7.4 Calculation of Percent Total Solids (Extractives-Free Biomass) 126
4.7.5 Calculation of Percent AIL from Extractives-Free Sample 126
4.7.6 Calculation of % ASL and %Lignin (Extractives Free) 127
4.7.7 Calculation of Percent Lignin (on As-Received Basis) 127
4.7.8 Calculation of Percent Sugars 127
4.7.9 Calculation of Percent Xylose Sugars 128
4.7.10 Biomass Compositional Analysis Summary 128
4.7.11 Biomass with the Highest Cellulose Content 128
4.7.12 Biomass with the Highest Lignin Content 128

4.7.13 Biomass with the Highest Hemicellulose Content ... 128
4.7.14 Whole Plant Biomass into Bioethanol ... 129
4.7.15 Animal Manure into Pure Methane ... 129
References ... 129
Assess Your Knowledge ... 130
Additional Learning Resources ... 132

Chapter 5 Biodiesel Production ... 133

Learning Objectives ... 133
5.1 Introduction ... 133
5.2 Vegetable Oil and Animal Fats Production in the United States ... 135
5.3 Vegetable Oil and Animal Fat Characteristics ... 136
5.4 Fatty Acid Composition ... 137
5.5 Other Basic Oil Properties ... 139
5.6 Oil Extraction Processes ... 139
5.7 Oil Refining Processes ... 141
5.7.1 Degumming ... 142
5.7.2 Neutralization ... 143
5.7.3 Dewaxing ... 144
5.7.4 Bleaching ... 145
5.7.5 Deodorization ... 146
5.7.6 Vegetable Oil Refining Calculations ... 148
5.8 Transesterification ... 151
5.9 ASTM Characterization ... 152
5.9.1 Flash Point ... 153
5.9.2 Kinematic Viscosity ... 154
5.9.3 Cloud Point ... 156
5.9.4 Water and Sediment ... 156
5.9.5 Total and Free Glycerin ... 157
5.9.6 ASTM Analysis of Various Oil Seed Methyl Esters ... 158
5.10 Engine Performance and Exhaust Emissions ... 159
5.11 Design of Biodiesel Plants in the United States ... 160
5.11.1 General Transesterification Processes ... 160
5.11.2 Crown Iron Works Design ... 160
5.11.3 Westfalia Biodiesel Plant Design ... 161
5.11.4 Hydrotreated Vegetable Oil ... 161
5.11.5 Economics of Biodiesel Facility ... 162
5.12 Conclusion ... 164
5.13 Problems and Discussion Issues ... 165
5.13.1 Various Types of Fatty Acids ... 165
5.13.2 Refined, Bleached and Deodorized (RBD) Oil ... 165
5.13.3 Transesterification Process ... 165
5.13.4 Oil Refining Processes ... 165

5.13.5 Acreage to Establish 100 MGY Biodiesel Plant.... 166
5.13.6 Refining Processes ... 166
5.13.7 Refining Efficiency Calculations ... 166
5.13.8 ASTM Fuel Properties ... 166
5.13.9 Energy Density of Biodiesel ... 166
5.13.10 Biodiesel Production Processes ... 166
5.13.11 Major Fatty Acid in Vegetable Oil Types ... 167
5.13.12 Density of Diesel Fuel Versus Vegetable Oil ... 167
5.13.13 Biodiesel with Lowest Flash Point ... 167
5.13.14 Acreage for Biodiesel Feedstock ... 167
5.13.15 Glycerin Production and Marketing ... 167
5.14 Case Study 5.1. Biodiesel Economics and Sustainability ... 167
References ... 169
Assess Your Knowledge ... 169
Additional Learning Resources ... 172

Chapter 6 Bioethanol Production ... 173

Learning Objectives ... 173
6.1 Introduction ... 173
6.2 Sugar Crops ... 175
6.2.1 Production of Ethanol from Sweet Sorghum ... 176
6.2.2 Ethanol Dehydration Methods ... 178
6.3 Starchy Crops ... 180
6.4 Cellulosic Biomass ... 184
6.5 Biomass Pretreatment Processes ... 185
6.5.1 Physical Pretreatment of Biomass ... 185
6.5.2 Chemical Pretreatment of Biomass ... 186
6.5.3 Enzymatic Pretreatment of Biomass ... 188
6.5.4 Combination of Various Pretreatment Methods ... 189
6.6 Pilot-Scale Ethanol Production Systems ... 189
6.7 Cellulosic Ethanol Studies ... 192
6.8 Commercial Cellulosic Ethanol Plants in the United States ... 198
6.9 Conclusion ... 198
6.10 Problems and Discussion Issues ... 200
6.10.1 Rules of Thumb in Ethanol Production ... 200
6.10.2 First- and Second-Generation Biofuels ... 200
6.10.3 Sugars and Ethanol Yield from Sweet Sorghum ... 200
6.10.4 Fermentation Efficiency and Ethanol Yields per Unit of Area ... 200
6.10.5 Pretreatment and Ethanol Production Process ... 200
6.10.6 Differences in First- and Second-Generation Ethanol ... 201
6.10.7 Enzyme and Its Role in Ethanol Production ... 201
6.10.8 Enzymatic Conversion of Lignocellulosics into Ethanol ... 201

6.10.9 Advantages and Disadvantages of Ethanol Production 201
6.10.10 Lignin Conversion and Uses 202
6.10.11 Improving Biofuel Production 202
6.10.12 Feedstock with the Highest Bioethanol Production Potential 202
6.10.13 Feedstock with the Most Carbohydrates 202
6.10.14 Crop with the Highest Residue to Crop Ratio 202
6.10.15 Acreage Required for a Bioethanol Refinery 202
6.10.16 Bioethanol from Whole Plant Biomass 202
6.11 Case Studies 202
References 203
Assess Your Knowledge 205
Additional Learning Resources 208

Chapter 7 Biogas Production 209

Learning Objectives 209
7.1 Introduction 209
7.1.1 Hydrolysis 209
7.1.2 Acidogenesis 210
7.1.3 Acetogenesis 211
7.1.4 Methanogenesis 211
7.2 Biomass/Waste Parameters Important in Anaerobic Digestion 217
7.2.1 Biochemical Oxygen Demand 217
7.2.2 Chemical Oxygen Demand 218
7.2.3 Total Organic Carbon 219
7.2.4 Total Solids and Volatile Suspended Solids 219
7.2.5 Typical Septic Tanks or Municipal Solid Wastes Effluent Data 221
7.3 Acid- and Methane-Forming Microbes 221
7.4 Advantages and Disadvantages of Anaerobic Digestion Processes 223
7.4.1 Advantages of Anaerobic Digestion Processes 223
7.4.2 Disadvantages of Anaerobic Digestion Processes 223
7.5 Biogas Conversion Process and Digester Designs 224
7.6 First- and Second-Generation Biogas Digesters 225
7.6.1 Chinese Dome Type 225
7.6.2 Indian Gobar Gas Plant 226
7.6.3 Taiwanese Rectangular Digester Design 226
7.6.4 Cylindrical Above-Ground Digesters 227
7.6.5 High-Rate Biogas Digesters (Second-Generation Biogas Digesters) 227
7.6.6 Anaerobic Filters 229
7.7 Design of Biogas Digester 230

7.8 Biogas Utilization ... 234
7.9 Conclusion ... 235
7.10 Problems and Discussion Issues ... 236
7.10.1 Acid- and Methane-Producing Microbes ... 236
7.10.2 Theoretical Acetic Acid Production ... 236
7.10.3 Calculation of BOD ... 236
7.10.4 Calculation of COD ... 236
7.10.5 Use of Laboratory Data to Estimate Digester Size ... 236
7.10.6 Energy Produced from Biogas Daily ... 237
7.10.7 Volume of Biogas Required for Electrical Power Production ... 237
7.10.8 Amount of Sludge Produced in an Anaerobic Digester ... 237
7.10.9 Theoretical COD from Biomass ... 237
7.10.10 Estimating Minimum Size of Digester ... 237
7.10.11 Feedstock with the Highest Methane Yield ... 237
7.10.12 Manure Type with the Highest Potential Methane Yield ... 238
7.10.13 Manure with the Highest Volatile Solids ... 238
7.10.14 Sizing of Anaerobic Digester ... 238
7.10.15 Power Production from Layer Biogas ... 238
7.11 Case Studies ... 238
References ... 240
Assess Your Knowledge ... 241
Additional Learning Resources ... 245

Chapter 8 Torrefaction ... 246

Learning Objectives ... 246
8.1 Introduction ... 246
8.2 Bio-Physico-Chemical Changes in Biomass during Torrefaction ... 247
8.3 Torrefaction Products ... 249
8.3.1 Solid Products ... 250
8.3.2 Liquid Condensibles ... 250
8.3.3 Gaseous Products ... 251
8.4 Physical Properties of Torrefied Biomass ... 252
8.4.1 Bulk Density ... 252
8.4.2 Grindability ... 254
8.4.3 Particle Size Distribution, Sphericity, and Surface Area ... 255
8.4.4 Moisture Content ... 255
8.4.5 Storage Properties ... 256
8.4.6 Hydrophobicity ... 256
8.4.7 Pelletability ... 257

8.5 Comparison between Torrefied Biomass versus Pelleted Biomass 258
8.6 Thermal Gravimetric Analysis Studies of Biomass 259
8.7 Chemical Composition Changes during Torrefaction of Biomass 261
8.7.1 Carbon Content 262
8.7.2 Hydrogen Content 263
8.7.3 Calorific Value 263
8.7.4 Degree of Carbonization or Conservation of Carbon 264
8.7.5 H/C Ratio 264
8.7.6 Nitrogen and Sulfur Content 265
8.8 Advantages and Disadvantages of Torrefaction Process 265
8.8.1 Advantages of Torrefaction as Pretreatment Process 266
8.8.2 Disadvantages of Torrefaction as Pretreatment Process 266
8.9 Conclusion 266
8.10 Problems and Discussion Issues 267
8.10.1 Energy Densification for Torrefied Product 267
8.10.2 Determining Energy and Mass Balances in Torrefaction 267
8.10.3 Bulk Density of Torrefied Biomass 267
8.10.4 Bulk Density of Biomass 268
8.10.5 Grinding Energy Usage 268
8.10.6 Moisture Content of Torrefied Biomass 268
8.10.7 Durability of Biomass Pellets 268
8.10.8 Bulk Density of Torrefied Biomass 269
8.10.9 Energy Density Changes in Torrefied Biomass 269
8.10.10 Van Krevelen Plot of Raw Biomass 269
8.10.11 Bulk Density Effect After Torrefaction 269
8.10.12 Grindability of Torrefied Biomass 270
8.10.13 Changes in Chemical Composition of Torrefied Biomass 270
8.10.14 Energy Farming for Torrefied Product 270
8.10.15 Biomass Pellets 270
References 270
Assess Your Knowledge 272
Additional Learning Resources 276

Chapter 9 Pyrolysis 277

Learning Objectives 277
9.1 Introduction 277
9.2 Various Pyrolysis Processes Based on Heating Rates 282
9.2.1 Slow/Conventional Pyrolysis 283

9.2.2 Fast/Flash Pyrolysis Systems ... 284
9.2.3 Flash Vacuum Pyrolysis ... 285
9.2.4 Ablative Pyrolysis ... 286
9.2.5 Microwave Pyrolyzer ... 286
9.3 Effects of Temperature on Product Yields from Pyrolysis of Microalgae ... 287
9.3.1 Microalgae Mass Balances ... 288
9.3.2 Microalgae Energy Balances ... 290
9.3.3 Overall Product Yields ... 295
9.4 Applications of Products from Fast Pyrolysis ... 295
9.5 Bio-Oil Characterization Processes ... 297
9.6 Bio-Oil Upgrade Processes ... 299
9.7 Studies on Pyrolysis of Various Biomass Resources ... 299
9.7.1 Cotton Gin Trash Pyrolysis (Aquino et al., 2010) ... 300
9.7.2 Switchgrass Pyrolysis Studies (Iman and Capareda, 2010) ... 300
9.7.3 Corn Stover Pyrolysis Studies (Capunitan and Capareda, 2012) ... 304
9.7.4 Sorghum Biomass Pyrolysis Studies (Santos and Capareda, 2012) ... 307
9.8 Conclusion ... 309
9.9 Problems and Discussion Issues ... 310
9.9.1 Effect of Temperature on Syngas Production in Pyrolysis ... 310
9.9.2 Pyrolysis Conditions that Favor Biochar and Bio-Oil Production ... 310
9.9.3 Advantages and Disadvantages of Pyrolysis Process ... 310
9.9.4 Problems and Issues of Pyrolysis Process at Extreme Temperatures ... 310
9.9.5 Energy Balance in Pyrolysis Process ... 310
9.9.6 Estimation of Residence Time in Fluidized Bed Reactor ... 311
9.9.7 Estimation of Residence Time in Auger Reactor ... 311
9.9.8 Calibration Curve for Augers ... 311
9.9.9 Calculation of HV of Syngas from Experiments ... 311
9.9.10 Calculating Yields per Unit Weight of Biomass ... 312
9.9.11 Ideal Circular Bioeconomy Feedstock ... 312
9.9.12 Ideal Feedstock for Bio-Oil Production ... 312
9.9.13 Accurate Syngas Measurement ... 312
9.9.14 Pyrolysis of Dairy Manure ... 313
9.9.15 Pyrolysis of Microalgae Biomass ... 313
9.10 Case Studies ... 313
References ... 314
Assess Your Knowledge ... 315
Additional Learning Resources ... 320

Chapter 10 Gasification 321

Learning Objectives 321
10.1 Introduction 321
10.2 Chemistry of Biomass Gasification 323
10.3 Various Types of Gasifiers 326
10.3.1 Updraft Gasifier 326
10.3.2 Downdraft Gasifier 328
10.3.3 Cross-Draft Gasifier 328
10.3.4 Fluidized Bed Gasifier 329
10.4 Applications of Biomass Gasifiers 333
10.4.1 Production of Thermal Sensible Heat 334
10.4.2 Electrical Power Generation 335
10.4.3 Mechanical Power Generation 336
10.4.4 Liquid Fuel Production from Synthesis Gas 337
10.5 Gasifier TDR and Throughput 337
10.6 Gasification Studies 339
10.7 Conclusion 345
10.8 Problems and Discussion Issues 346
10.8.1 Empirical Chemical Formula of Biomass 346
10.8.2 Air-to-Fuel Ratio of Biomass for Thermal Conversion 346
10.8.3 Air Requirement for Gasification 346
10.8.4 Volumetric Air Requirement in Fixed-Bed Gasifiers 346
10.8.5 Acreage Required to Build Power Plants 346
10.8.6 Equivalence Ratios Calculations in a Gasifier 346
10.8.7 Power from Manure Gasification 347
10.8.8 Superficial Velocity in Updraft Gasifiers 347
10.8.9 Syngas Requirement in Internal Combustion Engines 347
10.8.10 Design of Downdraft Gasifier 347
10.8.11 Exothermic Versus Endothermic Reactions 347
10.8.12 Gasification Reactor Types 348
10.8.13 Capacities of Gasifiers 348
10.8.14 Gasifier in a Cotton Gin 348
10.8.15 Gasifier in a Poultry Farm 348
10.9 Case Studies 348
References 349
Assess Your Knowledge 350
Additional Learning Resources 354

Chapter 11 Advanced Gasification 355

Learning Objectives 355
11.1 Introduction 355
11.2 Determining Average Particle Size of Bed Material 358

11.3 Minimum Fluidizing Velocity, Terminal Velocity, and Pressure Drop in Fluidized Bed Reactors 360
11.4 Operation of 0.3048 Fluidized Bed Gasifier 364
11.4.1 Fluidized Bed Gasifier Operational Protocol 367
11.4.2 Proper Use of Bed Material 370
11.4.3 Cold Fluidization Tests 371
11.5 Designing Dimensions of Fluidized Bed Gasifier 372
11.6 Designing Dimensions of Series Cyclone Char Removal System 373
11.6.1 Primary (Low-Efficiency) Cyclone 373
11.6.2 Secondary (High-Efficiency) Cyclone 376
11.6.3 Pressure Drop in Cyclones 377
11.7 Direct Use of Synthesis Gas for Heat and Steam Production 378
11.8 Electrical Power Production from Fluidized Bed Gasification 378
11.9 Conclusion 380
11.10 Problems and Discussion Issues 381
11.10.1 Calculation of Bed Material Average Particle Size 381
11.10.2 Determining Minimum Fluidizing Velocity 381
11.10.3 Terminal Velocity Calculations 381
11.10.4 Volumetric Flow Rate Calculations 382
11.10.5 Pressure Drop through Fluidized Bed Gasifier 382
11.10.6 Capacity of Screw Conveyor 382
11.10.7 Determining Dimensions of Fluidized Bed Gasifier 382
11.10.8 Determining Dimensions of 1D2D Cyclone 382
11.10.9 Determining Dimensions of 1D3D Cyclone 383
11.10.10 Predicting Pressure Drop in Cyclones 383
11.10.11 Conversion of Biomass into Electrical Power Using Fluidized Bed 383
11.10.12 Fluidization Issues 383
11.10.13 Factors Affecting Fluidization 383
11.10.14 Pressure Drop in Cyclone Particulate Biochar Removal 383
11.10.15 Cyclone Design 384
11.10.16 Sizing of Fluidized Bed Gasifier 384
References 384
Assess Your Knowledge 385
Additional Learning Resources 389

Chapter 12 Biomass Liquefaction 390

Learning Objectives 390
12.1 Introduction 390
12.2 Indirect Liquefaction Processes 391

12.2.1 Methanol Production from Synthesis Gas 392
12.2.2 Mobil Process 394
12.2.3 Fischer–Tropsch Process 396
12.2.4 China Lake Process 399
12.3 Direct Liquefaction Processes 402
12.3.1 BOM or PERC Process 402
12.3.2 LBL Process 404
12.4 Other Biomass Liquefaction Processes 407
12.4.1 Hydrothermal Liquefaction (HTL) Process........... 408
12.4.2 Hydrolytic and Hydrothermal Upgrading (HTU) Processes.. 409
12.5 Advantages and Disadvantages of Biomass Liquefaction Processes 410
12.5.1 Advantages of Direct Liquefaction Process........... 410
12.5.2 Disadvantages of Direct Liquefaction Process...... 411
12.5.3 Advantages of Indirect Liquefaction Process 411
12.5.4 Disadvantages of Indirect Liquefaction Process.... 411
12.6 Conclusion 411
12.7 Problems and Discussion Issues 412
12.7.1 Heating Value of Synthesis Gas Produced from Pyrolysis 412
12.7.2 Methanol from Biomass Conversion Efficiency.... 412
12.7.3 Mobil Process Conversion Efficiency................... 413
12.7.4 Fischer–Tropsch Syngas Calculations.................. 413
12.7.5 Heating Value of Syngas Produced for China Lake Process 413
12.7.6 BOM Process Efficiency Calculations 413
12.7.7 Determining Bio-Oil Yields by Hydrothermal Processes 414
12.7.8 Differences among BOM, PERC, and LBL Liquefaction Processes 414
12.7.9 Hydrogen-to-Carbon Ratios for Biomass Liquefaction Processes 414
12.7.10 Conversion Efficiency for HTU Process 414
12.7.11 Catalysts to Produce Methanol from Syngas......... 414
12.7.12 Catalysts for Fischer–Tropsch Process.................. 415
12.7.13 Catalysts for the Mobil Process............................ 415
12.7.14 Methanol Production from Fast-Growing Trees.... 415
12.7.15 Bio-Oil Production from Hydrothermal Liquefaction of Animal Manure 415
References 415
Assess Your Knowledge 417
Additional Learning Resources 420

Chapter 13 Biomass Combustion 421

Learning Objectives 421
13.1 Introduction 421
13.2 Types of Biomass Combustion Systems 423
13.2.1 Spreader Stoker System 424
13.2.2 Fluidized Bed Combustion System 426
13.2.3 Biomass Cookstoves 428
13.2.4 Pellet Stoves 432
13.3 Co-Combustion of Biomass and Co-Firing with Coal 433
13.4 Power Generation Case Studies in the United States 434
13.5 Slagging and Fouling Issues with Agricultural Biomass 436
13.5.1 Coal Slagging and Fouling Indices 437
13.5.2 Other Slagging and Fouling Indices 438
13.6 Determining Eutectic (or Melting) Point of Biomass Ash Pellets 441
13.7 Applications of Biomass Combustion Systems 444
13.7.1 Cooking, Heating, Drying, and Steam Production 445
13.7.2 Mechanical Power Generation 445
13.7.3 Electrical Power Generation 445
13.7.4 Combined Heat and Power Applications 446
13.8 Conclusion 447
13.9 Problems and Discussion Issues 448
13.9.1 Amount of CO_2 Produced for Every Metric Ton of Biomass Combusted 448
13.9.2 Amount of Air Required for Combustion of Sorghum Biomass 448
13.9.3 Biomass Combustion Efficiency and Acreage Needed 448
13.9.4 Boiling Water at Mile High Stadium 448
13.9.5 Biomass Co-Firing Efficiency 448
13.9.6 Conversion of Biomass into Small-Scale Electrical Power 449
13.9.7 Alkali Index for Animal Manure 449
13.9.8 Base-to-Acid Ratio in Animal Manure Ash Sample 449
13.9.9 Bed Agglomeration Index of Animal Manure Ash Sample 449
13.9.10 Compressive Strength of Animal Manure Ash Pellet Sample 449
13.9.11 Melting Materials During Combustion 449
13.9.12 Sticky Materials During Combustion 450
13.9.13 Slagging and Fouling Indices for Agricultural Biomass 450
13.9.14 Biomass Requirement for a Power Plant and Potential Revenue 450

13.9.15 Biomass Pellet Stoves 450
References 450
Assess Your Knowledge 452
Additional Learning Resources 456

Chapter 14 Biomass Sustainability Issues 457

Learning Objectives 457
14.1 Introduction 457
14.2 Well-to-Wheel (WTW) Approach 461
14.2.1 Feedstock Production 461
14.2.2 Feedstock Transport 461
14.2.3 Biofuel Production 462
14.2.4 Biofuel Distribution and Storage 462
14.2.5 Dispensing Biofuels 462
14.2.6 End Use of Biofuels 463
14.2.7 GHG Emissions of Biomass Combustion 466
14.3 Discussions of Software for LCA Analysis 467
14.4 Biofuel Economics 469
14.4.1 Major Components of Biofuel Economic Analysis 470
14.4.2 Economics of Production Processes for Major Biofuels 470
14.5 Sustainability of Biofuels Production 477
14.5.1 Measuring Sustainability of Biofuels 478
14.6 Circular Economy Concepts and Activities 481
14.6.1 Benefits and Example of Creating Circular Bioeconomy Systems 483
14.6.2 Carbon Capture and Sequestration 487
14.6.3 Carbon Footprint (CF) of Facilities 489
14.6.4 Carbon Intensity 491
14.7 Conclusion 492
14.8 Problems and Discussion Issues 493
14.8.1 Greenhouse Gas Equivalency Calculations 493
14.8.2 Gas Turbine Greenhouse Gas Equivalency Calculations 493
14.8.3 GREET Modeling Part 1 493
14.8.4 GREET Modeling Part 2 494
14.8.5 Infield Pyrolysis System for Immediate Biochar Field Application 494
14.8.6 Wood Bioethanol Economics 495
14.8.7 Pyrolysis Economics 495
14.8.8 Gasification Economics 495
14.8.9 Net Energy Ratio (NER) Calculations 496
14.8.10 Net Energy Balance (NEB) Calculations 496
14.8.11 Carbon Dioxide Equivalent Emissions per Unit of Energy 496

14.8.12 Best Biodiesel Type with Lowest Carbon Intensity....496
14.8.13 Capital Intensive Biofuel for Conversion Processes....496
14.8.14 Converting a Cotton Gin Operation from Linear to Circular Economy....496
14.8.15 Converting a Biodiesel Facility into Circular Bioeconomy....497
References....497
Assess Your Knowledge....500
Additional Learning Resources....508

Appendix A: Appropriate Units and Conversion....509

Appendix B: Glossary of Terms in Biomass Conversion Processes....513

Index....521

List of Figures

Figure 2.1 Various biomass energy conversion pathways 38

Figure 2.2 Simple schematic diagram to convert vegetable oil into biodiesel with glycerin 40

Figure 2.3 Various pathways to convert biomass high in sugar, starch, or lignocellulose into ethanol 41

Figure 2.4 Conversion pathway to break down complex biomass organics into methane and carbon dioxide 44

Figure 2.5 Schematic of various thermal conversion processes 46

Figure 2.6 Simple biomass gasifier schematic of basic component parts 51

Figure 2.7 Schematic of energy and mass balances in gasifier 52

Figure 2.8 Schematic of OSU pilot syngas fermentation system. (Adapted from Kundiyana, D. K., R. L. Hunke and M. R. Wilkins, Syngas fermentation in a 100-L pilot scale fermentor: Design and process consideration. *Journal of Bioscience and Bioengineering* 109, 492, 2010.) 57

Figure 3.1 Selected parameters for complete biomass characterization 74

Figure 3.2 System flow alternatives for biomass before being used in biorefinery 74

Figure 3.3 Tube furnace used for measurement of VCM 81

Figure 3.4 Slagging and fouling conditions in boiler 89

Figure 4.1 Problem-solving roadmap to establish compositional analyses of biomass 102

Figure 4.2 Plot of absorbance versus concentration of algae samples 117

Figure 4.3 Pretreatment pathway for conversion of lignocellulosic cotton gin trash into ethanol 124

Figure 5.1 Growth of the biodiesel industry in the United States by volume of production. (*Source*: USEIA. 2022. US Biodiesel plant production capacity. U.S. Energy Information Administration, Washington, DC. Available at: https://afdc.energy.gov/data/10325. Accessed March 12, 2022) 134

Figure 5.2 Overall biodiesel production and testing flow processes chart 134

Figure 5.3 Basic chemical configuration of triglyceride 137

Figure 5.4 Refining processes for some oils showing stages of refining........... 147

Figure 5.5 Various refining process efficiencies from three types of oil seed..........149

Figure 5.6 Schematic of a simple oil transesterification process....................... 151

Figure 5.7 Mass balance in transesterification process 151

Figure 5.8 Typical NO_x emissions for a John Deere engine fueled by soybean oil biodiesel, blends, and reference diesel fuel................... 159

Figure 5.9 Crown Iron Works Inc. biodiesel plant design 161

Figure 5.10 Wesfalia biodiesel plant design. (Under license of Archer Daniels Midland Company, ADM, Decatur, IL.)............................ 162

Figure 6.1 Schematic of ethanol production from various biomass components.. 174

Figure 6.2 Process flow chart for conversion of sorghum into ethanol 177

Figure 6.3 Dry and wet milling processes for conversion of starch into ethanol. (Adapted from Nichols, N. N., B. S. Dien, R. J. Bothast and M. A. Cotta, The corn ethanol industry. In: *Alcoholic Fuels* (S. Minteer, ed.). CRC Press, Boca Raton, FL. pp. 59–78, 2006.)...........182

Figure 6.4 Pretreatment economic analysis for various chemical pretreatment processes. (From Dale, B., E. Newton, M. Laser, S. Kim, L. Laureano-Perez, H. Alizadeh, H. Jin and L. Lynd, *Recent Improvements for the Ammonia Fiber Explosion (AFEX) Process and Resulting Reductions in Projected Ethanol Prices.* Great Lakes Bioenergy Research Center Report, Michigan State University, Ann Arbor, MI, 2009.).. 188

Figure 6.5 Schematic of 1,300 L (350 gal) ethanol production facility for grain sorghum and corn 189

Figure 6.6 Cooking and fermenting log for grain-to-ethanol production process.. 190

Figure 6.7 Glucose consumption and ethanol production during fermentation of sweet sorghum juice... 193

Figure 6.8 Plot of fermentation efficiency against fermentation time for ethanol production from sweet sorghum juice 194

Figure 6.9 Lignocellulosic conversion into glucose using sweet sorghum biomass.. 196

Figure 6.10 Effect of various enzymes used in this study on ethanol production 197

Figure 7.1 Schematic of anaerobic digestion process 210

Figure 7.2 Schematic of laboratory batch digestion system used to establish HRT or SRT.......213

Figure 7.3 Photo of simple laboratory setup to determine retention times in the digester.......214

Figure 7.4 Typical biogas production from anaerobic digestion of biomass and pure manure.......214

Figure 7.5 Schematic diagram of an anaerobic digestion system showing feedstock that may be used, products, and applications.......222

Figure 8.1 Schematic of torrefaction facility.......247

Figure 8.2 Physico-chemical changes in biomass during torrefaction. (Modified from Bergman, P. C. A., A. R. Boersma, R. W. H. Zwart and J. H. A. Kiel, Torrefaction for biomass cofiring in existing coal-fired power plants. Energy Research Center of the Netherlands (ECN), ECN Biomass, Petten, The Netherlands. ECN Report No. ECN-C-05–013, 2005.).......248

Figure 8.3 Products formed during torrefaction of biomass. (Modified from Bergman, P. C. A., A. R. Boersma, R. W. H. Zwart and J. H. A. Kiel, Torrefaction for biomass co-firing in existing coal-fired power plants. Energy Research Center of the Netherlands (ECN), ECN Biomass, Petten, The Netherlands. ECN Report No. ECN-C-05–013, 2005.).......250

Figure 8.4 Flow diagram for converting torrefied biomass into densified material or pellets.......257

Figure 8.5 Changes in heating values of biomass product as torrefaction temperature is increased. (Adapted from data by Bridgeman, T. G., J. M. Jones, I. Shield and P. T. Williams, *Fuel*, 87, 844, 2008.).......264

Figure 9.1 Schematic of pyrolysis process.......278

Figure 9.2 Process flow in continuous pyrolyzer to generate solid, liquid, and gaseous products.......281

Figure 9.3 Mass balance of microalgae pyrolysis at 400°C.......289

Figure 9.4 Mass balance of microalgae pyrolysis at 500°C.......289

Figure 9.5 Mass balance of microalgae pyrolysis at 600°C.......289

Figure 9.6 Energy balance of microalgae pyrolysis at 400°C.......291

Figure 9.7 Energy balance of microalgae pyrolysis at 500°C.......292

Figure 9.8 Energy balance of microalgae pyrolysis at 600°C.......292

Figure 9.9 Energy distribution of solid, liquid, and gaseous products from pyrolysis of microalgae samples.......295

Figure 9.10 Various applications of products of pyrolysis process.................... 296

Figure 9.11 Bio-oil characterization pathway for product identification and analysis.................... 297

Figure 9.12 Results of pyrolysis of switchgrass samples at various temperatures.................... 301

Figure 9.13 Proximate analyses of biochar with pyrolysis temperatures.................... 302

Figure 9.14 Yields of biochar, bio-oil, and syngas from batch pyrolysis of corn stover.................... 305

Figure 9.15 Non-condensible (NC) gas, bio-oil, biochar, and aqueous products as function of pyrolysis temperature.................... 308

Figure 10.1 Various conversion zones in a gasifier and corresponding temperature profile.................... 324

Figure 10.2 Schematic of updraft gasification process.................... 327

Figure 10.3 Temperature profile in an updraft gasifier and reactions occurring during gasification process.................... 327

Figure 10.4 Schematic of downdraft gasifier.................... 328

Figure 10.5 Temperature profile within a downdraft gasifier.................... 329

Figure 10.6 Schematic diagram of a fluidized bed gasifier.................... 330

Figure 10.7 Schematic of a patented fluidized bed gasifier.................... 331

Figure 10.8 Schematic diagram of turbo gasifier stove developed by Tom Reed.................... 334

Figure 10.9 Schematic diagram of a downdraft gasifier with gas cleanup system to generate power from internal combustion engines.................... 336

Figure 10.10 Photo of material found in fluidized bed combustor using animal manure.................... 339

Figure 10.11 Typical temperature profile during gasification, start-up, and operation.................... 342

Figure 10.12 Temperature profile when successive biomass feedstocks (woodchips, switchgrass, and animal manure) were gasified using a fluidized bed gasifier.................... 343

Figure 11.1 Behavior of fluidizing media in fluidized bed reactor. (Adapted from Kunii, D. and O. Levenspiel, *Fluidization Engineering*, Robert E. Krieger Publishing Company, Huntington, NY, 1977.).................... 357

Figure 11.2 Quality of fluidization as influenced by type of gas distribution plates. (Adapted from Kunii, D. and O. Levenspiel, *Fluidization Engineering*, Robert E. Krieger Publishing Company, Huntington, NY, 1977.).................... 357

Figure 11.3 Several examples of gas distribution systems for fluidized beds. (Adapted from Kunii, D. and O. Levenspiel, *Fluidization Engineering*, Robert E. Krieger Publishing Company, Huntington, NY, 1977.) 358

Figure 11.4 Capacity of screw conveyor. (Modified from Woodstock, C. R. and J. S. Mason, *Bulk Solids Handling: An Introduction to the Practice and Technology*, Chapman & Hall Inc., New York, 1987.) 365

Figure 11.5 Dimensions nomenclature for cyclone design 374

Figure 12.1 Schematic diagram of producing methanol from syngas. (Modified from Trifiro, F., From syngas to methanol and dimethylether. Lecture presented in Summer School 2009, University of Bologna, Spain. Sponsored by the International Center for Science and High Technology, UNIDO, Rome, Italy, 2009.) 394

Figure 12.2 Schematic diagram of FT process 397

Figure 12.3 Schematic diagram of the China Lake process 400

Figure 12.4 Schematic of BOM or PERC liquefaction process. (Adapted from Elliott, D. C, In: *Thermochemical Processing of Biomass: Conversion into Fuels, Chemicals and Power* (R. C. Brown, ed.). John Wiley and Sons, Ltd., Chichester, UK. pp. 200–231, 2011.) 403

Figure 12.5 Detailed schematic of LBL process. (Adapted from Elliott, D. C, In: *Thermochemical Processing of Biomass: Conversion into Fuels, Chemicals and Power* (R. C. Brown, ed.). John Wiley & Sons, Ltd., Chichester, UK. pp. 200–231, 2011.) 405

Figure 12.6 Schematic of HTL biomass conversion process. (Adapted from Holladay, J. E., A. J. Schmidt and D. C. Elliott, *Hydrothermal Liquefaction—A Route to Improved Biooils.* Pacific Northwest National Laboratory, Richland, WA. #PNNL-SA-78800, 2011. With credit to the National Advanced Biofuels Consortium funded by the USDOE through its Office of the Biomass Program.) 408

Figure 13.1 Schematic of biomass combustion process for electricity production 423

Figure 13.2 Schematic of spreader stoker biomass combustion system 424

Figure 13.3 Schematic of FBC system for biomass 427

Figure 13.4 Schematic of various zones in FBC system, including the option to recycle bed material in the reactor 429

Figure 13.5 Designs of cookstoves used for cooking in developing countries (Philippines) 430

Figure 13.6 Design of cookstove used for cooking in developing countries (Vietnam)........ 432

Figure 13.7 Graph showing eutectic points of manure and CGT biomass ash........ 443

Figure 13.8 Photo of about 18 kg (40 lb) of slag collected after full-day gasification of animal manure using a fluidized bed gasifier........ 444

Figure 14.1 Role of biomass for a sustainable future........ 458

Figure 14.2 LCA for conversion of coal into electricity. (Modified from Mann, M. K. and P. L. Spath, The net CO_2 emissions and energy balances of biomass and coal-fired power systems. NREL, USDOE Project Report NREL/TP-430–23076 and NREL/TP-570–25119. National Renewable Energy Laboratory (NREL) of the USDOE, Golden, CO, 1999.) 459

Figure 14.3 LCA for conversion of biomass materials into electricity. (Modified from Mann, M. K. and P. L. Spath, The net CO_2 emissions and energy balances of biomass and coal-fired power systems. NREL, USDOE Project Report NREL/TP-430–23076 and NREL/TP-570–25119. National Renewable Energy Laboratory (NREL) of the USDOE, Golden, CO, 1999.) 460

Figure 14.4 Diagram of LCA for corn ethanol........ 464

Figure 14.5 Stages covered in ANL GREET sustainability model........ 468

Figure 14.6 Breakdown of total operating cost for the production of biodiesel 471

Figure 14.7 Breakdown of conversion and processing cost for making biodiesel........ 472

Figure 14.8 Breakdown of capital cost for commercial biodiesel plant 472

Figure 14.9 Capital cost composition in complete commercial biodiesel facility 473

Figure 14.10 NERs of renewable and fossil fuel energy sources. (From Shapouri, H., M. Wang and J. A. Duffield, Net energy balancing and fuel-cycle analysis. In: *Renewable-Based Technology Sustainability Assessment* (J. Dewulf and H. Van Langenhove, eds.). John Wiley & Sons Ltd., New York, 2006.) 479

Figure 14.11 Linear economy for milk production in a dairy farm converted into a circular bioeconomy 483

List of Tables

Table 1.1 US Energy Summary over a Period of 11 Years (in Units of Quads) 3

Table 1.2 Breakdown of Energy Consumption in the United States for 11 Years (in Units of Quads) 4

Table 1.3 Wood as Percentage of Total Energy Use in Asian and African Countries 5

Table 1.4 Estimated Production of Agricultural Residue in United States Based from Current Crop Yield 9

Table 1.5 Forest and Energy per Capita Resources in Selected Developing Countries (FAO, 2012) 11

Table 1.6 Potential Ethanol Yields of Some Sugar and Starchy Crops in L/kg... 13

Table 1.7 Potential Ethanol Yields of Some Sugar and Starchy Crops in L/ha and gal/acre 14

Table 1.8 Various Seed Yields of Oil for Major Crops and Equivalent Biodiesel Production per Unit Area 15

Table 1.9 Various Yields of Oil for Major Crops in Units of kg/ha 15

Table 1.10 Manure Production of Some Major Farm Animals 18

Table 1.11 Yields of Dedicated Crops and Variability of Yields across the US States 23

Table 1.12 Emergent Aquatic Plants for Biomass Energy Systems Development..... 26

Table 2.1 Data Required to Estimate Conversion of Glucose into Ethanol on Unit Weight Basis 42

Table 2.2 Effect of Steam in Gasification of Biomass 63

Table 3.1 Table of Standard Sieve Screen Sizes 77

Table 3.2 Some Proximate Analysis of Common Biomass (as Received) 79

Table 3.3 List of ASTM Standards Used for Biomass Proximate Analysis for Wood Fuel, Coal, and Coke 80

Table 3.4 List of ASTM Standards Used for Biomass Characterization 84

Table 3.5 Some Ultimate Analyses of Common Biomass 85

Table 3.6 Ultimate Analysis Data for Selected Solid and Biomass Fuels 87

Table 3.7 Slagging and Fouling Indices for Coal Samples 90

Table 3.8 Ash Composition of Cotton Gin Waste and Sorghum Stalks 90

Table 3.9 Other Standard Methods for Biomass Analysis 92

Table 3.10 Fusibility of Some Biomass Ash 92

Table 4.1 Ranges of Values of Chemical Composition of Common Biomass 103

Table 4.2 LAPs for Biomass Characterization 105

Table 4.3 Data to Establish Absorptivity Constant for Algae Sample 116

Table 4.4 Example of Calibration Curve Data for Sugar Analysis 119

Table 4.5 Results of Sugar Analysis from Example 4.12 121

Table 4.6 Total Composition of *Nannochloropsisoculate* 122

Table 4.7 Summary of ASTM Standards for Compositional Biomass Characterization 123

Table 5.1 Vegetable Oil and Animal Fat Production in the United States in 2020 136

Table 5.2 Systematic, Common Name, and Structure of Most Common Fatty Acids 138

Table 5.3 Fatty Acid Composition of Some Common Edible Fats and Oils 138

Table 5.4 Common Properties of Oil and Diesel 140

Table 5.5 Oil Extraction Yield Data from Jatropha Seeds 141

Table 5.6 Results of Refining Studies for Three Types of Oil Seeds Showing Initial and Final Weights after Each Refining Step 148

Table 5.7 ASTM D6751–07: Standards Specification for Biodiesel Fuel 153

Table 5.8 Operating Conditions of GC Used to Measure Total and Free Glycerin Samples in Biodiesel 157

Table 5.9 Results of ASTM Analysis of Various Biodiesel Samples at BETA Lab 158

Table P6.1 Experimental Data for Glucose-to-Ethanol Conversion 201

Table 7.1 Volatile Fatty Acids in Anaerobic Digestion Processes 211

Table 7.2 Typical Characteristics of Septage Effluent 221

Table 7.3 Optimum Conditions for a Single-Stage Anaerobic Digestion Process 228

Table 7.4 Daily Manure Production from Various Animals 230

Table 7.5 Biogas Production Heating Values and Approximate Equivalents to Gasoline, Diesel, Natural Gas, and Propane 231

Table 7.6 Methane Yield of Various Woody Biomass Types 231

Table 8.1 Compositional Analysis of Some Selected Biomass and Wastes....... 249

Table 8.2 Bulk Density (Dry Basis) of *Eucalyptus grandis* as Affected by Torrefaction Temperature.. 252

Table 9.1 Conditions for Various Thermal Conversion Processes..................... 280

Table 9.2 Yields of Solid, Liquid, and Gaseous Products for Various Thermal Conversion Processes 280

Table 9.3 Initial Biomass Characterization for Microalgae Used for Pyrolysis Studies 287

Table 9.4 Yields of Solid, Liquid, and Gaseous Components of Microalgae Sample by Weight.. 288

Table 9.5 Biochar and Bio-Oil Analysis Results 290

Table 9.6 Gaseous Components of Typical Syngas from Pyrolysis Experiments by Volume Basis (%v/v) 290

Table 9.7 Parameters of Common Gases Used for Calculating HV of Synthesis Gas.. 291

Table 9.8 Types of Compounds Identified in Corn Stover Bio-Oil Sample....... 298

Table 9.9 Compounds Separated through Simple Distillation of Bio-Oil between 100°C and 180°C.. 299

Table 9.10 Syngas Composition as Function of Pyrolysis Temperature for Switchgrass Sample 302

Table 9.11 Elemental Composition, HV, Surface Area, and Total Pore Volume of Char.. 303

Table 9.12 Complete Corn Stover Biomass Characterization before Pyrolysis Studies 303

Table 9.13 Some Corn Stover Bio-Oil Properties.. 305

Table 9.14 Proximate, Ultimate, HV, and Structural Analyses of High-Tonnage Sorghum Biomass 307

Table 9.15 Proximate Analysis of Biochar as Function of Pyrolysis Temperature... 308

Table 9.16 Yields of Pyrolysis Coproducts as Function of Pyrolysis Temperature... 309

Table 10.1 Table of Calculations for Establishing Chemical Formula for Sorghum Stalks 321

Table 10.2 Data for Gasification of Various Fuels Using Fluidized Bed Gasifier... 331

Table 10.3 Composition of LCV Gas Using a 305-mm Gasifier 341

Table 11.1 Common Particle Size Data of Bed Material for Pilot-Scale Gasification Research 359

Table 11.2 Voidage at Minimum Fluidizing Conditions 361

Table 11.3 Bulk Density and Loading Factors for Various Biomasses Used for Gasification in a Fluidized Bed Gasifier 365

Table 11.4 Properties of Three Biomasses Used in Gasification Experiments 367

Table 11.5 Data for Gasification of Various Fuels Using a Fluidized Bed Gasifier 368

Table 11.6 Particle Size Analysis and Properties of Olivine Bed Material Used for Fluidized Bed Gasifiers 371

Table 11.7 Nomenclatures for Design of 1D2D Cyclone 375

Table 11.8 Nomenclatures for Design of 1D3D Cyclone 376

Table 13.1 Woody Biomass Power Development Case Studies 435

Table 13.2 Raw Data for CGT and Manure Ash for Calculating Various Slagging and Fouling Indices 439

Table 13.3 Measured Compressive Strengths of Treated Ash Pellets of Dairy Manure and CGT Measured Compressive Strengths of Treated Ash Pellets of Dairy Manure and CGT 442

Table 14.1 Emission Factors for Natural Gas and Diesel Drivers Based on Fuel Usage 465

Table 14.2 Comparative Results of Energy and GHG Emissions WTP for Two Biofuels Using GREET1 Software 469

Table 14.3 Relative Impact of Cost Components on Various Biofuels 470

Table 14.4 Energy Associated with Ethanol Production from Corn without Coproducts Allocation 478

Table 14.5 Energy Associated with Ethanol Production from Corn with Coproducts Allocation 481

Table 14.6 Emission Factors Used for Energy and Utility Usage 490

Table 14.7 Carbon Dioxide Emissions Coefficients by Fuels (USEIA, 2021 and Gelfand et al., 2020) 491

Table 14.8 Carbon Intensities of Various Biodiesel Types Compared with Fossil Fuel Carbon Intensity (Posada et al., 2012) 492

Foreword

Renewable energy has great potential as a supplement to traditional fuels (coal, oil, gas, etc.) and can reduce greenhouse gas emissions. This book is intended to serve as a textbook for courses in biomass energy conversion. It is a valuable addition to the extensive references already published and referenced herein.

Chapter 1 is an introduction and overview. It provides an estimation of the available biomass resources in the world while recognizing that biomass for energy may compete with food availability and may raise food prices but emphasizes that significant quantities of biomass may be available as a fuel source.

Chapter 2 provides an overview of all biomass conversion technologies. The standard procedures for characterization of biomass, an essential aspect of the process of biomass conversion, are comprehensively presented in Chapters 3 and 4. Chapters 5–13 describe in detail the various processes that may be used to convert biomass to more useful solid, liquid, and gaseous forms that may be used to power mobile vehicles (tractors, trucks, automobiles, etc.) or generate electrical power or valuable materials. Also, direct combustion of biomass may be useful to fire boilers in power plants (Chapter 13).

One of the most useful aspects of this book is the clear statement of learning objectives at the beginning of each chapter. The extensive numbers of examples throughout the chapters clarify the text material, and the large numbers of example problems complete with solutions are noted. Students will find these learning aids very useful.

Chapter 14 is perhaps the most important one in the book. It deals with sustainability and economics of biomass for fuels. Moving most any technology from the original concept in the research laboratories to full-scale commercialization is a long and arduous process that often requires several decades to complete. Many technologies fall by the wayside in this scale-up process. Many biomass conversion technologies have already achieved several decades of scale-up toward commercialization. Thus, Chapter 14, dealing with sustainability and economics, may be the key to successful competition with oil and gas prices. Lifecycle analysis is vital. The final paragraph is crucial, "Finally, since the world is still a fossil fuel–based economy, at present, the economics of biofuel production will always be compared with the price of crude oil." No one knows the price of oil next year or in future decades. It is governed by many factors other than supply and demand (politics, wars, etc.). Some sort of government intervention (laws, regulations, subsidies, etc.) may be necessary to stabilize the price of oil and provide all forms of renewable energy investors stable oil prices on which to base their business plans.

Again, this book will provide students with a valuable understanding of biomass energy conversion processes and an introduction to sustainability and economic issues.

B.A. Stout
Professor Emeritus
Biological and Agricultural Engineering Department
Texas A&M University

Preface

This textbook was written for various reasons. My main goal is to provide a primary textbook for a 15-week class on biomass energy conversion. The book has 14 chapters, enough to be covered in a one-semester, three-credit-hour course.

Much has been written concerning biomass conversions, but I could not seem to find one book that had all the information that I wanted to use for teaching a biomass conversions class. This book compiles all the topics and various references in the field of biomass conversions and related areas that I have used over the years as a student, instructor, and practitioner. In this textbook, I have included several solved example problems within each chapter and similar problems at the end of each chapter for students to work out. These were devised mostly by my graduate students based on actual problems that were encountered during their theses or dissertations.

Chapter 1 describes all the possible biomass resources available for conversion. I tried to incorporate what data I could gather on production tables, yields, and recent estimates. Chapter 2 provides an overview of all the major types of biomass energy conversion processes.

Biomass characterization procedures are discussed in Chapters 3 and 4. I had originally planned on covering this topic in one chapter, but because some of the analyses are quite complex, I decided to split the discussion and analyses of characterization procedures into those important in thermal conversion processes (Chapter 3) and those used before biological conversion processes (Chapter 4). For those interested in setting up a biomass energy conversion laboratory, they should find these two chapters useful. I have enumerated and shown all facilities and equipment needed for characterization and analyses, including all standard methods of analyses that are being used, mainly American Society for Testing and Materials or National Renewable Energy Laboratory, US Department of Energy, procedures. The remaining chapters discuss in detail the various biomass conversion processes and concepts.

Chapter 14 deals with biomass sustainability issues (economics and sustainability). It is common to find these topics discussed in separate chapters in other books, but I believe economics and sustainability issues are best discussed together.

Engineers, practitioners, and hobbyists should find this textbook useful because of the design calculation approach used in most chapters. With "biofuels" currently a popular buzzword, K–12 instructors may also use the book as reference for lectures on special topics to encourage renewable energy awareness in the youth.

This textbook comes with a solutions manual for all the problems given at the end of each chapter as well as a set of PowerPoint slides for use in class lectures. My future plan is to write a laboratory manual for a biomass bioconversion laboratory course. I have now adapted a set of laboratory exercises that we developed at the University of the Philippines at Los Baños into my classes at Texas A&M University.

I welcome any suggestions and feedback for the improvement of this textbook.

This second edition of the textbook addresses new research information concerning biomass energy conversions. There is an eBook version in this edition, where PowerPoint slides are available including additional multiple-choice questions (MCQs) at the end of each chapter. There are also additional MCQs at the end of each chapter and a couple of open-ended essay type questions with calculations required. Each major biofuel has comprehensive case studies focusing on economics and sustainability. Ready-made presentation materials were updated and available for teachers, including an updated solutions manual and recommended laboratory exercises. Modern fuels such as hydrogenated vegetable oils (HVO) were added. Numerous questions and discussions on electrical power production are included, including cogeneration systems.

The primary focus of the revision is concerning circular bioeconomy. Hence, numerous discussions have been covered on carbon capture and sequestration (CCS), carbon intensity (CI), carbon footprint work (CF), and introduced circular bioeconomy principles and evaluation protocol. Numerous examples were presented focused on converting a linear bioeconomy into a circular bioeconomy.

Acknowledgments

I would like to express my immense gratitude to the following individuals for sharing their time and creativity toward the completion of the first edition of this book. Dr. Bill Stout served as technical editor and provided invaluable comments and suggestions to improve the presentation of the chapters despite his busy schedule. I also appreciate the help of my former graduate students, Jewel Capunitan, Jerrson Placido Escobar, Tahmina Imam, Jinny Kongkasawan, Amado Maglinao, Monet Maguyon, and Bjorn Santos, for providing the sample problems and some tables and figures used in this book. I wish to thank Ryan Roessler, formerly an undergraduate student then graduate student in my group, for doing most of the Adobe Illustrator drawings.

I also acknowledge the following people for their key roles in the inspirational experiences that have directly or indirectly influenced this book project. Dr. Reynaldo M. Lantin, former dean of the College of Engineering and Agro-Industrial Technology at the University of the Philippines at Los Baños, who got me started on biofuels research when he hired me as a research assistant, fresh out of college, for his project on village-level ethanol production.

My first technical training on biomass conversion processes was provided by Dr. S. C. Bhattacharya, Professor at the Asian Institute of Technology in Bangkok, Thailand.

Drs. Wayne A. LePori and Calvin B. Parnell, Jr., have been very supportive mentors and colleagues at Texas A&M University. Dr. LePori taught me how to operate a fluidized bed gasifier. I owe my knowledge on and experience with the air quality aspects of agricultural industries to Dr. Parnell.

I wish to thank my daughter Alexa Jean (AJ) for finding time to proof this second edition, despite her busy schedule. I also wish to thank my graduate students El Jirie Baticados and Eunice Arzadon, who helped significantly with the eBook version. They both helped provide callouts and popups in each chapter including linking images and developing new audio/videos to help students improve their learning objectives for each chapter.

Finally, I wish to thank all who have helped in one way or another in this endeavor to elevate classroom instruction in biomass conversion and all who share my goal of disseminating knowledge on the relevance of biofuels in bio-based circular economy.

Author

Sergio Capareda is a professor at Texas A&M University, a licensed professional engineer in Texas, and an active member of the American Society of Agricultural and Biological Engineers. He earned a Bachelor of Science degree in agricultural engineering from the University of the Philippines at Los Baños, a Master of Engineering degree in energy technology from the Asian Institute of Technology in Bangkok, Thailand, and a PhD in agricultural engineering from Texas A&M University in the United States. He had additional training on remote sensing and geographic information systems at the Asian Institute of Technology and on high-rate biogas reactor design and development at the King Mongkut's Institute of Technology.

After obtaining his PhD in 1990, Capareda returned to the Philippines to embark on an academic career in the field of renewable energy at the University of the Philippines at Los Baños. With funding from the Philippine Department of Energy, he developed the Biomass Energy Laboratory at the University of the Philippines at Los Baños. As a consultant for the coconut industry, he helped develop an engine that ran on 100% coconut oil, in collaboration with German engineers (IDEA, GmbH). Capareda was program director for World Bank– and United States Agency for International Development–funded projects implemented by the Philippine Department of Energy on rural electrification and market assessment of renewable energy in the Philippines. With funding from the United States Department of Energy, Philippine Department of Energy, and United States Agency for International Development, he developed the *Biomass Atlas of the Philippines* in 2000. From 2000 to 2003, he served as director of the Institute of Agricultural Engineering at the University of the Philippines at Los Baños and program manager of the Agricultural Mechanization Development Program, a United Nations Development Program–funded program. After a 2-year stint at the Department of Biological and Agricultural Engineering, Texas A&M University, as a visiting scientist working on air quality research, Capareda was appointed faculty at Texas A&M University in 2005 and was tasked to develop the alternative energy program of the department. He developed and established the Texas A&M University BioEnergy Testing and Analysis Laboratory (http://betalab.tamu.edu) in that year.

Dr. Capareda has authored or co-authored more than 120 refereed journal publications since 2003 and four book chapters in the field of renewable energy and air quality. With some colleagues at Texas A&M University, he holds two patents—Integrated Biofuel Production System and Pyrolysis and Gasification System for Biomass Feedstock. He is recipient of three "Balik" (Returning) Scientist Awards from the

Philippines Government, serving as "balik" scientist in 2011, 2014, and 2019. Other awards include a 2016–2017 USAID-STRIDE Visiting Professorship Award, 2017 DOST-ERDT Philippines Visiting Professorship Award, 2016 Charles H. Barklay Jr., '45 Fellow for Engineering Award for Outstanding Contributions, 2015 Texas A&M AgriLife Research Vice Chancellors Award in Excellence for International Involvement, and recognition as a 2015 Texas A&M University AgriLife Research Faculty Fellow. He also received the 2010 Vice Chancellor's Award in Excellence at the College of Agriculture and Life Sciences as a member of the Bio-Resource RecCycling Team and the 2011 Award for Excellence in Research from the TAMU Department of Biological and Agricultural Engineering.

1 Biomass as Energy Source

LEARNING OBJECTIVES

On completion of this chapter, one should be able to

1. Provide an estimate of the available biomass energy resources in developed countries and the developing world.
2. Identify various biomass resources that may be used for energy production.
3. Discuss the advantages and disadvantages of using biomass as an energy resource.
4. Describe the different types and amounts of biomass that can be generated from various sources.
5. Learn the various units and conversions used for estimating energy equivalents of biomass resources.

1.1 INTRODUCTION

Biomass includes every living plant and animal. The most common form of biomass is plant biomass from agricultural food production. Other sources of biomass include forestry resources and residue, and animal manure from confined animal production operations. Each country has its own biomass residue accumulating in agricultural lands, such as rice straw from rice-producing countries and rice hulls accumulating in rice mills. Cotton-producing states such as Texas would have abundant cotton gin trash that accumulates in cotton gins. Plant biomass resources are produced through the process of photosynthesis, which makes use of the sun, arable land, carbon dioxide, and water. The conversion of solar energy into biomass per unit area is quite low and inefficient. However, because of the vastness of land and water bodies available for photosynthesis, some 1,260 EJ (10^{18} J) of energy is produced each year (Boyle, 2004). The world's projected energy requirement has been projected to be around 623 EJ by 2035 (Kopetz, 2013), which is about half of the energy stored in the form of biomass each year. The annual energy consumption of the United States in 2020 was around 98 EJ (92.9 quads, see Example 1.1 for unit conversion), representing slightly less than a quarter of the world's energy consumption of 418 EJ in 2019 (IEA, 2021). Most recently, the United States became a net exporter of energy because its annual energy production (101 EJ or 95.7 quads) exceeds consumption from all sources. About 11.6 quads (12.2 EJ) are coming from renewable energy. The contribution of biomass energy to this energy mix is around 5 quads (5.3 EJ), which is approximately 5.5% of its total energy production.

DOI: 10.1201/9781003294306-1

Biomass resources could play a significant role in meeting the future energy requirements of many countries. However, the approach to their utilization should be carefully analyzed, while considering the diverse cultural, socioeconomic, and technological factors in each locality. Agricultural and agro-industrial wastes can provide an inexpensive source of energy and an effective low-sulfur fuel. It could be processed into other fuels, thereby reducing environmental hazards (e.g., biomass from sewage). Also, there is a relative ease with which it could be gathered and generated. The extensive use of biomass resources for energy purposes will also lead to a scarcity of land and water for other uses. Additional land for biomass production is becoming scarce. The high moisture content (MC) of fresh biomass makes its collection and transport expensive. Thus, biomass energy conversion could be relatively inefficient if improperly done. There are biomass resources with low moisture (such as rice hull after milling), and these resources are ideal for direct utilization of energy without significant removal of water. Moreover, extensive utilization of these biomass waste resources may affect the price of the food from which the biomass was derived.

With depleting forest and agricultural resources for energy use, attention should be focused on new and emerging biomass technologies for their efficient conversion. There are several sources of energy for both rural or agricultural, and urban or industrial residues. There is a need to diversify the traditional resources for energy to meet local demands. These include, among others, (a) planting high-energy value crops, fast-growing trees, sugar and starchy crops, aquatic plants, oil and hydrocarbon crops, and (b) obtaining energy from municipal sewage and solid waste. For an in-depth understanding of biomass as a source of energy compared with traditional fossil fuel sources, some common properties of biomass resources will be reviewed in the subsequent chapters.

Example 1.1: World Energy Consumption in Other Units

Convert the US primary energy consumption of 98 EJ in units of Btu. Convert the Btu units into quads because Btu is a relatively smaller unit and is equivalent to 1,055 J.

Solution:

a. Btu

$$\text{Btu} = 98\,\text{EJ} \times \frac{\text{Btu}}{1{,}055\ \text{J}} \times \frac{1 \times 10^{18}\ \text{J}}{\text{EJ}} = 9.29 \times 10^{16}\ \text{Btu}$$

b. Quads

$$\text{Quad} = 9.29 \times 10^{16}\ \text{Btu} \times \frac{\text{quads}}{1 \times 10^{15}\ \text{Btu}} = 92.9\,\text{quads}$$

1.2 US ENERGY USE AND NEEDS

Table 1.1 summarizes the trend in the United States for primary energy production and consumption over a period of 11 years. The US energy consumption continues to rise but declined in 2020 due to the COVID pandemic. The breakdown of these energy sources is summarized in Table 1.2. Note that most energy consumed in the United States comes from petroleum, followed by natural gas. In the United States, there is a decline in the use of coal, which has a positive effect on the environment, and biomass consumption is still higher than that of most other renewable energies such as solar and wind.

The use of biomass energy in the United States comes primarily from wood and wood-derived fuels (52.5%) followed by ethanol and biodiesel (36.4%) (Nelson, 2011). In addition, energy is also derived from landfill gas and municipal solid waste (MSW) (11.1%). The use of biomass for commercial energy usage in the United States is still in its infancy, and more research is required for biomass resources to have a significant impact. The tracts of land available for biomass growth are quite vast; however, the significant necessity of water for widespread biomass growth may be an environmental and sustainability issue. The use of biomass energy in the United States must be viewed with purpose, and the issue of sustainability will always have to be considered. Renewable energy is defined as "energy resources that are replenished at the same rate that they are used." Any renewable energy resource will cease to be a renewable resource when it is used at a rate faster than it is produced.

TABLE 1.1
US Energy Summary over a Period of 11 Years (in Units of Quads)

Year	Production	Consumption	Exports	Imports
2010	74.9	97.5	8.2	29.9
2011	78.1	96.9	10.4	28.7
2012	79.2	94.4	11.3	27.1
2013	81.8	97.1	11.8	24.6
2014	87.7	98.3	12.3	23.2
2015	88.3	97.4	12.9	23.8
2016	84.3	97.3	14.1	25.4
2017	88.1	97.6	17.9	25.5
2018	95.8	101.2	21.2	24.8
2019	101.3	100.3	23.5	22.9
2020	95.7	92.9	23.5	20.0

Source: EIA, *US Energy Information Administration Annual Energy Review 2020*. Office of Energy Markets and End Use, US Department of Energy, Washington, DC, 2021.

TABLE 1.2
Breakdown of Energy Consumption in the United States for 11 Years (in Units of Quads)

Year	Coal	Natural Gas	Petroleum	Nuclear	Renewables	Totals
2010	20.8	24.6	35.3	8.4	8.3	97.4
2011	19.7	25.0	34.6	8.3	9.2	96.7
2012	17.4	26.1	33.8	8.1	8.8	94.2
2013	18.0	26.8	34.4	8.2	9.5	96.9
2014	18.0	27.4	34.7	8.3	9.7	98.1
2015	15.5	28.2	35.4	8.3	9.7	97.2
2016	14.2	28.4	35.7	8.4	10.4	97.1
2017	13.8	28.1	36.0	8.4	11.1	97.4
2018	13.3	31.2	36.9	8.4	11.3	101.0
2019	11.3	32.2	36.9	8.5	11.4	100.2
2020	9.2	31.5	32.2	8.2	11.6	92.8

Source: EIA, *US Energy Information Administration Annual Energy Review 2020*. Office of Energy Markets and End Use, US Department of Energy, Washington, DC, 2021.

1.3 BIOMASS ENERGY USAGE AND OVERALL ENERGY NEEDS IN OTHER COUNTRIES

Biomass is used differently in other countries. Most developing countries use biomass as a major source of energy for households and communities. Table 1.3 shows the significant use of fuel wood in many Asian and African countries reported as a percentage of their total energy usage. The country with the highest wood consumption is Nepal (98%) in Asia, followed by Mali (97%) in Africa. Today, biomass energy still produces, on average, around 30% of the energy consumed globally (Nelson, 2011). Most of this energy is used for cooking and heating, and people normally rely on fuelwood, charcoal, and animal manure (called dung in India) for these purposes.

1.4 FUTURE OF AGRICULTURE

The agricultural industry in the United States has been providing the growing population with food, feed, and fiber (the three "F's"). In the future, one additional "F" will be added to mean fuel (or biofuel if it comes primarily from biomass). Later, it will be shown that the amounts of biomass waste generated during agricultural production are usually more than the amounts of products produced and sometimes twice as much. Thus, these wastes can be converted into biofuels. It will be shown that the per capita energy consumption for human nutrition is also a small fraction of the energy produced from crop production. However, care must be taken with the extent of the use of biomass for making biofuels so that the agricultural land will still be productive or sustainable in the future. These biomass wastes are the primary source of nutrients for the next generation of crops and must remain in the soil for crop sustainability.

TABLE 1.3
Wood as Percentage of Total Energy Use in Asian and African Countries

Africa	%
Mali	97
Rwanda	96
Burkina Faso	95
Tanzania	95
Chad	85
Ethiopia	94
Central African Rep.	93
Somalia	91
Burundi	90
Niger	86
Benin	85
Nigeria	85
Cameroon	83
Sudan	82
Madagascar	80
Sierra Leone	76
Angola	75
Ghana	75
Mozambique	75
Kenya	70
Ivory Coast	46
Zambia	37
Asia	%
Nepal	98
Thailand	63
Sri Lanka	63
Pakistan	37
India	35
Malaysia	10

Source: Nelson, V., *Introduction to Renewable Energy*. CRC Press, Taylor & Francis Group, Boca Raton, FL, 2011.

In the future, one would expect the widespread use of all available biomass wastes for conversion into biofuels or chemicals. Biomass utilization will perhaps start with those resources that have the lowest costs, such as MSWs, sewage residues, and animal manure. The future of agriculture will demonstrate wastes recycled into energy, power, and liquid biofuels. MSWs may have to be segregated to separate the combustibles, which will be used to generate electrical power. Landfill gas is now being used to generate electrical power in many states. MSW is perhaps a biomass resource that has a negative feedstock cost, as MSW-processing companies are paid a tipping fee for collecting or accepting these wastes from municipalities. Municipal sludge is another biomass with a negative feedstock cost and is abundant in most cities and municipalities.

1.5 ADVANTAGES AND DISADVANTAGES IN USE OF BIOMASS AS ENERGY SOURCE

The inherent low energy density of biomass makes it difficult to gather and transport over long distances to a centrally located refinery. Most biomass resources have high moisture content (MC) on harvest, making on-site processing difficult. It has been reported that transportation of biomass such as corn stover to a biorefinery on an average distance of 64 km through rail may cost between \$55 and \$61 per dry megagram (or per tonne), whereas transportation by truck is more expensive at around \$71–\$77 per dry tonne (Turhollow and Sokhansanj, 2007). Some processes are now being used to convert biomass into high-energy-density material. This reduces moisture, and bulk density is increased. The amount of moisture in biomass is calculated by placing a wet biomass sample in a convection oven. The temperature is usually set to 103°C (217°F) (American Society for Testing and Materials Standard E 871–82 Reapproved 2006: Standard Test Method for Moisture Analysis of Particulate Wood Fuels) until the final weight of the material has stabilized, normally after 24 h. The MC, on a wet basis, is then calculated using Equation 1.1. Example 1.2 demonstrates the use of Equations 1.1 and 1.2:

$$\text{MC (wet basis)\%} = \frac{\text{Wet weight} - \text{Dry weight}}{\text{Wet weight}} \times 100\% \tag{1.1}$$

To calculate the MC on a dry basis, Equation 1.2 was used:

$$\text{MC (dry basis)} = \frac{\text{Wet weight} - \text{Dry weigh}}{\text{Dry weight}} \times 100\% \tag{1.2}$$

Example 1.2: Determining Moisture Contents of Biomass

Corn stover is harvested with around 25% moisture (wet basis). Determine the amount of moisture to be removed per 1,000 kg of raw corn stover if the ideal safe storage MC is 10%.

Solution:

a. The MC (wet basis) equation is as follows:

$$\text{MC (wet basis)\%} = \frac{\text{Wet weight} - \text{Dry weight}}{\text{Wet weight}} \times 100\%$$

b. The dry weight of the material after drying is calculated from the following equation:

$$25\% = \frac{1{,}000 \text{ kg} - \text{Dry weight}}{1{,}000 \text{ kg}} \times 100\%$$

$$\text{Dry weight} = 750 \text{ kg}$$

c. The new wet weight after drying at 10% MC is as follows:

$$10\% = \frac{\text{Wet weight} - 750 \text{ kg}}{\text{Wet weight}} \times 100\%$$

$$\text{New wet weight} = 833 \text{ kg}$$

d. Thus, the amount of moisture removed is 167 kg (1,000–833).

1.5.1 Advantages

As an energy resource, there are various advantages in the use of biomass as fuel energy. They are as follows:

1. It provides an effective source of low-sulfur fuel.
2. It is a relatively inexpensive source of energy.
3. It reduces environmental hazards (e.g., energy from sewage or animal manure).
4. It may generate additional coproducts that are valuable (as feed or as protein sources).

Clearly, the use of biomass has more advantages for energy and environmental sustainability concerns. Perhaps the only thing to remember is that to become a renewable energy resource, biomass must not be used at a much faster rate than it is produced by the sun, carbon dioxide, and water.

There are two distinct reasons why biomass is an ideal energy resource from the point of view of sustainability. First, it will not contribute to global warming because the carbon dioxide balance is always positive, and second, it will provide additional nutrients to the soil (improvement of soil structure) when byproducts of any conversion process are returned to the soil or ground.

The use of biomass near its production site is becoming popular to minimize transportation costs. The use of a local conversion process that would require minimal pretreatment of inputs is an important consideration for energy conversion.

1.5.2 Disadvantages

Although biomass could play a very important role in replacing nonrenewable sources of fuels, there are numerous disadvantages as listed below:

1. Many forms of biomass are available only seasonally (such as crop residues and in energy farms).

2. The conversion of sunlight into energy in the form of biomass is very low.
3. The concentration of biomass over a given area is also low, leading to high transport costs.
4. A scarcity of land for biomass cultivation exists in some regions of the world.
5. The average biomass has high MC.
6. The energy conversion of most processes is quite inefficient.

Biomass for energy must be grown in marginal land areas so it will not conflict with food sources. However, biomass grown on marginal lands will likewise produce marginal yields. The high MC makes this material difficult to transport over longer distances. Thus, the selection of a conversion process becomes very important. This is particularly evident if the biomass has properties that do not conform to a particular conversion process and would require expensive inputs and pretreatments.

1.6 SOURCES OF BIOMASS AVAILABLE FOR ENERGY USE

There are numerous sources of biomass for energy. This will be enumerated in the following sections. The most common biomass resources are those from agricultural and forestry operations. Some biomass resources are prolific in aquatic areas, whereas others are centrally produced for their specific components such as sugars or oils. A large majority of wastes that are not converted into fuel at present are found in animal farms. These wastes contribute to environmental degradation. However, there are various biomass resources available in many municipalities that have not yet been used in widespread energy conversion. Because of the necessity to generate energy locally, biomass resources such as MSW and sewage will become important in the future. Finally, renewable energy companies may grow biomass for energy. This will be termed "energy farming." This option will provide more freedom to produce biomass as quickly as possible. It will be possible to control some properties of biomass through breeding.

1.6.1 Crop Residues

In the pursuit of meeting the food requirements for increasing populations, humans have improved the yields of agricultural crops. For example, corn production in the 1940s was a dismal 1.5 tonnes/ha (0.67 tons/acre). Over the years, production has improved to its highest yield of more than 10.8 tonnes/ha (4.8 tons/acre) (NASS, USDA, 2021). The increase in yield also increases the amount of biomass residue produced. The waste biomass yield for most crops is greater than the yield of its primary product, as summarized in Table 1.4. Most biomass residues are returned to the field for field productivity and sustainability purposes. When using biomass for biofuel production, care should be taken to return some nutrients back to the field. There are many factors that should be considered to determine how much nutrient is removed from the soil if the biomass harvested is not returned to the field to maintain soil sustainability for some nutrients. For example, the phosphorous removal rate by harvesting corn will most likely be lower if only the corn grain is harvested. If a low

TABLE 1.4
Estimated Production of Agricultural Residue in United States Based from Current Crop Yield

Crop	Crop Yield (tons/acre/yr)	Crop Yield (tonnes/ha/yr)	Assumed Crop/Residue Ratio	Residue Production (tons/acre/yr)	Residue Production (tonnes/ha/yr)
Rice	3.8	8.5	1:2	7.6	17.1
Wheat	1.5	3.4	1:1.75	1.8	4.0
Corn	4.8	10.8	1:2.5	12.0	26.9
Sorghum	1.8	4.0	1:2.5	4.5	10.1
Barley	1.9	4.3	1:1.75	3.3	7.5
Cotton	0.4	0.9	1:2	0.8	1.8
Soybean	1.5	3.4	1:2	3.0	6.8

Source: NASS, USDA,2021. *Crop Production 2020 Annual Summary*. USDA, Washington, DC. Published January 21, 2021. Also available at: https://downloads.usda.library.cornell.edu/usda-esmis/files/k3569432s/w3764081j/5712n018r/cropan21.pdf. Accessed 7 March 2022.

yield of 9 tons/ha dry biomass (4 tons of dry biomass/acre) is harvested at 2.5 kg/tonne of P_2O_5 (5 lb P_2O_5/ton), the removal would be about 22.5 kg/ha of P_2O_5 (20 lb P_2O_5/acre) (Fernandez, 2012). The amount will be different for other nutrients. Thus, 100% of crop residues should not be used to produce biofuels owing to nutrient depletion.

Table 1.4 summarizes a classic estimate of crop yields for major agricultural crops, including their biomass residue yields. Thus, one may use the information found in Table 1.4 to make a preliminary estimate of the biomass residue produced for a specified area of land (see Example 1.3). In developing countries where rice is the staple food, rice production wastes (such as rice hulls and rice straw) become the most abundant biomass resource for bioenergy production. Rice, wheat, and corn are still the top three crops produced worldwide.

Example 1.3: Determining Biomass Wastes Given Acreage

Determine the amounts of biomass waste that can be generated from 500 ha (1,235 acres) of corn (assumed stover) each year. Compare this with that from rice production (assuming straw) on similar acreage. Use Table 1.4 for your data. Convert to English units.

Solution:

a. The amount of waste generated is based on Table 1.4 as follows:

$$\text{Corn stover (tonnes)} = 500\text{ ha} \times \frac{27\text{ tonnes}}{\text{ha-yr}} = 13{,}500\frac{\text{tonnes}}{\text{yr}}$$

$$\text{Corn stover (tons)} = \frac{13{,}500\ \text{tonnes}}{\text{yr}} \times \frac{1.1\ \text{tons}}{\text{tonne}} = 14{,}850\,\frac{\text{tons}}{\text{yr}}$$

b. The amount of waste generated from rice farming is calculated as follows:

$$\text{Rice straw (tonnes)} = 500\ \text{ha} \times \frac{17\ \text{tonnes}}{\text{ha} - \text{yr}} = 8{,}500\,\frac{\text{tonnes}}{\text{yr}}$$

$$\text{Rice straw (tons)} = \frac{8{,}500\ \text{tonnes}}{\text{yr}} \times \frac{1.1\ \text{tons}}{\text{tonne}} = 9{,}350\,\frac{\text{tons}}{\text{yr}}$$

c. Thus, there is more corn residue than rice per unit area.

1.6.2 Fuelwood

Fuelwood in most developed countries, such as the United States, is used mainly for industrial (72.8%) and housing needs (16.7%) (EIA, 2010). Fuelwood is seldom used for biofuels or power production in these countries. However, in the United States, some 8.4% of the total wood supply has been used for power generation (EIA, 2010). In developing countries, fuelwood may be one primary source of fuel for household needs. Unfortunately, the use of fuelwood in developing countries has depleted forestry resources, resulting in widespread flooding and erosion. Thus, replanting programs must be initiated in conjunction with the use of wood for energy and fuel. Few countries, such as Finland and Sweden, have significant use of fuelwood for energy production. Table 1.5 summarizes the forest area remaining in several selected developing countries and the per capita forest reserve. Assuming a sustainable forest energy yield of about 15 GJ/ha (Krausmann et al., 2007), a sustainable wood yield may be estimated. This value was compared with the nutrition requirements in selected countries to show that the forest reserves of some developing countries may not be enough to provide energy for human nutrition alone. Countries such as Nepal, the Philippines, India, and Kenya have per capita energy consumption greater than the sustainable wood yield. The US per capita energy consumption is 5.8 GJ/head (hd)/day (i.e., to satisfy daily nutrition only). This figure is higher than those reported in developing countries and is near the world average of 6 GJ/capita/yr. Most forest resources are still used for cooking needs such as charcoal and fuelwood. For example, in Tanzania, about 91% of total energy consumption comes from fuelwood sources (FAO, 2012).

In the United States, some states have embarked on power production from fuelwood, such as California in the early 1990s. However, because of widespread forest fires in California, there is hardly any excess biomass for power generation available. Recently, the City of Austin invested in a 115 MW biomass-fueled facility located in Sacul, Texas. This project is expected to yield 490 jobs in the area (Xu et al., 2008). Yields from wood waste have been constant for most years. The US total fuelwood resource is in the order of 1.5 million tonnes each year after taking out the moisture of about close to 50%. Thus, fuelwood is still an important biomass resource that

TABLE 1.5
Forest and Energy per Capita Resources in Selected Developing Countries (FAO, 2012)

Country	Population (10^6)	Forest Area Remaining (10^6ha)	Forest Area per Capita (ha)	Sustainable Wood Yield (GJ/Capita/yr)	Energy per Capita (GJ/Capita/yr)
Congo	4.0	22.4	5.6	84	3.9
Brazil	194.9	519.5	2.7	40.5	4.8
Argentina	40.4	29.4	0.73	10.9	4.6
Belize	0.30	1.4	4.7	70.5	4.1
Myanmar	50.1	31.8	0.63	9.5	4.4
Tanzania	44.8	33.4	0.75	11.3	3.1
Thailand	69.1	19.0	0.27	4.1	3.9
Nepal	30.0	3.6	0.12	1.8	3.6
Mali	15.4	12.5	0.81	12.2	3.9
Philippines	93.3	7.7	0.08	1.2	3.9
Zambia	13.1	49.4	3.8	57	2.9
India	1,224.6	68.4	0.05	0.75	3.6
Kenya	40.5	3.5	0.09	1.4	3.1
Bangladesh	148.7	1.4	0.009	0.14	3.5
Total	1,969.2	803.4	0.41	6.2	3.7
World	10,174	4,030	0.40	6.0	4.3

may be used for biofuel or power production. Note that for mill residue, about 97% of these forest biomass resources are being utilized by the forest product industry for production of steam, electricity, and other uses (SECO, 2005).

1.6.3 Aquatic Biomass

As early as 1982, the US Department of Energy (USDOE) established the Aquatic Species Program with the goal of implementing a plan to develop, demonstrate, and commercialize aquatic biomass energy systems for maximum contribution to US energy supplies (Kresovich et al., 1982). The yields of some fresh and salt-tolerant varieties of aquatic plants are very high under optimal conditions. The final report for the US algae project suggests more than 140,234 L/ha (15,000 gal/acre) per year of oil. Previous studies have not yet investigated saltwater biomass in nearby shores and oceans, such as giant kelp and seaweeds. Freshwater aquatic plants include water hyacinth (*Eichornia crassipes*), duckweed (*Lemma*), cattails (*Typha latifolia*), and others. More recently, funding toward investigating the potential of microalgae species for biofuel production has increased. Perhaps the only difficulty with biomass propagated from large bodies of water is the appropriate harvesting and extracting methods. In addition, when dry materials are required for the conversion process, the high MC of aquatic plants would add drying and material-handling costs. Aquatic biomass will become an important energy resource in the future, considering that

75% of the earth is composed of water systems. The list of potential aquatic species will be discussed in Section 1.6.8.

1.6.4 Sugar Crops

There are numerous sugar crops that are potential sources of energy for conversion into biofuels. These sugar crops have been used for hundreds of years because certain strains of yeast were discovered to convert sugary materials into alcohol through fermentation. The most popular strain of yeast is *Saccharomyces cerevisiae*, which can convert glucose ($C_6H_{12}O_6$) into ethyl alcohol (C_2H_5OH) and carbon dioxide (CO_2), as shown in Equation 1.3.

$$C_6H_{12}O_6 + \text{Yeast} \rightarrow 2C_2H_6O + 2CO_2 + \text{Heat} \tag{1.3}$$

In the batch ethanol conversion process, the substrate is diluted to a sugar content of about 20% by weight, acidified to a pH of between 4 and 5. This pH ensures that only this yeast will survive and ferment for 3–5 days. Beer (with about 10% alcohol) will be produced at the end of this fermentation process. The beer will then be distilled to generate a higher percentage of alcohol for fuel use. About 3.785 L (21,257 kcal) may be obtained from about 5.85 kg of pure sugar (21,840 kcal) so that there is almost no energy loss in the fermentation process.

The sugar from sugarcane is still one of the most popular sources of feedstock for ethanol production, such as in the large ethanol program in Brazil. Many developing countries are now investigating the use of sweet sorghum for similar goals.

When a starchy material, such as corn, grain sorghum, or barley, is used as a substrate, the starch must be converted first into fermentable sugar before yeast fermentation. The decomposition of large molecules of starch requires the catalytic action of certain enzymes also produced by microorganisms. The most popular microbe for this step is *Aspergillus niger*, a mold that produces or secretes those enzymes. Crops that are potential substrates to produce bioethanol in units of L/kg are listed in Table 1.6. The potential ethanol yields of some sugar and starchy crops in units of L/ha (or gal/acre) are summarized in Table 1.7. Jerusalem artichoke has the highest yield of ethanol per hectare, primarily because this is accomplished by harvesting three heads per year. The primary source of ethanol produced in the United States is corn. Example 1.4 shows how this table may be used to make a preliminary estimate of large-scale ethanol production. About 810 ha (2,000 acres) may be needed to generate 3.785 million liters of ethanol each year (or 1 million gallons per year [1 MGY]).

Example 1.4: Determining Acreage to Generate a 3.785 MLY Ethanol Plant (1 MGY)

Determine the number of hectares required to establish a 3.785 million liters per year (MLY) (~1 MGY) ethanol plant based from sugarcane in the state of Hawaii with a yield of 8,000 L/ha-yr of ethanol. Compare this with that from sorghum cane ethanol production, with a yield of 4,674 L/ha-yr. Convert units in the English system.

TABLE 1.6
Potential Ethanol Yields of Some Sugar and Starchy Crops in L/kg

Crop	Residue/ Crop Ratio	Dry Matter (%)	Lignin (%)	Carbohydrates (%)	Ethanol Yield (L/kg Dry Biomass)
Barley	1.2	88.7	2.90	67.10	0.41
Barley straw		81.0	9.00	70.00	0.31
Corn	1.0	86.2	0.60	73.70	0.46
Corn stover		78.5	18.69	58.29	0.29
Oat	1.3	89.1	4.00	65.60	0.41
Oat straw		90.1	13.75	59.10	0.26
Rice	1.4	88.6		87.50	0.48
Rice straw		88.0	7.13	49.33	0.28
Sorghum	1.3	89.0	1.40	71.60	0.44
Sorghum straw		88.0	15.00	61.00	0.27
Wheat	1.3	89.1		38.85	0.40
Wheat straw		90.1	16.00	54.00	0.29
Sugarcane		26.0		67.00	0.50
Bagasse	0.6	71.0	14.50	67.15	0.28

Source: Kim, S. and B. E. Dale, Global potential bioethanol production from wasted crops and crop residues. *Biomass and Bioenergy* 26(2004):361–375, 2004.

Solution:

a. The acreage required can be easily calculated as follows:

$$\text{Acreage (ha)} = \frac{3.785\times10^{6}\ \text{L}}{\text{yr}} \times \frac{\text{ha}-\text{yr}}{8{,}000\ \text{L}} = 473\ \text{ha}\left[1{,}169\ \text{acres}\right]$$

b. The acreage required for sweet sorghum is as follows:

$$\text{Acreage (ha)} = \frac{3.785\times10^{6}\ \text{L}}{\text{yr}} \times \frac{\text{ha}-\text{yr}}{4{,}674\ \text{L}} = 810\ \text{ha}\left[2{,}000\ \text{acres}\right]$$

c. Note the tremendous increase in acreage requirements when the yield is very low.

1.6.5 Oil Crops

Oil crops are primarily cultivated for their use as food; however, there are excess materials that can be used for biofuel production. Unfortunately, the use of materials that are primarily intended for food will affect their use as biofuel feedstock, thereby creating a food versus fuel dilemma. In the United States, soybean oil has been the

TABLE 1.7
Potential Ethanol Yields of Some Sugar and Starchy Crops in L/ha and gal/acre

Crop Material	Yield (L/ha)	Yield (gal/acre)
Jerusalem artichoke	11,219	1,200
Sugarcane (HI)	8,311	889
Sugarcane (LA)	5,189	555
Sorghum cane	4,674	500
Sugar beets	3,852	412
Potatoes	2,795	299
Corn	2,001	214
Sweet potatoes	1,776	190
Rice rough	1,636	175
Apples	1,309	140
Dates, dry	1,178	126
Grain Sorghum	1,169	125
Carrots	1,131	121
Raisins	951	101.7
Yams	879	94
Grapes	845	90.4
Peaches	785	84
Barley	776	83
Prunes, dry	774	82.8
Wheat	739	79
Pineapples	729	78
Oats	533	57
Rye	505	54
Pears	461	49.3
Molasses	421	45
Apricots	383	41
Buckwheat	320	34.2
Figs, fresh	294	31.5
Figs, dry	276	29.5
Plums (non-prunes)	204	21.8

Source: Adapted from Mother Earth News. 1980. Mother earth alcohol fuel. Chapter 3. Available at: http://www.motherearthnews.com, with permission.

predominant feedstock for biodiesel production (an ester of oil) when the price per gallon was still below the diesel cost. However, in 2006, the price of soybean oil per gallon exceeded $4, whereas the cost of diesel stayed below $3 per gallon. In this scenario, it was not economical to produce the biofuel if the feedstock cost exceeded the product cost (unless subsidies are in place). As a result, there was a massive shift in the use of low-cost oil feedstock, such as animal fats and oil (e.g., chicken oil as well as waste oil from restaurants). Various crops are potential sources of oil for biofuel production. Table 1.8 summarizes the potential seed yields of oil for major crops in units of kg/ha (or lb/acre). The estimated biodiesel production yields in L/ha (gal/acre) are also summarized in the table. Table 1.9, on the other hand, presents some yields of oil for some major crops in units of kg/ha. Tropical palm and coconut trees are still the largest producers of oil per unit area of land. To convert the weight unit into volume, the density of oil must be known. The reported density of soybean oil, for example, is about 916 kg/m^3 (7.6 lb/gal) (Noureddini et al., 1992).

TABLE 1.8
Various Seed Yields of Oil for Major Crops and Equivalent Biodiesel Production per Unit Area

Plant	Seed Yield (lb/acre)	Biodiesel (gal/acre)	Seed Yield (kg/ha)	Biodiesel (L/ha)
Corn	7,800	18	8,757	168
Oats	3,600	23	4,042	215
Cotton	1,000	35	1,123	327
Soybean	2,000	48	2,245	449
Mustard	1,400	61	1,572	570
Camelina	1,500	62	1,684	580
Crambe	1,000	65	1,123	608
Safflower	1,500	83	1,684	776
Rice	6,600	88	7,410	823
Sunflower	1,200	100	1,347	935
Peanut	2,800	113	3,144	1,056
Rapeseed	2,000	127	2,245	1,187
Coconut	3,600	287	4,042	2,683
Oil Palm	6,251	635	7,018	5,937

Source: Collins, H. P., A. Hang, S. Fransen and P. Wanderschnieder, Biofuel Variety Trials Factsheet. USDA-ARS Publications, Washington State University, Prosser, WA. Also available at http://www.usda.prosser.wsu.edu. Accessed August 28, 2012. Bioenergy Washington, Washington's Website for BioEnergy Information, 2012.

TABLE 1.9
Various Yields of Oil for Major Crops in Units of kg/ha

Material	Yield (kg/ha)
Corn	145
Cashew nut	148
Oat	183
Palm	190
Lupine	195
Rubber seed	218
Kenaf	230

(Continued)

TABLE 1.9 (*Continued*)
Various Yields of Oil for Major Crops in Units of kg/ha

Material	Yield (kg/ha)
Calendula	256
Cotton	274
Hemp	305
Soybean	376
Coffee	387
Linseed	403
Hazelnut	405
Euphorbia	441
Pumpkin	450
Coriander	451
Mustard	482
Camelina	491
Sesame	586
Crambe	591
Safflower	657
Buffalo gourd	667
Rice	697
Tung oil tree	792
Sunflower	802
Cocoa	865
Peanut	891
Opium poppy	980
Rapeseed	1,003
Olive tree	1,022
Piassava	1,115
Gopher	1,122
Castor	1,190
Bacuri	1,199
Pecan	1,508
Jojoba	1,531
Babasu	1,544
Jatropha	1,593
Macadamia	1,891
Brazil nut	2,014
Avocado	2,222
Coconut	2,265
Oiticia	2,525
Buriti palm	2,748
Pequi	3,148
Macauba	3,782
Palm oil	5,011

Source: Kemp, W. H.,*Biodiesel Basics and Beyond: A Comprehensive Guide to Production and Use for the Home and Farm*. Aztext Press, Ontario, Canada, 2006.

Example 1.5 illustrates how to estimate the acreage required to produce several million liters (or gallons) of oil per year.

Example 1.5: Determining Acreage to Generate a 15 MLY Biodiesel Plant (4 MGY)

Determine the number of hectares required to establish a 15 million liters per year (~4 MGY) biodiesel plant based on Jatropha having a yield of 1,600 kg/ha. Compare this with that from soybean oil with a yield of 376 kg/ha-yr. The density of Jatropha oil is 903 kg/m³ (Akbar et al., 2009), whereas that of soybean oil is 920 kg/m³. Convert units in the English system.

Solution:

a. The acreage required for Jatropha can be easily calculated as follows:

$$\text{Acreage (ha)} = \frac{15\times10^6\ \text{L}}{\text{yr}} \times \frac{\text{ha}-\text{yr}}{1{,}600\ \text{kg}} \times \frac{903\ \text{kg}}{\text{m}^3} \times \frac{1\ \text{m}^3}{1{,}000\ \text{L}}$$

$$= 8{,}466\ \text{ha}\left[20{,}910\ \text{acres}\right]$$

b. The acreage required for soybean is as follows:

$$\text{Acreage (ha)} = \frac{15\times10^6\ \text{L}}{\text{yr}} \times \frac{\text{ha}-\text{yr}}{376\ \text{kg}} \times \frac{920\ \text{kg}}{\text{m}^3} \times \frac{1\ \text{m}^3}{1{,}000\ \text{L}}$$

$$= 36{,}702\ \text{ha}\left[90{,}654\ \text{acres}\right]$$

c. Note the tremendous increase in acreage requirements when yield is very low.

1.6.6 Animal Manure

The domestication of animals and the production of these animals in confined feeding operations in developed countries lead to the production of an abundant amount of manure in one location. Thus, installation of larger scale conversion systems is appropriate to take advantage of economies of scale. In developing countries, small modular manure conversion systems have been established, with anaerobic digestion as the most common conversion process to produce biogas (methane and carbon dioxide). Table 1.10 summarizes the estimated manure production by domesticated animals. A more detailed estimation procedure for manure production is discussed in the anaerobic digestion or biogas production in Chapter 7. Note that in Table 1.10, it is assumed that these animals are the mature and finishing types. The amount of manure varies with the weight of the animal as well as the stage of growth. Estimation of manure production should be made. There is a significant difference between the volume of manure excreted by the animal and that which can be practically recovered. As summarized in Table 1.10, recovery varies from 33% (beef cattle) to as high

TABLE 1.10
Manure Production of Some Major Farm Animals

Animal	Manure Production (kg/hd/day)	Manure Production (lb/hd/day)	Manure Production (kg/hd/day)	Manure Production (lb/hd/day)
	As Excreted		As Removed	
Beef cattle	22	50	7.5	16.5
Dairy	68	150	35	77
Horse	25	59	8.5[a]	18.7
Layer	0.088	0.19	0.03	0.07
Swine	5.0	11	3.5	7.7

Source: Adapted from ASABE D384.2 Standards, *Manure Production and Characteristics*. ASABE, St. Joseph, MI, 2010.

[a] Assumed 1/3 recovery.

as 70% (for swine). Manure collection from confined animal-feeding operations is higher than from scattered feedlots. For dairies, about 51% of the manure excreted by the animals may be recovered for energy purposes. Note that the MC of these excretions is rather high. It could range from a low of 75% in chicken litter to as high as 90% in swine. Thus, it is likely that conversion systems selected for animal manure would be those that would not require drying of the material to reduce preprocessing costs. Example 1.6 makes an estimate of the amount of manure generated in a cattle feedyard facility.

Example 1.6: Determining the Amount of Manure Generated in a Feedyard Facility

Determine the amount of manure (tonnes/day) (tons/day) generated each day in a 40,000-head cattle facility. The estimated practical recoverable manure is around 2.3 kg/hd/day (5 lb/hd/day). If the energy content of this manure is 19.73 MJ/kg (8,500 Btu/lb), how much energy is contained in this amount of manure each day in MJ? If this manure is to be converted into energy in 1 day, how much power may be released in MW?

Solution:

a. This is a simple conversion calculation as shown below:

$$\text{Manure}\left(\frac{\text{tonnes}}{\text{day}}\right) = 40{,}000\ \text{hd} \times \frac{2.3\ \text{kg}}{\text{hd-day}} \times \frac{\text{tonne}}{1{,}000\ \text{kg}}$$

$$= 92\ \text{tonnes}\left[101\,\text{tons}\right]$$

b. The energy contained in the manure is calculated as follows:

$$\text{Energy (MJ)} = 92 \text{ tonnes} \times \frac{1{,}000 \text{ kg}}{1 \text{ tonne}} \times \frac{19.73 \text{ MJ}}{\text{kg}}$$

$$= 1{,}815{,}160 \text{ MJ}\left[1{,}721 \text{ Million Btu}\right]$$

c. If this is to be converted into power in 1 day, the power calculation is as follows:

$$\text{Power (MW)} = \frac{1{,}815{,}160 \text{ MJ}}{1 \text{ day}} \times \frac{1 \text{ day}}{24 \text{ h}} \times \frac{1 \text{ h}}{3{,}600 \text{ s}} \times \frac{\text{s} - \text{MW}}{\text{MJ}} = 21 \text{ MW}$$

1.6.7 Municipal Solid Wastes

Each day, each household generates abundant waste, which is dumped in landfills in every country. Some communities segregate recyclables. Recyclables include bottles, metals, or plastics, whereas wastes with no significant value are directly brought to landfills. There are facilities that convert these organic, biodegradable materials into methane gas, which is subsequently converted into electrical power. The US Environmental Protection Agency (USEPA) has collected and reported data on waste generation and disposal in the United States for more than 50 years. This information is used for waste reduction and recycling programs. In 2018, the United States generated more than 265.8 million tonnes (292.4 million tons) of MSW and recycled and composted over 85.5 million tonnes (94 million tons) of this material. The per capita waste generated increased from 1.22 kg/hd/day in 1960 to 2.23 kg/hd/day in 2018 (2.68–4.9 lb/hd/day) (USEPA, 2022).

Waste generation in other countries varies according to their level of development. Many developing countries do not have sophisticated handling systems for household waste, and such waste would be difficult to collect and convert. In addition, waste generation may also vary significantly according to the level of development in each country. To estimate the waste generation in each country, one would have to base the calculation on the waste generation based on the different stages of the economy. The estimate would be from 1 to 2 kg/person/day. The heating value of segregated MSW varies. The heating value of MSW is higher compared with agricultural biomass. Analysis of segregated waste from a municipality such as Dallas, Texas, shows a heating value (as received) of about 24.9 MJ/kg (10,729 Btu/lb). Example 1.7 shows how to estimate power from MSW.

Example 1.7: Power from Municipal Solid Wastes

An analysis of segregated MSWs (paper and combustibles only) showed that the energy content as received was found to be 24.67 MJ/kg (10,632 Btu/lb). In a municipality with a population of 2.2 million people (e.g., Houston) and a per capita waste generation rate of 2 kg/person/day, how much MSW is generated per day in tonnes, and how much energy is contained in the MSW generated each day in million MJ? If this is to be converted into power in 1 day, how much power may be released in MW assuming 10% conversion efficiency?

Solution:

a. The amount of MSW generated each day is calculated as follows:

$$\text{MSW (tonnes)} = 2.2\times10^6\ \text{hd}\times\frac{2\ \text{kg}}{\text{hd-day}}\times\frac{\text{tonne}}{1{,}000\ \text{kg}}$$

$$= 4{,}400\ \text{tonnes}\left[4{,}840\ \text{tons}\right]$$

b. The energy contained in this MSW is calculated as follows:

$$\text{Energy}\left(10^6\ \text{MJ}\right) = 4{,}400\ \text{tonnes}\times\frac{1{,}000\ \text{kg}}{1\ \text{tonne}}\times\frac{24.67\ \text{MJ}}{\text{kg}} = 108.5\times10^6\ \text{MJ}$$

c. The calculation of power in 1 day, with a conversion efficiency of only 10%, is as follows:

$$\text{Power (MW)} = \frac{108.5\times10^6\ \text{MJ}}{1\ \text{day}}\times\frac{1\ \text{day}}{86{,}400\ \text{s}}\times\frac{\text{s}-\text{MW}}{\text{MJ}}\times0.10 = 125.6\ \text{MW}$$

d. This is quite a lot of power, but it is very small compared with their electrical power consumption.

1.6.8 Energy Farming

Biomass could be grown specifically for the purpose of producing energy for heat, electrical, or mechanical power generation. This category would include the following items:

1. Fast-growing trees
2. Sugar and starchy crops
3. Oil and hydrocarbon crops
4. Herbaceous crops
5. Aquatic plants

Wood energy is already a significant contributor to the overall biomass consumption in the world. However, most of the wood resources are traditional crops already grown for housing needs and are not suitable for energy farming, because of their slow growth and low energy density. It is logical to grow new varieties of wood species that are fast growing, highly tolerant of harsh weather with very high productivity per unit of area, and with minimal contributions for growth, such as nutrients and water. Some breeding programs must be initiated to address these issues, and it may take years before a potential biomass crop could become a candidate as an energy crop.

There are also additional environmental aspects in growing new breeds of biomass plants to solve the eroding forest cover, improve erosion characteristics, and retain moisture in addition to being used as an energy crop. Conventional forests may only produce between 2 and 7 dry tonnes equivalent (DTE) per hectare per year, but new energy crops such as high-tonnage sorghum could reach a potential yield of as much as 10–15 DTE/ha/yr (Rooney et al., 2007).

Plants can be grown specifically for energy production. A number of options are available to address the diverse nature of biomass as well as its low-energy yields from photosynthetic conversion of energy from the sun. Thus, "energy farming" could be initiated where biomass resources are produced specifically for biofuel production such as fast-growing trees, sugar and starchy crops, oil and hydrocarbon crops, herbaceous crops, and aquatic plants such as microalgae.

1.6.8.1 Fast-Growing Trees

There are species of trees that qualify as energy crops. These species can be grown on a very short-rotation basis, providing yields that are better than conventional forest varieties. An example is the fast-growing *Camelina* species. Others include the following:

1. *Leucaena leucocephala*, also known commonly as giant "ipil-ipil"
2. Hybrid poplar
3. Tulip poplar
4. Weeping willow
5. Thuja green giant
6. Eucalyptus

The hybrid poplar is perhaps one of the fastest growing kinds of trees in the North American continent and can provide shade for a one-story house with 3 years' growth. This variety is wind and insect resistant. The weeping willow is probably the fastest growing shade tree. This plant will grow as much as 2.5–3 m (8–10 ft) a year under ideal conditions. *Leucaena leucocephala* is a leguminous tree and can fix nitrogen from the atmosphere and has the effect of improving the soil it grows in. This tree is also noted for producing a very high protein seed and foliage that can be used as animal feed. It was reported that yields as high as 30 DTE/ha/yr are possible (Bhattacharya, 1983).

For energy purposes, the criteria for these fast-growing trees are as follows:

1. Rapid growth even in poor soil
2. No adverse effect on the environment
3. Minimal management requirements
4. Disease and pest resistance
5. Ability to survive droughts and other ecological stresses
6. Ability to resprout from the stump
7. Low planting cost
8. Ability to improve soil erosion properties
9. High heating value
10. Suitability for multiple uses

11. Short rotation period of between 3 and 7 years
12. High yield per unit area

Fuelwood comprises most of the energy mix for most developed countries, because of high yields and growth over wide areas. This could also be the best option to generate enormous amounts of energy for biofuel or bio-power production. There are a number of fast-growing trees that could be grown that exceed conventional yields of from 2 to 7 DTE/ha/yr. Research on identifying high-yielding and fast-growing species of wood are underway in many developed countries. Candidates include various species of *Camelina*, giant "ipil-ipil" trees (*L. leucocephala*), and others. It appears that these trees can have yields as high as 30 DTE/ha/yr. *Leucaena leucocephala* species, for example, have yields from 2 to 20 DTE/ha/yr. This is a nitrogen-fixing species because it is considered a legume and can achieve 6 m (20 ft) in only 3 years (Bhattacharya, 1983).

The amount of land required to establish refineries will vary depending on the productivity of various tree species. For example, to build a 50 MW thermal power plant from biomass, around 20,000 ha (49,400 acres) is required if the yield of crops is >12 DTE/ha/yr. It will also depend on the process used for conversion. The conversion efficiencies for power production may range from a low of 15% to as high as 50%. In the previous illustration, the efficiency will be around 30%, as shown in Example 1.8.

There are also several varieties of short-rotation woody crops that may be grown for energy such as cottonwood, silver maple, black locust, and poplar. The average yield of these woody crops in Michigan is around 9–11.2 tonnes/ha/yr or 4–5 dry tons/acre/yr (Launder, 2002).

Example 1.8: Efficiency of Conversion of Biomass into Electrical Power

Determine the required conversion efficiency for producing 50 MW electrical power output using biomass. The heating value of the biomass (dry) is 20 MJ/kg. The biomass yield iis 12 tonnes (dry) per hectare per year. The total area devoted to this project is 20,000 ha. Assume the power plant is operated for 333 days/year and 24 h/day.

Solution:

a. This is a simple conversion process using the efficiency calculation introduced in this chapter as follows:

$$\%\ \text{Eff} = 50\ \text{MW} \times \frac{\text{kg}}{20\ \text{MJ}} \times \frac{\text{MJ}}{\text{MW}-\text{s}} \times \frac{\text{ha-yr}}{12\ \text{tonnes}}$$

$$\times \frac{1}{20{,}000\ \text{ha}} \frac{333\ \text{days}}{\text{yr}} \times \frac{3{,}600\ \text{s}}{\text{h}} \times \frac{24\ \text{h}}{\text{day}} \times \frac{\text{tonne}}{1{,}000\ \text{kg}} \times 100\% = 30\%$$

b. Thus, conversion technologies that can achieve a conversion efficiency of at least 30% are appropriate for this application.

1.6.8.2 Sugar and Starchy Crops

There are countries with available tracts of land that can be devoted to energy farming that do not compete with food production and can be grown with new varieties of sugar and starchy crops. Texas A&M University is considered the home of hybrid sorghum, where various varieties of the sorghum crop have been modified togenerate the most sugar, and a sorghum grain variety with starch more readily available than that of corn (Rooney et al., 2007). Both new hybrids have potential for growing crops specifically for energy. In developing countries, several varieties of root crops, such as cassava, have also been genetically modified and improved to produce high yields in a shorter time. Other root crops include sweet potatoes, hybrid corn, rice, and cereals; however, the latter varieties are more important for food than for energy. Sorghum crops are noted for being highly drought tolerant and high yielding.

There are also new breeds of sugar and starchy crops that have the potential for biofuel production. Hybrid sorghum, for example, can have higher yields in developing countries (>35 tonnes/ha/yr) than in arid regions of the world. Table 1.11 summarizes some yields reported across the US states for sugar, starchy, and dedicated lignocellulosic biomass.

The yields for dedicated biomass resources will depend on favorable soil and inputs for production (fertilizer and irrigation water). The yields will vary for marginal lands and areas that are deficient in irrigation water. Note that most of the bioethanol produced in the Unites States comes from corn, primarily a starchy material that has limited yields. Sugarcane still produces the most ethanol per hectare.

1.6.8.3 Oil and Hydrocarbon Crops

Several plants, such as various species of *Euphorbia*, can be grown just for their oil, or to some extent some hydrocarbon substances as part of their natural metabolism. *Euphorbia* is a highly diverse plant genus containing about 2,000 currently accepted species. It is perhaps one of the most diverse genera in the plant kingdom with a very important milky white latex product that is a potential feedstock for energy use. This product, when extracted properly, may produce some terpene-like products and

TABLE 1.11
Yields of Dedicated Crops and Variability of Yields across the US States

Crop	Ames, IA	Chariton, IA	Bushland, TX	Jay, FL
	(tonnes/ha/yr)			
Sweet sorghum	17.5	18.0		31.8
Forage sorghum	15.7	15.5	23.8	27.0
Photoperiod sorghum			31.5	
Corn	14.0	10.6	23.5	
Switchgrass	10.6	10.3		

Source: Rooney, W. L., J. Blumenthal, B. Bean and J. E. Mullet, Designing sorghum as a dedicated bioenergy feedstock. *Biofuels, Bioproducts and Biorefining* 1: 147, 2007.

esters that may be easily converted into biofuel. We have already enumerated some traditional oil crops, but there are superior oil crops that have excellent potential as energy crops. Table 1.9 summarizes several varieties of palm or coconut trees that can be included in this category. There have been varieties of hybrid coconut trees that are propagated with a short rotation period: although dwarfish, they produce a more decent yield than conventional varieties. Coconut and palm trees yield oil from 1.5 to 5.0 tonnes/ha/yr under good conditions. The highest yields have been reported to be from 3.3 to 8.4 tonnes/ha/yr (Bhattacharya, 1983). The oil yields from ground crops such as sunflower, castor, soybean, and ground nut only range from 0.6 to 1.2 tonnes/ha/yr, with a growing season lasting between 120 and 160 days.

There are also crops that have very high hydrocarbon yields and are suitable, not only for biofuels but also for biochemicals. Hydrocarbon plants are also called petroleum plants. These crops produce certain terpenoids as their secondary metabolites. It may be economically feasible to convert these terpenoids into gasoline-like fuels. Nobel laureate Melvin Calvin has investigated the terpene biosynthetic pathways of several species of plants to study and quantify the properties of the compounds. Various species of *Euphorbia* were used in this study.

There are quite a few species of plants that produce hydrocarbon or hydrocarbon-like substances, including high yields of oil. This is normally part of the plant's metabolism. These compounds are terpenoids and come from latex-producing plants such as *Euphorbiaceae* and rubber species. For these feedstocks, all that is needed is a simple extraction procedure and purification. The resulting fuel can be readily used in many conversion facilities such as commercial engines and boilers. The production of high-quality oil from specialized species of microalgae such as *Botryococcus brunii* has also received considerable attention lately because of the production of near-hydrocarbon-type compounds. Thus, to be economical, these plant species must be cheap to produce and high yielding, and must require the least amount of investment for harvesting, extraction, and production in a biorefinery.

Hydrocarbon crops are also quite rare. Consequently, their propagation and cultivation are not well understood. The other examples of oil and hydrocarbon crops are as follows:

1. "Hanga" plant
2. Milk weed (*Asclepiad speciosa*)
3. Tree legume (*Copaifera multijuga*)

Extensive research must be conducted to make hydrocarbon production from these species economical. For example, the growth rate of *B. brunii* is slower than most microalgae species because they double their weight in about a week. On the other hand, high-yielding algae species double their weight in a few hours, assuming adequate nutrient adjustment and feeding.

1.6.8.4 Herbaceous Crops

In the United States, some bioenergy and bioproduct efforts have been focused on the growing of herbaceous grasses and short-rotation woody crops that can be used for fuel production. All high-yielding grasses are potential candidates for fuel production as well as the basis for cellulosic feedstocks. Of the herbaceous crops, switchgrass has received the greatest attention (*Panicum virgatum*). The major focus of the European community has been *Miscanthus*, which has also begun attracting attention in the United States. Other potential candidates include the invasive species *Arundo donax* and reed canary grass. Switchgrass may be one of the largest sources of ethanol in the United States. The highest switchgrass yield in the United States ranged from 11.5 to 14.4 dry tonnes/ha (5.1–6.4 dry tons/acre) (Walsh et al., 2007). If switchgrass is to be used for ethanol production from lignocellulosic biomass, and to satisfy the US requirement for ethanol by the year 2015, around 69 million dry tonnes (76 million dry tons) of switchgrass would need to be produced. Other cellulosic biomasses that will have to be produced would include 86 million dry tonnes (95 million dry tons) of crop residues and 48 million dry tonnes (53 million dry tons) of forest residues (Walsh et al., 2007).

Other examples of herbaceous crops include napier grass, also called elephant grass. A hybrid napier grass (Sordan 7a) is the outstanding short-rotation grass with an average yield of 27 DTE/ha/yr. There are prospects of increasing yields of up to 90 DTE/ha/yr. It is likely that fertilizer, water, and agronomic requirements for their growth would be less than those for more sensitive crops, such as sugarcane (Bhattacharya, 1983).

A great number of grass species have the potential for biofuel production. The USDOE has reported studies on the potential of switchgrass for biofuel production. The other species are as follows:

1. Napier grass
2. Miscanthus
3. Bluestem
4. Elephant grass
5. Wheat grass

These grass species (e.g., switchgrass) are perennials. Consequently, they do not need to be planted each year and would require less water and fertilizer for growth. In addition, they have an extensive root system that prevents soil erosion. They are resistant to pests and diseases and can grow up to 3 m in one season (Launder, 2007). The average yield in Michigan is 11.2 tonnes/ha/yr (5 tons/acre/yr).

1.6.8.5 Aquatic Plants

The yields of some freshwater or saltwater plant species could be as high as the yields of terrestrial crops. Consequently, they are also potential energy crops. Some reported yields are in the order of about 50 DTE/ha/yr (see Table 1.12) and may exceed some land-based crops. Some issues concerning these plants are their harvesting methods as well as the high MC of the plant species, including high salt content for marine types.

A common example of a rapid growing sea species is *E. crassipes*, also known as water hyacinth. Water hyacinth is a very prolific and fast-growing aquatic plant, and in some instances, it is an invasive and damaging species in waterways and canals.

TABLE 1.12
Emergent Aquatic Plants for Biomass Energy Systems Development

Biomass Type	Yield (dry tonnes/ha/yr)
Arundo donax L.	12–38
Cyperus papyrus L.	30–50
Phragmites communis Trin.	
Aboveground biomass	15–35
Belowground biomass	1–14
Saccharum spontaneoum L.	15–40
Spartina alterniflora Loisel	
Aboveground biomass	5–22
Belowground biomass	1–8
Typha latiflora L.	
Aboveground biomass	11–33
Belowground biomass	14–26

Source: Kresovich, S., C. K. Wagner, D. A. Scanland, S. S. Groet and W. T. Lawton, The utilization of emergent aquatic plants for biomass energy systems development. USDOE Report Under Contract No. EG-77-C-01–4042. USDOE, Golden CO, 1982.

Others include duckweed, hydrilla, water lilies, and cattails. A more recent invasive and near-aquatic species is the alligator grass (*Alternanthera philoxeroides*), which thrives under water-logged conditions in mining areas.

Alligator grass is a potential prolific biomass feedstock for biofuel production. This biomass resource is considered an invasive species and will not compete with agricultural food or feed production. It can generate more than 11.7 tonnes/ha (wet) or 11.7 tonnes/ha/yr (dry at 85% moisture) of biomass continuously throughout the year (Schooler et al., 2008). Alligator grass can be harvested in a 45-day rotation to provide biomass continuously, thereby improving the practical issues of harvest and transport. This biomass resource is more of an aquatic plant rather than a terrestrial biomass: it is best suited to produce methane gas through anaerobic digestion, while converting the sludge into other biofuels using thermochemical conversion processes such as pyrolysis and gasification. This feedstock can be processed using readily available equipment. Common grass/hay harvesting equipment can be used for harvesting the material. The high MC of the freshly harvested material makes it an ideal feedstock for anaerobic digestion processes to produce biogas ($CH_4 + CO_2$).

Some saltwater and freshwater plants can produce higher biomass yields than land-based crops, assuming optimal conditions of good sunlight, available nutrients, and other environmental conditions for growth. The most popular of those plants are the water hyacinth, duckweed, hydrilla species, cattails, and other fresh-growing plants such as alligator grass. Most of these aquatic species are invasive in nature. Consequently, the removal and harvesting of such plants will be beneficial to waterways and tributaries.

Saltwater species are also prolific, such as giant kelp, a seaweed. The only major issues with these resources are difficulty in the harvesting processes and the need to remove the high moisture, if a relatively dry feedstock is needed.

The USDOE conducted a study several decades ago (Kresovich et al., 1982), which identified various emergent aquatic plants for biomass energy production systems. The yield data are summarized in Table 1.12. The contribution of both below- and aboveground biomass could be as high as 50 DTE/ha/yr (e.g., *Cyperus papyrus* L.). Some aquatic plants have about the same amount of biomass produced below and above ground, emphasizing the prolific production of roots in aquatic systems.

Thus far, we have enumerated various biomass resources that may be used as energy crops for biofuel production. However, biomass characteristics are quite diverse in terms of heating values, MC, and ash components. Chapter 2 explains biomass conversion systems, followed by ways to characterize specific biomass for various applications. One will have to be familiar with the conversion units used. A review and discussion of conversion units is included in the Appendices of this textbook. Some examples are provided in Section 1.7.

1.7 UNITS AND CONVERSIONS

Throughout this textbook, the metric system of units (or SI, Systems International) is used, and the students should start remembering the basic units of conversion from the SI system to the English system of units. Appendix A enumerates some conversion units to use. The first important practice that a student should remember is the proper use of appropriate prefixes for some units such as megawatts (MW), gigawatts (GW), megajoules (MJ), or gigajoules (GJ). Some units are too large or too small, so prefixes are very important to remember. Example 1.9 shows how these prefixes are to be used properly in reporting answers to problems.

Example 1.9: Use of Appropriate Prefix

Power plants are normally reported in megawatts (1×10^6 W) because 1 W is a relatively small unit of power. If a 50 MW biomass powered facility operates for 24 h a day and 360 days a year, how much energy (in MWh or GWh, whichever is appropriate) is produced during this period? Report the units with the appropriate prefix.

Solution:

a. Energy is power multiplied by time. Thus, the estimated numbers, when these values are multiplied, are as follows:

$$\text{Energy}(\text{MWh}) = 50\,\text{MW} \times \frac{24\text{ h}}{\text{day}} \times \frac{360\text{ days}}{\text{yr}} = 432{,}000\,\text{MWh}$$

b. To keep the values within three significant digits, the appropriate prefix should be divided by 1,000 to report a much better value with minimal zeros. The energy should be reported as 432 GWh (gigawatt-hours).

Another important thing to remember with units and conversions is the order of magnitude of some fuels that are being produced from biomass or other sources. In the oil industry, the use of barrel as a unit (which is equivalent to 42 US gallons) is very common—conventional fossil fuels such as crude oil as well as transport fuels such as gasoline and diesel are used abundantly, and the fuel industry uses the unit of barrel instead of the typical units of volume reported and used by the consumers (i.e., liters or gallons). Thus, students must be aware of these orders of magnitude in reporting fuel production. One of the most common conversion factors for fuel production is that 3.785 L is equivalent to a gallon. Example 1.10 illustrates these orders of magnitude of unit used for biomass fuel production.

Sizes of biofuel plants or biorefineries are also reported in millions of liters (or gallons) a year, and thus, the proper prefixes must also be used because a liter of a gallon is such a small unit to describe these facilities.

Example 1.10: Barrels of Crude Oil

In 2010, the United States consumed around 20 million barrels of crude oil per day. Convert this unit into liters and gallons. Use appropriate prefixes to report your answer.

Solution:

a. The simple conversion factor to remember is that there are 42 US gallons in a barrel and there are 3.785 L/gal. Thus, to convert this unit into liters, the solution is as follows:

$$\text{Liters (L)} = \frac{20 \times 10^6 \text{ barrels}}{\text{day}} \times \frac{42 \text{ US gal}}{1 \text{ barrel}}$$

$$\times \frac{3.875\text{L}}{1\,\text{gal}} = 3{,}179{,}400{,}000 \text{ L} \left[3.2 \text{ gigaliters, GL}\right]$$

b. To convert this into gallons, a similar procedure is done. The giga (G) and mega (M) prefixes are appropriate because liters and gallons are very small numbers when it comes to describing daily crude oil consumption of a huge country such as the United States.

$$\text{Gallons}(\text{gal}) = \frac{20 \times 10^6 \text{ barrels}}{\text{day}} \times \frac{42 \text{ US gal}}{1 \text{ barrel}} = 840{,}000{,}000 \text{ gal} \left[840 \text{ M gal}\right]$$

1.8 CONCLUSION

There are numerous biomass resources available for conversion into heat energy and biofuel, or to produce power. Several biomass resources are underutilized and may be used in the developing world as fossil fuels become scarce and expensive. Biomass resources are produced during crop production for the delivery of food and fiber, and most are simply returned to the field for recycling of nutrients. However, it is shown that these resources are very high in energy potential and can contribute to each country's energy mix. Although there are numerous advantages in the use of biomass as energy resources, there are also many disadvantages. Foremost is the high MC of most biomass as well as its dispersed distribution across a large area of land. Thus, transportation of biomass and related logistical issues must be considered. In the subsequent chapters, the characteristics of biomass will be discussed, including methods for analysis followed by a discussion of various conversion technologies.

There is a rapid development of technologies for the conversion of biomass into heat energy, fuel, and power. Countries should take advantage of these rapidly developing technologies. However, as more conversion technologies are developed, the next constraint will be on biomass resources. If biomass energy is used inappropriately, it cannot be considered a renewable energy resource. Thus, methods to diversify the biomass resource base must be developed in conjunction with the use of emerging technologies for conversion. Sources of biomass must be diversified beyond traditional crop residues, livestock manure, and fuelwood to include culturing energy crops such as aquatic biomass (e.g., algae), while recovering energy from MSW and sewage.

Attempts must be made to make use of more efficient equipment and technologies for energy utilization. In anaerobic digestion, for example, many countries are still using ancient first-generation anaerobic reactors, although high-rate biogas reactors have gained popularity. In implementing new and emerging technologies, lessons from the past must be considered. Many of these technologies require highly qualified and skilled manpower, more advanced monitoring techniques, and equipment and materials that many developing countries may not have. Government support would be ideal, such as information campaigns, training and improving local expertise in the use of advanced materials, and process equipment for biomass conversion into energy and fuel.

Finally, to reverse the trend of depletion of agriculture and forestry resources, massive reforestation programs must be implemented, along with developing technologies for harvesting, preprocessing, and storage of biomass. This should be implemented in conjunction with infrastructure development for efficiently transporting biomass to where it is needed, or to develop technologies that can be brought to areas where biomass resources are abundant.

1.9 PROBLEMS AND DISCUSSION ISSUES

1.9.1 Simple Definitions

P1.1 Define or explain the following terms and list their common English and metric units: energy, power, and heating value.

1.9.2 Energy Units

P1.2 Convert the world's energy consumption of 418 EJ into units of Btu. Convert the Btu units into quads because Btu is a relatively smaller unit.

1.9.3 Net Energy Trends in United States

P1.3 Plot the net energy use in the United States from 2010 to 2020 using data from Table 1.1 (consumption minus production). Describe the trends in net energy use in the United States.

1.9.4 Contribution of Energy Sources to the US Economy

P1.4 Make a pie chart showing the percentage contribution of each energy source consumed in the United States in 2020. Discuss the sources that comprised the majority of the use. Please use Table 1.2 for this problem. What was the percentage of use for renewables?

1.9.5 Moisture Removal in Biomass

P1.5 Algae biomass is harvested with around 87% moisture (wet basis). Determine the amount of moisture to be removed per 1,000 kg of algae if the ideal safe storage MC is 13%.

1.9.6 Energy from Imported Crude Oil

P1.6 In 2021, the United States consumed an average of 19.8 million barrels of oil per day. How many quads does this represent for the year? How many exajoules per year? Assume 1 gal of crude oil contains 138,095 Btu. There are 42 gal to a barrel.

1.9.7 Pure Sugar to Ethanol Conversion

P1.7 Determine the amount of pure sucrose sugar (in kg) to generate a liter of ethanol using the idealized equation shown. The density of ethanol is 0.8 kg/L. Convert this unit into its English equivalent (lb/gal).

$$\underset{\text{Sucrose +}}{C_{12}H_{22}O_{11}} + \underset{\text{water +}}{H_2O} + \underset{\text{yeast} \rightarrow}{\text{yeast} \rightarrow} \underset{\text{ethanol + carbon dioxide + heat}}{4C_2H_5OH + 4CO_2 + \text{heat}}$$

1.9.8 Biogas Utilization in Households

P1.8 In 2020, the average annual electricity consumption for a US residential utility customer was 10,715 kWh or an average of 893 kWh per month. Biogas will be used to fuel an engine generator to produce the needed electrical power for this household. The thermal efficiency of the engine is 30% and the electrical to mechanical efficiency is 85%. Thirty liters (1 ft^3) of biogas with a methane content of 65% will provide 0.63 MJ (600 Btu) of energy. Determine the volume of biogas required (in cubic meters and cubic feet per day) for this purpose. Note that kWh = 3.6 MJ.

1.9.9 Acreage Needed to Build Power Plants

P1.9 A 100 MW biomass-based power plant is being operated in Nacogdoches County, Texas, fueled by wood residue. This power plant will be operated continuously (24 hrs/day) for three months in a year (90 days) to handle peak summer-month loads from the City of Austin, Texas. The heating value of the wood waste is approximately 22 MJ/kg (9,480 Btu/lb). Assume an overall conversion efficiency of 45%. How many lbs/day and tons/yr of wood waste is needed every year of operation? If the wood waste is to be gathered from wood trimmings from a forest land with an average yield of 25 tonnes/ha/yr, how many hectares (or acres) should be devoted to this project?

1.9.10 Air-to-Fuel Ratios for the Combustion of Fuel

P1.10 Determine the air-to-fuel ratio (kg/kg) for the complete combustion of methane (CH_4) gas in air. Note that air contains 79% nitrogen and 21% oxygen by volume. The main products are water and carbon dioxide.

1.9.11 Ideal Biomass Feedstock for Methane Production

P1.11 Identify the ideal feedstock for methane production.
a. Manure b. sugarcane c. corn stover d. rice straw

1.9.12 Ideal Biomass Feedstock for Bioethanol Production

P1.12 Identify the ideal feedstock for bioethanol production.
a. Manure b. sweet sorghum c. microalgae d. wood waste

1.9.13 Ideal Biomass Feedstock for Biodiesel Production with the Highest Yield

P1.13 Identify the ideal feedstock for biodiesel production.
a. Sunflower b. Lupine c. Kenaf d. Palm Oil

1.9.14 Biomass Feedstock Production and Delivery Project

P1.14 You are tasked to establish, produce, and deliver biomass feedstock to 3.785 million liters of bioethanol refinery. Outline a comprehensive logistics for this facility by identifying a feedstock, acreage, and enumerate equipment needed for processing and delivery.

1.9.15 Circular Bioeconomy Strategies for a Dairy Farm

P1.15 Convert a simple dairy farm (milk production) into a circular bioeconomy by enumerating all possible interventions to recycle all water usage, use of manure for energy or power, and other farm wastes that can be converted into valuable materials.

REFERENCES

Akbar, E., Z. Yaakob, S. K. Kamarudin, M. Ismail and J. Salimon. 2009. Characteristic and composition of *Jatropha curcas* oil seed from Malaysia and its potential as biodiesel feedstock. *European Journal of Scientific Research* 29 (3): 396–403.

ASABE D384.2 Standards. 2010. *Manure Production and Characteristics*. ASABE, St. Joseph, MI.

Bhattacharya, S. C. 1983. Lecture Notes in "Renewable Energy Conversion" Class. Asian Institute of Technology, Bangkok, Thailand.

Boyle, G. 2004. *Renewable Energy: Power for a Sustainable Future*. Oxford University Press, Oxford, UK.

Collins, H. P., A. Hang, S. Fransen and P. Wandershnieder. 2012. Biofuel Variety Trials Factsheet. USDA-ARS Publications, Washington State University, Prosser, WA. Available at https://www.usda.prosser.wsu.edu. Accessed August 28, 2012. Bionergy Washington, Washington's Website for BioEnergy Information.

EIA. 2010. *US Energy Information Administration Annual Energy Review 2009*. Office of Energy Markets and End Use, US Department of Energy, Washington, DC.

EIA. 2021. *US Energy Information Administration Annual Energy Review 2020*. Office of Energy Markets and End Use, US Department of Energy, Washington, DC.

FAO. 2012. *Food and Agricultural Organization Production Yearbook*. Rome, Italy.

Fernandez, F. 2012. Nutrient removal with drought-stressed corn. *The Bulletin* Issue No. 17, Article 5, July 27, 2012. The University of Illinois Extension Publications, The University of Illinois, Urbana-Champaign, IL.

IEA (International Energy Agency). 2021. Key World Energy Statistics 2021. Statistics Report, September 2021. IEA Publications International Energy Agency. Website: www.iea.org. Contact information: www.iea.org/about/contact.

Kemp, W. H. 2006. *Biodiesel Basics and Beyond: A Comprehensive Guide to Production and Use for the Home and Farm*. Aztext Press, Ontario, Canada.

Kim, S. and B. E. Dale. 2004. Global potential bioethanol production from wasted crops and crop residues. *Biomass and Bioenergy* 26 (2004): 361–375.

Kopetz, Heinz. 2013. Build a biomass energy market. *Nature* 494: 29–31. 7 February 2013. Macmillan Publishers.

Krausmann, F., H. Schandl and R. P. Sieferle. 2007. Socio-ecological regime transitions in Austria and the United Kingdom. CSIRO Working Papers on SocioEconomics and Environment. ISSN 1834-5638. CSIRO Sustainable Ecosystems, Canberra, Australia.

Kresovich, S., C. K. Wagner, D. A. Scanland, S. S. Groet and W. T. Lawton. 1982. The utilization of emergent aquatic plants for biomass energy systems development. USDOE Report Under Contract No. EG-77-C-01-4042. USDOE, Golden, CO.

Launder, K. 2002. Energy Crops and Their Potential Development in Michigan. Michigan Biomass Energy Program Report. Michigan Department of Consumer and Industry Services, Energy Office, Biomass Energy Program, Lansing, MI. August, 2002.

Launder, K. 2007. *Energy Crops and Their Potential Development in Michigan*. Michigan Department of Consumer and Industry Services, Energy Office, Michigan Biomass Energy Program.

Mother Earth News. 1980. Mother earth alcohol fuel. Chapter 3. Available at: https://www.motherearthnews.com.

NASS, USDA. 2021. *Crop Production 2020 Annual Summary*. USDA, Washington, DC. Published January 21, 2021. Also available at: https://downloads.usda.library.cornell.edu/usda-esmis/files/k3569432s/w3764081j/5712n018r/cropan21.pdf. Accessed 7 March 2022.

Nelson, V. 2011. *Introduction to Renewable Energy*. CRC Press, Taylor & Francis Group, Boca Raton, FL.

Noureddini, H., B. C. Teoh and L. D. Clements. 1992. Densities of vegetable oils and fatty acids. *Journal of the American Oil Chemists' Society* 69 (12): 1184–1188.
Rooney, W. L., J. Blumenthal, B. Bean and J. E. Mullet. 2007. Designing sorghum as a dedicated bioenergy feedstock. *Biofuels, Bioproducts and Biorefining* 1: 147–157.
Schooler, S., T. Cook, A. Bourne, G. Prichard and M. Julien. 2008. Selective herbicides reduce alligator weed (*Alternanthera philoxeroides*) biomass by enhancing competition. *Weed Science* 56: 259–264.
SECO. 2005. Texas renewable energy resource assessment. State Energy Conservation Office Publication, Austin, TX. Report Prepared by Frontier Associates, LLC for Seco, TX.
Turhollow, A. F. and S. Sokhansanj. 2007. Costs of harvesting, storing in a large pile and transporting corn stover in a wet form. *Applied Engineering in Agriculture* 23 (4): 439–448.
USEPA. 2022. National Overview: Facts and Figures on Materials, Wastes and Recycling. USEPA, Washington, DC. Available at: https://www.epa.gov/facts-and-figures-about-materials-waste-and-recycling/national-overview-facts-and-figures-materials. Accessed March 12, 2022.
Walsh, M. E., D. G. De La Torre, B. C. Ugarte, K. J. English, C. Hellwinckel, R. J. Menard and R. G. Nelson. 2007. Agricultural impacts of biofuels production. *Journal of Agricultural and Applied Economics* 39 (2): 365–372.
Xu, W., Y. Li and A. B. Carraway. 2008. *Estimation of Woody Biomass Availability for Energy in Texas*. Texas Forest Service, College Station, TX. 96 p.

ASSESS YOUR KNOWLEDGE

1. Name an example of a fast-growing trees
   a. Hybrid poplar
   b. Jatropha
   c. Crambe
   d. Coconut

2. Which crop has the higher potential bioethanol yield?
   a. Grapes
   b. Sugar beets
   c. Corn
   d. Grain sorghum

11. Estimate the number of hectares required to supply the feedstock for an 11.355 million liters biodiesel plant that will come from sunflower oil?
    a. 199
    b. 377
    c. 11,219
    d. 9,960

12. Determine the total amount of manure that can be gathered from a 500-head dairy farm (as removed, assuming 1/3 recovery) in a year. Report your units in tonnes.
    a. 4,015
    b. 3,850
    c. 2,605.6
    d. 6,387.5

3. Which oil crop has the higher oil yield per unit of area?
   a. Linseed
   b. Pumpkin
   c. Calendula
   d. Sesame

4. Which dedicated crop has the highest reported yield in the United States?
   a. Forage sorghum
   b. Corn
   c. Switchgrass
   d. Sweet sorghum

5. California consumes 6,923 trillion Btu of energy. Convert this energy usage unit into Quads.
   a. 6.923
   b. 0.0069
   c. 7.303
   d. 0.0073

6. California consumes 6,923 trillion Btu of energy. Convert this energy usage unit into EJ.
   a. 6.923 Quads
   b. 7.304 Quads
   c. 0.0069 Quads
   d. 0.0073 Quads

13. The household population of California is around 13.1 million. Determine the municipal solid waste (MSW) production in the state of California on a yearly basis if the per capita MSW production is around 3 kg/household/day. Report your units in metric tonnes.
   a. 39,300
   b. 54,256
   c. 35,125
   d. 25,220

14. Determine the required efficiency of conversion to produce a 1 MW (output) biomass power plant (operated 350 days a year and 24 hr a day) if the heating value of the biomass (dry) is 19.4 MJ/kg. The biomass yield in tonnes (dry) per hectare per year in dry tonnes is about 5. The total area devoted to this project was 2,000 ha.
   a. 31.2%
   b. 15.6%
   c. 46.8%
   d. 62.4%

15. Which emergent aquatic plant has the highest yield in dry tonnes/ha/yr?
   a. *Arundo donax* L.
   b. *Typha latiflora* L.
   c. *Cyperus papyrus* L.
   d. *Spartina alterniflora* L.

16. Which developing country had the highest forest energy per capita?
   a. Congo
   b. India
   c. Mali
   d. Brazil

7. A metric tonne of newly harvested softwood has a moisture content of 45% (wet basis). How much water should be removed (in kg) if the goal is to bring this material to 10% moisture (wet basis)?
   a. 450
   b. 611.1
   c. 388.9
   d. 550

8. Estimate the wastes biomass in a year from rice harvested from a 200 ha plot. Report the units in tonnes.
   a. 1,800
   b. 5,300
   c. 8,500
   d. 2,380

9. Estimate the waste biomass in a year from cotton harvested from a 1,000 ha plot. Report the units in tonnes.
   a. 1,900
   b. 2,380
   c. 5,300
   d. 8,500

10. Estimate the number of hectares required to supply the feedstock for a 3.785 million liters bioethanol plant that will come from Jerusalem artichoke?
    a. 199
    b. 377
    c. 11,219
    d. 9,960

17. Which developing country has an energy per capita greater than its sustainable wood energy yield?
    a. Philippines
    b. Brazil
    c. Argentina
    d. Mali

18. Which developing country has a reported sustainable wood energy yield higher than its energy per capita?
    a. Philippines
    b. Brazil
    c. Nepal
    d. India

19. Which crop has the lowest reported potential ethanol yield?
    a. Carrots
    b. Potatoes
    c. Wheat
    d. Barley

20. Which crop has the lowest reported potential oil yield?
    a. Cashew nut
    b. Linseed
    c. Cotton
    d. Corn

Answers: 1a, 2b, 3c, 4d, 5a, 6b, 7c, 8d, 9a, 10b, 11c, 12d, 13a, 14b, 15c, 16d, 17a, 18b, 19c, 20d

ADDITIONAL LEARNING RESOURCES

- Access the Lecture Notes at the Instructor and Student Resources: https://routledgetextbooks.com/textbooks/instructor_downloads/

2 Biomass Conversion Processes

LEARNING OBJECTIVES

Upon completion of this chapter, one should be able to

1. Describe an overview of the various biomass conversion processes.
2. Differentiate between chemical, biological, and thermal conversion processes.
3. Recognize the various units and terminologies used for estimating efficiencies of different biomass energy conversion processes.
4. Compare new and advanced biomass conversion processes, including combinations and hybrids.
5. Enumerate all applications of various conversion processes and list important products and coproducts produced.

2.1 INTRODUCTION

As discussed in Chapter 1, there are numerous ways to convert biomass resources into useful energy, chemicals, fuels, or mechanical and electrical power. In some ways, the conversion process is dependent on specific properties of the biomass, such as moisture and ash content. However, some conversion processes are affected not so much by the biomass properties but rather by the energy content of the biomass resource. There are also various ways to categorize these conversion processes—by the final products of the process or by the specific conditions used during the conversion process, such as temperature or heat. Biomass resources that have very low moisture upon harvest and have high energy density are mostly suitable for thermal conversion processes, because removing the moisture will significantly affect the overall efficiency of the conversion process. Biomass resources very high in moisture are traditionally used in the biological conversion pathway, primarily because this conversion process relies on living microorganisms that normally thrive under high-moisture conditions. Finally, there are biomass conversion processes that use neither heat nor microorganisms but rely on simple chemicals or catalysts to convert extracted biomass constituents into useful energy-related products.

In this textbook, biomass conversion processes are broken down into physical, chemical, biological, and thermal conversions. Some conversion processes make use of a combination of these conditions and may be categorized as hybrid conversion systems. The order of discussion of the biomass conversion processes is arranged

DOI: 10.1201/9781003294306-2

according to this system, whereby the first set of conversion processes uses only primarily physical and chemical factors, followed by the use of biological microorganisms, and finally by the use of heat or thermal energy for conversion. All these conversion processes, however, are evaluated the same way when it comes to their efficiency. The overall efficiency of conversion is a function of the energy from the biomass input and the energy of the primary output of the process, as illustrated in Equation 2.1.

$$\text{Energy conversion efficiency } (\%) = \frac{\text{energy output}}{\text{energy input}} \times 100 \tag{2.1}$$

One may argue about what could be included in the energy output group; however, it should be clear that if one were to produce ethyl alcohol (ethanol) from biomass, the efficiency of conversion should only be based on the energy content of ethanol and not include the energy from the distiller's grain that is co-produced. If energy is further extracted from the distiller's grain, the overall energy conversion efficiency will be higher (and can be defined appropriately). In some thermodynamics textbooks, this calculation is called the second law of efficiency (thermodynamics efficiency). The efficiency of conversion could be as low as 10% in some thermal conversion systems. However, the efficiency could be as high as 90% in other thermal conversion systems, if heat is the only useful energy required for the process.

Example 2.1: Conversion Efficiency Calculations from Biomass

Determine the efficiency of converting a certain biomass for heat generation given the following data: the amount of biomass converted is 1 metric ton, and the heating value is 18 MJ/kg. The heat energy produced was used to power a boiler that has a heat output of 16,500 MJ in the form of steam.

Solution:

a. The efficiency equation as presented in Equation 2.1 is as follows:

$$\text{Efficiency } (\%) = \frac{\text{output}}{\text{input}} \times 100$$

b. Substituting the values with the correct units to estimate efficiency is as follows:

$$\text{Efficiency } (\%) = \frac{16{,}500 \text{ MJ}}{18 \text{ MJ/kg}} \times \frac{1}{1 \text{ metric ton}} \times \frac{1 \text{ metric ton}}{1{,}000 \text{ kg}} \times 100 = 91.7\%$$

2.2 OVERVIEW OF BIOMASS CONVERSION PROCESSES

The development of conversion technologies for the utilization of biomass resources for energy has grown at a fast pace in recent years. However, conversion of biomass into useful energy has been present since humans first made use of fuelwood for personal heat and energy needs many thousands of years ago. Figure 2.1 shows the various methods of converting biomass into useful energy, chemicals, fuels, and mechanical or electrical power.

Biomass conversion into heat energy is still the most popular and perhaps the most efficient process using combustion (Chapter 13). However, not all energy requirements are in the form of heat. Biomass resources can be converted into chemicals, liquid fuels, or electrical or mechanical energy, in order to have widespread use. These products take the form of solid fuel, such as charcoal; liquid fuel, such as ethanol; or gaseous fuel, such as methane. These fuels can be used in a wide range of energy conversion devices to satisfy diverse energy needs. The relatively high cost of transporting low-energy density biomass makes conversion into some other form a necessity.

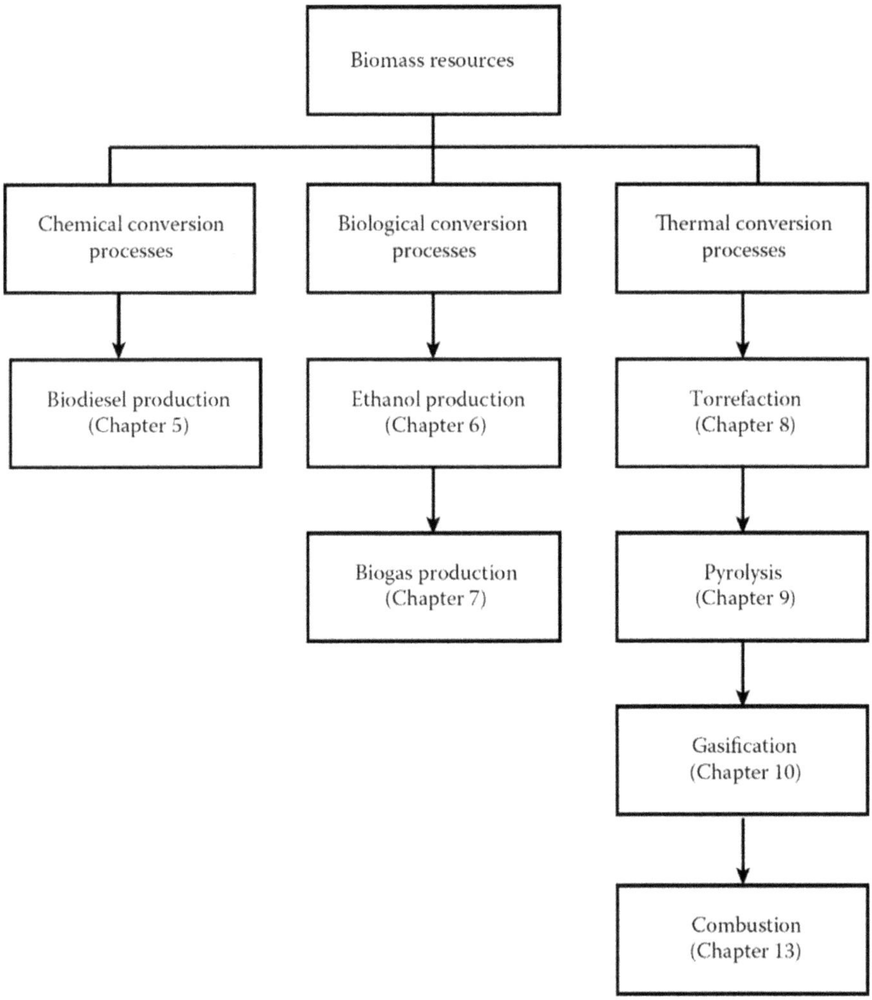

FIGURE 2.1 Various biomass energy conversion pathways.

Each biomass conversion process will be described in detail in subsequent chapters. This chapter presents an overview of all possible products of the several processes. Biodiesel production from biomass or oil seed crops will fall under the simple chemical conversion processes (Chapter 5). All that is needed for converting extracted oil from crops into biodiesel (an ester of oil that is compatible with conventional diesel fuels for diesel engines) is a chemical catalyst (such as methanol and sodium hydroxide). Some form of heat is also needed, but the temperature is normally below 100°C (212°F), which does not alter the chemical composition of the primary components of the biomass.

The next set of conversion processes is biological in nature, relying on microorganisms to convert biomass into fuels. Chapter 6 deals with ethanol production using various types of microbes. Chapter 7 deals with biogas (methane and carbon dioxide) production using different sets of microbes.

The remaining chapters deal with thermal conversions (Chapters 8 through 13). The selection of a conversion process for a particular biomass is not straightforward, and there are no set criteria that a particular biomass should use a specific conversion route. As mentioned earlier, the properties of each biomass play a significant role in deciding which particular biomass conversion is appropriate. With a clear understanding of the various biomass properties (Chapters 3 and 4), one may be able to select the best conversion route for the biomass in question. Some conversion pathways are easy to determine simply by assessing the main component of a biomass. If the biomass has very high sugar content, such as sugarcane or sweet sorghum, ethanol conversion is the best conversion pathway. Thermally converting sugar crops into biofuels by exposing them to high-temperature conditions (i.e., thermal means) is a challenge. Imagine putting pure cane sugar into a pan and heating it slightly. Even before the temperature reaches the boiling point of water, one will observe a viscous black substance that solidifies shortly upon cooling, an irreversible state that makes it even more difficult to convert the biomass into a hydrocarbon-like fuel.

2.3 CHEMICAL CONVERSION PROCESSES

If the major component of the biomass is oil, or triglycerides, then the most logical step is to form esters of this oil. One of the most popular biofuels from biomass is biodiesel, which is simply the ester of the oil. Biodiesel can be blended with diesel fuel easily and may be used in conventional diesel engines without engine modification. Diesel fuel may be blended with biodiesel and the blending ratio will depend on the engine warranty. Some engine manufacturers can warrant the use of as much as 20% biodiesel in the blend. This blending terminology is commonly called B20 (i.e., 20% biodiesel). Thus, a B100 fuel description means that the fuel is 100% biodiesel. When sulfur in diesel fuel was eliminated by the fuel industry to improve emissions and reduce smog (i.e., sulfur dioxides, SO_2), there was a search for a replacement additive. It turns out that biodiesel is an excellent lubricant and qualifies as a blending agent to reduce smog. A 5% blend, or B5, provides the lubrication, which is equivalent to that of the sulfur originally used to improve lubrication in diesel engines.

The process of making biodiesel is very simple. Figure 2.2 shows the chemical equation and the simple product balance for the biodiesel conversion process. For

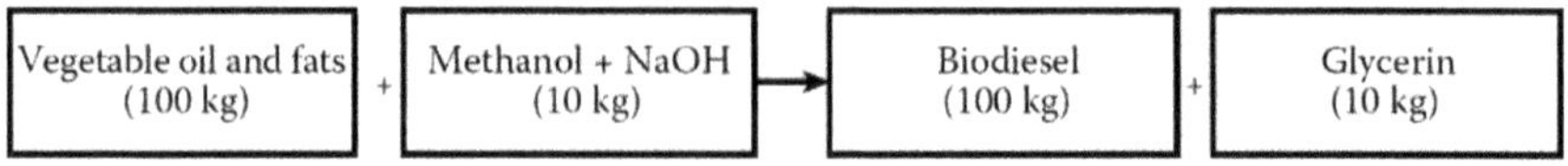

FIGURE 2.2 Simple schematic diagram to convert vegetable oil into biodiesel with glycerin.

every 100 kg of refined vegetable oil used, about 10 kg of methanol is used with some catalysts and about 100 kg of biodiesel is formed with about 10 kg of glycerin as a coproduct.

The process is technically called transesterification of oils and fats, requiring only alcohol and a catalyst for conversion (a simple chemical conversion process). However, the overall biofuel production process is complex, as elaborated in Chapter 5 on biodiesel production. The raw oil products directly extracted from crops have inherent components that must be removed before biodiesel production: gum, free fatty acids, and other impurities, including colors. Commercial biodiesel producers would normally begin with a refined, bleached, and deodorized (or RBD) oil. Before oil seed extraction, the seed kernels must be cleaned and dried, or the hulls must be removed. More importantly, the biodiesel produced must pass American Society for Testing and Materials (ASTM) Standards before it can be blended with diesel. This standard includes more than 15 parameters to which the biodiesel product must conform before it can be used in engines. The conversion efficiency for this process is quite high, nearly 100% (see Example 2.2 on comparing input and output energy from the process). However, the pre- and post-treatment processes must be strictly followed to achieve this high conversion efficiency. This procedure would include ensuring good-quality refined oil before the transesterification process, as well as excellent downstream processes after the chemical catalytic conversion process. These refinement processes, as well as adhering to set criteria, will be discussed in detail in Chapter 5.

Example 2.2: Comparing Energy Content of Oil and Its Ester

About 100 L (26.4 gal) of refined canola oil was converted into biodiesel following the yield formula, as shown in Figure 2.2. Compare the energy content of the ester product (in GJ or million Btu) with that of the refined oil. The energy content of the refined oil was 27.87 MJ/L (100,000 Btu/gal), whereas that of the resulting ester was 33.45 MJ/L (120,000 Btu/gal). Convert the units into the English system.

Solution:

a. Energy content of refined oil is as follows:

$$\text{Energy (Btu)} = 100\ \text{L} \times \frac{27.87\ \text{MJ}}{\text{L}} = 2.8\ \text{GJ}$$

$$\text{Energy (Btu)} = 26.4\ \text{gal oil} \times \frac{100{,}000\ \text{Btu}}{\text{gal}} = 2.64\ \text{million Btu}$$

b. Energy content of the esters of oil is as follows:

$$\text{Energy (Btu)} = 100\ \text{L} \times \frac{33.45\ \text{MJ}}{\text{L}} = 3.3\ \text{GJ}$$

$$\text{Energy (Btu)} = 26.4\ \text{gal oil} \times \frac{120{,}000\ \text{Btu}}{\text{gal}} = 3.17\ \text{million Btu}$$

c. Note that there is more energy contained in the ester product than in the refined oil. Methanol and heat energy were also added as a reactant containing additional amounts of energy. Thus, this calculation does not imply that one would produce more energy in the product than in the reactant.

2.4 BIOLOGICAL CONVERSION PROCESSES

Biological conversion processes involve the use of various naturally occurring microorganisms to convert biomass organic material into various high-energy liquids or gaseous fuels. The two most important biological conversion processes are ethanol fermentation and anaerobic digestion processes. Fermentation is the conversion of sugars from crops into alcohol, whereas anaerobic digestion is the conversion of a complex biomass into methane gas and carbon dioxide (termed "biogas") by the action of anaerobic microorganisms. However, complexities occur in the conversion processes if the major component of the biomass is not pure sugar. Shown in Figure 2.3 are the various pathways to generate ethanol from sugar, starchy, and lignocellulosic biomass. All of these conversion processes require the use of various types of microorganisms.

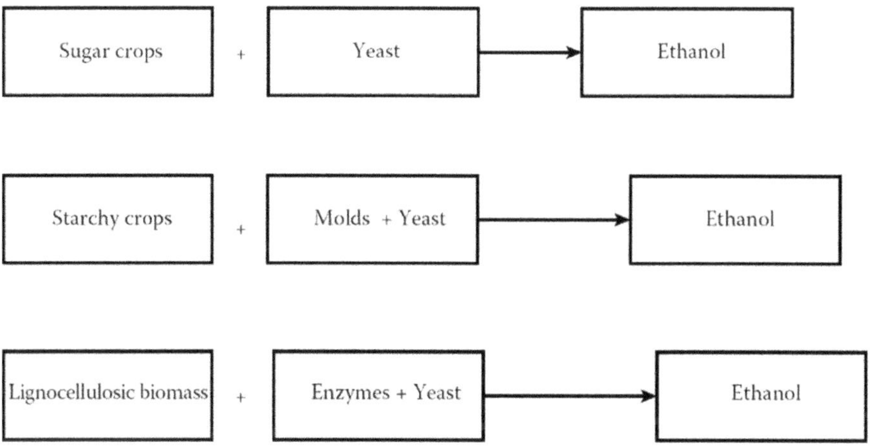

FIGURE 2.3 Various pathways to convert biomass high in sugar, starch, or lignocellulose into ethanol.

2.4.1 Ethanol Fermentation

The raw materials used in ethanol production include biomass components, such as sugars, starches, or their cellulose or hemicellulose components. The only component in the biomass that microbes are unable to convert into useful fuels is the lignin fraction. Sugars (such as cane sugar or sweet sorghum juice) can be easily converted into ethanol using yeast. The most commonly used yeast is *Saccharomyces cerevisiae*. However, there are other microbes or enzymes produced by microbes that can convert sugars into ethanol or use other substrates besides a sugar component. The starchy and cellulosic components of biomass must first undergo saccharification or hydrolysis (convert these components into simple sugars) before the yeast can convert the sugars into ethanol. Molds such as *Aspergillus niger* produce enzymes that would break down starches into simple sugars, whereas enzymes from *Trichoderma reesi* break down the cellulose components into simple sugars by enzymatic hydrolysis, which would then be converted into ethanol with the action of yeast. Some plants (e.g., sorghum) would have these three components (sugars, starch, and lignocellulosics), and nearly the entire plant may be converted into ethanol component materials. Only the lignin component would remain unconverted into biofuel.

Majority of ethanol in the United States is produced from corn, or the starch component of corn. However, in Brazil, majority of ethanol is produced from sugarcane. The major component of sugarcane is sugar, chemically represented by sucrose or glucose. The conversion of pure glucose compound into ethanol can be illustrated by the idealized equation introduced in Chapter 1 (Equation 1.3), in which 1 mol of glucose can be converted into 2 mol of ethanol and 2 mol of carbon dioxide, upon the action of yeast.

In terms of weight basis, one would simply use the molecular weight of the chemical formula in Equation 1.3 to determine how much ethanol is produced per weight of pure glucose. Table 2.1 summarizes the basic data required to accomplish this task; Example 2.3 shows how this is calculated. From this example, one would be able to make an informed estimate of the amount of material required to produce a certain volume of biofuel. The theoretical amount of pure sugar required to generate a liter of ethanol is about 1.6 kg/L (if the density of ethanol is assumed to be 0.8 kg/L). In the English system, the use of 15 lb/gal (1.8 kg/L) of ethanol is quite common.

TABLE 2.1
Data Required to Estimate Conversion of Glucose into Ethanol on Unit Weight Basis

Element	Symbol	Atomic Number	Atomic Weight	Common Weight Used
Carbon	C	6	12.011	12
Hydrogen	N	1	1.008	1
Oxygen	O	8	14.007	14
Nitrogen	N	7	15.9994	16
Sulfur	S	16	32.06	32
Phosphorus	P	15	30.974	31

For conversion of corn into ethanol, the current state-of-the-art production rate is between 0.40 and 0.43 L/kg (2.7–2.9 gal/bushel) of corn grain. Note that a bushel of corn with a 15% moisture content weighs about 56 lb (25.5 kg).

The conversion of lignocellulosic biomass into ethanol has not yet been established, although preliminary data has been published in this area. Perhaps the "holy grail" in ethanol production will come from the conversion of lignocellulosic biomass feedstock, which is the most abundant biomass resource on Earth. The primary barrier to commercialization of this approach is the cost of enzymes used in the conversion as well as the various pretreatment processes that must be adopted. This process will be discussed in detail in Chapter 6.

Example 2.3: Conversion of Glucose into Ethanol by Weight

Determine the amount of pure glucose sugar (in kg) to generate a liter of ethanol using the idealized Equation 1.3. The density of ethanol is 0.8 kg/L. Convert this unit into its English equivalent (lb/gal).

Solution:

a. The mass balance for Equation 1.3 (rewritten here) is as follows:

$$C_6H_{12}O_6 + \text{yeast} \rightarrow 2C_2H_5OH + 2CO_2 + \text{heat}$$

$$\underbrace{\left[6^*12+12^*1+6^*16\right]}_{180\,\text{kg}} \rightarrow \underbrace{2^*\left[2^*12+6^*1+1^*16\right]}_{92\,\text{kg}} + \underbrace{2^*\left[1^*12+2^*16\right]}_{88\,\text{kg}}$$

b. Thus, for every 180 kg of glucose, 92 kg of ethanol is produced; if these units are converted into volume units, we have the following:

$$\frac{180 \text{ kg sugar}}{92 \text{ kg ethanol}} \times \frac{0.8 \text{ kg ethanol}}{\text{L ethanol}} = 1.6 \text{ kg sugar/L ethanol}$$

$$\frac{180 \text{ kg sugar}}{92 \text{ kg ethanol}} \times \frac{0.8 \text{ kg ethanol}}{\text{L ethanol}} \times \frac{2.2 \text{ lb}}{1 \text{ kg}} \times \frac{3.785 \text{ L}}{\text{gal}} = 13 \frac{\text{lb sugar}}{\text{gal ethanol}}$$

c. Converted into English units, this is about 13 lb of glucose sugar to generate a gallon of ethanol. For the rules of thumb (in the English system of units), the use of 15 lb of sugar per gallon of ethanol is easy to remember.

2.4.2 Biogas Production

The production of methane gas and carbon dioxide (termed "biogas") is also an old biomass conversion process. Ruminant animals have been converting cellulosic grasses into energy for their own metabolic uses. There are numerous anaerobic microbes present in the stomach of ruminants responsible for such a novel conversion process. However, no one has yet been able to emulate these conversion processes as they happen in an animal's stomach in large-scale refineries. The best that can be done is to collect the manure samples immediately after excretion and place them in a reactor, thereby ensuring there is strictly no entry of any oxidant (oxygen or air). This conversion process is also called the anaerobic digestion process. The digestion or breakdown process occurs under strict anaerobic conditions (i.e., no oxygen). The microbes responsible for the conversion are called anaerobic microbes.

Figure 2.4 shows the pathway for the conversion of complex organic components in biomass into methane gas. These complex organics are broken down into higher organic acids before being converted into acetic acid and hydrogen. The hydrogen and acetic acid react to form methane (CH_4) with carbon dioxide (CO_2) as a respiration by-product. Also shown in the figure are the general process stages as follows: stage 1 is the hydrolysis and fermentation step; stage 2 is the acetogenesis and dehydrogenation step; and stage 3 is the methane fermentation step.

Villages in China and India have been utilizing methane gas from animal manure for quite some time (NAS, 1976). A few animals would be enough to satisfy a household's cooking and lighting needs. The amount of biogas produced from animal manure is approximately 150–200 L/kg of dry matter (or 2.4–3.2 ft^3/lb) (Stout et al., 1979). The combustion of about 30 L of biogas (1 ft^3) releases the amount of energy equivalent to lighting a 25 W bulb for about 6 h (NAS, 1976). This is equivalent to about 0.54 MJ of energy (512 Btu). Example 2.4 shows a simple way to determine the volume of biogas required to satisfy a household's electrical requirements. More details of this process will be discussed in Chapter 7.

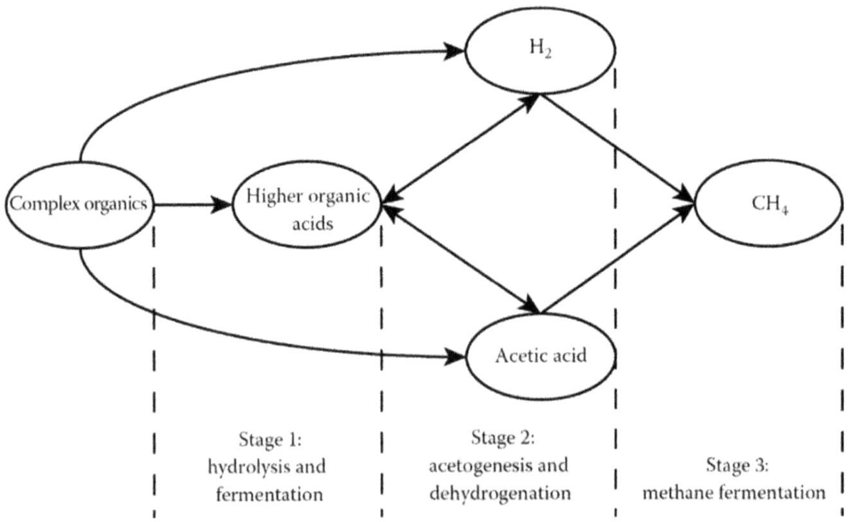

FIGURE 2.4 Conversion pathway to break down complex biomass organics into methane and carbon dioxide.

Example 2.4: Volume of Biogas Required for Electrical Power Production

A household in India requires 100 kWh of electricity per day. Biogas will be used to fuel an engine generator to produce the required electrical power. The thermal efficiency of the engine is 25% and the electrical to mechanical efficiency is 80%. Thirty liters (1 ft^3) of biogas with a methane content of 55% will provide 0.58 MJ (550 Btu) of energy. Determine the volume of biogas required (in cubic meters and cubic feet per day) for this purpose.

Solution:

a. From the efficiency equation, we have the following energy required from the biogas:

$$\% \text{ Efficiency} = \frac{\text{output}}{\text{input}}$$

$$\text{Input} = \frac{100 \text{ kWh}}{(0.25) \times (0.80)} = 500 \text{ kWh}$$

b. The biogas required can be easily calculated as follows:

$$\frac{500 \text{ kWh}}{\text{day}} \times \frac{3.6 \text{ MJ}}{1 \text{ kWh}} \times \frac{30 \text{ L}}{0.58 \text{ MJ}} \times \frac{\text{m}^3}{1{,}000 \text{ L}} = 93 \frac{\text{m}^3}{\text{day}} \left[3{,}285 \frac{\text{ft}^3}{\text{day}} \right]$$

2.5 THERMAL CONVERSION PROCESSES

When biomass materials are exposed to heat (normally above 300°C, 572°F), an irreversible chemical change occurs and the biomass is transformed into various solid, liquid, or gaseous components. Perhaps the most common practice using wood biomass is to convert it into a higher energy density material, namely charcoal. Subsequent uses, such as cooking, will provide the required heat energy with minimal volatiles (or smoke). The range of products produced will depend on the temperature used for the process, the rate at which the biomass is exposed to this temperature, and the presence or absence of an oxidant (oxygen or air). The amount of solid, liquid, or gaseous coproducts will also depend on the same parameters but will be complicated using various catalysts or chemical additives.

The various thermochemical biomass conversion processes to be discussed in this textbook include the following:

1. Torrefaction—a biomass thermal pretreatment process
2. Pyrolysis—thermal process in complete absence of oxygen
3. Gasification—thermal process with incomplete amounts of oxygen
4. Combustion—thermal conversion with excess amounts of oxidant

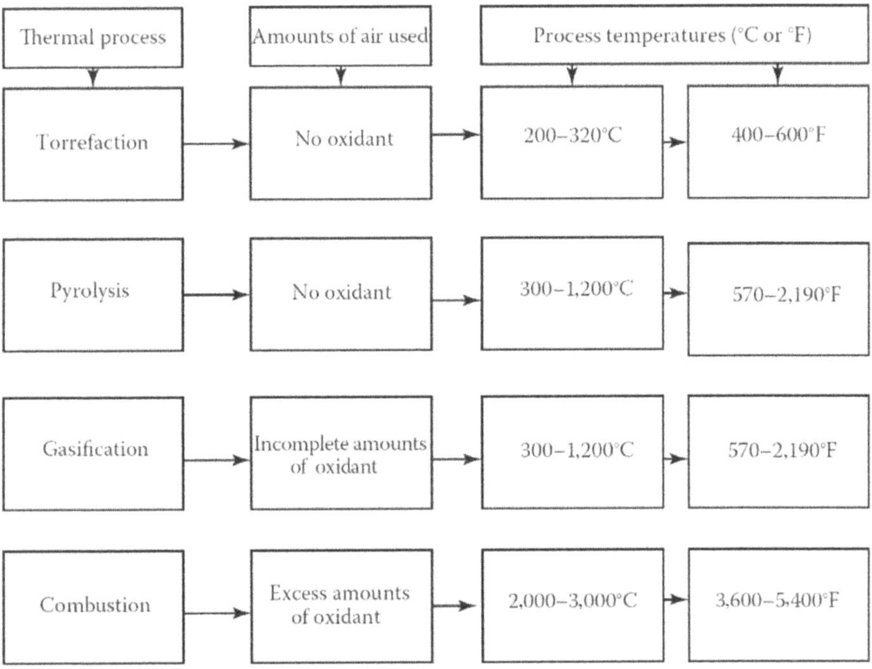

FIGURE 2.5 Schematic of various thermal conversion processes.

Figure 2.5 shows the relationships among the various thermal conversion processes used for biomass in terms of the relative amount of air used during the conversion process as well as the range of temperatures.

Torrefaction and pyrolysis are oxygen-free conversion processes, whereas gasification and combustion require the use of an oxidant. Gasification uses oxidants in small amounts, whereas combustion uses stoichiometric amounts of air or, frequently, excess amounts of air.

2.5.1 Torrefaction

Thermal conversion processes vary in the amount of oxidant used for conversion as well as the temperature used for conversion. Even when biomass is heated gradually, there are still irreversible chemical changes. Torrefaction, although strictly speaking not a real biomass conversion process, is now being used as a pretreatment process for biomass that will undergo a thermal conversion process. The growing literature about torrefaction clearly defines this process by the temperature that is commonly reported between 200°C and 320°C. A biomass sample is usually placed in a tube reactor and heated in an oxidant-free electronically controlled furnace to between 200°C and 320°C. Before heating, a nitrogen purge system is commonly used to evacuate any air that is trapped inside the reactor. The gaseous products are collected and analyzed. At the end of the torrefaction process, the remaining solid materials are analyzed or processed further. The products or coproducts that are recovered are thought to be improvements on the original biomass material. This conversion process is also described as mild thermal pyrolysis of biomass.

The main advantage in using this process is that a higher energy density biomass material results in reduced moisture and improved bulk density. In some instances, the biomass will undergo pelleting to further improve the bulk density. There is additional energy used for the process, and this energy may have to be justified in the overall economics of the conversion process. In addition, numerous authors writing about torrefaction claim to have improved the composition of the biomass in its hydrophobic behavior (repels water), eliminated further biological degradation, and improved grindability if the particle size is to be further reduced (Tumuluru et al., 2011).

2.5.2 Pyrolysis

Pyrolysis is an irreversible thermal conversion process whereby the biomass undergoes conversion at elevated temperatures (normally higher than 300°C) in the complete absence of an oxidant. It is also called "destructive distillation." There are three main products for this process: condensible liquids (termed bio-oil), noncondensible gases (also called producer gas or synthesis gas [syngas]), and char or charcoal (product with high carbon content). The biomass is placed in a completely enclosed biomass bin. Before starting the process, the system is purged with inert gas, usually nitrogen gas, to remove any oxygen trapped in the enclosed system. The biomass is moved through the reactor by a simple auger.

This auger is normally calibrated for a particular biomass to relate the revolutions per minute of the auger shafting to the biomass feed rate. The tube where the auger is located is also the reactor and is normally made of thick stainless steel or black iron pipe. This tube is normally passed through an electrically heated or programmable tube furnace. The tube furnace provides the external heat required to carry out the process. The biomass bin is filled with the feedstock to be tested; it is then sealed and purged. Finally, the tube furnace is set to provide the desired reactor temperature. As soon as the reaction temperature is reached, the biomass is fed through the reactor. Biochar is collected at the lower biochar bin, while syngas is passed through a chilled condenser to recover all condensible materials. The gaseous products that are not condensed are then sampled, measured, and completely analyzed. The energy associated with the operation of the pyrolyzer includes the biomass energy input, the electrical energy input to run the auger, and the electrical energy needed by the tube furnace. The product outputs include energy contained in the biochar, bio-oil, and syngas. The overall energy and mass balance is shown in Equation 2.2.

$$E_b + E_a + E_f = E_{bc} + E_{sg} + E_{bo} + E_l \tag{2.2}$$

where

E_b = energy of biomass

E_a = electrical energy supplied to the electric motor that drives the auger

E_f = electrical energy supplied to the furnace
E_{bc} = energy contained in the biochar
E_{sg} = energy contained in the synthesis gas
E_{bo} = energy contained in the bio-oil sample produced
E_l = energy lost through the system

Note that a similar mass balance may be established from the pyrolysis process, as shown by Equation 2.3. The components in the mass balance include the mass of the biomass converted (normally the difference between the initial weight of biomass before the test and the mass of the biomass unconverted) and the individual masses of the solid biochar, liquid bio-oil, and syngas.

$$M_b = M_{bc} + M_{sg} + M_{bo} + M_l \tag{2.3}$$

where
M_b = net mass of biomass used during conversion
M_{bc} = gravimetric mass of biochar (after cooling)
M_{sg} = total mass of the synthesis gas collected
M_{bo} = mass of the bio-oil collected
M_l = loss of mass in the system

Although the pyrolysis process is the same as that for making charcoal, the state-of-the-art pyrolysis reactor is complicated. The reactor is designed with the ability to control operational temperature and heating rates, how the coproducts are recovered, and how they are further converted into beneficial products. The ideal operation of the pyrolyzer includes the following guidelines:

1. Biochar is maximized when the pyrolyzer is operated at very low temperatures and longer retention times.
2. More syngas is produced at higher temperatures.
3. Bio-oil is maximized when the reactor is operated at medium temperatures of between 400°C and 600°C (750°F–1,115°F) and at the highest heating rates and reduced contact time possible.

A simple energy balance for an auger-based pyrolyzer is provided in Example 2.5. Note that the energy conversion efficiency may be calculated differently; it may be based on a particular product of interest, or it may be based on all the products produced. It will also be observed in the example that an auger-based pyrolyzer generates more char than any other coproducts. Consequently, the conversion efficiency of char will be higher than that of the other coproducts. In this case, about 27% of the energy in the biomass is contained in the char coproduct (6/22.3). The overall conversion efficiency is around 83% if one considers energy from the char, syngas, and bio-oil. Chapter 9 provides a more detailed discussion of the pyrolysis process.

Example 2.5: Simple Energy Balance in Auger-Based Pyrolyzer

Determine the energy balance for the pyrolysis of sorghum biomass using the following data (12-min test):

a. The amount of sorghum biomass used is 1.23 kg, with a heating value of 18.1 MJ/kg.
b. The auger electrical motor is rated 5 A at 220 V.
c. The furnace rating is 2,400 W.
d. The amount of char produced is 0.5 kg, with a heating value of 24 MJ/kg.
e. The energy content of the syngas produced is 100 MJ/kg, and about 0.06 kg of syngas was collected.
f. The amount of bio-oil produced was 0.1 kg, with a heating value of 26.2 MJ/kg.

Determine the overall conversion efficiency, assuming all products are utilized for energy purposes and the energy losses.

Solution:

a. The energy balance in units of MJ is calculated as follows:

$$E_b + E_a + E_f = E_{bc} + E_{bo} + E_{sg}$$

First, the energy input is calculated as follows:

$$1.23\ \text{kg} \times \frac{18.1\ \text{MJ}}{\text{kg}} + \frac{5 \times 220\ \text{W}}{\text{W}-\text{s/J}} \times \frac{1\ \text{MJ}}{10^6\ \text{J}} \times \frac{720\ \text{s}}{\text{W}-\text{s/J}} \times \frac{2{,}400\ \text{W}}{10^6\ \text{J}} \times 720\ \text{s}$$

$$22.3\ \text{MJ} + 0.79\ \text{MJ} + 1.73\ \text{MJ} = 24.8\ \text{MJ}$$

b. Next, the energy output is calculated as follows:

$$0.5\,\text{kg} \times \frac{24\,\text{MJ}}{\text{kg}} + 0.1\ \text{kg} \times \frac{26.2\ \text{MJ}}{\text{kg}} + 0.06\,\text{kg} \times \frac{100\,\text{MJ}}{\text{kg}}$$

$$12\,\text{MJ} + 2.62\,\text{MJ} + 6\,\text{MJ} = 20.62\,\text{MJ}$$

c. The overall conversion efficiency is calculated as follows:

$$\text{Overall conversion efficiency} = \frac{E_{bc} + E_{bo} + E_{sg}}{E_b + E_a + E_f} \times 100\%$$

$$\text{Overall conversion efficiency (\%)} = \frac{20.62\ \text{MJ}}{24.8\ \text{MJ}} \times 100\% = 83.1\%$$

d. The system losses are calculated as follows:

$$\text{Losses} = \text{output energy} - \text{input energy} = 24.8\ \text{MJ} - 20.62\ \text{MJ} = 4.18\ \text{MJ}$$

2.5.3 Gasification

Gasification is a thermal conversion process that utilizes some amounts of oxidants or oxygen, but well below the stoichiometric requirements. In the combustion of fuels such as methane, the stoichiometric amounts of air necessary for complete combustion may be calculated by completing a balanced combustion equation, as shown in Equation 2.4.

$$CH_4 + 2\left(O_2 + 3.76\ N_2\right) \rightarrow CO_2 + 2H_2O + \text{heat} \quad (2.4)$$

The equation shows that for every mole of methane, 2 mol of air is required for complete combustion. This combination generates 1 mol of carbon dioxide and 2 mol of water (or steam) and some heat that is equal to the heat of combustion of methane. Note that to have a balanced equation, the number of elements on the left-hand side of the equation must be equal to that on the right-hand side. If biomass is to be thermally converted, a similar process could be used to determine the stoichiometric amounts of air required for complete combustion. Note that the 3.76 mol of nitrogen per mole of oxygen used is based on the molar ratio of nitrogen and oxygen in air (79% N_2/21% O_2 = 3.76). In Chapter 3, a procedure will be discussed that establishes the empirical chemical formula for a biomass feedstock to set up a similar combustion stoichiometric equation.

In the gasification process, some amount of oxidant is required for the reaction to proceed, but well below the stoichiometric value. If an incomplete amount of air is used, intermediate products are produced, and most of these gaseous products are combustible. Normally around 30%–70% of the stoichiometric amount of air is used, but not 100%. The primary gaseous product of the gasification reaction is called synthesis gas (also called "syngas" or "producer gas"), which is composed of hydrogen gas and carbon monoxide ($CO + H_2$).

The products of the gasification process are very similar to the products of the pyrolysis process: solid biochar and gaseous syngas. The liquid product, sometimes called tar, is minimized and is not enhanced or collected during the conversion process. Syngas is normally used directly for heat and power generation. In most gasification systems, conversion efficiencies are frequently based on how well the carbon of the biomass is used. A common term is carbon conversion efficiency, which is the ratio of all the carbon in the product divided by carbon products of the input biomass. Example 2.6 shows how carbon efficiency is calculated for a gasifier. Figure 2.6 is a simple schematic of a fixed-bed gasification system where biomass is loaded into a reactor (normally a 200 L metal drum) (52.84 gal).

A screen is placed near the bottom of the drum to contain the biomass and to allow a small space (called the plenum) where air can be introduced. A small axial-flow blower introduces controlled and limited amounts of air. The biomass is first ignited by combusting part of it at the bottom of the reactor. Thus, an access port (firing door in the figure) must be available to accomplish this task. Once a substantial

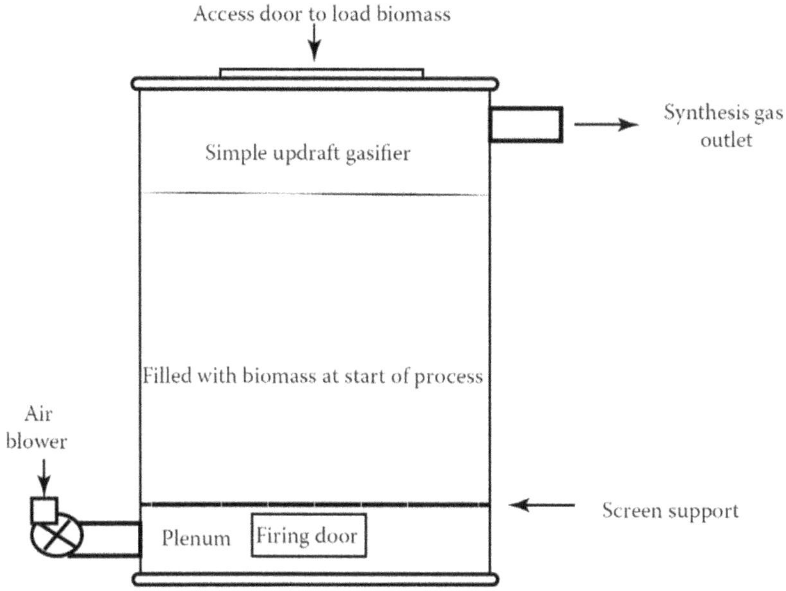

FIGURE 2.6 Simple biomass gasifier schematic of basic component parts.

portion of the biomass is burned or lit, air is introduced and controlled, such that a thick brownish-black smoke is produced at the syngas outlet near the top of the reactor. As the reaction temperature is increased, the quality of the gaseous product will improve to a point where it can be sustainably ignited or burned. This gaseous product may be combusted immediately or stored for later use. When all the biomass materials have been completely gasified, syngas production ceases, biochar is removed, and fresh biomass material is added. The process is then repeated for every batch of biomass. Note that the biomass must be dried to about 10%–15% moisture for the process to run smoothly.

To accomplish energy and mass balance in a gasifier reactor, the procedure is similar to that discussed in the pyrolysis process and shown schematically in Figure 2.7. The mass and heating value of the biomass must first be measured on a given test run. At the end of the test run and after cooling the system, the amount of biochar produced must be weighed and the heating value measured. The amount of syngas produced must also be measured. Then, using the average temperature during the test, a complete chemical compositional analysis can be done. If possible, a complete ultimate analysis must be done on the biomass and the resulting biochar, defined as the amounts of carbon (C), hydrogen (H), oxygen (O), nitrogen (N), and sulfur (S) in the material, plus the ash. To estimate carbon efficiency, the difference between the carbon content in the biomass and the carbon content in the biochar may be assumed to be that produced in the syngas. This is easily accomplished in a batch gasifier by measuring the net amounts of biomass utilized during the run, the amount of biochar produced, and the syngas generated. To complete a meaningful energy and mass balance, all data must be corrected at standard temperature (0°C) and pressure (1 atm),

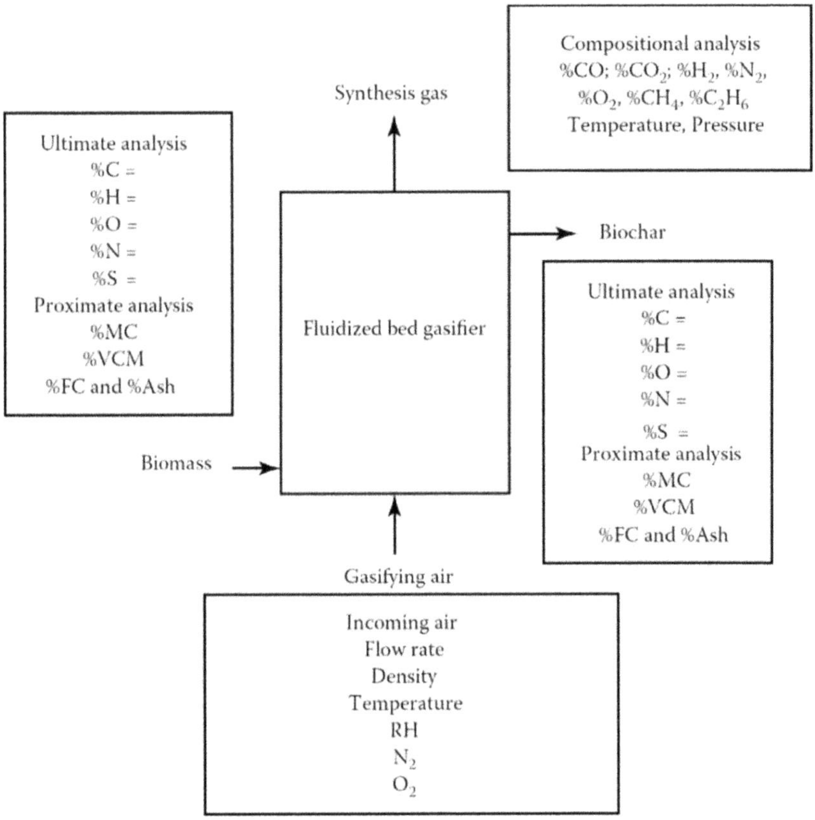

FIGURE 2.7 Schematic of energy and mass balances in gasifier.

or STP. Alternatively, use normal temperature (20°C) and pressure (1 atm), or NTP. Likewise, the conversion efficiency of the process must be calculated accordingly. As biochar is not used for much of the process, in most cases, the efficiency is based on the energy content of the syngas divided by the energy content of the biomass feedstock used. The gasification process is self-sustaining after the initial preheating. Thus, the energy input is simply the biomass used plus the energy used during the start-up procedure. During the start-up procedure, part of the biomass is utilized for the initial heating. Thus, the initial energy requirement is usually negligible, especially if the process is continuous. There are many variations of this gasification reactor. These variations will be discussed in detail in Chapter 10. An advanced biomass gasification system using a fluidized bed will be discussed in Chapter 11.

One simple efficiency calculation for gasification is the carbon conversion efficiency defined in Equation 2.5.

$$\text{Carbon conversion efficiency} = \frac{\text{carbon in syngas}}{\text{carbon in original biomass}} \times 100\% \tag{2.5}$$

This equation is the ratio of the carbon in the syngas to the carbon in the biomass that is being gasified. Thus, the syngas must be completely analyzed and the percentage of carbon calculated for all of the gaseous compounds produced. The original biomass must also undergo ultimate analysis to determine the percentage of carbon. This is normally done on a per weight basis. Unfortunately, a complete analysis of all the gaseous compounds cannot be done. Consequently, the usual procedure is to do a complete analysis of the biochar produced. Then, one assumes the difference in the carbon content of the biomass less the carbon content of the biochar is the carbon content of the syngas. Example 2.6 illustrates how carbon conversion efficiency in a gasification process is calculated.

Example 2.6: Carbon Efficiency Calculations in Gasification

Determine the carbon efficiency of a gasification conversion process wwusingww the following data:

The ultimate analysis of sorghum biomass has the following results:

$$C = 40.27\%, H = 5.47\%, O = 39.63\%, N = 0.40\%, S = 0.07\%, \text{and ash} = 14.16\%.$$

Thus, the important parameter is 40.27% carbon in the feedstock.

The biomass was gasified at a rate of 1.16 kg/min.

The ultimate analysis of the char is as follows:

$$C = 31\%, H = 2.1\%, O = 40.4\%, N = 1.0\%, S = 0.01\%, \text{and ash} = 25.4\%.$$

The char production rate was measured at 0.31 kg/min.

Assume that tar production is negligible.

Solution:

a. The carbon in syngas is calculated from the difference between the fuel and the char as follows:

$$C_{syngas} = C_{biomass} - C_{char}$$

$$C_{syngas} = \frac{1.16 \text{ kg}}{\text{min}} \times 0.4027 - \frac{0.31 \text{ kg}}{\text{min}} \times 0.31 = 0.371 \text{ kg/min}$$

b. Carbon conversion efficiency will be calculated based on the carbon in the syngas divided by the carbon in the biomass as follows:

$$\text{Carbon conversion efficiency} = \frac{\text{carbon in syngas}}{\text{carbon in original biomass}} \times 100\%$$

$$\text{Carbon conversion efficiency} = \frac{0.371 \text{ kg/min}}{1.16 \times 0.4027 \text{ kg/min}} \times 100\% = 79.4\%$$

2.5.4 Combustion

Combustion is the thermal conversion of biomass with excess air as the oxidant and with the production of heat. The main products of conversion are heat, water, and carbon dioxide. The energy produced during the process is called the heat of combustion of the fuel, or in this case, the heat of combustion of the biomass feedstock. This heat of combustion is synonymous with the heating value of the biomass fuel that was combusted. Combustion of biomass, specifically that of fuelwood, has been done because people make use of fuelwood for domestic heating, cooking, and lighting. Fuelwood is still the most important biofuel in the developing world. In general, fuelwood consists of cellulose, hemicellulose, and lignin; proportions vary for different wood or biomass species. In general, hardwood contains about 43% cellulose, 35% hemicellulose, and 22% lignin. Softwood contains about 43% cellulose, 28% hemicellulose, and 29% lignin (Bhattacharya, 1983). The heating value of holocellulose (cellulose and hemicellulose) is about 17.5 MJ/kg, whereas that of lignin is about 26.65 MJ/kg.

Although fuelwood is still a popular feedstock for combustion systems, agricultural residues are now becoming a popular source of heat and energy. However, there is one significant difference between fuelwood and agricultural residues. Some agricultural residues have a very high ash content. Consequently, these high-ash residues are inappropriate for combustion systems. The inorganic components of some agricultural residues (ash), particularly animal manure, are quite high. Combustion systems using high-ash residues for conversion will experience slagging and fouling issues. The ash in these types of biomass melts at high combustion temperatures. Upon cooling, hardened agglomerates form that clog the conveying surfaces. In addition, incineration of agricultural residues is now banned in most countries because of atmospheric pollution issues. As such, not all biomass resources are appropriate for combustion systems.

Chapter 13 will present a thorough discussion of biomass combustion as a conversion process and identify biomass resources that are suitable for combustion systems. In some cases, the best option is to simply mix biomass with other conventional fuels that are already being used for combustion systems, such as coal. Co-combustion or co-firing of biomass with coal is an excellent option and should be further investigated. If mixed in proper proportions, the low-sulfur content of biomass will balance the high-sulfur content of coal. This combination may be more economical than the use of after-burner treatment systems to reduce smog formation. There are still numerous biomass-to-electrical power plants operating around the world. The majority of feedstock comes from either hard or soft woods. Example 2.7 shows the growing area required to generate several megawatts of electrical power from biomass.

Example 2.7: Conversion of Biomass into Electrical Power

How many acres would be required to build a 50 MW (energy output) biomass power plant (operated 365 days a year and 24 h a day) if the heating value of the biomass (dry) is 7,000 Btu/lb? The biomass yield in dry tons per acre per year is about 10. Assume a conversion efficiency of 35% and using the formula for percentage efficiency as follows:

$$\% \text{ Efficiency} = \text{energy output/energy input} \times 100\%$$

Solution:

a. The energy input must be equal to 50 MW/0.35 = 142.9 MW.
b. This is a simple conversion process using dimensional analysis to the correct units of acres as follows:

$$142.9\text{ MW} \times \frac{1\times10^6\text{ W}}{\text{MW}} \times \frac{\text{J}}{\text{W}-\text{s}} \times \frac{\text{Btu}}{1{,}055\text{ J}} \times \frac{\text{lb}}{7{,}000\text{ Btu}} \times \frac{3{,}600\text{ s}}{\text{h}}$$

$$\times \frac{24\text{ h}}{\text{day}} \times \frac{365\text{ days}}{\text{yr}} \times \frac{\text{ton}}{2{,}000\text{ lb}} \times \frac{\text{acre}}{10\text{ tons}} = 30{,}500\text{ acres}$$

c. Answer = 30,500 acres

2.6 HYBRID CONVERSION PROCESSES

There are now biomass conversion processes that rely on a combination of various chemical, biological, or thermal conversion systems to generate suitable biofuels. Some of these mature and emerging technologies will be discussed in the succeeding section (Sections 2.6.1 and 2.6.2). Perhaps the most common technology is the combination of biological conversion on the front-end of the conversion pathway, followed by thermal conversion of the remaining biomass that was not converted. One of the most popular patented designs is the MixAlco™ process, developed at Texas A&M University by Dr. Mark Holtzapple.

2.6.1 Biological and Thermal Conversions

There are various advantages to combining the biological and thermal conversion processes. This is primarily because biological conversion results in the generation of large amounts of sludge, which, after the conversion process, may be as much as 40% of the initial biomass by weight. The MixAlco process is basically a combination of the anaerobic digestion process on the "front-end" of the process and thermal conversion on the "back end." Dr. Holtzapple calls this process the "carboxylate platform" (Granda et al., 2009).

The MixAlco process begins with pretreatment of biomass with lime, during which the biomass is aerated and some moisture is added. The system could be a simple pile of biomass on the ground with gravel and lining at the bottom. The material is seeded with a mixture of microorganism populations taken from animal rumens (cattle, sheep, or deer), including anaerobic sewage digesters, swamp microorganisms, and termite guts. The pile is then covered so that the anaerobic digestion process can proceed. The process takes several weeks or months, during which organic acids are produced. Organic acid production can be enhanced by using higher reaction temperatures, thus resulting in more acetic and butyric acids. Methane production is inhibited by the addition of a methane suppressant such as iodoform, which inhibits the proliferation of methane-forming microorganisms.

Next, the organic acids (acetic, propionic, butyric, caproic, etc.) are dewatered and the dry carboxylate salts undergo thermal conversion at high temperature with no oxygen to generate ketones. The calcium acetates are then converted into acetones, while calcium propionate is converted into diethyl ketone, and the higher molecular weight calcium butyrate is converted into dipropyl ketone. These ketones are hydrogenated (addition of hydrogen) to generate higher alcohols (i.e., acetone to isopropanol, methyl-ethyl ketone to 2-butanol, and diethyl ketone to 3-pentanol). The reactions are thus quite complex. The alcohols are then oligomerized using suitable novel catalysts (such as zeolite-based ZSM-5) to produce hydrocarbon fuels. For example, the isopropanols are converted into propylene and then into hexene. The hexene can then be saturated to produce hexane (Landoll, 2012).

The advantages of this system are as follows: minimal solids handling, low capital cost, and wide range of feedstock that can be used. The energy efficiency is high if sludge is utilized. Likewise, there are many chemical products that can be produced. The overall system could be designed in a distributed fashion as well.

Perhaps a few disadvantages include a longer conversion period that results in a larger area required (or footprint) for the overall system, as well as the production of enormous amounts of sludge that, after the anaerobic digestion process, should be utilized to further improve the economics of the system. Although the whole process could take several weeks or months, several reactors could be utilized in a round-robin system to make the overall process continuous throughout the year. The remaining sludge can be dried and pyrolyzed to generate the hydrogen that is required during the hydrogenation step.

2.6.2 Thermal and Biological Conversions

The technology being developed at Oklahoma State University's (OSU) Biobased Products and Energy Center is the reverse of a process discussed in the previous section (Section 2.6.1). The OSU is developing an efficient thermal conversion process to produce syngas, which is utilized to produce alcohol. This process is a combination of thermal conversion on the front-end of the system and biochemical conversion on the finishing stage for liquid fuel production. Fermentation of syngas has several advantages compared with many other chemical catalytic processes for biomass to hydrocarbon fuel conversion. *Clostridium* strain P11 (Kundiyana et al., 2010) is being utilized in a strict anaerobic condition of syngas to generate various alcohol compounds such as isopropyl alcohol and butanol, two very important chemicals in the industry.

The front-end of the system could be a fixed-bed or fluidized-bed gasifier to generate carbon monoxide and hydrogen gas (the main components of syngas). The syngas is cleaned and purified using a syngas scrubber system; then, these gaseous materials are placed in syngas storage tanks for introduction into the fermentor, as shown in

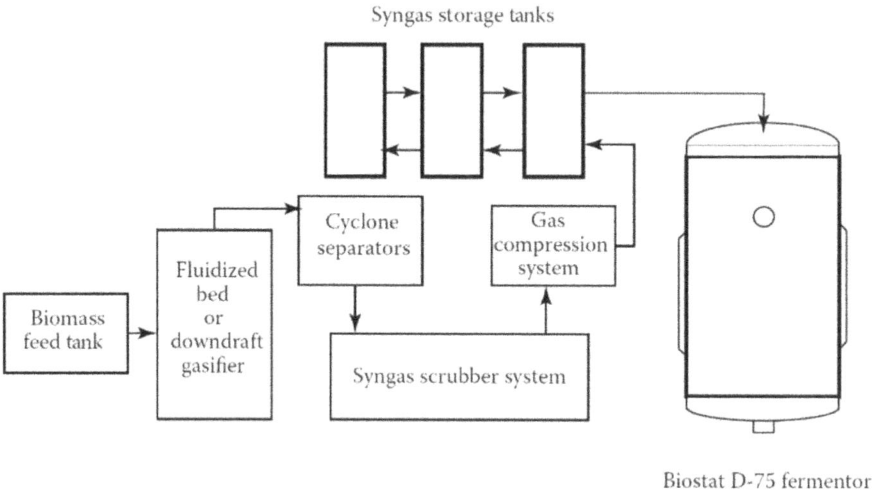

FIGURE 2.8 Schematic of OSU pilot syngas fermentation system. (Adapted from Kundiyana, D. K., R. L. Hunke and M. R. Wilkins, Syngas fermentation in a 100-L pilot scale fermentor: Design and process consideration. *Journal of Bioscience and Bioengineering* 109, 492, 2010.)

Figure 2.8. The typical product concentration recorded during the fermentation run in the OSU pilot study revealed an ethanol concentration of 25.26 g/L, 2-propanol concentration of 9.25 g/L, acetic acid concentration of 4.82 g/L, and 1-butanol concentration of 1.13 g/L (Kundiyana et al., 2010). To make the syngas fermentation more efficient, the medium used for the growth of *Clostridium* species must be inexpensive, and the feeding rate must be optimized to increase the syngas conversion into ethanol. A study was conducted to use corn-steep liquor as a nutrient replacement for yeast extract in syngas fermentation (Maddipati et al., 2011). About 32% more ethanol (1.7 g/L) was produced.

2.7 APPLICATIONS OF BIOMASS CONVERSION PRODUCTS

Biomass energy conversion systems provide a wide range of products and coproducts and can be used with many applications. The most common and direct application of biomass conversion is the production of heat, such as the use of fuelwood for heating and cooking using simple combustion systems. Liquid fuels may also be produced using various conversion processes, such as ethanol from sugar, starch, or cellulosic biomass. Another liquid fuel, biodiesel, may be derived from oil seed crops. Biomass may also be used for the generation of mechanical or electrical power, aside from biofuels. The next section (Sections 2.7.1–2.7.7) will discuss various applications of biomass conversions, including technical knowledge for evaluating a particular application.

2.7.1 Heat Energy

In many developing countries, the primary use of biomass (e.g., wood fuel) is for heating and cooking. Heat energy is perhaps the most important use of biomass resources. Wood fuel and animal manure are still being used as cooking fuel in many other countries. For large-scale applications, however, biomass may not be easily used to generate heat, because of the high percentage of ash present. In large-scale applications and those with high heat requirements, biomass ash will melt at elevated temperatures forming slag, which may affect some heat-conveying components of the system. In subsequent chapters, one will learn that there are only a few types of biomass that may be combusted in excess air because of slagging and fouling problems. The simplest way to generate energy from biomass is by using fuelwood to boil water. Equation 2.5 is the most used heat energy conversion equation to estimate the heat required to raise the temperature of water from an initial temperature of t_i to a final temperature of t_f. The other components needed are the mass (m) and the specific heat of the material being heated. For water, the value of C_p at standard conditions is 4.190 MJ/kg-°C. Thus, it takes about 300 kJ of energy to bring a liter of water starting at 30°C to a boil (100°C) (see Example 2.8). This example makes use of Equation 2.6.

$$Q(\text{MJ}) = mC_p\Delta t = mC_p(t_f - t_i) \tag{2.6}$$

Example 2.8: Energy Required to Heat or Boil a Liter of Water

How much energy is required to raise the temperature of 1 L of water from an initial temperature of 30°C to a final temperature of 100°C (boiling point of water) at standard temperature and pressure (STP)? Compare this energy to the energy content of a kilogram of wood waste with an energy content of 20 MJ. Assume the density of water at STP is 1.0.

Solution:

a. Equation 2.5 is used to arrive at the following values (note that 1 L of water is equal to 1 kg of water at STP):

$$Q(\text{MJ}) = 1\text{ kg} \times \frac{4.19\text{ kJ}}{\text{kg} - {}^\circ\text{C}} \times (100 - 30)^\circ\text{C} = 293.3\text{ kJ} = 0.2933\,\text{MJ}$$

b. Thus, 1 kg of wood has several times more energy than that required to boil 1 L of water. In short, this is only equivalent to 1.5% of the energy of the fuelwood. One kilogram of wood is more than enough to boil 1 L of water at STP.

2.7.2 Electrical Energy

Biomass may also be used to generate electrical power. In fact, a small portion of the fuelwood consumed in the United States is used to produce electrical energy.

There are various ways of generating electricity from biomass. The most popular power conversion pathway from biomass in the United States is the conversion of methane gas in landfills into electrical power. Large municipal or industrial landfills produce gas that can be tapped to generate electricity. The methane gas is used to run internal combustion engines that are compatible with this gas, to which an electrical generator has been attached. In 1995, California had 56 landfill gas recovery facilities: 14 collected only gases and 42 collected gases to produce electricity (CEC, 2012). These electricity-generating facilities have an installed capacity of about 246 MW.

There are internal combustion engines currently being designed to use landfill gases. For example, General Electric's (GE) gas engines, ranging from 0.12 to 9.5 MW, are designed for stationary, continuous duty operations using landfill gases. GE's Jenbacher gas engines (J420 gas engines) are reported to power China's largest landfill gas power generation project in Laogang, China (Business Wire, 2012). Seven of GE's Jenbacher J420 gas engines provide about 10 MW of electrical power. Each J420 engine combusts 2.7 million cubic liters of methane each year. Each engine can generate 1.4 MW of electricity. The calorific value of the landfill gas is reported to be around 5 kWh/Nm^3. Example 2.9 shows the estimated conversion efficiency of this system, which is around 50%.

Example 2.9: Conversion Efficiency of Landfill Gas to Power

Determine the conversion efficiency of landfill gas for a power generation facility with the following data: Electrical power output is 1.4 MW, and gas consumption is 2.7 million cubic meters of methane each year. Landfill gas has approximately 60% methane. Assume about 8,000 h of operation each year with a heating value of 5 kWh/Nm^3.

Solution:

a. The total amount of biogas used assuming 60% is methane is calculated from the following:

$$\text{Biogas }(\text{m}^3) = \frac{2.7\times10^6\,\text{m}^3}{0.6} = 4.5\times10^6\,\text{m}^3$$

b. The energy efficiency equation is used to estimate efficiency as follows:

$$\text{Efficiency }(\%) = \frac{\text{output}}{\text{input}}\times100\% = \frac{1.4\text{MW}}{4.5\times10^6\,\text{m}^3}\times\frac{\text{m}^3}{5\,\text{kWh}}$$

$$\times\frac{1{,}000\text{ kW}}{1\,\text{MW}}\times8{,}000\,\text{h}\times100\% = 49.78\%$$

c. Thus, the conversion efficiency is approximately 50%.

2.7.3 Combined Heat and Power

There are applications where the combination of heat and power may be necessary. Biomass resources converted into electrical power may be more economical when excess heat is utilized for other applications within the vicinity of the power plant. Facilities that utilize both electrical power and heat are called "cogeneration facilities," which are used for combined heat and power production. Biomass cogeneration is the sequential production of either heat or steam and electrical power. There are two types of cogeneration systems: topping cycles and bottoming cycles. Topping cycles refer to systems that utilize biomass feedstock first for the electrical output of the system (e.g., production of electricity) and then use the excess heat for other thermal applications. Bottoming cycles, on the other hand, refer to facilities that use the biomass first for heat or thermal energy generation, and the rejected heat is then used for production of electricity.

Cogeneration systems are currently being used for decentralized power generation systems utilizing biomass. For example, many poultry or turkey facilities require heat or steam to process the animals for market sale and distribution, while electricity is always a ubiquitous facility requirement. Thus, it is likely that the animal manure will be used to generate the required heat energy and the excess heat used to generate a small amount of electricity. This electricity may be used by the facility or might be sold to the nearby electrical utility for what is termed an "avoided cost." Avoided cost is simply the cost between the selling cost of the local utilities and the production cost of the decentralized power generation facility. Thus, if the selling cost for a local utility is 15 cents/kWh and the generation cost for the local biomass power generation facility is 10 cents/kWh, the local utility and the decentralized power generation facility will be negotiated up to 5 cents/kWh. In this case, both parties have "avoided any loss" in the sale of electricity. The local utilities are bound or obliged by law to purchase electricity. The power output is normally on a smaller scale and normally between 50 and 80 MW. This law is contained in the Public Utilities Regulatory Policies Act of 1978 (PURPA). Of course, the decentralized power generation system must be qualified to negotiate with the local utilities. Qualifications include certain overall conversion efficiency (e.g., >42.5% overall) or in terms of facilities size (<80 MW) and must demonstrate a cogeneration facility as well.

2.7.4 Mechanical Energy

Biomass may be converted into some form of energy that in turn will generate mechanical power. For example, bioethanol or biodiesel may be blended with gasoline and the blended mixture combusted in an internal combustion engine to generate mechanical power. Likewise, the syngas produced from gasification of biomass may be used to power internal combustion engines in the same manner. The same syngas may also be burned in a combustor, where the products of combustion are used to generate steam or superheated gas, which in turn runs steam or gas turbines to generate mechanical power. Various parameters may be used to describe, evaluate, and report the performance of a particular system. Thus, we can define several conversion efficiencies that may be used. The simplest will be the combustion efficiency, which may be defined as the ratio of ideal fuel-to-air ratio to the actual air-to-fuel ratio used (Sonntag and Van Wylen, 1991) as shown in Equation 2.7.

$$\eta_{\text{comb}}(\%) = \frac{\text{FA}_{\text{ideal}}}{\text{FA}_{\text{actual}}} \times 100\% \tag{2.7}$$

If, on the other hand, the biofuel is used in a boiler to generate steam, the efficiency may be defined as the ratio of the heat transferred to the steam to the higher heating value of the fuel. If the fuel is biomass, the heating value can be measured using an appropriate bomb calorimeter. Thus, the efficiency may be defined as that shown in Equation 2.8.

$$\eta_{\text{steam generator}}(\%) = \frac{\text{heat transferred to steam (MJ/kg)}}{\text{higher heating value of the fuel (MJ/kg)}} \times 100\% \tag{2.8}$$

In internal combustion engines, the efficiency may be defined as the ratio of actual amount of work done by the engine to the heating value of the fuel used (Equation 2.9). This efficiency is called the second law efficiency in most thermodynamics textbooks, or also called the thermal efficiency of the process. Testing to estimate the amount of work done by an engine is normally done with a suitable engine dynamometer. Example 2.10 illustrates how engine conversion efficiency is calculated.

$$\eta_{\text{th}}(\%) = \frac{\text{work done by the engine per fuel weight }\left(\text{MNm/kg}^{-s}\right)}{\text{heating value of the fuel (MJ/kg)}} \times 100\% \tag{2.9}$$

Example 2.10: Engine Conversion Efficiency

A biomass (wood) power plant uses steam to generate electrical power. Around 320,000 kg of water per hour enters the boiler at a pressure of 12.5 MPa and a temperature of 200°C. Steam leaves the boiler at 9 MPa and at 500°C. The power output of the turbine is 60,000 kW. The rate of biomass input is 55,000 kg/h with a heating value of 20 MJ/kg. Determine the efficiency of steam generator and the overall thermal efficiency of the plant.

Solution:

a. The heat transferred to the water to generate steam can be calculated by the enthalpy changes at various pressure and temperature combinations. From steam tables, the enthalpy at 12.5 MPa and 200°C was 857.1, whereas the enthalpy at the other conditions was 3386.1:

$$\text{Energy transferred to water }\left(E_{\text{H}_2\text{O}}\right) = m^*\left(h_{9\,\text{MPa},500^\circ\text{C}} - h_{12.5\,\text{MPa},200^\circ\text{C}}\right)$$

$$E_{\text{H}_2\text{O}} = 320{,}000\text{ kg/h}^*(3.3861 - 0.8571)\text{MJ/kg} = 809{,}280\text{ MJ/h}$$

$$E_{\text{fuel}} = 55{,}000\text{ kg/h}^*20\text{ MJ/kg} = 1{,}100{,}000\text{ MJ/h}$$

b. Equation 2.8 may then be used to calculate the efficiency of the steam generator as follows:

$$\eta_{\text{steam generator}}\,(\%) = \frac{809{,}280\ \text{MJ/h}}{1{,}100{,}000\ \text{MJ/h}} \times 100\% = 73.6\%$$

c. Equation 2.9 will be used to estimate the overall thermal efficiency of the system as follows:

$$\eta_{\text{th}}\,(\%) = \frac{60{,}000\ \text{kW}}{20\ \text{MJ/kg}} \times \frac{\text{h}}{55{,}000\ \text{kg}} \times \frac{3.6\ \text{MJ}}{1\ \text{kWh}} \times 100\% = 19.6\%$$

2.7.5 Liquid Biofuels or Bio-Oil Production

Biomass can be converted directly into transport fuel blends such as ethanol and biodiesel. However, one cannot use 100% of these biofuels directly in current transport vehicles, primarily because of warranty issues on the engines. All major original equipment manufacturers in the United States support at least B5 or lower blends provided the biodiesel meets ASTM D6751 standards (Standard Specification for Biodiesel Fuel Blend Stock [B100] for Middle Distillate Fuels). Most engines, for example, may be blended with 5% biodiesel (B5) without limitations because of the lubricity behavior of biodiesel. However, some engine manufacturers, such as Caterpillar, Chrysler, Cummins, Ford, GM, John Deere, and Yanmar, limit their blending to 20% (B20), whereas few manufacturers, such as Case IH, Fairbanks Morse, and New Holland, support B100 use (NBB, 2011). More than 60% of US manufacturers now support B20 or higher blends in at least some equipment, according to the National Biodiesel Board.

Ethanol also has blending limits for use in gasoline engines. Pure ethanol use in internal combustion engines is only possible if that engine has been designed or modified for this purpose, such as the flex fuel engines. Ethanol fuel mixtures have "E" designations in the fuel mixture by volume. Thus, E85 is 85% ethanol and 15% gasoline. In the United States, E10 is the most common and already in the gasoline blend. Since 2011, E10 has been mandated by law. Ethanol is mixed in gasoline engine to improve the octane rating of the gasoline. This is because the octane rating of pure ethanol is around 110. Thus, blending ethanol with gasoline will definitely improve the enhanced fuel's octane rating. Perhaps only Brazil has engines that can use 100% ethanol (called neat ethanol vehicles). E85 is the highest blend available in the United States and is intended for use only in flex fuel vehicles.

There is also a growing trend in the use of bio-oil from the fast pyrolysis of biomass and blending with diesel fuel. Most reports are from research data only and the fuel has not yet been utilized in commercial engines. However, because of bio-oil's high heating value, its use in oil burners is becoming very common. When blended with conventional hydrocarbon fuels, the heating values may range between 32.5 and 41.2 MJ/kg, approaching that of diesel or gasoline fuel (Weerachanchai et al., 2009). However, bio-oil does not blend well with diesel fuel, and some solvents may be needed to have a homogeneous mixture. When bio-oil is used in boilers, some modifications may be necessary, mainly for fuel nozzles and fuel-conveying systems. Some bio-oils are quite viscous and may require special nozzles to direct their flow.

2.7.6 Syngas Production and Use

Gasifiers may be used solely to produce syngas. Syngas is composed mainly of carbon monoxide and hydrogen gas, with typical compositions of 15% and 10%, respectively (LePori and Soltes, 1985). These typical gas compositions are rather low if syngas is to be used for subsequent liquid fuel production. In some gasification systems, steam is used to improve the quality of syngas. In a study made on the gasification of the carbon in the charcoal using a downdraft gasifier (Cruz, 1985), it was shown that the amount of hydrogen is improved from 7.5% to 21.5% as steam is increased from 10% to 40%, respectively. The syngas yield was also increased from 5.2 m^3/kg C (84 ft^3/lb C) to 6.3 m^3/kg C (101 ft^3/lb C) at those increased steam rates. Table 2.2 summarizes the data illustrating the effect of steam in the gasification of biomass.

Syngas may be used directly in internal combustion engines to generate mechanical and electrical power. However, care must be taken when the syngas is cooled because of the formation of tar during this cooling process. There are now various low-Btu engines that have been used in directly combusting syngas. Depending on the engine requirements, particulates must be removed and the syngas must be cooled down. Thus, a series of tar removal systems must be in place. There are also

TABLE 2.2
Effect of Steam in Gasification of Biomass

% Steam	10%	20%	30%	40%
Blast temp (°C)	47	60	70	76
Reaction temp (°C)	795	699	646	601
Gas Composition (%)				
CO	34.1	26.7	19.1	12.3
H_2	7.5	13.6	18.3	21.5
CO_2	1.7	7.0	11.8	15.6
H_2O	0.4	2.2	5.7	10.7
CH_4	–	0.1	0.2	0.5
N_2	56.3	50.4	44.9	39.4
GCV (Btu/ft^3)	142	139	130	123
Gas yield ft^3/lb-°C	84	89	96	101
Heat available (Btu/lb-°C)	11,899	12,268	12,537	12,521

Source: Cruz, I. E., Producer-gas technology for rural applications. FAO Agricultural Services Bulletin No. 61. FAP of the United Nations, Rome, Italy, 1985.

companies that sell biofuel and retrofit conversion system to handle two types of fuel. The primary fuel could be diesel or methane gas and the secondary fuel could be syngas. An example is the GTI-Bifuel system from Altronics (Girard, OH). These systems were originally intended for the combustion of conventional fuels such as natural gas and diesel. Now their use has expanded to include biogas or syngas with diesel engines. These bifuel system retrofits will not require any engine modifications and will allow the use of natural gas in a diesel engine by as much as 70%, as claimed by Altronics.

Bifuel operation is achieved using a fumigated gas-charge design, whereby the natural gas is premixed with engine intake air and delivered to the combustion chamber through the air intake valve. The air–gas mixture is ignited by a reduced amount of diesel sprayed in the combustion chamber of the engine. The diesel fuel becomes the "pilot" fuel and acts as the ignition source for the primary air–gas mixture. Another company, ComAp Systems, LLC (Roscoe, IL), developed a fully automated bifuel control system that blends natural gas with diesel fuel in many engines: Caterpillar (150–2,000 kW), Cummins (400–1,900 kW), Deutz (all outputs), Man (400 kW), Mercedes (440 kW), Perkins (590 kW), Volvo (225–520 kW), and ComAp (ComAp Brochure, 2012).

These applications are currently being used for biomass-generated biofuel in internal combustion engines. The control system includes the following components:

1. A fully integrated bifuel control panel
2. Comprehensive knocking detection system
3. Diesel monitoring system
4. Gas train, made according to local and site standards
5. Gas actuator
6. Complete gas feed and air manifold vacuum feedback hoses
7. Gas/air blenders

In the future, there will be a wide range of systems that would utilize the gaseous products from biomass conversion systems in conventional internal combustion engines. Note that syngas has about 10%–20% of the heating value of natural gas and would require more volume compared with natural gas. Thus, engines are normally operated well below their rated capacity.

2.7.7 Biochar Production

Biochar, or charcoal, is a major coproduct of thermal conversion systems such as pyrolysis and gasification. This material has a higher heating value than the original biomass, and the rates of production are between 10% and 40% of the original biomass. This coproduct must be utilized to improve the overall system economics. There are numerous applications of biochar, as enumerated below:

1. Charcoal or briquettes fuel
2. Activated carbon pollutant absorption or adsorption
3. Soil amendment and recycling

4. Bedding material
5. Fertilizer material
6. Co-firing fuel
7. Carbon capture and sequestration
8. Graphene or graphene inks
9. Capacitor energy storage (super capacitors)
10. Black carbon

The various applications of biochar necessitate extensive research or post-processing. The higher heating value of biochar makes it an ideal material for making charcoal briquettes or co-firing fuel. However, the cost of producing pellets or briquettes is normally in the range of $30–$40 per tonne. Consequently, a guaranteed market for the product must be available above its production cost. The value-use of biochar as a fertilizer needs further research, including an understanding of the nutrients available following thermal conversion. At elevated temperatures of conversion, most inorganic fertilizer materials oxidize. They may be rendered unavailable to plants. Thus, some post-processing, such as acidulation, may be required. Some microbial conversion processes must also be investigated to hasten the release of nutrients upon application of char to the agricultural soil. Optimistically, the carbon in the char will be of value in the future carbon trade.

2.8 CONCLUSION

Technologies for the conversion of biomass into heat energy, fuels, and power are undergoing rapid development. Countries should be fully aware of and take advantage of these rapidly developing technologies.

This chapter provided a general overview of various biomass conversion processes, from simple physico-chemical conversion to biological and thermal conversion systems. Combinations of the conversion pathways and hybrid systems were also presented. Detailed discussion of each conversion technology will be made in subsequent chapters. All the basic products and coproducts of the conversion systems were enumerated, and possible future applications were presented. The efficiencies of conversion for most systems and how to do simple mass and energy balance calculations were described.

Applications of biomass conversion processes range from simple heat energy to fundamental electrical power production. Perhaps the most important use of biomass conversion processes is the production of liquid fuels or liquid hydrocarbon fuels. These liquid products, being of high energy density, may be stored for future use and transported over longer distances. The other important coproducts are gaseous fuels, such as syngas or biogas, that may be used on-site; solid products such as biochar that may be returned to the agricultural land for improved land sustainability; and other coproducts such as distiller's grain that may be used as animal feed or supplement. The overall efficiency of conversion will depend on how well these coproducts are further utilized.

2.9 PROBLEMS AND DISCUSSION ISSUES

2.9.1 Heat Energy Conversion Efficiency

P2.1 One tonne (1.1 ton) of dried manure is converted into heat energy. Determine the conversion efficiency of converting this biomass into heat if the heat produced (energy output) during the process is 15,000 MJ. The heating value of the manure was reported at 19.7 MJ/kg (8,500 Btu/lb).

2.9.2 Energy of Feedstock and Biofuel Product

P2.2 One hundred kilograms (220 lbs) of refined soybean oil were converted into biodiesel. Assuming 80% conversion efficiency by weight, that is, 80 kg of biodiesel was produced, determine the energy of the resulting biodiesel product in units of MJ. The energy content of refined soybean oil was found to be 27.87 MJ/L (100,000 Btu/gal). The resulting energy content of the ester was 32.94 MJ/L (118,170 Btu/gal). The density of biodiesel product was 888.7 kg/m^3 (7.4 lbs/gal), and the density of refined soybean oil was 912.7 kg/m^3 (7.6 lbs/gal). Likewise, determine the overall energy conversion efficiency of this biodiesel production process.

2.9.3 Ethanol Yield per Unit of Area

P2.3 A certain variety of sweet sorghum has a sugar content 15.9% by weight, and the biomass yield is 57.8 tonnes/ha (25.74 tons/acre) and a stalk sugar yield of around 9.2 tonnes/ha (4.1 tons/acre). Estimate the ethanol yield per hectare based on the rules of thumb presented in this chapter. Compare this with the 5,000 L/ha (535 gal/acre) yield reported in some literature reports.

2.9.4 Number of Animals Required for Power Generation

P2.4 An irrigation pump is to be powered by biogas using a methane-driven internal combustion engine. About 540 W of power is required for the pump. The biogas engine has an overall conversion efficiency of 24% (thermal and mechanical), and the irrigation must be completed in 10 h each day. Estimate the number of swine animals required for this application. Assume the energy content of biogas is 19.3 MJ/m^3 (550 Btu/ft^3). Further, assume that each mature swine will produce 164 L biogas per day.

2.9.5 Pyrolysis Conversion Efficiency

P2.5 Determine the energy loss (MJ and MMBtu) and overall conversion efficiency (%) for the pyrolysis of 1 tonne (1.1 ton) of switchgrass using the following resulting data. The energy content of the switchgrass is 19.77 MJ/kg (8,520 Btu/lb). Convert the answers into English units.

- The amount of biochar produced is 227 kg (500 lb), with a heating value of 28.97 MJ/kg (12,480 Btu/lb).
- The amount of bio-oil produced is 227 L (60 gal), with a heating value of 30.66 MJ/L (110,000 Btu/gal).
- The amount of syngas produced is 110.5 m^3 (3,900 ft^3), with an energy content of 11.17 MJ/m^3 (300 Btu/ft^3).
- The pyrolyzer power requirement per hour was measured to be about 10.5 kW. The experiment occurred for about 6 h.

2.9.6 Gasification Carbon Efficiency

P2.6 Determine the syngas carbon conversion efficiency for manure gasification conversion process using the following data:

- The ultimate analysis of manure biomass revealed that carbon content is 45% and manure char carbon content is 31%.
- The biomass was gasified at a rate of 4.5 kg/min.
- The char production rate was measured at 1.14 kg/min.
- The tar production was measured to be 0.51 kg/min, with 0.15% C by weight.

Provide a comment on the syngas carbon conversion efficiency.

2.9.7 Area Required for a Power Plant

P2.7 How many hectares (and acres) would be required to build a 50 MW (energy output) biomass power plant (operated 365 days a year and 24 h a day) if the heating value of the biomass (dry) is 17.4 MJ/kg (7,500 Btu/lb)? The biomass yield is 9 tonnes/ha/yr (4 tons/acre/yr). Assume an overall conversion efficiency of 50%.

2.9.8 Energy Required to Raise Water Temperature

P2.8 How much energy is required to raise the temperature of 1 L of water from an initial temperature of 25°C to a final temperature of 100°C (boiling point of water) at standard temperature and pressure (STP). Compare this energy to the energy content of 1 kg of rice hull waste, with an energy content of 17 MJ/kg. Assume the density of water at STP is 1.0 kg/L.

2.9.9 Biogas Conversion Efficiency

P2.9 Determine the conversion efficiency of a biogas for power generation facility with the following data: Electrical power output is 300 kW and gas consumption is around 0.96 million cubic meters of methane each year. Landfill gas has approximately 60% methane. Assume about 8,000 h of operation each year, with a heating value of 5 kWh/Nm^3. The letter N stands for "normal" in this heating value unit.

2.9.10 Steam Cycle Conversion Efficiency

P2.10 A biomass (wood) power plant uses steam to generate electrical power. Around 160,000 kg of water per hour enters the boiler at a pressure of 12.5 MPa and a temperature of 200°C. Steam leaves the boiler at 9 MPa and at 500°C. The power output of the turbine is 40,000 kW. The rate of biomass input is 30,000 kg/h, with a heating value of 22 MJ/kg. Determine the efficiency of the steam generator and the overall thermal efficiency of the plant. Hint: In calculating the energy absorbed by the steam, one may use the enthalpy equation $E = m \times (h_2 - h_1)$. The enthalpy values are found in steam tables.

2.9.11 Major Gases in Gasification Systems

P2.11 Identify the major gases in gasification systems.
a. $CO{+}H_2$ b. $CO_{2+}H_2O$ c. $CO_{2+}H_2$ d. CH_{4+H2}

2.9.12 Major Gases in Anaerobic Digestion Process

P2.12 Identify the major gases in Anaerobic Digestion Process.
a. $CO{+}H_2$ b. $CO_{2+}H_2O$ c. $CO_{2+}H_2$ d. CH_{4+CO2}

2.9.13 Thermal Conversion Systems

P2.13 Which thermal conversion system uses the most oxidants?
a. Torrefaction b. Pyrolysis c. Gasification d. Combustion

2.9.14 Advantages of Thermal Conversion Systems versus Biological Conversions

P2.14 Discuss the advantages of thermal conversion systems versus biological conversion systems when it comes to making biofuels for commercial use.

2.9.15 Topping versus Bottoming Cycle

P2.15 In cogeneration systems, which pathway (topping or bottoming cycle) do you think is best when the heating value of biomass is rather high? Provide a defense for each case.

REFERENCES

Bhattacharya, S. C. 1983. Lecture Notes in "Renewable Energy Conversion" Class. Asian Institute of Technology, Bangkok, Thailand.

Business Wire. 2012. GE gas engine technology to power China's largest landfill gas project. February 23, 2012 News Release. A Berkshire Hathaway Company, New York.

California Energy Commission (CEC). 2012. The California Energy Almanac. The California Energy Commission, Sacramento, CA. Also available at: https://energyalmanac.ca.gov.

ComAp Systems, LLC. 2012. COMAP bifuel (dual fuel)-The practical solution brochure for high speed and low speed engines. Roscoe, IL.

Cruz, I. E. 1985. Producer-gas technology for rural applications. FAO Agricultural Services Bulletin No. 61. FAP of the United Nations, Rome, Italy.

Granda, C. B., M. T. Holtzapple, G. Luce, K. Searcy and D. L. Mamrosh. 2009. Carboxylate platform: The MixAlco™ process. Part 1: Comparison of three biomass conversion platforms. *Applied Biochemistry and Biotechnology* 156 (156): 525–536.

Kundiyana, D. K., R. L. Hunke and M. R. Wilkins. 2010. Syngas fermentation in a 100-L pilot scale fermentor: Design and process consideration. *Journal of Bioscience and Bioengineering* 109 (5): 492–498.

Landoll, M. 2012. Ketone production from the thermal decomposition of carboxylate salts. PhD Dissertation, Chemical Engineering Department, Texas A&M University, College Station, TX.

LePori, W. A. and E. J. Soltes. 1985. Thermochemical conversion for energy and fuel. In: *Biomass Energy: A Monograph* (E. A.Hiler and B. A.Stout, eds.). Texas Engineering Experiment Station, The Texas A&M University Press, College Station, TX. pp. 9–75.

Maddipati, P., H. K. Atiyeh, D. D. Bellmer and R. L. Hunke. 2011. Ethanol production from syngas by *Clostridium* strain P11 using corn steep liquor as a nutrient replacement to yeast extract. *Bioresource Technology* 102 (2011): 6494–6501.

National Academy of Sciences. 1976. *Methane Generation from Human, Animal and Agricultural Wastes*. National Research Council, Washington, DC.

National Biodiesel Board (NBB). 2011. *Original Equipment Manufacturers (OEM) Engine Warranty Positions*. NBB, Jefferson City, MO.

Sonntag, R. E. and G. J. Van Wylen. 1991. *Introduction to Thermodynamics: Classical and Statistical*. John Wiley & Sons, New York.

Stout, B. A., C. A. Myers, A. Hurand and L. W. Faidley. 1979. *Energy for World Agriculture*. FAO Agriculture Series. Food and Agriculture Organization of the United Nations, Rome, Italy.

Tumuluru, J. S., S. Sokhansanj, C. T. Wright, R. D. Boardman and R. J. Hess. 2011. Review on biomass torrefaction process and product properties and design of moving bed torrefaction system model development. ASABE Technical Paper No. 1110459. ASABE, St. Joseph, MI.

Weerachanchai, P., C. Tangsatthitkulchai and M. Tangsatthitkulchai. 2009. Phase behaviors and fuel properties of biooil-diesel-alcohol blends. *World Academy of Science, Engineering and Technology Journal* 56: 387–393.

ASSESS YOUR KNOWLEDGE

1. Which of the following biomass conversion process is ideal for relatively dry biomass?
   a. Gasification
   b. Anaerobic digestion
   c. Fermentation
   d. Enzymatic hydrolysis

2. Oils from vegetable crop is an ideal feedstock for what conversion process?
   a. Bioethanol production
   b. Gasification
   c. Biodiesel production
   d. Biogas production

3. Name the most common microbe used for conversion of sugar into bioethanol.
   a. *Aspergillus niger*
   b. *Saccharomyces cerevisiae*
   c. *Aspergillus awamori*
   d. *Trichoderma reesi*

4. Name a common microbe that can breakdown cellulose into sugars.
   a. *Aspergillus awamori*
   b. *Aspergillus niger*
   c. *Saccharomyces cerevisiae*
   d. *Trichoderma reesi*

11. Which biomass thermal conversion process is endothermic?
    a. Gasification
    b. Pyrolysis
    c. Combustion
    d. Torrefaction

12. Which thermal conversion process requires the least oxidant?
    a. Gasification
    b. Combustion
    c. Anaerobic digestion
    d. Pyrolysis

13. Two metric tonnes of biomass with a heating value of 15 MJ/kg was converted thermally for 24 h and the resulting power output was 50 kW. Determine the conversion efficiency for this process.
    a. 30%
    b. 15%
    c. 45%
    d. 60%

14. Determine the conversion efficiency for using 100 kg of rapeseed oil (or canola) with a heating value of 27.87 MJ/L and converting this into its ester form (rapeseed methyl ester, a biodiesel) with a resulting heating value of 39.8 MJ/kg. Around 73.7 kg of rapeseed methyl ester was produced. The density of rapeseed oil is 0.922 kg/L.
    a. 97%
    b. 90%
    c. 85%
    d. 75%

5. Name a common mold used to convert starch into sugars.
   a. *Aspergillus awamori*
   b. *Trichoderma reesi*
   c. *Aspergillus niger*
   d. *Saccharomyces cerevisiae*

6. Majority of bioethanol in the United States comes from what feedstock?
   a. Corn
   b. Soybean
   c. Sugarcane
   d. Jerusalem artichoke

7. What is the theoretical amount of pure glucose sugar (kg) needed to produce a liter of bioethanol?
   a. 15
   b. 0.40
   c. 1.6
   d. 2.9

8. What is the current state-of-the art yield of bioethanol for every bushel of corn?
   a. 0.40
   b. 1.6
   c. 15
   d. 2.9

9. What are the primary gases contained in pyrolysis gas?
   a. $CO + H_2$
   b. $CH_4 + CO_2$
   c. $CO_2 + H_2O$
   d. $CH_4 + H_2$

15. How many moles of carbon dioxide is released for every mole of glucose consumed by the yeast to produce bioethanol?
    a. 1
    b. 2
    c. 3
    d. 4

16. Determine the conversion efficiency in using 30 liters of biogas (energy content is 24.2 MJ/m^3) for lighting purposes in a rural household. The equivalent light energy is about equal to the intensity of a 25 W bulb that lasted for 6 h.
    a. 35.5%
    b. 74.4%
    c. 50.5%
    d. 90%

17. This is another name for the MixAlco process developed at Texas A&M University by Dr. Mark Holtzapple?
    a. Torrefaction
    b. Hybrid anaerobic digestion
    c. Anaerobic fermentation
    d. Carboxylate platform

18. T/F. Steam, when used in gasification system will increase the heating value of synthesis gas.
    a. False
    b. True

19. Which of the following is not an application of biochar?
    a. Activated carbon
    b. Graphene
    c. Resistor
    d. Capacitor

10. Which thermal conversion process requires the most oxidant?
    a. Pyrolysis
    b. Gasification
    c. Combustion
    d. Torrefaction

20. Name the most popular chemicals used for biodiesel production?
    a. Yeast
    b. Enzymes
    c. Bleach
    d. Alcohol and catalyst

Answers: 1a, 2c, 3b, 4d, 5c, 6a, 7c, 8d, 9a, 10c, 11a, 12d, 13b, 14a, 15b, 16b, 17d, 18b, 19c, 20d

ADDITIONAL LEARNING RESOURCES

- Access the Lecture Notes at the Instructor and Student Resources: https://routledgetextbooks.com/textbooks/instructor_downloads/

3 Biomass Properties for Thermal Conversion

LEARNING OBJECTIVES

Upon completion of this chapter, one should be able to

1. Describe the various properties of biomass that are important for thermal conversion.
2. Describe the various methods of characterizing the physical and thermal properties of biomass.
3. Specify the various equipment used for characterizing various biomass properties.
4. Enumerate and describe the different standard methods for characterizing biomass.
5. Estimate some biomass properties, given basic data and empirical equations.

3.1 INTRODUCTION

This chapter deals with important biomass properties that are commonly used for thermal conversion processes. These properties play a significant role in the efficiency of thermal conversion systems. Biomass materials are so diverse in physical, chemical, and biologically related properties that there is not a single property that is used for a particular process. Almost always, biomass needs to undergo pretreatment or preprocessing. Numerous biomass characterization parameters will be discussed, beginning with the physical properties and equipment required for characterization. The physical and thermal properties discussed in this chapter are shown in Figure 3.1. Other compositional analyses of biomass, and those that are important for the biological conversion processes, are discussed in Chapter 4.

Important physical properties include particle and bulk density, particle size, and particle size distribution (PSD). Other important properties in biomass handling and storage include angle of repose, friction, heat capacity, and heat conductivities. Major thermally related properties include proximate analysis, ultimate analysis, and heating values. Others include the eutectic point or the fusion temperature of ash in the biomass. This parameter is important to note in evaluating slagging and fouling problems in biomass thermal conversions. The US Department of Energy (USDOE) has also developed various laboratory analytical procedures for compositional analyses of biomass that include lignin, carbohydrates, and extractives. These are important in the biological or biochemical conversion of biomass, particularly for ethanol production.

DOI: 10.1201/9781003294306-3

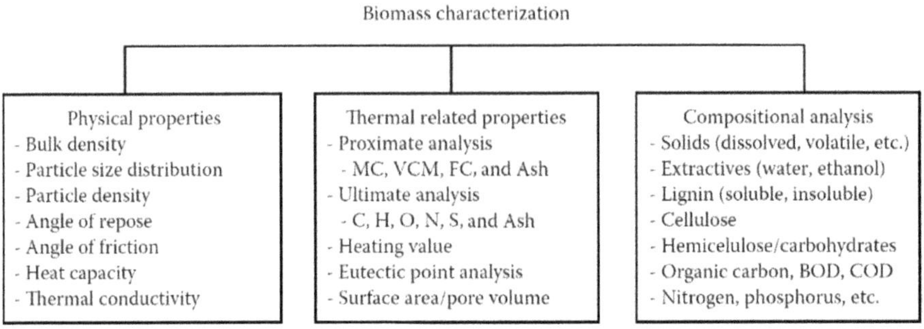

FIGURE 3.1 Selected parameters for complete biomass characterization.

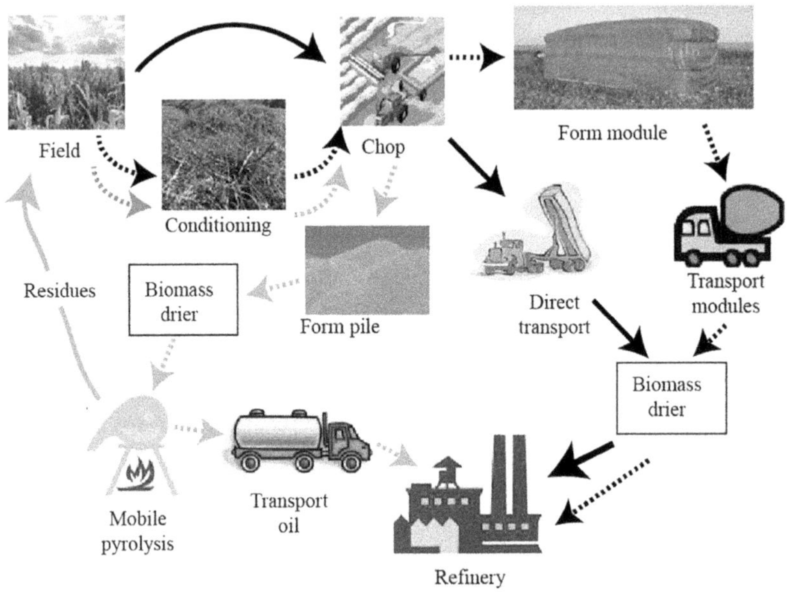

FIGURE 3.2 System flow alternatives for biomass before being used in biorefinery.

One should observe that as the biomass is harvested, the system flow alternatives greatly vary with the type of conversion system used. Figure 3.2 shows an example of system flow alternatives for biomass before it is transported to a refinery (Searcy, 2010). The field biomass may have very high moisture and would therefore require some conditioning in the field (the dotted black route) or may require some particle size reduction (or chopping). It may be formed in modules as shown before being transported by truck or rail. Biomass properties may vary at each stage of the process and thus should be evaluated. Refineries may require further reduction in moisture, and the biomass may be dried appropriately before being converted into biofuel. One direct route to refineries (the solid black route) may be direct reduction in size after being harvested in the field, followed by direct transport in large trucks; however, this may still require moisture reduction before the delivery to biorefineries. One other option (the dotted gray route) is the harvest of biomass that has been conditioned in the field before being chopped into coarse particles and ground to reduce the size further. This ground biomass is dried before being converted in a mobile flash pyrolyzer, near where the biomass is harvested (to reduce the transportation cost). This biomass will be pyrolyzed on-site to generate a high-density bio-oil that can then be transported by tankers to refineries for the generation of hydrocarbon fuels. The char by-product (the solid gray route) will be returned back to the field, thereby minimizing all biomass transport costs. Note that in all stages of the process, the biomass will have changed in characteristics and must be known for ease of transport and processing.

3.2 PHYSICAL PROPERTIES OF BIOMASS

3.2.1 Particle Density

Particle density is the measure of the biomass weight per unit volume (kg/m^3 or lb/ft^3). The most common device used for measurement of particle density is the pycnometer from Micromeritics (Model Accupyc 1330, Norcross, GA). The principle for particle density measurements is quite intuitive. A sample of ground and dried biomass is weighed accurately and placed in a receptacle of known and calibrated volume. This receptacle is then placed in an enclosed system where all void spaces are filled with an inert gas (normally helium) and the volume occupied by the biomass is calculated. Thus, this parameter is the measure of the biomass particle weight per unit of volume that the biomass particle has occupied. Sometimes, this is also referred to as the true particle density of the material. The measurement procedure is very fast and can be easily replicated with great accuracy.

3.2.2 Bulk Density

Bulk density, on the other hand, has the same units as particle density (kg/m^3); however, the volume unit includes the void spaces occupied by the biomass. The standard method for measuring bulk density of biomass is contained in the American Society for Testing and Materials (ASTM) E 873–82 (2006) entitled "Standard Test Method for Bulk Density of Densified Particulate Biomass Fuels." The bulk density of biomass is easily measured using a simple box of nominal size equal to 305 × 305 × 305 mm^3 (12 × 12 × 12 in^3) with handles. The volume of this box should be within 16.39 cm^3 (1 in^3). A scale is needed with the capacity to weigh the box sample to within 100 g. The quantity of the sample must be large enough to be representative, but not less than 45.45 kg (100 lb).

The tare weight of the box is first determined (to within 100 g, 0.22 lb), and a certain amount of biomass is poured into this container from a height of about 610 mm (2 ft). The box is dropped about five times from a height of about 150 mm (6 in) to a nonresilient surface to allow settling. Additional samples are added, and excess samples are removed from the top edge. The box is then weighed to within 100 g (0.22 lb). The bulk density is calculated using Equation 3.1.

$$\text{Bulk density}\left(\frac{\text{g}}{\text{cm}^3}\right) = \frac{(\text{weight of box and sample}) - (\text{weight of box})}{(\text{volume of box})} \tag{3.1}$$

The volume occupied by the biomass is the denominator of the bulk density equation. The numerator is the weight of the biomass. The bulk density value varies with how well the biomass is packed in the container. Biomass with a high bulk density is suitable for transport because it will require less storage volume. The average bulk density of wood is in the range of 300–550 kg/m^3 (18.7–34 lb/ft^3). For charcoal, the reported range is 200–300 kg/m^3 (12.5–18.7 lb/ft^3) (Lata and Mande, 2009). Rice straws and rice hulls have a lower bulk density of about 70 kg/m^3 (4.4 lb/ft^3), whereas densified products (or briquettes) may have bulk densities as high as 700–800 kg/m^3 (43.6–50 lb/ft^3).

Woody biomass, depending on the species, has a wide range of densities. A solid block of dry wood will weigh from 400 to 866 kg/m^3 (25–54 lb/ft^3). Chipped green wood has roughly 353 kg/m^3 (22 lb/ft^3). Dry wood shavings may only have 80–96 kg/m^3 (5–6 lb/ft^3). Thus, trucks hauling biomass may be limited by volume instead of weight. Wood for energy is usually sold by weight. Consequently, the relatively low weight that can be transported, because of volume limitations, can make hauling wood quite expensive (Badger and Pullammanappallil, 2011).

3.2.3 Particle Size and Particle Size Distribution

Biomass materials, after being harvested from agricultural crops, will need to be reduced in size if they are to be used for any conversion process. Moisture must be removed, and some important components must be separated. Freshly harvested biomass must first be dried, chopped into small particles, and ground into finer particles. When the biomass materials are already in ground form, they will be further characterized for physical dimensions, so they will be suitable for appropriate storage. Storing bulky materials is always costly.

The particle size or PSD of the biomass material is measured using various standard sieves. Biomass materials are normally ground to less than 10 mm in size for various conversion processes. The PSD is estimated using a sieve shaker.

Table 3.1 summarizes the list of standard sieve screen sizes from 37 to 3,360 microns. Ground samples of biomass are placed on different sets of sieve combinations from higher pore space on top to smaller screen size at the bottom. A known amount of biomass is placed on the top sieve and then subjected to several minutes of shaking (normally about 15–20 min). A mechanical sieve shaker is more appropriate for this purpose.

The biomass collected at each size range is weighed. The percentage of each size range is then calculated based on the original weight of the biomass used for the test. For particle sizes less than 100 microns, a different set of equipment is used. The Beckman Coulter Counter Multisizer 3 (Beckman Coulter, Brea, CA) and the Malvern Mastersizer 3000 (Malvern Instruments, Worcestershire, UK) are the most common PSD analysis equipment.

In the Beckman Coulter Counter unit, the biomass particles are dispersed using a sonicator in a solvent to break apart each component particle. Then, each particle is passed through an aperture to measure the particle size, and the PSD is plotted. The Malvern Mastersizer is operated similarly and has both dry and wet options for measuring the PSD of the material. In both cases, the particle equivalent spherical diameter (ESD, in microns) is first reported by the instrument. The particle aerodynamic

TABLE 3.1
Table of Standard Sieve Screen Sizes

Tyler Mesh No.	US Standard ASME 11–61	British Standard BSS 410 1969	Aperture Inch	Aperture Microns (μ)
6	6	5	0.131	3360
7	7	6	0.111	2830
8	8	7	0.0937	2380
9	10	8	0.0787	2000
10	12	10	0.0661	1680
12	14	12	0.0555	1410
14	16	14	0.0469	1190
16	18	16	0.0394	1000
20	20	18	0.0331	840
24	25	22	0.0280	707
28	30	25	0.0232	595
32	35	30	0.0197	500
35	40	36	0.0165	420
42	45	44	0.0138	354
48	50	52	0.0117	297
60	60	60	0.0098	250
80	80	85	0.0070	177
100	100	100	0.0059	149
150	140	150	0.0041	105
200	200	200	0.0029	74
250	230	240	0.0025	63
400	400	400	0.0015	37

equivalent diameter (AED, in microns) of the material is then calculated by multiplying the ESD with the square root of the particle density (g/cm^3), as shown in Equation 3.2.

$$AED = ESD \times \sqrt{\rho} \tag{3.2}$$

The particle densities are those values measured using the AccuPyc 1330 Pycnometer. Particle sizes in this range are normally reported in AED, because biomass particles are not spherical in nature. However, most instruments approximate the particle sizes as spheres.

3.2.4 Angle of Repose

The angle of repose is the steepest angle of descent or the slope relative to the horizontal plane when biomass material is poured on a horizontal surface. For example, when granular biomass materials are poured from a funnel and dropped on a horizontal surface, a mound or conical pile is formed. The angle or slope of the mound is called the angle of repose. This angle is technically defined as the internal angle between the surface of the pile and the horizontal surface. The factors affecting this angle are the density of the material, the surface area and shape of the particles, and the coefficient of friction of the material. This parameter is important in the design of equipment for biomass material storage, particularly silos and hoppers for biomass-conveying systems. If a belt conveyor is used to transport the biomass material, the size or width of this belt is dependent on the angle of repose for transporting a given volume of mass of the material without overflowing along the belt.

3.2.5 Angle of Friction

The angle of friction, on the other hand, is the angle with which a pile of biomass will slide when raised from the horizontal. This term is also synonymous with the angle of friction of the biomass material on a given surface. One will observe that if the surface is smooth, the biomass slides at a lower angle than when the surface is rather rough. This parameter is also important in the design of conveying systems and hoppers for biomass fuel bins. Proper use of angle of friction prevents bridging from biomass feed bins or conveying systems.

3.2.6 Heat Capacity and Thermal Conductivity of Biomass

The heat capacity of a material is defined as the ratio of the amount of heat energy transferred to the material to the resulting increase in temperature of this material. In the SI system of units, the heat capacity is given in units of joules per kelvin. There is currently limited research and few publications about the heat capacity of biomass materials. Information is lacking because heat capacity is an extensive property, that is, its physical characteristics vary with the scale of the system. In some applications, specific heat capacity is more practical to use because this is the heat energy per unit of mass at a given temperature. This parameter is now being used in the modeling of some biomass processes.

Thermal conductivity, on the other hand, is the ability of the material to conduct or transfer heat. Heat transfer of the material with high thermal conductivity is higher than that with lower thermal conductivity. Consequently, materials with high thermal conductivity are used to store heat energy (as heat sink), whereas those with low conductivity are used as insulation. Most biomass materials are more suitable for insulation than as heat sinks. The units of thermal conductivity are W/m-K (Btu/h-ft). Likewise, thermal conductivity is used for modeling biomass in storage systems to measure their ability to conduct heat through the enclosure as well as their ability to store heat energy.

3.3 IMPORTANT THERMAL RELATED PROPERTIES

There are standard test methods for biomass characterization developed by the ASTM. There are also methods developed by the USDOE National Renewable Energy Laboratory. The ASTM standards originated from the characterization of wood, coal, or coke. The three sets of standards commonly used for biomass characterization as it relates to its thermal conversion properties are the proximate, ultimate, and heating value analysis. For further thermal conversion, analysis of the eutectic point of biomass becomes important, and the test methods used originally were for the evaluation of coal. The test methods described here may be used to establish the rank of biomass fuels and to show the ratio of combustibles to incombustibles, and sometimes to provide the basis for marketing the biomass.

3.3.1 Proximate Analysis

Proximate analyses include the measurement of moisture content (MC), volatile combustible matter (VCM), fixed carbon (FC), and ash. Moisture is important for biomass transport to biorefineries and would also influence the type of conversion process used. The MC could be as high as 99% for some aquatic biomass, such as algae, or as dry as 10% for some field crops. Drying biomass requires enormous amounts of energy. Almost always, the most economical method for drying uses the sun. Table 3.2 shows some proximate analysis of biomass measured at the BioEnergy Testing and Analysis Laboratory (Beta Lab) at Texas A&M University (https://betalab.tamu.edu). The reported high heating value (HHV) described in the table is the reported dry-basis heating value of these biomass. The "as-received" biomass heating value is lower because of the presence of moisture (H_2O), which does not have any energy content.

Table 3.3 lists the various ASTM standards for the measurement of proximate analysis. Both the coal and the wood industries have their own standards, and the standards used for wood fuels are those that are closely related to the biomass. However, the differences in the procedures are quite minor.

TABLE 3.2
Some Proximate Analysis of Common Biomass (as Received)

Biomass	MC (%)	VCM (%)	FC (%)	Ash (%)	HHV (MJ/kg)
Rice straw (CA)	8.2	67.5	10.5	13.8	15.8
Rice straw (TX)	8.7	61.6	13.4	16.3	15.4
Corn stover	9.9	70.7	13.7	5.7	15.6
Woodchips (softwood)	10.0	78.5	11.2	0.3	19.4
Switchgrass	9.4	76.3	10.7	3.6	19.8
Cotton gin trash	10.7	66.7	15.7	6.9	17.9
Peat	14.3	58.8	24.1	2.8	19.0
Wheat straw	8.7	69.4	11.1	10.8	17.5
Rice husks (or hulls)	10.0	60.3	11.9	17.8	14.4
Poultry manure	10.7	48.7	11.6	29.0	11.8
Almond shells (outer)	9.5	65.9	18.9	5.7	18.7
Almond hulls (inner)	9.3	73.5	15.3	1.9	19.6
Alligator grass	11.3	64.4	10.1	14.2	17.1
Coconut shells	11.2	72.7	15.3	0.8	20.6
Sorghum biomass	12.8	65.4	14.4	7.4	17.0
Jatropha shells	9.8	64.5	21.6	4.1	19.2
Arundo donax	9.3	73.2	12.3	5.2	18.7

TABLE 3.3
List of ASTM Standards Used for Biomass Proximate Analysis for Wood Fuel, Coal, and Coke

ASTM Designation	Description
E 871–82 (reapproved 2006)	MC for particulate wood fuels
E 872–82 (reapproved 2006)	VCM for particulate wood fuels
D 1103–84 (reapproved 2007)	Ash analysis of wood
D 3172–07a	Proximate analysis of coal and coke
D 3173–03 (reapproved 2008)	MC for coal and coke
D 3174–04	Ash in coal and coke
D 3175–07	VCM for coal and coke

3.3.1.1 Moisture Content

To measure the biomass moisture, the sample (as received) is normally ground to reduce the particle size. Replicated amounts are placed in a dry container in a forced convection oven with a temperature setting of 103°C (ASTM E871–82 [reapproved 2006] for wood). This is normally done overnight, or for at least 16h, until the weight of the biomass (with the container) becomes constant. The equation to calculate the MC for both wet and dry bases was presented in Chapter 1. A similar standard is used for coal or coke samples (ASTM D 3173-03 [reapproved 2008]). The difference is that the temperature used for coal and coke is between 104°C and 110°C. An alternate form shown in Equation 3.3 may be used to measure the MC of biomass.

$$\text{Moisture content } (\%\text{MC}) = \frac{m_\mathrm{i} - m_\mathrm{f}}{m_\mathrm{i}} \times 100\% \tag{3.3}$$

In Equation 3.3, m_i is the initial weight of the biomass sample before placing in the drying oven and m_f is the final constant weight after drying. This MC, if the biomass is freshly harvested, is what is referred to as the "as-received" basis. Moisture is very important in the transportation of biomass and thermal conversion. If the MC of biomass is quite high, additional energy is required to remove the water, and this will affect the conversion efficiency.

The equipment for the measurement of biomass MC is a simple gravimetric forced convection dryer. The unit must have the ability to accurately control and maintain the temperature within the range of 104°C–110°C. In the oven, the door should contain a hole of approximately 3.2mm (1/8 in) in diameter near the bottom to permit the free flow of air through the oven space.

3.3.1.2 Volatile Combustible Matter

To measure the VCM, the bone-dry biomass sample is placed in a platinum crucible with a close-fitting cover and placed in a tube furnace (like that shown in Figure 3.3). The temperature setting is 950°C ± 20°C (ASTM E872–82 [reapproved 2006] for wood). The material is exposed for about 7min, cooled, and reweighed.

FIGURE 3.3 Tube furnace used for measurement of VCM.

The loss in weight is from the VCM of the biomass. Likewise, a similar procedure is used for coal or coke samples (ASTM D 3175–07). The temperature used is the same (950°C ± 20°C), including the time of exposure (7 min). Equation 3.4 may be used to measure the VCM of the biomass.

$$\text{Volatile combustible matter}\ (\%\text{VCM}) = \frac{v_i - v_f}{v_i} \times 100\% \tag{3.4}$$

In Equation 3.4, v_i is the initial weight of the biomass sample before it is placed in the tube furnace and v_f is the final constant weight after cooling and removal of volatiles. Volatiles are important in thermal conversion, as they lead to more combustible gases during the conversion. The ASTM standard calls for the use of specially made platinum crucibles for the test. However, these crucibles are quite expensive. VCM determinations by some laboratories use alternate nickel–chromium alloy crucibles, which are less expensive. However, if they are used, they must be calibrated with a platinum crucible to determine if relative bias exists. If a bias is identified, it will be used to correct the results when using the alternative crucibles. Note in the figure that a contraption to hold the crucible properly must be made in order to lower the sample appropriately into the tube furnace.

3.3.1.3 Fixed Carbon and Ash

After the measurement of the VCM, the biomass material is placed in a muffle furnace for about an hour, with the cap removed and the temperature set to 600°C. After an hour, the crucible is placed in a desiccator until the material is cooled. Then, it is reweighed. Ensure that there are no carbons left in the material after it is taken out of the furnace. Otherwise, one may add another 30 min until the color of the material becomes grayish white. The material left on the crucible is the ash present in the biomass. The difference between the ash content and the VCM is the FC. The ASTM standard for wood is designated as ASTM D 1102–84 (reapproved 2007). The equivalent ash analysis for coal and coke is designated as ASTM D 3174-04. The main difference between these standards is the temperature used for ash analysis. For coal samples, the recommended temperature is between 700°C and 750°C. For coke samples, the recommended temperature is 950°C. The total time for the ash analysis of coal and coke is about 4 h. It is very important to place the material in a desiccator to avoid absorption of moisture. In addition, the material must be cooled appropriately before weighing, to remove the effects of temperature on the material being weighed. Example 3.1 shows how proximate analysis is calculated and reported. The method should be replicated at least three times in order to calculate the mean. Equation 3.5 may be used to measure the ash content of the biomass.

$$\text{Ash (\% ash)} = \frac{f_i - f_f}{f_i} \times 100\% \tag{3.5}$$

In Equation 3.5, f_i is the initial weight of the biomass sample before placing in the muffle furnace and f_f is the final constant weight after cooling and removal of the FC components.

The FC content is calculated by taking the difference between 100% and the sum of the VCM and ash percentages of the bone-dry sample, as shown in Equation 3.6. The ash component is important in thermal conversion because high-ash-content biomass will tend to form slag and may cause fouling problems in many thermal conversion-conveying parts.

$$\text{\% Ash} = 100 - \text{\%VCM} - \text{\%FC} \tag{3.6}$$

FC is also important for conversion. Carbon is the prime element useful for making liquid biofuels, and, if matched with hydrogen, will generate hydrocarbon fuels.

Example 3.1: Proximate Analysis Calculations

Ground Jatropha waste samples after coming off the oil seed press were used to estimate the proximate analysis. The following are the data from the experiment:

- Moisture content (MC) measurement. An initial sample weighing 4.9122 g (m_i) was used and placed in the drying oven set at 103°C overnight until the final constant weight became 4.6197 g (m_f).

- Volatile combustible matter (VCM) measurement. That same bone-dry sample weighing 4.6197 g ($m_f = v_i$) was placed in the tube furnace set at 950°C for 7 min, and the resulting weight of the product was 0.8694 g (v_f).
- Ash content measurement. The same sample weighing 0.8694 g ($f_i = v_f$) was then placed in a muffle furnace set at 600°C for 4 h with a resulting final weight of 0.7936 g (f_f).

Calculate the proximate analysis for this sample. Report your data on "as-received" basis and dry basis.

Solution:

a. The moisture content is calculated from the following equation to report an MC of 5.95%:

$$\%\text{MC} = \frac{m_i - m_f}{m_i} \times 100\% = \frac{4.9122 - 4.6197}{4.9122} \times 100\% = 5.95\%$$

b. The volatile combustible matter is calculated from the following equation resulting in a VCM value of 81.18%:

$$\%\text{VCM} = \frac{v_i - v_f}{v_i} \times 100\% = \frac{4.6197 - 0.8694}{4.6197} \times 100\% = 81.18\%$$

c. The %Ash is calculated from the equation below to obtain an ash content of 8.72%:

$$\%\,\text{Ash} = \frac{f_i - f_f}{f_i} \times 100\% = \frac{0.8694 - 0.7936}{0.8694} \times 100 = 8.72\%$$

d. Finally, fixed carbon (FC) is calculated by difference as shown below:

$$\%\text{FC} = 100 - \%\text{VCM} - \%\text{Ash} = 100 - 81.18 - 8.72 = 10.1\%$$

e. Hence, the proximate analysis on dry basis is as follows: VCM = 81.18%; FC = 10.1%; and ash = 8.72% to total 100%.
f. If the proximate values are to be reported on an "as-received" basis, the following are the results: MC = 5.95%; VCM = 76.35%; FC = 9.5%; and Ash = 8.2%, and the resulting total is 100%. This is done by multiplying the VCM, FC, and Ash values by 94.05% (the dry weight of the sample).

3.3.2 Ultimate Analysis

The ultimate analysis consists of measuring the major elemental components of the biomass, such as carbon (C), hydrogen (H), oxygen (O), nitrogen (N), sulfur (S), and ash. The ash should have been established during the proximate analysis.

TABLE 3.4
List of ASTM Standards Used for Biomass Characterization

Designation	Title
ASTM E 777–08	Standard Test Method for Carbon and Hydrogen in the Analysis Sample of Refuse-Derived Fuel
ASTM E 775–87 (2008) e1	Standard Test Methods for Total Sulfur in the Analysis Sample of Refuse-Derived Fuel
ASTM E 778–08	Standard Test Methods for Nitrogen in the Analysis Sample of Refuse-Derived Fuel

The individual elements in the ultimate analysis also have their individual ASTM standard as listed in Table 3.4. Carbon and hydrogen are integrally analyzed in accordance with ASTM E 777; the measurement of sulfur content follows ASTM E 775; and the measurement of nitrogen content follows ASTM E 778. The oxygen content of biomass is a calculated value and is the difference between 100% and the sum of C, H, N, S, and ash. This is a photo of one type of an ultimate analyzer (Elementar Analyzer, Exeter Analytical, North Chelmsford, MA). Other companies manufacture ultimate analyzers, such as the TruSpec Micro Series CHNS/O Analyzer (Leco Corporation, St. Joseph, MI). The more expensive units are those that can handle biomass sample at the gram level rather than the microgram level. The micro-level ultimate analyzers must have a dedicated microbalance that must be accurate to the microgram level, or lower to provide accurate readings. Some ultimate analyses of common biomass are summarized in Table 3.5. Note that the carbon contents vary from a low of around 27% (poultry manure) to as high as 49.5% for peat. The carbon content of coconut shell is very close to peat at 46.9%, followed by softwood (46.1%). The hydrogen content for most agricultural biomass is around 5%–6% as shown. A couple of important parameters that may be calculated from the ultimate analysis are the atomic carbon-to-hydrogen ratio and the atomic oxygen-to-carbon ratio. Example 3.2 shows how to calculate the atomic C/H and O/C ratios important to thermal conversion systems. This will be discussed in subsequent chapters.

Example 3.2: Atomic Hydrogen-to-Carbon Ratios for Biomass

Determine the atomic carbon-to-hydrogen ratio for corn stover. Compare this with softwood. Also, calculate the atomic O/C ratios. Use the ultimate analysis data from Table 3.5.

Solution:

a. The atomic C/H ratios are calculated by first determining the number of moles of carbon and hydrogen from data in the table (assuming say, 100 g sample) as follows:

$$\text{Corn stover carbon (mol)} = 40/12\ \text{g/mol} = 3.33$$

$$\text{Corn stover hydrogen (mol)} = 5.9/1\ \text{g/mol} = 5.9$$

TABLE 3.5
Some Ultimate Analyses of Common Biomass

Biomass	C (%)	H (%)	O (%)	N (%)	S (%)	Ash (%)
Rice straw (CA)	37.1	5.7	42.2	1.0	0.2	13.8
Rice straw (TX)	39.9	6.0	36.3	1.3	0.2	16.3
Corn stover	40.0	5.9	46.9	1.3	0.2	5.7
Woodchips (softwood)	46.1	6.6	46.7	0.2	0.1	0.3
Switchgrass	42.0	6.1	47.8	0.4	0.1	3.6
Cotton gin trash	41.6	5.8	44.4	1.0	0.3	6.9
Peat	49.5	5.6	40.4	1.3	0.4	2.8
Wheat straw	37.1	5.6	45.7	0.5	0.3	10.8
Rice husks (or hulls)	35.0	5.2	41.4	0.5	0.1	17.8
Poultry manure	27.6	4.6	33.4	4.1	1.3	29.0
Almond shells (outer)	43.1	6.5	43.9	0.7	0.1	5.7
Almond hulls (inner)	40.9	6.1	50.2	0.7	0.2	1.9
Alligator grass	36.2	5.7	42	1.5	0.4	14.2
Coconut shells	46.9	6.1	45.7	0.3	0.2	0.8
Sorghum biomass	38.1	5.6	47.8	0.9	0.2	7.4
Jatropha shells	40.8	5.9	46.7	2.2	0.3	4.1
Arundo donax	38.0	5.6	50.3	0.6	0.3	5.2

Thus, the atomic C/H ratio is calculated as C/H = 3.33/5.9 = 0.56.

The number of moles of each element is in the same ratio as the number of atoms.

b. The atomic O/C ratios are calculated by first determining the number of moles of oxygen and carbon from data in the table as follows:

$$\text{Corn stover oxygen (mol)} = 46.9/16\ \text{g/mol} = 2.9$$

$$\text{Corn stover carbon (mol)} = 40/12\ \text{g/mol} = 3.33$$

Thus, the atomic O/C ratio is calculated as O/C = 2.9/3.3 = 0.88.

c. The atomic C/H ratio for softwood is calculated the same way as follows:

$$\text{Softwood carbon (mol)} = 46.1/12\ \text{g/mol} = 3.8$$

$$\text{Softwood hydrogen (mol)} = 6.6/1\ \text{g/mol} = 6.6$$

Thus, the atomic C/H ratio is calculated as C/H = 3.8/6.6 = 0.58.

d. The atomic O/C ratio is calculated as follows:

$$\text{Softwood oxygen (mol)} = 46.7/16\ \text{g/mol} = 2.9$$

$$\text{Softwood carbon (mol)} = 46.1/12\ \text{g/mol} = 3.8$$

Thus, the atomic O/C ratio is calculated as O/C = 2.9/3.8 = 0.76.

3.3.3 Heating Value Analysis

The most important characteristic of biomass is its heating value or calorific value. The heating value measures the energy content of the biomass and is measured by a calorimeter. This test procedure combusts the biomass in pure oxygen, and the energy released during the combustion process is measured to indicate the heating value. The heating value of biomass is highly dependent on the presence of moisture in the biomass. Moisture has no heating value. Consequently, one should be very careful in reporting heating value numbers. When the biomass heating value is reported and noted as "as received," the MC must be provided. Otherwise, the heating value must be reported on either a higher or lower heating value basis as affected by moisture. In some disciplines, as in coal combustion, the heating value of feedstock is reported in either the dry basis (excludes moisture) or the dry ash-free basis (excludes moisture and ash). Example 3.3 shows the distinction between these two parameters. Equation 3.7 shows the calculation of HHV as a function of moisture.

$$\text{HHV}(\text{dry basis}) = \frac{\text{HV}(\text{wet basis})}{1 - \text{moisture content (fraction)}} \tag{3.7}$$

The heating value calculation on an ash-free basis is given by Equation 3.8.

$$\text{HHV (dry ash -free basis)} = \frac{\text{HHV (dry basis)}}{1 - \text{ash content (fraction)}} \tag{3.8}$$

The primary reason for using the HHV of feedstock on a dry ash-free basis is because most power plants, such as coal power plants, are fed on a heating value basis and not on a mass basis, to ensure that the output is always constant.

Example 3.3: Heating Value Calculations

The heating value of a certain biomass was measured by a bomb calorimeter and reported as 17.8 MJ/kg "as received." The moisture content of this biomass was 12%. Convert this heating value numbers into its HHV equivalent. The ash content of the biomass on a dry basis was found to be 13%. Determine the heating value on a dry ash-free basis as well.

Solution:

a. Use Equation 2.1 for this calculation with the moisture content transformed into decimal equivalent:

$$\text{HHV (dry basis)} = \frac{\text{HHV (wet basis)}}{1 - \text{moisture}} = \frac{17.8}{(1 - 0.12)} = 20.2 \text{ MJ/kg}$$

b. Use Equation 2.2 for this calculation:

$$\text{HHV (dry ash-free basis)} = \frac{\text{HHV (dry basis)}}{1 - \text{ash content (fraction)}} = \frac{20.2}{1 - 0.13} = 23.2 \text{ MJ/kg}$$

The standard for the measurement of heating value is designated as ASTM E 711 (Standard Test Method for Gross Calorific Value of Refuse-Derived Fuel by Bomb Calorimeter). Approximately 1 g of relatively dry biomass sample is needed for the analysis. Note that the MC of the biomass sample to be tested must be known in order to convert the units into its HHV (bone-dry sample) or lower heating values (as-received) basis. The protocol being followed at the BioEnergy Testing and Analysis Laboratory at Texas A&M University (Beta Lab) is to use biomass with around 10% moisture to be consistent. Misfiring in the bomb calorimeter may occur if the biomass material is too dry (or too wet).

If a calorimeter is not available, there are various ways of estimating the heating value of biomass that are based on other compositional parameters, as reported in the literature. Table 3.6 shows ultimate analysis data for a variety of biomass fuels

TABLE 3.6
Ultimate Analysis Data for Selected Solid and Biomass Fuels

	Elements (% Dry Weight)						
Material	C	H	O	N	S	Ash	HV (MJ/kg)
Pittsburgh seam coal	75.5	5.0	4.9	1.2	3.1	10.3	31.7
West Kentucky coal	74.4	5.1	7.9	1.5	3.8	7.3	31.3
Utah coal	77.9	6.0	9.9	1.5	0.6	4.1	32.9
Wyoming Elkol coal	71.5	5.3	16.9	1.2	0.9	4.2	29.5
Lignite	64.0	4.2	19.2	0.9	1.3	10.4	24.9
Charcoal	80.3	3.1	13	0.2	0.0	3.4	57.4
Douglas fir	52.3	6.3	40.5	0.1	0.0	0.8	21.0
Douglas fir bark	56.2	5.9	36.7	0.0	0.0	1.2	22.1
Pine bark	52.3	5.8	38.8	0.2	0.0	2.9	20.4
Western hemlock	50.4	5.8	41.4	0.1	0.1	2.2	20.0
Redwood	53.5	5.9	40.3	0.1	0.0	0.2	21.0
Beech	51.6	6.3	41.5	0.0	0.0	0.6	20.4
Hickory	49.7	6.5	43.1	0.0	0.0	0.7	20.1
Maple	50.6	6.0	41.7	0.3	0.0	1.4	19.9
Poplar	51.6	6.3	41.5	0.0	0.0	0.6	20.7
Rice hulls	38.5	5.7	39.8	0.5	0.0	15.5	15.4
Rice straw	39.2	5.1	35.8	0.6	0.1	19.2	15.2
Sawdust pellets	47.2	6.5	45.4	0.0	0.0	0.9	20.5
Paper	43.4	5.8	44.3	0.3	0.2	6.0	17.6
Redwood waste wood	53.4	6.0	0.1	0.1	39.9	0.5	21.3
Alabama oak (waste)	49.5	5.7	41.3	0.2	0.0	3.3	19.2
Animal waste	42.7	5.5	31.3	2.4	0.3	17.8	17.2
Municipal solid waste	47.6	6.0	32.9	1.2	0.3	12.0	19.9
Cotton gin trash	42.0	5.4	36.2	1.4	0.5	14.5	15.5
Sorghum stalks	40.0	5.2	40.7	1.4	0.2	12.5	15.4
Feedlot manure (fresh)	45.5	5.4	31.9	1.0	0.3	15.9	17.4
Feedlot manure (aged)	33.0	4.9	19.0	0.7	0.8	41.6	15.1
Corn cobs	46.2	7.6	42.3	1.2	0.3	2.4	26.3
Rice hulls	41.3	8.4	30.98	1.0	0.02	18.3	16.5
Sawdust	49.7	6.2	42.0	0.7	0.7	0.7	20.0

Source: LePori, W. A. and E. J. Soltes, Thermochemical conversion for energy and fuel. In: *Biomass Energy: A Monograph* (E. A. Hiler and B. A. Stout, eds.). Texas Engineering Experiment Station, The Texas A&M University Press. College Station, TX, 1985.

and solids. The data are from an earlier publication (LePori and Soltes, 1985), which reported various approaches for ultimate analysis of biomass. The reported heating values shown in in the table may be used for comparative purposes.

3.3.3.1 Heat of Combustion

When biomass is converted thermally in the presence of excess amounts of air (i.e., combustion), the total energy released in the form of heat is called its heat of combustion (or heating value/calorific value). The heating value of the biomass is reported in units of Btu/lb or kJ/kg. Gasoline used as fuel for running internal combustion engines has a reported heating value of about 47 MJ/kg (20,250 Btu/lb), and diesel has a heating value of about 45 MJ/kg (19,400 Btu/lb). Biomass materials, on the other hand, may have heating values ranging from 15 to 30 MJ/kg (6,500–13,000 Btu/lb). In the absence of devices to measure heating values of any biomass waste or residue, several methods are used. Primary information is required to estimate heating values. There are two categories of information: first, based on ultimate analysis, that is, the percentage of carbon, hydrogen, oxygen, nitrogen, and sulfur or CHONS (carbon, hydrogen, oxygen, nitrogen and sulfur), and second, based on proximate analysis data (i.e., MC, VCM, FC, and ash).

3.3.3.2 Stoichiometry of Reactions

In the analysis of thermal conversion processes, one must be able to write a complete reaction of carbon, hydrogen, and other combustible elements in the fuel that are in the presence of oxygen. For example, all carbon present is assumed to be burnt to carbon dioxide and all the hydrogen converted into water. No oxygen is present in the products of combustion. The complete combustion of methane is shown in Equation 3.9.

$$CH_4 + 2O_2 \rightarrow CO_2 + H_2O + \text{heat} \tag{3.9}$$

The heat released, if measured, is called the heat of combustion of the fuel, the basic premise in calorimetry measurements. To balance the equation, begin by balancing the carbon, followed by hydrogen, and balance the oxygen last. Note that if air is used as the oxidant, it is convenient to assume air is composed of 21% O_2 and 79% N_2; thus, the stoichiometric equation is presented as Equation 3.10.

$$CH_4 + 2O_2 + 2(3.76)N_2 \rightarrow CO_2 + H_2O + 7.52\ N_2 + \text{heat} \tag{3.10}$$

3.3.3.3 Eutectic Point Analysis

The fusion (or melting) temperature of the ash integrally contained in the biomass sample is also an important parameter used for its suitability for thermal conversion. At the high temperature of the thermal conversion process, the ash components will melt and upon cooling will form heavy slag; these particles will stick through the walls of the conveying surfaces and create a blockage. If the biomass has a very high content of ash, the problem becomes severe, causing a complete blockage of the conveying conduits, which leads to fouling. Figure 3.4 shows a conduit that is filled with slag, which may cause fouling if the conduit is not cleaned.

The two most common indices for evaluating the potential of a biomass to agglomerate are the slagging (R_s) and fouling (R_f) factors. These factors are primarily used

FIGURE 3.4 Slagging and fouling conditions in boiler.

for coal samples but are sometimes used for biomass materials. Mathematically, these factors are defined in Equations 3.11 and 3.12.

$$R_s = \left(\frac{\text{Base}}{\text{Acid}}\right) \times \%\text{S on dry coal} \tag{3.11}$$

$$R_f = \left(\frac{\text{Base}}{\text{Acid}}\right) \times \%\text{Na}_2\text{O} \tag{3.12}$$

The total acid in coal ash is the sum of the percentages of SiO_2, TiO_2, and Al_2O_3, whereas the total base in coal ash is the sum of the percentages of Fe_2O_3, CaO, MgO, Na_2O, and K_2O. The percentage of sulfur in coal is also measured independently so that it can be used on the slagging factor, or it could be based on the SO_3 components. The range of values for R_s and R_f is summarized in Table 3.7, with expected four types of slagging and fouling conditions. Severe slagging is expected if the calculated slagging factor is greater than 2.6, while severe fouling is expected if the fouling factor is greater than 1.0. Unfortunately, these factors are not suitable for biomass or lignite samples (Maglinao and Capareda, 2010). The following are other indices that are also used to evaluate the slagging and fouling potential of biomass samples:

1. Base-to-acid ratio
2. Agglomeration index
3. Alkali index

TABLE 3.7
Slagging and Fouling Indices for Coal Samples

Slagging/Fouling Type	R_s	R_f
Low	<0.60	<0.2
Medium	0.6–2.0	0.2–0.5
High	2.0–2.6	0.5–1.0
Severe	>2.6	>1.0

Source: Winegartner, E. C. (ed.), Coal fouling and slagging parameters. Report prepared by the American Society of Mechanical Engineers (ASME) Research Committee on Corrosion and Deposits from Combustion Gases. Copyright 1974 by the American Society of Mechanical Engineers (ASME), Three Park Avenue, NY, 1974.

TABLE 3.8
Ash Composition of Cotton Gin Waste and Sorghum Stalks

Inorganic Ash	Cotton Gin Trash (%)	Sorghum Stalks
SiO_2	41.8	73.2
K_2O	10.5	8.4
CaO	10.8	5.0
MgO	3.3	1.5
Na_2O	0.6	0.4
P_2O_5	2.6	1.1
Fe_2O_3	0.7	1.0
Al_2O_3	3.1	5.1
SO_3	5.9	0.5

A more detailed discussion about how these indices are calculated will be presented in Chapter 13. The slagging and fouling indices for coal samples are not appropriate for biomass samples that are high in inorganic components. The sample analysis starts by producing ash samples from the biomass in question. Then, a complete compositional analysis is made to determine the basic and acid components and applied to Equations 3.16 and 3.17. Example 3.9 shows how the slagging and fouling factors are calculated. Data for ash composition of agricultural wastes, such as cotton gin trash and sorghum stalks, is listed in Table 3.8. These values may be used to make a preliminary estimate of slagging and fouling indices.

Example 3.4: Slagging and Fouling Factor

Determine the slagging (R_s) and fouling (R_f) factors for cotton gin trash ash data summarized in Table 3.7. Discuss the slagging and fouling potential based on these indices.

Solution:

a. The acid components include the following: SiO_2 (41.8%), Al_2O_3 (3.1%), and TiO_2 (0%), and amount to 44.9%.
b. The basic components include the following: Fe_2O_3 (0.7%), CaO (10.8%), MgO (3.3%), Na_2O (0.6%), and K_2O (10.5%), and amount to 25.9%.
c. Thus, the slagging factor is calculated from the equation shown. Note that the percentage of sulfur in SO_3 is 40% based on its molecular weight (i.e., 1 mol sulfur = 32 g/mol and 3 mol oxygen = 16 × 3 = 48 g/mol to obtain 32/(32 + 48) × 100% = 40% sulfur in SO_3) and 40% of 5.9% = 2.36%.

$$R_s = \left(\frac{\text{Base}}{\text{Acid}}\right) \times \%S = \left(\frac{25.9\%}{44.9\%}\right) \times 2.36\% = 1.36$$

d. The fouling factor is calculated in the same manner.

$$R_f = \left(\frac{\text{Base}}{\text{Acid}}\right) \times \%\text{Na}_2\text{O} = \left(\frac{25.9\%}{44.9\%}\right) \times 0.6\% = 0.35$$

e. Based on Table 3.7, this material has medium potential for slagging and medium potential for fouling.
f. It will be shown in a later chapter (Chapter 13) that this material has less tendency for slagging and fouling using the other indices suited for biomass wastes.

3.4 OTHER STANDARD METHODS FOR BIOMASS ANALYSIS

There are numerous methods for characterizing biomass, all of which are not discussed in detail in this chapter. However, Table 3.9 summarizes some of the additional methods that may be used for other applications. Most of the standard methods were developed by the ASTM. However, some methods were developed by other organizations such as the American Society for Agricultural and Biological Engineering (ASABE), formerly the American Society of Agricultural Engineering (ASAE).

The fusibility of biomass ash is an important parameter. Table 3.10 lists some biomass types and reports the temperatures for their initial deformation, softening, formation of hemispherical conditions, and transition to a fluid.

These fusibility values are important for thermal conversion such as gasification, where the temperature of the reaction is (normally) slightly below 1,000°C. Thus, there are biomass materials listed in Table 3.10 whose initial deformation temperature is below 1,000°C. One should be aware of slagging and fouling issues when converting these biomass materials under extremely high reaction temperatures. Problems may not happen at the conversion reactor but may occur when these materials are cooled downstream from the reactor.

In using the biomass for thermal conversion, such as those for power generation, the simple rule of thumb is illustrated in Example 3.5.

TABLE 3.9
Other Standard Methods for Biomass Analysis

Property/Designation		Title
Heating value	ASTM D 2015	Gross Calorific Value of Solid Fuel by the Adiabatic Bomb Calorimeter
Particle size distribution	ASTM E 828	Designing the Size of RDF-3 from Its Sieve Analysis
	ASAE S 319	Expressing Fineness of Feed Materials by Sieving
Specific gravity	ASTM D 2395	Specific Gravity of Wood and Wood-Based Materials
Proximate analysis	ASTM D 3172	Proximate Analysis of Coal and Coke
Moisture content	ASTM D 2016	Moisture Content of Wood
VCM	ASTM D 3175	Volatile Matter in the Analysis Sample of Coke and Coal
Ash	ASTM D 3174	Ash in the Analysis Sample of Coal and Coke
Ultimate analysis	ASTM D 3176	Ultimate Analysis of Coal and Coke
Thermal properties	ASTM C 351	Mean Specific Heat of Thermal Insulation
	ASTM C 687	Thermal Resistance of Low-Density Fibrous Loose Fill-Type Building Insulation
Mechanical properties	ASTM D 2555	Establishing Clear Wood Strength Values

TABLE 3.10
Fusibility of Some Biomass Ash

	Temperature (°C)			
Biomass Type	**Initial Deformation**	**Softening**	**Hemispherical**	**Fluid**
Alfalfa seed straw	700		1,550	
Almond shell	790		1,440	
Barley straw	925		1,100	
Bean straw	900		1,150	
Corn cobs	900		1,020	
Corn stalks	820		1,091	
Cotton gin trash	1,010		1,380	
Grape prunings (oxidizing)	1,313	1,368	1,374	1,424
Grape prunings (reducing)	1,310	1,360	1,371	1,382
Madrone wood (reducing)	1,271	1,330	1,352	1,404
Manzanita wood	1,268	1,299	1,310	1,321
Olive pits	1,080		1,400	
RDF pellets	850		1,480	
RDF (oxidizing)	890	1,092	1,130	1,193
RDF (reducing)	1,065	1,063	1,131	1,182
Rice hull	1,024		1,097	
Rice straw	1,439		>1,650	
Rice straw (oxidizing)	1,060	1,421	1,250	>1,500
Rice straw (reducing)	1,027	1,410		>1,500
Safflower straw	770		1,430	
Tan oak wood (oxidizing)	1,390	1,440	1,449	1,457

(Continued)

TABLE 3.10 (*Continued*)
Fusibility of Some Biomass Ash

	Temperature (°C)			
Biomass Type	**Initial Deformation**	**Softening**	**Hemispherical**	**Fluid**
Tan oak wood (reducing)	1,377	1,438	1,446	1,454
Walnut shell	820		1,225	

Source: Jenkins, B. M., Physical properties of biomass. In: *Biomass Handbook* (O. Kitani and C. W. Hall, eds.). Gordon and Breach Science Publishers, New York, 1989.
Note: The materials are exposed to inert (nitrogen) atmosphere unless noted.

Example 3.5: Rules of Thumb: Wood to Power

As a rule of thumb, 0.9 tonne/h (1 dry ton/h or 2 green tons/h) of wood fuel can generate 1 MW of electrical power. Freshly harvested wood would normally have 40%–50% moisture when harvested in summer. Prove this calculation, assuming the energy content of wood fuel as 20 MJ/kg (8,616 Btu/lb) and an efficiency of 20%.

Solution:

a. The power from wood each hour is calculated as follows:

$$\text{Power (kW)} = \frac{0.9\ \text{tonne}}{\text{h}} \times \frac{20\,\text{MJ}}{\text{kg}} \times \frac{1{,}000\ \text{kg}}{\text{tonne}}$$

$$\times \frac{1\times10^{6}\,\text{J}}{1\,\text{MJ}} \times \frac{1\,\text{h}}{3{,}600\,\text{s}} \times \frac{\text{W}}{\text{J/s}} \times \frac{\text{MW}}{1\times10^{6}\,\text{W}} = 5\,\text{MW}$$

b. Then, the energy output may be calculated from the efficiency equation presented in Chapter 2 as follows:

$$\text{Output power (MW)} = \text{Input power} \times \text{Efficiency} = 5\,\text{MW} \times 0.20 = 1\,\text{MW}$$

3.5 CONCLUSION

This chapter has described certain parameters for biomass characterization that are important for thermal conversion. Physical parameters, such as bulk density, particle density, angle of repose, and angle of friction, including related thermal properties, are important in biomass transport and storage, including conveying systems. The properties of biomass that are important for thermal conversion are heating value, proximate analysis, and ultimate analysis. Proximate analysis includes the determination of MC, VCM, FC, and ash. Ultimate analysis includes elemental analysis of carbon, hydrogen, oxygen, nitrogen, sulfur, and ash. Perhaps the most important parameters for thermal conversion are those associated with the tendency of ash in biomass to melt at high temperatures and agglomeration and formation of slag upon cooling. The slag may block conveying surfaces, leading to fouling of those conveying surfaces and ultimately clogging the pathways for biomass coproducts.

Heating value is another very important biomass parameter. It describes the energy content of the biomass in units of energy per unit of weight (MJ/kg or Btu/lb). This parameter is most important in determining the process efficiency when converting the biomass to a particular product. Although the biofuels from biomass may have higher energy content than the biomass feedstock itself, the overall energy input must not be greater than the energy output. Otherwise, this biomass conversion process will not be economically justifiable. Thus, care must be taken in reporting the energy content of the biomass and its products. One should now be familiar with reporting energy units on a dry basis (higher heating value), on an "as-received" basis (with a specification of MC), or with the dry ash-free basis. Other conversion systems will be discussed in future chapters (Chapters 5–13), the biomass may be fed into a conversion system in either of those units. One will have to become familiar with shifting from one unit of measure to another.

All of the properties important to thermal conversion of biomass that are introduced in this chapter have corresponding ASTM standard for measurements. For a clear understanding of all the methods, one should refer to these standards for specific step-by-step details of the measurement procedures.

3.6 PROBLEMS AND DISCUSSION ISSUES

3.6.1 Proximate Analysis

P3.1 Enumerate the biomass properties measured using proximate analysis. The sum of all these properties must be equal to 100%. How do you report the proximate analysis on an "as-received basis," dry basis and dry ash-free basis? The data from rice straw in California (CA) show the different values of proximate analysis components. Please use the data from Table 3.2 for the calculations.

3.6.2 Reporting Heating Values of Biomass

P3.2 What is the difference between low heating value and high heating value of biomass? Use the data from rice straw (CA) from Table 3.2 to estimate the low heating value of this biomass. Discuss the practical consequence of these two values in operating a biomass thermal conversion system.

3.6.3 Approximate Chemical Formula of Biomass

P3.3 Determine the estimated chemical formula for charcoal. Please use the data from Table 3.6 for your calculation.

3.6.4 Stoichiometric Combustion and A/F Calculations for Biomass

P3.4 Determine the estimated chemical formula for alligator grass using the data from Table 3.5 and set up the stoichiometric combustion equation for this biomass using atmospheric air. In addition, determine the air-to-fuel ratio by weight (kg/kg) for this biomass during combustion to produce CO_2, H_2O, and heat.

3.6.5 Sustainability of Biomass Use

P3.5 List and discuss the major factors that limit the removal of crop residues from the land.

3.6.6 Proximate Analysis Calculations

P3.6 Ground Jatropha hull samples were used to estimate the proximate analysis. The following are the data from the experiment:

- An initial sample ("as-received" basis) weighing 2.9657 g was used and placed in the drying oven set at 103°C overnight until the final weight became 2.6938 g.
- About 0.9101 g of bone-dry samples were placed in the tube furnace set at 950°C for 7 min, and the resulting weight of the product was 0.2659 g.
- The same sample was then placed in a muffle furnace set at 600°C for 4 h with a resulting final ash weight of 0.0142 g.

Calculate the proximate analysis for this sample. Likewise, report the values on an "as-received" basis.

3.6.7 Air-to-Fuel Ratio Calculations

P3.7 Calculate the air-to-fuel ratio for the complete combustion of biogas comprising 65% methane (CH_4) and 35% CO_2 by volume. Estimate the air-to-fuel ratio by weight (kg/kg) for the complete combustion of this fuel in air.

3.6.8 Heating Value Equivalents

P3.8 How much cotton gin trash (CGT) (in kg and in lb) is required to provide the same heating value as a pound of gasoline? Assume the heating value of CGT to be 15.5 MJ/kg and that of gasoline to be 47 MJ/kg.

3.6.9 Biomass Requirements for Power Generation

P3.9 A gasifier is used to generate 1 MW of electrical power from municipal solid waste (MSW). Determine the amount of MSW required per day (tonnes/day) if the conversion efficiency is around 15%. The unit will be operated for 350 days in a year at 24 h per day. The heating value of MSW is 24.9 MJ/kg.

3.6.10 Heating Value Calculations

P3.10 The heating value of animal manure was reported as 19.7 MJ/kg (8,500 Btu/lb) on a dry ash-free basis. Convert this value on an "as-received" basis and dry basis if the MC of this biomass was 15% and the ash was 20%. Convert the units to the English system.

3.6.11 Slagging and Fouling Factors

P3.11 Determine the slagging (R_s) and fouling (R_f) factors for sorghum stalks ash data shown in Table 3.8. Discuss the slagging and fouling potential based on these indices.

3.6.12 Biomass Feedstock with the Highest Carbon Content

P3.12 Which among the biomass has the highest carbon content by weight?
a. poultry manure b. rice husks c. coconut shells d. switchgrass

3.6.13 Biomass with Lowest Ash Content

P3.13 Which among the biomass has the lowest ash content?
a. wood chips b. poultry manure c. coconut shells d. rice straw

3.6.14 Biomass with Highest Energy Content

P3.14 Which among the biomass has the highest energy content by weight?
a. sorghum stalks b. rice hulls c. corn cobs d. rice straw

3.6.15 Cotton Gin Trash for Electrical Power

P3.15 Develop a project to convert the energy content of cotton gin trash (CGT) into electrical power by gasification. Assume a cotton gin with capacity of 60 bales per hour (bph) processing stripped cotton with about 320 kg [700 lbs] waste per bale. A bale weighs around 455 kg [1,000 lbs]. The cotton gin uses 50 kWh/bale of electrical power. Will the energy from the CGT be enough to satisfy the power requirement of this cotton gin using an overall conversion efficiency of 15%?

3.6.16 Dairy Manure for Electrical Power

P3.16 You are given a 500-head dairy farm. You will use the animal manure in an anaerobic digestion system that will generate the biogas that will be used to generate electrical power. The electrical energy consumption of this facility is around 450 kWh/head/day. Will the energy from the dairy manure be enough to satisfy the power requirement of this dairy facility using an overall conversion efficiency of biogas in an internal combustion engine of 35%?

REFERENCES

Badger, P. and P. Pullammanappallil. 2011. Biomass conversion to energy and fuels. In: Wood to Energy: Using Southern Interface Fuels for Bioenergy (C. Staudhammer, A. L. Hermansen-Baez, D. R. Carter and E. A. Macie, eds.). USDA Forest Service General Technical Report GTR-132.

Lata, K. and S. P. Mande. 2009. Bioenergy resources. In: *Renewable Energy Engineering and Technology: Principles and Practice* (V. V. N. Kishore, ed.). Earthscan Publishers, London.

LePori, W. A. and E. J. Soltes. 1985. Thermochemical conversion for energy and fuel. In: *Biomass Energy: A Monograph* (E. A. Hiler and B. A. Stout, eds.). Texas Engineering Experiment Station, The Texas A&M University Press, College Station, TX.

Maglinao, A. L. and S. C. Capareda. 2010. Predicting fouling and slagging behavior of dairy manure (DM) and cotton gin trash (CGT) during thermal conversion. *Transactions of the ASABE* 53 (3): 903–909.

Searcy, 2010. Biomass system flow alternatives. Biomass Logistics Lecture Notes. Biological and Agricultural Engineering Department, Texas A&M University, College Station, TX.

Winegartner, E. C. (ed.). 1974. Coal fouling and slagging parameters. Report prepared by the American Society of Mechanical Engineers (ASME) Research Committee on Corrosion and Deposits from Combustion Gases. Copyright 1974 by the American Society of Mechanical Engineers (ASME), Three Park Avenue, NY.

ASSESS YOUR KNOWLEDGE

1. A biomass material is placed in a 305 × 305 × 305 mm^3 box [0.305 m each side]. It is dropped about five times from a height of around 610 mm, and the sample is refilled to the brim and weighed. The weight is around 5.5 kg. Determine its bulk density in units of kg/m^3.
   a. 300
   b. 250
   c. 150
   d. 450

2. What are the items analyzed for proximate analysis of biomass samples?
   a. Moisture, volatiles, fixed carbon, and ash
   b. Cellulose, hemicellulose, and lignin
   c. Moisture and ash
   d. Heating value and elemental analysis

11. Identify which biomass sample has the highest heating value?
   a. Alligator grass
   b. Coconut shells
   c. Rice straw
   d. Rice husks

12. A freshly harvested rice straw was weighed and dried. The initial weight of the sample was 100 kg, and after drying the final weight of the bone-dry material was 75 kg. Determine the moisture of this sample on dry basis.
   a. 33.3%
   b. 75%
   c. 90%
   d. 25%

3. About 67.5 metric tonnes of waste biomass sample having a bulk density of 450 kg/m^3 is to be transported using a standard container van having a dimension of 2.43 m width × 2.59 m height, and 12.2 m length. How many containerized van(s) is/are required to transport this biomass sample?
   a. 1
   b. 2
   c. 3
   d. 4
4. A freshly harvested rice straw was weighed and dried. The initial weight of the sample was 100 kg, and after drying the final weight of the bone-dry material was 75 kg. Determine the moisture of this sample on wet basis.
   a. 10%
   b. 75%
   c. 90%
   d. 25%
5. What are the items analyzed for ultimate analysis of biomass samples?
   a. Moisture, volatiles, fixed carbon, and ash
   b. Elemental C, H, O, N, S, and ash
   c. Cellulose, hemicellulose, and lignin
   d. Heating value and elemental analysis

13. Identify the biomass sample with the lowest heating value?
   a. Alligator grass
   b. Coconut shells
   c. Poultry manure
   d. Rice husks
14. Determine the chemical formula for coconut shells. Hint: Please use Table 3.5 for calculations.
   a. $CH_{6.1}\ O_{3.5}\ N_{0.5}\ S_{.005}$
   b. $C_{3.91}H_{6.1}O_{2.86}\ N_{0.02}\ S_{0.025}$
   c. $C_{3.91}H_{6.1}\ O_{2.86}\ N_{0.3}\ S_{0.8}$
   d. $C_{3.91}H_{3.05}\ O_{3.5}\ N_{0.3}\ S_{0.05}$
15. Identify a slagging and fouling index used for coal.
   a. Base-to-acid ratio
   b. Agglomeration index
   c. Alkali index
   d. Rf

6. This is defined as the angle with which a pile of biomass will slide when the sample is raised from the horizontal plane.
   a. Slope
   b. Angle of repose
   c. Azimuth angle
   d. Angle of friction

7. This is defined as the steepest angle of descent or dip relative to the horizontal plane to which the biomass material can be piled without slumping.
   a. Angle of friction
   b. Slope
   c. Angle of repose
   d. Azimuth angle

8. This is an example of a particle size distribution (PSD) analyzer for biomass particles less than 10 microns. Name an example of a commercial PSD equipment.
   a. Malvern Multisizer
   b. Pycnometer
   c. Tumbling box
   d. Quadrat sampler

9. The particle size distribution of biomass sample is measured using what device(s)?
   a. Brix meter
   b. Biomass tumbler
   c. Pycnometer
   d. Tyler sieves

16. A biomass sample with 25% moisture and 10% ash content was evaluated for heating value (as received), and the results showed it had a heating value of 15.5 MJ/kg. Determine the dry ash-free heating value of this sample in units of MJ/kg.
    a. 12.5
    b. 15.5
    c. 20.7
    d. 18.5

17. A biomass sample with 25% moisture and 10% ash content was evaluated for heating value (as received), and the results showed it had a heating value of 15.5 MJ/kg. Determine the high heating value of this sample in units of MJ/kg.
    a. 12.5
    b. 15.5
    c. 20.7
    d. 18.5

18. Which among the choices has the lowest reported heating value?
    a. Charcoal
    b. Rice straw
    c. Lignite
    d. Poplar

19. Which among the choices has the highest reported heating value?
    a. Charcoal
    b. Rice hull
    c. Pine bark
    d. Maple

10. This parameter is defined as the diameter of an irregularly shaped object that is the same as the diameter of a perfect sphere and used to enable a simplified representation of the particle size distribution of a biomass sample. Name this parameter.
    a. Geometric standard deviation
    b. Sample mean
    c. Equivalent spherical diameter
    d. Sphere-to-diameter ratio

20. Name a slagging and fouling index used for agricultural biomass.
    a. Rs
    b. Rf
    c. Agglomeration Index
    d. Ash melting point

Answers: 1a, 2a, 3b, 4d, 5b, 6d, 7c, 8a, 9d, 10c, 11b, 12a, 13c, 14b, 15d, 16d, 17c, 18b, 19a, 20c

ADDITIONAL LEARNING RESOURCES

- Access the Lecture Notes at the Instructor and Student Resources: https://routledgetextbooks.com/textbooks/instructor_downloads/

4 Biomass Properties for Biological Conversion

LEARNING OBJECTIVES

Upon completion of this chapter, one should be able to

1. Describe the various compositional properties of biomass that are important for biological conversion.
2. Describe the various methods of characterizing biomass based on various laboratory analytical procedures of the National Renewable Energy Laboratory of the US Department of Energy.
3. Specify the various types of equipment used for characterizing biomass properties that are important for biological conversion.
4. Enumerate the various standard methods associated with characterizing biological biomass.
5. Enumerate some pretreatment procedures for biomass feedstock prior to biological conversion.

4.1 INTRODUCTION

Chapter 3 dealt with biomass properties that are important in thermal conversion. Preparing biomass for biological conversion also requires specific characterization procedures. These procedures lead to suitable biomass conversion pathways. The National Renewable Energy Laboratory (NREL) of the US Department of Energy (USDOE) has developed several laboratory analytical procedures (LAPs) for categorizing biomass, specifically for biological conversion and particularly for ethanol production. These LAPs are available online and are free. The procedures include the preparation of samples for compositional analysis and all the other associated experimental protocols for enzymatic saccharification of lignocellulosic biomass and basic sugar fermentation. More importantly, LAPs will keep track of the important compositional analysis parameters such as carbohydrates (or sugars), cellulose, and lignin contents in biomass. Unlike the parameters for thermal conversion, most of the procedures for compositional analysis are very time consuming and require more sophisticated facilities and equipment. These procedures will be outlined in this chapter, including step-by-step characterization and the use of examples in each step.

To guide the reader with complete compositional analysis, a problem-solving roadmap was developed (Figure 4.1). The example used illustrates the compositional analysis of a microalgae, *Nannochloropsis oculata*. This species was selected primarily because calibration curves were not yet established for this biomass resource. As shown in the figure, particle size characterization is the first step, followed by determination of proper moisture content (usually below 10% moisture). The particle

DOI: 10.1201/9781003294306-4

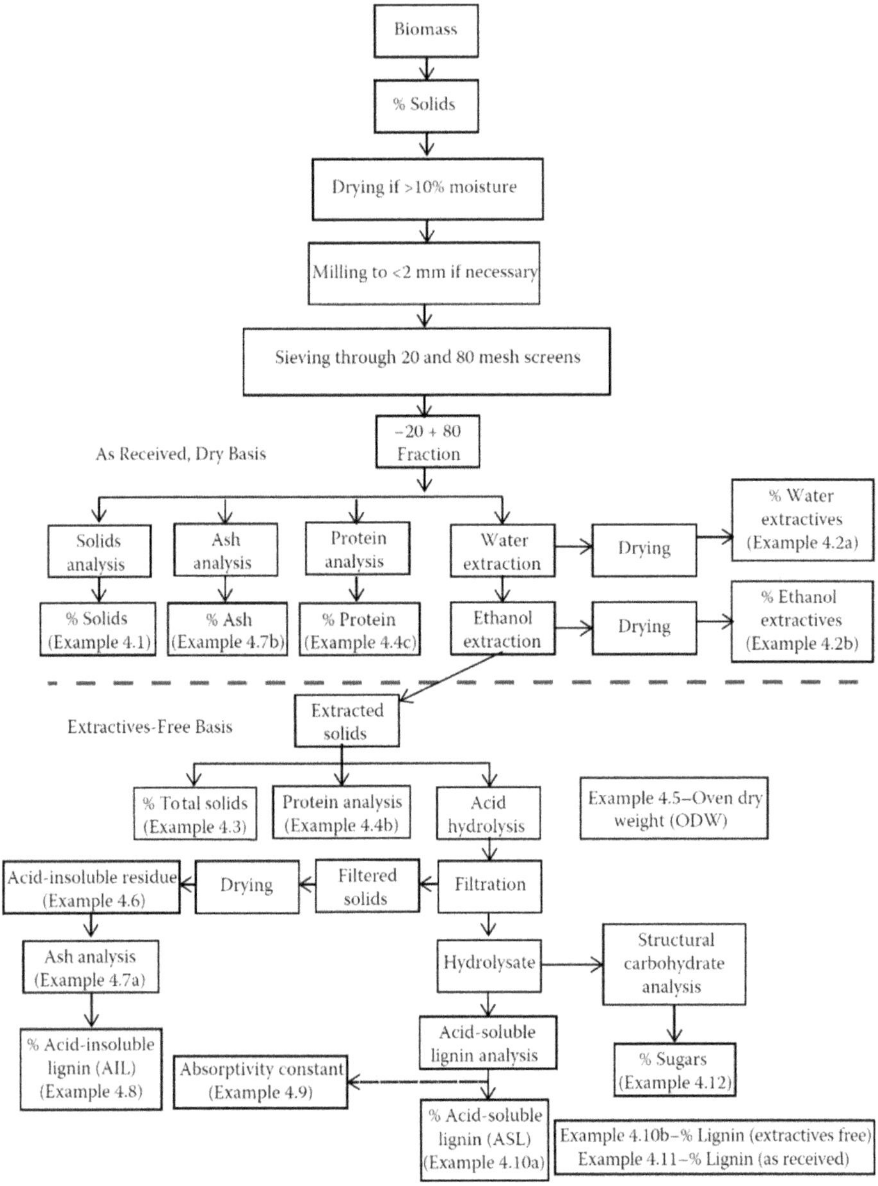

FIGURE 4.1 Problem-solving roadmap to establish compositional analyses of biomass.

size must be milled as necessary. Once this is done, particle sieving is accomplished. The top half of the figure is for calculations based on dry material, and this may be easily converted into "as-received" basis if the moisture content is known. The water and ethanol extractives are then removed and measured. The bottom half of the figure is for the analysis based on an extractive-free basis. This is where the amounts of lignin and sugars are measured, including protein concentration. The lignin is further classified into acid-soluble and acid-insoluble components, whereas the sugar components are broken down into specific sugar compounds (glucose, xylose, galactose, etc.). Finally, the mass closure must be calculated to ensure all components are properly quantified. Depending on the ash and moisture content, these components may be reported on an "as-received" or an "ash-free" basis.

4.2 PROPERTIES OF BIOMASS IMPORTANT FOR BIOLOGICAL CONVERSION

A vital step in the use of diverse biomass resources for conversion, aside from various physical analyses, is the compositional analysis based on macro-components (i.e., cellulose, hemicellulose, and lignin). Of these three major components, lignin is perhaps the most difficult component to convert biologically; thus, it will have to be removed from the biomass matrix to improve the process efficiency. In this case, the pretreatment of the biomass before conversion is necessary. The three conversion pathways discussed in the previous chapters have different ways of treating the biomass feedstock before they are used in the final conversion step. In the production of biodiesel, for example, the raw extracted oil from seeds must be refined properly before the transesterification process is conducted. Most biomass undergoing biochemical conversion processes must be pretreated, such as with the more popular ammonia fiber expansion (AFEX) processes, or with caustic soda in addition to reducing particle size. For thermal conversion processes, the most common pretreatment process is particle size reduction and removal of moisture, as the three major components (cellulose, hemicellulose, and lignin) are all converted into other products. Table 4.1 summarizes the relative composition of cellulose, hemicellulose, and lignin for common biomass.

TABLE 4.1
Ranges of Values of Chemical Composition of Common Biomass

Biomass	Cellulose (%)	Hemicellulose (%)	Lignin (%)
Hardwood stem	40–50	24–40	18–25
Nut shells	45–50	35–45	25–35
Corn cobs	45	35	15
Grasses	25–40	35–50	10–30
Wheat straw	33–40	20–25	15–20
Rice straw	40	18	5.5
Leaves	15–20	80–85	0
Coastal Bermuda grass	25	35.7	6.4
Switchgrass	30–50	10–40	5–20
Solid cattle manure	1.6–4.7	1.4–33	2.7–5.7
Swine waste	6.0	28	–
Paper	85–99	0	0–15
Newspaper	40–55	25–40	18–30

Source: Prasad, S., A. Singh and H. C. Joshi, *Resources, Conservation and Recycling* 50, 1, 2007.

Cellulose is a linear polysaccharide consisting of several hundred or thousand β(1→4) linked d-glucose units. It is the structural component of the cell wall of biomass or microalgae. As summarized in the table, the typical cellulose component of most wood wastes ranges from 40% to 50% to less than 10% for cattle manure or swine waste. The cellulose components of some plant parts, such as cotton fiber, are typically very high.

Hemicellulose is composed of groups of complex carbohydrates such as sugars and starch. Agricultural crops producing large fractions of free sugars such as xylose or mannose are ideal for conversion into ethanol, as are those crops that are high in starch, such as corn and cassava. Crops high in sugars include sugarcane and sweet sorghum. Crops that are high in hemicellulose content are ideal for biochemical conversion and are considered first-generation feedstocks. However, those crops that are high in cellulose and can also be converted into ethanol are normally considered second-generation feedstocks. The microbial population does not have the ability to convert the lignin component of biomass at all; they are useless for biochemical conversion. Thus, biochemical conversion processes rely heavily on pretreatment processes, which will be discussed in detail in the biological conversion chapters.

Lignin is a very important component of biomass because it binds the cellulosic compounds to keep the plant erect. Lignin is considered the mortar that keeps the cellulosic blocks rigid. Trees can grow to several hundred meters of height. The structure of lignin is not yet fully understood. Lignin's complexity makes it highly resistant to attack by microorganisms. Thus, biomass resources high in lignin are not suitable for fermentation or anaerobic digestion.

Other important biomass properties include protein content and extractables. The determination of protein is important in evaluating the nutrient value of biomass when they are used for feed. Some biomass contain protein and nitrogenous compounds that are essential for life. The economics of some biomass conversion processes are dependent on whether any of the by-products can be sold for additional revenue. Such is the case for ethanol production, where the distiller's grain offers enough value as animal feed, making the overall economics even more profitable.

Extractables are biomass components that are soluble in water, ether, or ethanol. These compounds include resins, waxes, chlorophylls, and fatty acids. With minor chemical pretreatments, these compounds can still find value as fuel or additional value-added chemicals and must be measured.

Table 4.2 summarizes the various NREL LAPs that have already been developed. The titles and authors are also provided in the table. Most of these procedures will be outlined in this chapter, except for those related to enzymatic saccharification of lignocellulosic biomass, simultaneous saccharification and fermentation, experimental protocols, measurement of cellulose activities, and starch determination by high-performance liquid chromatography (HPLC).

TABLE 4.2
LAPs for Biomass Characterization

LAP No.	Test Procedure Title	Authors
42617	Determination of Acid-Soluble Lignin Concentration Curve by UV-VIS Spectroscopy	Hyman, D., A. Sluiter, D. Crocker, D. Johnson, J. Sluiter, S. Black, and C. Scarlata, 2008
42618	Determination of Structural Carbohydrates and Lignin in Biomass	Sluiter, A., B. Hames, R. Ruiz, C. Scarlata, J. Sluiter, D. Templeton, and D. Crocker, 2008f
42619	Determination of Extractives in Biomass	Sluiter, A., R. Ruiz, C. Scarlata, J. Sluiter, and D. Templeton, 2008a
42620	Preparation of Samples for Compositional Analysis	Hames, B., R. Ruiz, C. Scarlata, A. Sluiter, J. Sluiter, and D. Templeton, 2008b
42621	Determination of Total Solids in Biomass and Total Dissolved Solids in Liquid Process Samples	Sluiter, A., B. Hames, D. Hymann, C. Payne, R. Ruiz, C. Scarlata, J. Sluiter, D. Templeton, and J. Wolfe, 2008d
42622	Determination of Ash in Biomass	Sluiter, A., B. Hames, R. Ruiz, C. Scarlata, J. Sluiter, and D. Templeton, 2008a
42623	Determination of Sugars, By-products, and Degradation Products in Liquid Fraction Process Samples	Sluiter, A., B. Hames, R. Ruiz, C. Scarlata, J. Sluiter, and D. Templeton, 2008a
42624	Determination of Starch in Solid Biomass Samples by HPLC	Sluiter, A. and J. Sluiter, 2008
42625	Determination of Protein Content in Biomass	Hames, B., C. Scarlata, and A. Sluiter, 2008a
42626	Rounding and Significant Figures	Michener, B., C. Scarlata, and B. Hames, 2008
42627	Determination of Insoluble Solids in Pretreated Biomass Material	Sluiter, A., D. Hymann, C. Payne, and J. Wolfe, 2008e
42628	Measurement of Cellulase Activities	Adney, B. and J. Baker, 2008
42629	Enzymatic Saccharification of Lignocellulosic Biomass	Selig, M., N. Weiss, and Y. Ji, 2008
42630	SSF-Experimental Protocols—Lignocellulosic Biomass Hydrolysis and Fermentation	Dowe, N. and J. McMillan, 2008

Setting up an analysis laboratory is a major undertaking. Such a facility would include a variety of physical resources, equipment, and instruments. Some of the items included in the following list have been, or will be, enumerated in this textbook:

1. Analytical balance, accurate to 0.1 g
2. Diode array ultraviolet and visible (UV-VIS) spectrometer
3. Vacuum solvent evaporation system, such as a Buchi Rotavapor

4. Convection drying oven with temperature control of 105°C ± 3°C
5. Muffle furnace, equipped with thermostat or with ramping program
6. Water bath
7. Autoclave, suitable for autoclaving liquids
8. Filtration setup equipped with vacuum source and vacuum adaptors for crucibles, such as the Buchner funnel filters with glass microfiber filter paper, 30 μm or less pore size
9. Desiccator with desiccants
10. HPLC system with refractive index detector and the following columns: Shodex sugar SP0810 or BioradAminex HPX-87P column or equivalent with ionic form H^+/CO_3^- deashing guard column and BioradAminex HPX-87H column (or equivalent) equipped with an appropriate guard column
11. Freeze-dryer system
12. Standard laboratory knife mill such as the Wiley Mill size No. 4 with a 2 mm screen or intermediate Wiley Mill with a 1 mm screen
13. Sieve shaker that provides motion in both horizontal and vertical axes
14. Sieve set: No. 20 (859 μm) and No. 80 (180 μm) stackable sieves with lid and bottom pan
15. Riffle sampler with pans
16. pH meter, accurate to 0.01 pH unit
17. Vortex mixer
18. Centrifuge, capable of 3,000 rpm with 15 mL centrifuge vial slots
19. Kjeldahl nitrogen measurement digestion facility
20. Glucose analyzer or equivalent glucose quantification method
21. Incubator-shaker capable of maintaining 50°C ± 1°C
22. Laminar fume hood or biosafety cabinet
23. Freezer capable of storing frozen yeast cultures (down to −70°C)
24. Cell counting chamber slide for yeast cell counts
25. Infrared ovens for determining percent dry solids

4.3 STANDARD METHODS FOR ANALYSIS AND EXAMPLES

Aside from the actual procedures developed by NREL, workbooks were also designed by the USDOE for use in conjunction with the above LAPs to assist the researcher with automatic calculations of compositional analysis and mass closure. Workbooks are available for woody biomass (hardwood and softwood), corn stover, biomass hydrolysate, corn stover intermediates, and nitrogen-to-protein factor calculations. However, there is no algae workbook currently available. Some additional calibration curves must be made if this has not yet been developed by NREL.

Figure 4.1 also shows how closure is made for all analyses. Step-by-step examples are provided in this section for complete analysis. The equations for calculations are provided in the discussions before the examples. The first step in the gamut of LAPs is the preparation of samples for compositional analysis.

4.3.1 Preparation of Samples for Compositional Analysis (LAP No. 42620)

This procedure is very similar to ASTM E 1757-01 (Standard Practice for Preparation of Biomass for Compositional Analysis). This procedure describes methods for drying, size reduction, obtaining samples with uniform particle size, and representative sampling of biomass samples. There are three options for drying biomass samples: air-drying, convection oven-drying, and freeze-drying. The most common drying method uses a convection oven capable of maintaining 45°C ± 5°C (called Method B). For Method A, one would only need a large table or drying rack for the purpose. Freeze-drying (also called "lyophilization"; Method C) is suitable for very wet biomass that is at risk of microbial growth during drying. This drying method is suitable for small samples of biomass weighing less than 20 g. One should refer to the published standards for a detailed description of Methods A and C.

Note that an analytical balance with a sensitivity of 0.1 g is necessary for drying methods. The drying procedure outlined in Chapter 3 is the same and should be followed. The dried biomass is then milled with a 2 mm screen. A Wiley Mill No. 4 with a 2 mm screen is suitable for samples >20 g; the intermediate model Wiley Mill, with a 1 mm screen, is suitable for samples <20 g that will not be sieved. Equivalent knife mills are also acceptable.

After milling, the next step is to sieve the biomass samples using a suitable sieve shaker that provides motion in both the horizontal and vertical axes. Sieve set No. 20 (850 μm) and No. 80 (180 μm) stackable sieves with lids and bottom pans should also be used. Sieves and bottom pans should be at least 8.9 cm (3 1/2 in) in height. Sieves should conform to ASTM Specification E 11–09 (Standard Specification for Woven Wire Test Sieve Cloth and Test Sieves).

Finally, a manual sample divider is required that splits the milled biomass into two or more equivalent subsamples. Riffle divisions should be between 6.4 and 12.7 mm (1/4 to 1/2 in) with at least 24 riffle openings. The feed chute and riffles should have a slope of at least 60°. Collection pans are required to pour samples onto the riffler and to collect the subsamples.

The following equations are used to report the data during the preparation of samples for compositional analysis. To calculate the fraction percent of –20/80 mesh fractions, use Equation 4.1:

$$\text{Fraction}_{20/80}(\%) = \left(\frac{W_{t20/80}}{W_{t20/80} + W_{t80}}\right) \times 100\% \tag{4.1}$$

where

$W_{t20/80}$ = weight of –20/80 mesh fractions, g
W_{t80} = weight of fines fraction, g

To calculate the fraction percent of 80 mesh fractions, use Equation 4.2:

$$\text{Fraction}_{80}(\%) = \left(\frac{W_{t80}}{W_{t20/80} + W_{t80}}\right) \times 100\% \tag{4.2}$$

To calculate the percent of total solids obtained by drying at 45°C (Method B), use Equation 4.3:

$$\%\ T_{45} = \left(\frac{W_f - W_t}{W_i - W_t}\right) \times 100\% \tag{4.3}$$

where

$\%\ T_{45}$ = percent total solids of a sample oven-dried at 45°C
W_t = tare weight of container, g
W_i = initial weight of container and sample, g
W_f = final weight of container and sample, g

4.3.2 Determination of Percent Solids in Biomass

The first example (Example 4.1) is how to calculate the percentage of solids in the biomass sample. In this step, the biomass is totally dried if the moisture content is more than 10%, followed by milling to less than 2 mm if necessary. The material is then sieved through 20 and 80 mesh screens, and the fractions in the −20 to +80 range are used for subsequent tests. This "as-received" and "dry basis" biomass is then analyzed for percent solids (Example 4.1), percent ash (Example 4.7b), and percent protein (Example 4.4b). The sample then undergoes water extraction and drying to determine the percent water extractives (Example 4.2a), followed by ethanol extraction and drying to determine the percent ethanol extractives. Equation 4.4 is used to calculate the percent total solids.

$$\%\text{Total solids} = \frac{\left(\text{Weight}_{\text{dry pan + dry sample}} - \text{Weight}_{\text{dry pan}}\right)}{\text{Weight}_{\text{sample as received}}} \times 100\% \tag{4.4}$$

where

$\text{Weight}_{\text{drypan+dry sample}}$ = weight of the pan and dry sample, g
$\text{Weight}_{\text{dry pan}}$ = tare weight of the dry pan, g
$\text{Weight}_{\text{sample as received}}$ = weight of the sample used for analysis, g

Example 4.1: Percent Total Solids

A sample of *N. oculata* was used to establish the percent total solids. The following data were gathered:

Weight of pan = 1.2762 g
Weight of pan + sample = 3.1599 g
Weight of pan + dry sample = 2.9976 g
Determine the percent total solids.

Solution:

a. The equation to calculate percent total solids is as follows:

$$\%\text{Total solids} = \frac{\left(\text{Weight}_{\text{dry pan+dry sample}} - \text{Weight}_{\text{dry pan}}\right)}{\text{Weight}_{\text{sample as received}}} \times 100\%$$

$$\%\text{Total solids} = \frac{(2.9976 - 1.2762)}{(3.1599 - 1.2762)} \times 100\% = 91.38\%$$

b. Thus, the percent total solids is 91.38%.

4.3.3 Determination of Extractives in Biomass

It is necessary to remove the nonstructural material from biomass before analysis to prevent interference with future analytical steps. As shown in Figure 4.1, the water and ethanol extractives are segregated. This procedure is the same as ASTM E 1690–08 (Standard Test Method for Determination of Ethanol Extractives in Biomass). Water-soluble materials include the inorganic portions, the nonstructural sugars, and the nitrogenous materials, primarily coming from the soil or fertilizer. The ethanol-soluble materials include chlorophyll, waxes, oils, and others. The test results are normally reported on a dry-weight basis. Thus, the oven dry weight (ODW) of the extracted solids is first established and will be used throughout the succeeding analysis. Note the upper portion of the pathway presented in Figure 4.1 is on an "as-received and dry basis," while the bottom half of the analysis pathway refers to an "extractives-free basis." Note that some biomass would only require ethanol extraction, whereas others would require both extraction procedures. For example, corn stover would require both extraction processes because large amounts of water extractives are usually present. An extraction facility is required, usually equipped with a Soxhlet apparatus and a rotary evaporator for solvent removal.

The ODW of the biomass sample is first established, and this is done by using Equation 4.5:

$$\text{ODW} = \frac{\text{Weight}_{\text{air dry sample}} \times \%\ \text{Total solids}}{100} \tag{4.5}$$

where

$\text{Weight}_{\text{air dry sample}}$ = the weight of the air-dried sample used, g
% Total solids = the total solids of the biomass sample, %

The extraction procedure would normally take a minimum of 12 h, where the material is placed above the boiling flasks. For water extraction, about 190 ± 5 mL of HPLC-grade water is added to the tared receiving flask. The heating mantles must be able to provide a minimum of four to five siphon cycles per hour. The reflux is normally done for 6–24 h, depending on the biomass material used. For corn stover, this is normally about 8 h. The amounts of extractives in the sample, on a percent dry-weight basis, is given in Equation 4.6.

$$\%\text{Ethanol extractives} = \frac{\text{Weight}_{\text{flask+ extractives}} - \text{Weight}_{\text{flask}}}{\text{ODW}} \times 100\% \tag{4.6}$$

After extraction, the biomass material is totally dried.

Example 4.2: Percent Extractives

Dried, ground, and sieved samples of *N. oculata* were used to measure water and ethanol extractives. The following data were gathered:

Water extraction data
Weight of thimble = 8.1547 g
Weight of thimble + sample = 18.1881 g
Weight of flask = 145.7160 g
Weight of flask + sample after extraction = 147.9881 g
% Total solids from Example 4.1 = 91.38%
Ethanol extractives data
Weight of thimble = 8.1547 g
Weight of thimble + sample = 18.1881 g
Weight of flask = 153.8718 g
Weight of flask + sample after extraction = 156.0089 g
% Total solids from Example 4.1 = 91.38

Determine the (a) percent water extractives, (b) percent ethanol extractives, and (c) percent total extractives, all on a dry basis.

Solution:

a. The ODW of the sample is calculated from Equation 4.4 because the percent total solids have already been established in Example 4.1 as follows:

$$\text{ODW} = \frac{(18.1881 - 8.1547) \times 91.38}{100} = 9.1685\,\text{g}$$

b. The percent water extractives (dry basis) is then calculated as follows:

$$\%\,\text{Water extractives} = \frac{(147.9881 - 145.7160)}{9.1685} \times 100\% = 24.78\%$$

The percent ethanol extractives calculation is shown here next:

$$\%\,\text{Ethanol extractives} = \frac{(156.0089 - 153.8718)}{9.1685} \times 100\% = 23.31\%$$

The percent total extractives will be 24.78 + 23.31 = 48.09%.

4.3.4 Percent Total Solids of Extractives-Free Biomass

After water and ethanol extraction (bottom half of Figure 4.1), the extracted solids undergo a series of analysis steps. The first step is the calculation of percent total solids of the extractives-free biomass. This was already done and shown in Example 4.1 for the "as-received" basis and again illustrated in Example 4.3 for the extractives-free biomass. The equation used is the same as that used in the sample illustrated in Example 4.1.

Example 4.3: % Total Solids (of the Extractives-Free Biomass)

The extractives-free biomass sample of *N. oculata* was used to establish the percent total solids. The following data were gathered:

Weight of pan = 1.3016 g
Weight of pan + sample = 2.3299 g
Weight of pan + dry sample = 2.2376 g
Determine the percent total solids of the extractives-free biomass.

Solution:

a. The equation to calculate percent total solids is as follows:

$$\%\text{ Total solids} = \frac{(2.2376 - 1.3016)}{(2.3299 - 1.3016)} \times 100\% = 91.02\%$$

b. Thus, the percent total solids of the extractives-free biomass is 91.02%.

4.3.5 Percent Protein of Extractives-Free, ODW Biomass

Protein and other nitrogenous components of biomass must also be measured to have a comprehensive biomass characterization. This parameter is also difficult to measure directly. For protein analysis, the NREL LAP 42625 is used (NREL/TP-510-42625). The biomass sample must be digested completely to remove and determine the protein content. The Kjeldahl method is most used, or some other combustion of methods, to measure the elemental nitrogen and to correlate this to the protein content using an appropriate nitrogen factor.

The Kjeldahl method consists of heating the biomass sample with sulfuric acid, which will decompose the organic substances by oxidation, extracting the nitrogen in the form of ammonium sulfate. In this procedure, potassium sulfate is added to increase the boiling point of the medium from 169°C to 189°C (refer to AOAC 984.13 Standard). Chemical decomposition is complete when the sample becomes clear and colorless, from a darker solution. The solution is then distilled with sodium hydroxide, which converts ammonium salt into ammonia. The amount of ammonia present is determined by back titration. The end of the condenser is normally dipped into a solution of boric acid. The ammonia reacts with the acid, and the remainder of the acid is then titrated with a sodium carbonate solution with a methyl orange pH indicator. Currently, the Kjeldahl method is largely automated and makes use of specific catalysts (i.e., mercury oxide or copper sulfate) to accelerate the decomposition (Persson et al., 2008).

Whatever procedure is used, the percent protein is recalculated according to the extractives-free and ODW bases as well as on an "as-received" basis. The equations used are shown in Equations 4.7 and 4.8. Example 4.4 shows how the results of protein analysis are recalculated according to the extractives-free and ODW, as well as on an "as-received" basis.

$$\%\text{ Protein}_{\text{ext-free, ODW}} = \frac{\%\text{ Protein}_{\text{ext-free, air dried}}}{\%\text{ Total solids}_{\text{ext-free}}} \times 100\% \tag{4.7}$$

$$\% \text{Protein}_{\text{as received}} = \frac{\% \text{Protein}_{\text{ext-free,ODW}} \times (100 - \% \text{Total extractives})}{100} \quad (4.8)$$

Example 4.4: Percent Protein (Extractives Free, Oven Dry Weight), Percent Protein (As Received)

The extractives-free and air-dried algae samples were sent to a laboratory for protein analysis, and the results revealed that the percent protein was 41.8%. Recalculate the percent protein content according to the extractives-free and ODW bases, as well as on an "as-received" basis. Use the percent total solids results from Example 4.3 and the total extractives results from Example 4.2.

Solution:

a. The % $\text{Protein}_{\text{ext-free, air-dried}} = 41.80$ and the % Total $\text{solids}_{\text{ext-free}}$ from Example 4.3 = 91.02, and the % Total extractives from Example 4.2 = 48.09%.
b. Thus, substitutions in Equations 4.7 and 4.8 are as follows:

$$\% \text{Protein}_{\text{ext-free, ODW}} = \frac{41.80}{91.02} \times 100\% = 45.92\%$$

c. The % Protein on an as-received and dry basis is likewise calculated as follows:

$$\% \text{Protein}_{\text{as received}} = \frac{45.92 \times (100 - 48.09)}{100} = 23.84\%$$

4.3.6 ODW of Extractives-Free Sample

The ODW of the extractives-free biomass samples must again be established to estimate the remaining important compositional analysis of sugars and lignin. Thus, Equation 4.5 will be modified slightly to reflect this recalculation of the ODW of the extractives-free sample as shown in Equation 4.9. Example 4.5 will illustrate this procedure.

$$\text{ODW}_{\text{ext-free}} = \frac{\text{Weight}_{\text{ext-free sample}} \times \% \text{Total solids}_{\text{ext-free}}}{100} \quad (4.9)$$

Example 4.5: Percent ODW Extractives-Free Sample

An extractives-free biomass sample weighing 0.3059 g from *N. oculata* was used to establish the $\text{ODW}_{\text{ext-free}}$. Calculate the ODW (of extractives free) if the percent total solids of the extractives-free sample was 91.02% (Example 4.3).

Solution:

a. Equation 4.9 is used for this calculation as follows:

$$\text{ODW} = \frac{0.3059 \times 91.02}{100} = 0.2784 \text{ g}$$

b. Thus, the ODW is 0.2784 g.

4.3.7 Percent Acid-Insoluble Residue

After the analysis of total solids and protein, the water- and ethanol-extracted solids will undergo acid hydrolysis to separate the lignin fraction and the carbohydrate (sugar) fractions. Carbohydrates and lignin comprise the major portion of the biomass samples after extractives are removed. As shown in Figure 4.1, the hydrolysate will be analyzed for sugar components as well as acid-soluble lignin (ASL). The filtered solids should be dried, and the acid-soluble residue must be determined.

Lignin content is an important biomass property in biochemical conversion processes because microbes are not able to convert them into biofuels. The upper limit in biomass conversion efficiency depends on the percentage of lignin in a sample. The NREL procedure employs a two-step acid hydrolysis to separate the acid-soluble and the acid-insoluble fractions. The soluble fractions can be easily measured using ultraviolet-visible (UV-VIS) spectroscopy, whereas the insoluble fractions (containing ash and protein) can be measured gravimetrically (by weight). The range of lignin content for switchgrass varies between 14% and 19% (Mann et al., 2009).

To calculate the lignin content of biomass, the sample must first be devoid of extractives (extractives-free sample) such that the calculations are based on an extractives-free basis (as shown in a series of examples above). The sample is then broken down into the acid-insoluble residue (% AIR) and the acid-insoluble lignin (% AIL) by hydrolysis of the sample. The ODW of the extractives-free sample should be established; the percent AIR is then calculated by treating the remaining biomass with acid to determine its AIR content as shown in Equation 4.10. The determination of percent AIR has also a similar ASTM Standard E 1721-01 (reapproved 2009) entitled "Standard Test Method for Determination of Acid-Insoluble Residue in Biomass."

$$\% \text{ AIR} = \frac{\left(\text{Weight}_{\text{crucible+residue}} - \text{Weight}_{\text{crucible}}\right)}{\text{ODW}_{\text{sample}}} \times 100\% \tag{4.10}$$

Example 4.6 shows the procedure for calculating percent AIR. The NREL LAPs for this process can be found in Technical Report No. NREL/TP-510-42618 (2008).

After the percent AIR is established, ash analysis is then performed, followed by the determination of percent AIL.

Example 4.6: Percent Acid-Insoluble Residue (% AIR)

Determine the percent acid-insoluble residue (% AIR) from the algae sample data shown below:

Weight of crucible = 43.1369 g
Weight of crucible + residue = 43.2950 g
ODW from Example 4.5 = 0.2784 g
Required: % AIR

Solution:

a. The % AIR is calculated from Equation 4.10 and shown as follows:

$$\% \text{ AIR} = \frac{(43.2950 - 43.1369)}{0.2784} \times 100\% = 56.79\%$$

b. Thus, the % AIR is 56.79%.

The ash content of the extractives-free biomass may be measured by a similar procedure described in Chapter 3, and the estimation of the percent ash (of the extractives-free sample) is shown in Equation 4.11. The muffle furnace is usually set at 575°C ± 25°C for the ashing procedure.

$$\% \text{Ash}_{\text{ext-free}} = \frac{\text{Weight}_{\text{dry crucible+ash}} - \text{Weight}_{\text{dry crucible}}}{\text{ODW}_{\text{sample}}} \times 100\% \tag{4.11}$$

The percentage of ash based on the "as-received" basis may also be calculated and the equation is given in Equation 4.12:

$$\% \text{Ash}_{\text{as received}} = \frac{\% \text{Ash}_{\text{ext-free}} \times (100 - \% \text{Total extractives})}{100} \tag{4.12}$$

Example 4.7: Ash Analysis of Extractives-Free and "As-Received" Basis

Determine the ash content of an extractives-free sample and "as-received" sample given the data below:

Weight of crucible = 43.1369 g
Weight of crucible + ash = 43.1578 g
ODW from Example 4.5 = 0.2784 g
% Total extractives from Example 4.2 = 48.09%
Required: Percent ash (extractives free), percent ash (as received)

Solution:

a. Percent ash (extractives free) is calculated using Equation 4.11:

$$\% \text{Ash}_{\text{ext-free}} = \frac{43.1578 - 43.1369}{0.2784} \times 100\% = 7.51\%$$

b. Percent ash (as received) is calculated using Equation 4.12:

$$\% \text{Ash}_{\text{as received}} = \frac{7.51 \times (100 - 48.09)}{100} = 3.90\%$$

4.3.8 Percent AIL (Extractives Free)

The percent AIL is measured by using the filtered solids after acid hydrolysis of the extracted solids sample, drying the sample to establish the AIR, which takes place at 105°C ± 3°C. This will be followed by ashing at 575°C ± 25°C for 24 ± 6 h. A furnace with temperature ramping may also be used with the following furnace temperature-ramping program enumerated as follows:

- Ramp from room temperature to 105°C.
- Hold at 105°C for 12 min.
- Ramp to 250°C at 10°C/min.
- Hold at 250°C for 30 min.
- Ramp to 575°C at 20°C/min.

- Hold at 575°C for 180 min.
- Allow the temperature to drop to 105°C.
- Hold the sample at 105°C until samples are removed.

The percent AIL may be calculated from Equation 4.13:

$$\%\,\text{AIL} = \frac{\left(\text{Wt}_{\text{crucible+AIR}} - \text{Wt}_{\text{crucible}}\right) - \left(\text{Wt}_{\text{crucible+ash}} - \text{Wt}_{\text{crucible}}\right) - \text{Wt}_{\text{protein}}}{\text{ODW}_{\text{sample}}} \times 100 \quad (4.13)$$

Note that the percent protein is also required for the calculation of percent AIL. The weight of protein may be calculated from Equation 4.14, which is based on the ODW and percent protein of the extractives-free sample data.

$$\text{Weight}_{\text{protein}} = \text{ODW} \times \left(\frac{\%\,\text{Protein}_{\text{ext-free, ODW}}}{100}\right) \quad (4.14)$$

Example 4.8 shows how this calculation is done.

Example 4.8: Percent Acid-Insoluble Lignin (% AIL) (Extractives Free)

Determine the % AIL of extractives-free algae samples from the data given below:
Weight of dry crucible = 43.1369 g
Weight of dry crucible + acid-insoluble residue (AIR) = 43.2950 g
Weight of dry crucible + ash = 43.1578 g
% Protein (extractives free, ODW) from Example 4.4a = 45.92%
ODW from Example 4.5 = 0.2784 g

Solution:

a. The weight of the protein sample is first calculated as follows:

$$\text{Weight}_{\text{protein}} = 0.2784 \times \left(\frac{45.92}{100}\right) = 0.1278\text{ g}$$

b. The % AIL is then calculated as follows:

$$\%\,\text{AIL} = \frac{(43.2950 - 43.1369) - (43.1578 - 43.1369) - 0.1278}{0.2784} \times 100$$

c. The % $\text{AIL}_{\text{ext-free}}$ = 3.38%

Also note that percent AIR (answer in Example 4.6) = percent protein (Example 4.4a) + percent ash (Example 4.7a) + percent AIL (Example 4.8).

4.3.9 Percent ASL (Extractives Free)

The ASL is measured by UV-VIS spectroscopy. As shown in Figure 4.1, the extracted solids first undergo hydrolysis and the hydrolysate is brought to ASL analysis. Sulfuric acid is used for acid hydrolysis with a concentration of 72% w/w (specific gravity 1.6338 at 20°C). A water bath is normally used and set at 30°C ± 3°C, and this is incubated for 60 ± 5 min. After this hydrolysis and incubation process, the material is diluted to a 4% concentration by adding deionized water. This sample is then brought to the UV-VIS spectrometer to analyze the percent ASL. This step must be done within 6h of hydrolysis. The recommended absorptivity constants for ASL measurements for selected biomass are listed in the publications by Sluiter and colleagues (Sluiter and Sluiter, 2008; Sluiter et al., 2008a–f). This publication was based on an NREL LAP no. 42618 entitled "Determination of Carbohydrates and Lignin in Biomass." The recommended wavelength for corn stover sample was 320nm, the absorptivity at the recommended wavelength was 30 L/g cm, and the λ_{max} was 198 and the absorptivity at this λ_{max} was 55. Unfortunately, the absorptivity constants for newer biomass samples such as microalgae have not yet been established. There is a need to develop such calibration for newer samples.

The percent ASL may be calculated on an extractives-free basis, as shown in Equation 4.15:

$$\%\,\text{ASL} = \frac{\text{UV}_{\text{abs}} \times \text{Volume}_{\text{filtrate}} \times \text{Dilution}}{\varepsilon \times \text{ODW}_{\text{sample}} \times \text{Path Length}\,(\text{cm})} \times 100\% \qquad (4.15)$$

where

UV_{abs} = average UV-VIS absorbance for the sample at 320nm

Volume of hydrolysis liquor = volume of filtrate, normally 87mL

$$\text{Dilution} = \frac{\text{Volume}_{\text{sample}} + \text{Volume}_{\text{diluting solvent}}}{\text{Volume}_{\text{sample}}}$$

ε = absorptivity of biomass at specific wavelength

If the absorptivity of the biomass is not known, this will have to be determined. Table 4.3 summarizes the data for algae samples used to generate this absorptivity constant. The data show the different concentrations of the biomass and the various absorbances at 240, 280, and 320nm. Beer's law was used to determine the absorptivity of this particular biomass at the various wavelengths used and given in Equation 4.16.

$$A_{\lambda} = \varepsilon \times b \times c \qquad (4.16)$$

TABLE 4.3
Data to Establish Absorptivity Constant for Algae Sample

Standard No.	Concentration (mg/mL)	Absorbance		
		240nm	280nm	320nm
1	0.1	1.670	1.103	0.780
2	0.05	0.827	0.540	0.374
3	0.02	0.385	0.262	0.187
4	0.015	0.228	0.141	0.089
5	0.01	0.112	0.070	0.043
6	0.006	0.089	0.053	0.032
7	0.002	0.029	0.018	0.011

where

A_λ = average UV-VIS absorbance at specific wavelength
ε = absorptivity constant at specific wavelength in L/g-cm
b = path length through the sample in cm (~1 cm)
c = concentration of a single analyte in mg/mL

From the above equation, one can simply plot the average absorbance, A_λ, with concentration, c, to estimate the slope of the line: this slope is equal to the absorptivity constant. The path length is constant for a specific UV-VIS unit. The data in Figure 4.2 are used to determine the absorptivity for this algae sample at various wavelengths. Example 4.9 shows how this is calculated.

The percent lignin on an extractives-free basis is the sum of the percent AIL (extractives free) and the percent ASL (extractives free) as shown in Equation 14.17. Example 4.10 shows how this is calculated.

$$\%\,\text{Lignin (extractives-free)} = \%\,\text{AIL}_{\text{ext-free}} + \%\,\text{ASL}_{\text{ext-free}} \tag{4.17}$$

Finally, the percent lignin on an "as-received" basis can be calculated using Equation 4.18. Example 4.11 illustrates this calculation as an example.

$$\%\,\text{Lignin}_{\text{as received}} = \left(\%\,\text{Lignin}_{\text{ext-free}}\right) \times \frac{\left(100 - \%\,\text{Extractives}\right)}{100} \tag{4.18}$$

Example 4.9: Determination of Absorbance for Algae Sample

Determine the various absorptivity constants for algae samples at various wavelengths of 240, 280, and 320 nm. Use Figure 4.2 linear regression equations for calculations.

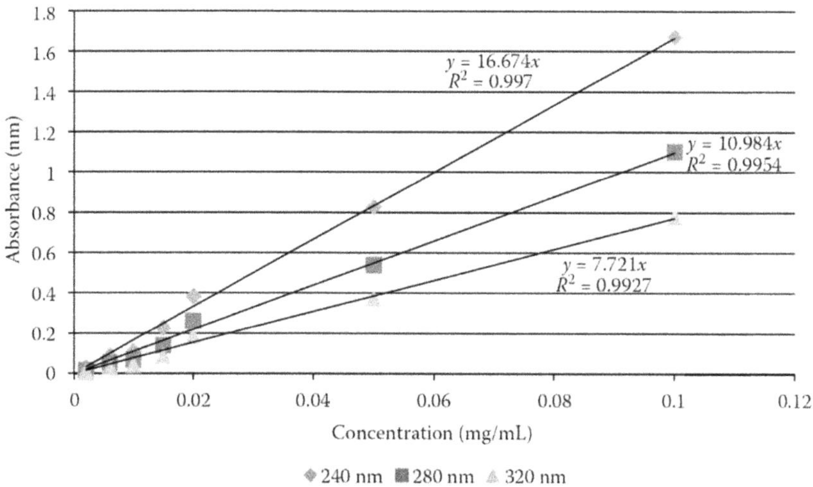

FIGURE 4.2 Plot of absorbance versus concentration of algae samples.

Solution:

a. From Figure 4.2, one can directly use the linear regression equation to establish the following absorptivity constants:
b. At 240 nm: $A_\lambda = 16.675\ c \rightarrow \varepsilon = 16.674$ L/g-cm
c. At 280 nm: $A_\lambda = 10.985\ c \rightarrow \varepsilon = 10.984$ L/g-cm
d. At 320 nm: $A_\lambda = 7.7249\ c \rightarrow \varepsilon = 7.721$ L/g-cm

Example 4.10: Determination of Acid-Soluble Lignin (ASL) (Extractives Free) and Percent Lignin (Extractives Free)

Determine the % ASL (extractives free) and percent lignin (extractives free) from the data given below:

ODW from Example 4.5 = 0.2784 g
UV absorbance = 0.825
Volume of filtrate = 87 mL
ε (320 nm) from Example 4.9 = 7.7249 L/g-cm
Dilution = 1.6667
% AIL (extractives free) from Example 4.8 = 3.38

Solution:

a. The % ASL (extractives free) is calculated using Equation 4.15 as follows:

$$\%\,\text{ASL} = \frac{0.825 \times (87/1000) \times 1.6667}{7.7249 \times 0.2784} \times 100\% = 5.56\%$$

b. The percent lignin (extractives free) is calculated as follows:

$$\%\,\text{Lignin (extractives-free)} = \%\,\text{AIL}_{\text{ext-free}} + \%\,\text{ASL}_{\text{ext-free}}$$

$$\%\,\text{Lignin (extractives-free)} = 3.38 + 5.56 = 8.94\%$$

Example 4.11: Determination of Percent Lignin (As Received)

Calculate the percent lignin (as received) of the algae sample from the following data:

% Lignin (extractives free) from Example 4.11b = 8.94%
% Total extractives from Example 4.2 = 48.09%

Solution:

a.

$$\%\,\text{Lignin}_{\text{as received}} = \left(\%\,\text{Lignin}_{\text{ext-free}}\right) \times \frac{\left(100 - \%\,\text{Extractives}\right)}{100}$$

$$\%\,\text{Lignin}_{\text{as received}} = (8.94) \times \frac{(100 - 48.09)}{100} = 4.64\%$$

b. Note the very low lignin content of algae samples compared with agricultural biomass.

4.3.10 Carbohydrates in Biomass

The carbohydrate content of biomass samples can be structural or nonstructural. Structural carbohydrates are bound in the matrix of the biomass, whereas the nonstructural carbohydrates can be removed using the extraction or washing steps shown above. This LAP for carbohydrates is very similar to ASTM E 1758-01 (Standard Test Method for Determination of Carbohydrates in Biomass by High-Performance Liquid Chromatography). Looking back at Figure 4.1, observe that the hydrolysate after the acid hydrolysis of extracted solids must also be analyzed for the presence of structural carbohydrates or for the cellulose and hemicellulose contents. The sugar contents would include analysis of cellobiose, glucose, xylose, galactose, arabinose, and mannose. Thus, during sugar analysis, sets of high-purity standards are required for this method; the procedure also calls for a set of sugar recovery standards (SRS), including their suggested sugar concentration ranging from 0.1 to 4.0 mg/mL. This standard will also be used to correct for losses due to destruction of sugars during the dilute acid hydrolysis process. A fresh set of standards is not necessary for each analysis. A second set of high-purity standards is also required and called the calibration verification standards (CVS), which is normally sourced from a different manufacturer.

1. d-Cellobiose
2. d(+)Glucose
3. d(+)Xylose
4. d(+)Galactose
5. d(+)Arabinose
6. d(+)Mannose

The HPLC system is used for all sugar analyses using a Shodex sugar SP0810 or BioradAminex HPX-87P column equipped with the appropriate guard column.

Table 4.4 summarizes an example of calibration data for sugar standards. The slope of the calibration curve has already been calculated to use in the subsequent calculation for sugar concentration. The CVS recovery must also be calculated as shown in Equation 4.19, while the percentage of each sugar is calculated using Equation 4.20:

TABLE 4.4
Example of Calibration Curve Data for Sugar Analysis

Concentration (mg/mL)	Area				
Sugar	Arabinose	Mannose	Galactose	Xylose	Glucose
0.1	8,577	11,871	9,386	7,066	9,565
1.0	73,390	77,362	68,975	67,215	75,011
4.0	283,032	299,153	276,062	267,346	291,818
Retention time	11.916	12.307	10.874	10.122	9.394
Slope	70,922	74,965	69,030	66,861	73,089

Note: For slope: ($y = mx$; y = area, x = concentration).

$$\%\,\text{CVS recovery} = \frac{\text{concentration detected using HPLC, mg/mL}}{\text{known concentration of standard, mg/mL}} \tag{4.19}$$

$$\%\,R_{\text{sugar}} = \frac{\text{concentration detected using HPLC, mg/mL}}{\text{known concentration of sugar before hydrolysis, mg/mL}} \tag{4.20}$$

Use the percent hydrolyzed sugar recovery values, using the value of percent R_{sugar} in Equation 4.20, to correct the corresponding sugar concentration values obtained using HPLC for each of the hydrolyzed samples ($C_{\text{cor/sample}}$), accounting for any dilution made before HPLC analysis. Equation 4.21 may then be used for this step.

$$C_x = \frac{C_{\text{HPLC}} \times \text{dilution factor}}{\%\,R_{\text{ave.sugar}}/100} \tag{4.21}$$

where
- C_{HPLC} = concentration of sugar as determined using HPLC, mg/mL
- $\%\,R_{\text{ave.sugar}}$ = average recovery of specific SRS component
- $C_x = C_{\text{cor/sample}}$, concentration in mg/mL of a sugar in the hydrolyzed sample after correction for loss on 4% hydrolysis

Calculate the concentration of the polymeric sugars from the concentration of the corresponding monomeric sugars, using an anhydro correction of 0.88 (or 132/150) for C5 sugars (xylose and arabinose) and a correction of 0.90 (or 162/180) for C6 sugars (glucose, galactose, and mannose) using Equation 4.22.

$$C_{\text{anhydro}} = C_{\text{corr}} \times \text{Anhydro correction} \tag{4.22}$$

Then, calculate the percentage of each sugar on an extractives-free basis using Equation 4.23.

$$\%\,\text{Sugar}_{\text{ext-free}} = \frac{C_{\text{anhydro}} \times V_{\text{filtrate}} \times \dfrac{1\,\text{g}}{1{,}000\,\text{mg}}}{\text{ODW}_{\text{sample}}} \times 100 \tag{4.23}$$

where V_{filtrate} is the amount of filtrate used. The percentage of sugars on an "as-received" basis may also be calculated using Equation 4.24.

$$\%\,\text{Sugar}_{\text{as received}} = \left(\%\,\text{Sugar}_{\text{ext-free}}\right) \times \frac{\left(100 - \%\,\text{Extractives}\right)}{100} \tag{4.24}$$

where the percent extractives are the percent extractives in the prepared biomass sample, as determined in the LAP "Determination of Extractives in Biomass." Note that additional tests for acetyl content may be necessary for some samples. This is important for biomass containing hemicellulose with xylan backbone but not biomass containing mannan backbone. Example 4.12 illustrates these sugar calculations, and the results of calculations are summarized in Table 4.5.

Example 4.12: Determination of % Sugars ("As-Received" Basis)

The following table summarizes the data from sugar analysis in an HPLC. Use the calibration data given in Table 4.4 to calculate all of the percent sugars (on "as-received" basis). The ODW is 0.2784 g (as calculated in Example 4.4) and the percent total extractives from Example 4.2 is 48.09%. The volume of filtrate used ($V_{filtrate}$) was 87 mL.

Sugar	Area from HPLC	% Recovery
Glucose	6,048	98.41
Xylose	620	97.58
Galactose	4,850	94.44
Arabinose	0	98.13
Mannose	22,777	94.00

Solution: (sample calculation for glucose only)

a. The glucose concentration is calculated as follows:

$$C_{\text{glucose, HPLC}} = \frac{\text{Area}}{\text{Slope of calibration curve}} = \frac{6048}{73,089} = 0.083\,\text{mg/mL}$$

b. The corrected sugar concentration is as follows:

$$C_{\text{corr}} = \frac{C_{\text{HPLC}} \times \text{dilution factor}}{\% R_{\text{sugar}}} \times 100 = \frac{0.083 \times 1}{98.41} \times 100 = 0.084\ \text{mg/mL}$$

c. The correction for anhydro sugar is as follows:

$$C_{\text{anhydro}} = C_{\text{corr}} \times \text{Anhydro correction} = 0.084 \times 0.90 = 0.076\,\text{mg/mL}$$

d. Thus, the percent glucose (extractives free) is calculated as follows:

$$\%\,\text{Glucose}_{\text{ext-free}} = \frac{C_{\text{anhydro}} \times V_{\text{filtrate}} \times \frac{1\,\text{g}}{1,000\ \text{mg}}}{\text{ODW}_{\text{sample}}} \times 100$$

$$= \frac{0.076 \times 87 \times 1/1,000}{0.2784} \times 100 = 2.375\%$$

TABLE 4.5
Results of Sugar Analysis from Example 4.12

Sugar	C_{HPLC}, mg/mL	C_{corr} mg/mL	$C_{anhydro}$, mg/mL	% Sugars (Ext. Free)	% Sugar (As Received)
Glucose	0.0827	0.0841	0.0757	2.36	1.23
Xylose	0.0093	0.0095	0.0084	0.26	0.14
Galactose	0.0703	0.0744	0.0670	2.09	1.09
Arabinose	0.000	0.0000	0.0000	0.00	0.00
Mannose	0.3038	0.3232	0.2909	9.09	4.72
Total				13.81	7.17

e. The percent glucose (as received) is calculated as follows:

$$\%\,\text{Glucose}_{\text{as received}} = \left(\%\,\text{Glucose}_{\text{ext free}}\right) \times \frac{\left(100 - \%\,\text{Extractives}\right)}{100}$$

$$\%\,\text{Glucose}_{\text{as received}} = (2.375\%) \times \frac{(100 - 48.09)}{100} = 1.23\%$$

f. The other sugars were calculated similarly, and the results are summarized in Table 4.6.

4.3.11 Total Composition of *Nannochloropsisoculata*

Table 4.6 summarizes the results of one replication for the compositional analysis of *N. oculata* to demonstrate the total composition and mass closures. Note that mass closures vary with samples. One should attempt to get this as close as possible to above 90%.

4.4 SUMMARY OF ASTM PROCEDURES FOR COMPOSITIONAL ANALYSIS

Table 4.7 summarizes the various related ASTM Standard Procedures for biomass compositional analyses. Note the similarities with most NREL LAP procedures. There are various versions published, and the earliest and most recent updated versions should be used. Some methods are cross-referenced with the NREL methods discussed earlier.

TABLE 4.6
Total Composition of *Nannochloropsisoculate*

Component	% w/w (As Received, Dry Basis)
Extractives (Example 4.2)	48.09
Water (Example 4.2a)	24.78
Ethanol (Example 4.2b)	23.31
Protein (Example 4.4b)	23.84
Ash (Example 4.7b)	3.90
Lignin (as received) (Example 4.11)	4.64
Acid soluble (extractives free) (Example 4.10a)	5.56
Acid insoluble (extractives free) (Example 4.8)	3.38
Lignin (extractives free) (Example 4.10b)	8.94
Structural carbohydrates (Example 4.12)	7.17
Glucan	1.23
Xylan	0.14
Galactan	1.09
Arabinan	0.00
Mannan	4.72
Total	87.64

TABLE 4.7
Summary of ASTM Standards for Compositional Biomass Characterization

ASTM	Test Procedure Title
E11-09	Standard Specification for Woven Wire Test Sieve Cloth and Test Sieves
E873-82 (reapproved 2006)	Standard Test Method for Bulk Density of Densified Particulate Biomass Fuels
E1757	Standard Practice for Preparation of Biomass for Compositional Analysis
E1690-08	Standard Test Method for Determination of Ethanol Extractives in Biomass
D1107	Test Method for Ethanol–Toluene Solubility of Wood
E1756	Test Method for Determination of Total Solids in Biomass
E1721-01 (reapproved 2009)	Standard Test Method for Determination of Acid-Insoluble Residue in Biomass
E1758-01 (reapproved 2007)	Standard Test Method for Determination of Carbohydrates in Biomass by High-Performance Liquid Chromatography
D1193	Specification for Reagent Water
E1755-01 (reapproved 2007)	Standard Test Method for Ash in Biomass
E1534	Test Method for Determination of Ash Content of Particulate Wood Fuels

4.5 BIOMASS PRETREATMENT

One would think that, after a tedious compositional analysis, biomass conversion could proceed readily. However, this is not the case for most biomass. Knowing the composition leads to strategies for further biomass processing, pretreatment procedures, and removing components from the process that are not necessary for conversion. The lignin content, for example, may interfere with the biological process that may be the next step in the conversion, and as much as possible must be removed from the matrix. One will realize that pretreatment of biomass is also a ubiquitous step in biological conversions. From the biomass characterization processes just described, some steps in the analytical procedures are actually pretreatment steps that are also used before biomass conversion. One may also delineate these pretreatments according to whether they are done physically, chemically, or with the use of heat or pressure. These biomass pretreatment procedures will be discussed in the chapter that deals with biological conversion processes such as ethanol and biogas production (Chapters 6, 7, respectively). The major pretreatment processes are categorized according as follows:

1. Physical
2. Chemical
3. Biological
4. Thermal
5. Combination of 1, 2, and 3

The physical pretreatments include grinding, milling, micro-milling, and the use of ultrasonication (ultrasound techniques). The latter would bring the biomass particles down to micron level. However, lengthy physical breakdown would also mean additional energy and costs for the process. Chemical pretreatments would include mild or strong acid digestion and the use of basic chemicals on the other end of the pH spectrum, such as lime and other strong bases. The physical pretreatments may also use high pressure and, when combined with high temperatures, would lead to thermal pretreatment procedures. Biological pretreatments include the use of microorganisms to break down the matrix of the biomass into simpler and manageable components. Finally, a combination of all these pretreatment procedures may be used. The pretreatment processes done at the BioEnergy Testing and Analysis Laboratory (Beta Lab) at Texas A&M University use the combination of these. This is schematically shown in Figure 4.3 for a procedure that converts cotton gin trash into ethanol. This procedure will be discussed in detail in the chapter on biomass cellulosic conversion into ethanol (Chapter 6).

The process begins with extensive size reduction, followed by steam explosion and ultrasonication. Minimal lime pretreatment is combined with the use of lignolytic enzymes. Chemical hydrolysis is also being investigated with the use of low to medium temperature and, less costly, mild to strong acids. To generate the biofuels of choice, all processes end with saccharification and fermentation. Results of some initial studies will be presented in later chapters.

4.6 CONCLUSION

This chapter has described some biomass characterization procedures that are important in biological conversion. The compositional analysis leads to the determination of the lignin content level in biomass as well as the breakdown of cellulose, hemicellulose, or carbohydrates. These are the properties of biomass that are important for biological conversion, such as ethanol and biogas production. Perhaps the most important parameters to note are the lignin content of both the acid-soluble and the acid-insoluble fractions. Knowledge of the lignin fraction levels, as well as the sugar level, will lead to

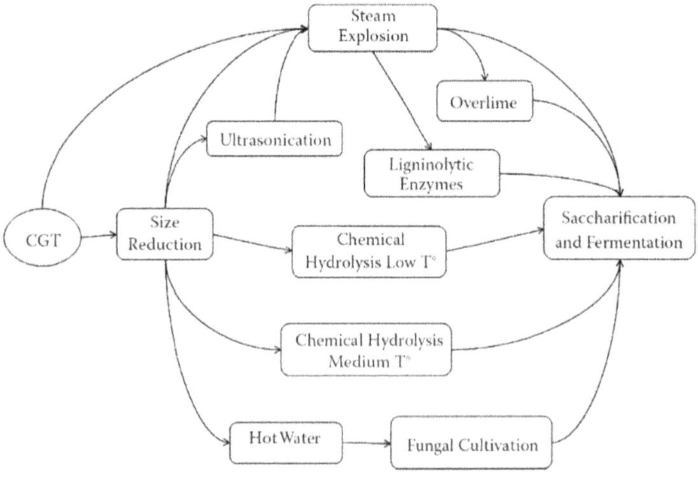

FIGURE 4.3 Pretreatment pathway for conversion of lignocellulosic cotton gin trash into ethanol.

better designs of subsequent conversion pathways. Note that microbes have not yet been able to degrade or convert the lignin component in biomass. The lignin must either be removed from the biomass or its concentration determined for reporting efficiencies and yield. Several pretreatment processes have been designed to remove or convert the lignin into some form that can be easily managed or separated during conversion.

Note the lengthy procedures for complete compositional analyses, when they are compared with methods used in thermal conversions. These procedures also require the use of sophisticated equipment supplements by analytical skills in the methodology. For example, the use of an HPLC and the UV-VIS near-infrared spectrometer is essential and requires special skills for analysis. Moving further into the conversion processes, additional equipment may be necessary, such as for fuels and product characterization. When developing facilities for biomass conversion, the choice of equipment depends on its importance for conversion pathway that may be adopted. This chapter presented most of the equipment and facilities required and presented a step-by-step procedure for complete compositional analyses. For better understanding of all components, one should be able to report the summary table of closure for the compositional analysis and have a closure as close as possible to 100% or at least within 90%.

Observe also that the reporting process for compositional analysis may depend on the moisture content of the material as well as whether it is extractives free or not. One should also be able to switch from one method of reporting to another. Details of the procedures discussed in this chapter may be sourced from Table 4.2 LAPs.

The example presented in this chapter is only a single experimental run. To obtain data of better quality, these experiments must be made in triplicate and the values averaged, with reports of standard deviations or standard errors. The sample problems at the end of this chapter were from a similar analysis that used corn stover feedstock.

4.7 PROBLEMS AND DISCUSSION ISSUES

The problems in Section 4.7.1 are patterned after the examples presented in this chapter with particular focus on the use of corn stover as the feedstock. Figure 4.1 will also be used to guide the reader of each step of the process. The data used are actual values from laboratory experiments so that the reader will be familiar with the range of weights and samples that will be used.

4.7.1 Calculation of Percent Total Solids

P4.1 Determine the percentage of total solids in a corn stover sample. Refer to Figure 4.1 for the flow chart for analysis. The weight of the empty crucible used for the analysis was 18.1119 g. A corn stover sample was placed in this crucible and was reweighed and found to have a total weight of 19.2048 g. The sample was then placed in the forced convection dryer oven until the weight became constant. The final weight after drying was 19.1189 g.

4.7.2 Calculation of Percent Ash Content (Dry Basis)

P4.2 A bone-dry corn stover sample used in Problem 4.1 was placed in a crucible (which was dried completely) and then placed in a muffle furnace with a temperature setting of 575°C. The ODW of the sample was already established to be 0.9845 g. Determine the percent ash of this sample (dry basis) if the weight of the crucible that is dry was 14.832 g and the weight of this crucible after exposure to the oven at 550°C for several hours was 14.8702 g.

4.7.3 Calculation of Percent Extractives

P4.3 Determine the following:

- Percent water extractives
- Percent ethanol extractives
- Percent total extractives from the following laboratory data:

  Part A. Water Extraction Data

  1. Weight of sample = 10.7204 g
  2. Weight of flask = 154.9616 g
  3. Weight of flask + sample after extraction = 156.55 g
  4. The percent total solids from Example 3.1 was established to be = 92.13%

  Part B. Ethanol Extraction Data

  1. Weight of sample = 10.7204 g
  2. Weight of flask = 154.5637 g
  3. Weight of flask + sample after extraction = 154.7824 g

4.7.4 Calculation of Percent Total Solids (Extractives-Free Biomass)

P4.4 Calculate the percent total solids from an extractives-free biomass sample. The following are the laboratory data:

Weight of dry pan = 1.3425 g
Weight of dry pan + sample = 1.9707 g
Weight of dry pan + dry sample = 1.925 g

4.7.5 Calculation of Percent AIL from Extractives-Free Sample

P4.5 Determine the percent AIL from the extractives-free sample with the following given laboratory data:

- Weight of sample = 305.3 mg or 0.3053 g
- Percent total solids (extractives free) already established from Problem 4.4, equal to 92.73%
- Weight of dry crucible + AIR = A = 43.229 g Weight of dry crucible = 43.1811 g (B)
- Weight of dry crucible + ash = 43.1852 g (C)
- Protein content = 4.2% (as received)

4.7.6 Calculation of % ASL and %Lignin (Extractives Free)

P4.6 Determine (a) the percent ASL and (b) the percent lignin (extractives free) from the laboratory data given as follows:

- Weight of sample = 305.3 mg or 0.3053 g
- Percent total solids (extractives-free biomass) (established from Example 3.4) = 92.73%
- UV absorbance = 0.991
- Volume of filtrate = 86.73 mL
- Absorptivity at 320 nm, ε = 30 L/(g-cm)
- Dilution = 1

4.7.7 Calculation of Percent Lignin (on As-Received Basis)

P4.7 Calculate the percent lignin if the extractives were found to be 18.44% and lignin 15.77% (as established from Problems 4.3 and 4.6).

4.7.8 Calculation of Percent Sugars

P4.8 You are given the following calibration curve data for various sugar standards and at various concentrations. The retention time for each sugar sample is also given as well as the areas under the curve (or peak) and the slope of the calibration curve for each sugar standards. The % extractives is 18.44%, ODW = 0.2832 g, and $V_{filtrate}$ is 86.73 mL. Determine the percent glucose (as received) for this sample. The data for the injected sample in an HPLC are shown in the succeeding table:

Calibration Curve Data

Concentration (mg/mL)	**Area**				
Sugar	**Arabinose**	**Mannose**	**Galactose**	**Xylose**	**Glucose**
0.16		18,022	12,100	12,062	17,027
0.5	37,092	42,955	35,034	35,104	42,724
1	70,770	77,349	70,145	69,942	76,843
2	140,300	147,187	140,477	138,829	146,944
4	276,747	286,068	281,520	275,603	290,733
Retention time	11.88	12.26	10.86	10.11	9.39
Slope ($y = mx; y =$ area, $x =$ concentration)	69,501	72,400	70,345	69,069	73,218

Data for Samples

Sugar	Area from HPLC	% Recovery
Glucose	103,541	98.41
Xylose	60,332	97.58
Galactose	2,916	94.44
Arabinose	7,120	98.13
Mannose	8,053	94.00

4.7.9 Calculation of Percent Xylose Sugars

P4.9 You are given the calibration curve data given in Problem 4.8 for various sugar standards and at various concentrations. The retention time for each sugar sample is given as well as the areas under the curve (or peak) and the slope of the calibration curve for each sugar standards. Determine the percent xylose sugars (as received) for this sample. The data for the injected sample in an HPLC is also shown in the previous problem. Complete also the analysis for the remaining sugars and tabulate your results including totals.

4.7.10 Biomass Compositional Analysis Summary

P4.10 Summarize the total composition of the corn stover sample from Problems 4.1 through 4.9 and calculate the overall total composition. Note that the percentage of protein is 5.8%. Break down the percent lignin (as received) into its acid-soluble and acid-insoluble components.

4.7.11 Biomass with the Highest Cellulose Content

P4.11 Which of the following biomass has the highest cellulose content?
a. swine waste b. rice straw c. cattle manure d. nut shells

4.7.12 Biomass with the Highest Lignin Content

P4.12 Which of the following biomass has the highest lignin content?
a. paper b. hardwood stem c. rice straw d. cattle manure

4.7.13 Biomass with the Highest Hemicellulose Content

P4.13 Which of the following biomass has the highest hemicellulose content?
a. leaves b. rice straw c. paper d. manure

4.7.14 Whole Plant Biomass into Bioethanol

P4.14 Develop a project to convert all components of sweet sorghum plant biomass into bioethanol. The sorghum juice is first extracted and converted into bioethanol followed by starch from its panicle and all the other cellulosic components into bioethanol as well. Document assumptions properly.

4.7.15 Animal Manure into Pure Methane

P4.15 Propose a project to produce 100% pure natural gas or methane from anaerobic digestion of all animal manure from a 1,000-head dairy facility. The pure natural gas will be sold as a commercial fuel. Enumerate all the technical barriers for this process.

REFERENCES

Badger, P. and P. Pullammanappallil. 2011. Biomass conversion to energy and fuels. In: Wood to Energy: Using Southern Interface Fuels for Bioenergy (C. Staudhammer, A. L. Hermansen-Baez, D. R. Carter and E. A. Macie, eds.). USDA Forest Service General Technical Report GTR-132.

Lata, K. and S. P. Mande. 2009. Bioenergy resources. In: *Renewable Energy Engineering and Technology: Principles and Practice* (V. V. N. Kishore, ed.). Earthscan Publishers, London.

LePori, W. A. and E. J. Soltes. 1985. Thermochemical conversion for energy and fuel. In: *Biomass Energy: A Monograph* (E. A. Hiler and B. A. Stout, eds.). Texas Engineering Experiment Station, The Texas A&M University Press, College Station, TX.

Maglinao, A. L. and S. C. Capareda. 2010. Predicting fouling and slagging behavior of dairy manure (DM) and cotton gin trash (CGT) during thermal conversion. *Transactions of the ASABE* 53 (3): 903–909.

Searcy, 2010. Biomass system flow alternatives. Biomass Logistics Lecture Notes. Biological and Agricultural Engineering Department, Texas A&M University, College Station, TX.

Winegartner, E. C. (ed.). 1974. Coal fouling and slagging parameters. Report prepared by the American Society of Mechanical Engineers (ASME) Research Committee on Corrosion and Deposits from Combustion Gases. Copyright 1974 by the American Society of Mechanical Engineers (ASME), Three Park Avenue, NY.

ASSESS YOUR KNOWLEDGE

1. What are the major constituents of compositional analysis for biomass samples?
   a. Cellulose, hemicellulose, and lignin
   b. MC, VCM, FC, and ash
   c. Moisture and ash contents
   d. C, H, O, N, S, and ash

2. What does AFEX stand for in compositional analysis and retreatment of biomass samples?
   a. Associated foreign exchange
   b. Ammonia fiber extraction
   c. Advanced fiber extraction process
   d. Associated fiber extrusion process

3. This biomass component is an important organic polymer, which is also abundant in cell wall, but it binds cellulose together making the biomass rigid and woody. It is synonymous with mortar binder in concrete.
   a. Lignin
   b. Cellulose
   c. Hemicellulose
   d. Ash

4. This biomass component is composed of groups of complex carbohydrates such as sugars and starch. They indicate potential bioethanol yield using simple yeasts as microbe for conversion.
   a. Hemicellulose
   b. Lignin
   c. Cellulose
   d. Ash

11. If one is asked to select a material with high cellulose content, which among the choices should quality?
    a. Cattle manure
    b. Hardwood stem
    c. Leaves
    d. Coastal Bermuda grass

12. If one is asked to select a material with high cellulose content, which among the choices should qualify?
    a. Cattle manure
    b. Hardwood stem
    c. Leaves
    d. Coastal Bermuda grass

13. What particle size will a sieve set No. 20 on the Tyler scale pass through this sieve?
    a. Smaller than 840 microns
    b. Greater than 840 microns
    c. Greater than 200 microns
    d. Smaller than 900 microns

14. What is another name for freeze-drying process?
    a. Solidification
    b. Solid convection
    c. Immobilization
    d. Lyophilization

5. This biomass component is the major structural component of biomass specifically of the plant or microalgae's cell wall.
   a. Hemicellulose
   b. Lignin
   c. Cellulose
   d. Sugars

6. What is a riffle sampler?
   a. Device to shot biomass into feed bin
   b. Device to condense biomass samples
   c. Device to divide sample into equal half parts
   d. Device to reduce biomass size

7. A 500 g biomass sample was prepared for compositional analysis. The first step was to sieve the sample in Tyler sieve sets No. 20 (top) and 80 (middle) and pan (botttom), and the results are as follows: The amount of material captured in the middle sieve was 350 g and the pan sample weighs 50 g. Determine the 20/80 fraction and the 80 fraction. For this question, what is the Fraction_{80} component (in %)?
   a. 12.5
   b. 80
   c. 70
   d. 87.5

15. What particle size will a sieve set No. 80 on the Tyler scale pass through this sieve?
   a. Greater than 177 microns
   b. Smaller than 177 microns
   c. Smaller than 800 microns
   d. Greater than 200 microns

16. A 500 g biomass sample was prepared for compositional analysis. The first step was to sieve the sample in Tyler sieve sets No. 20 (top), No. 80 (middle) and pan (Bottom), and the results are as follows: The amount of material captured in the middle sieve was 350 g and the pan sample weighed 50 g. Determine the 20/80 fraction and the 80 fraction. For this question, what is the 20/80 fraction (in %)?
   a. 12.5
   b. 80
   c. 70
   d. 87.5

17. These materials include inorganic portions of the biomass, the nonstructural sugars, and the nitrogenous materials, primarily coming from the soil or fertilizer.
   a. Water-insoluble
   b. Water-soluble
   c. Ethanol-soluble
   d. Ethanol-insoluble

8. What does HPLC stands for?
   a. High-performance liquid chromatography
   b. Homogeneous particles dissolved in liquid chromatography
   c. High-pressure lyophilization contraption
   d. High-performance linear chromatography
9. Which apparatus is used for solvent extraction?
   a. Distillation system
   b. Boiler
   c. Freeze dryer
   d. Soxhlet apparatus
10. Which device is commonly used for protein determination?
    a. Nitrogen extractor
    b. Kjeldahl method
    c. Nitrogen oxidizer
    d. Soxhlet apparatus
18. These materials in biomass include chlorophyll, waxes and oils.
    a. Water-insoluble
    b. Water-soluble
    c. Ethanol-soluble
    d. Ethanol-insoluble
19. True or False. Can micro degrade lignin?
    a. No
    b. It depends
    c. No idea
    d. Yes
20. What is AOAC 984.13 Standard?
    a. Official method for analyzing lignin
    b. Official method for analyzing hemicellulose in biomass
    c. Official method for determining protein in plant tissue
    d. Method for determining cellulose in biomass

Answers: 1a, 2b, 3a, 4d, 5c, 6c, 7a, 8a, 9d, 10b, 11b, 12b, 13a, 14d, 15b, 16d, 17b, 18c, 19d, 20c

ADDITIONAL LEARNING RESOURCES

- Access the Lecture Notes at the Instructor and Student Resources: https://routledgetextbooks.com/textbooks/instructor_downloads/

5 Biodiesel Production

LEARNING OBJECTIVES

Upon completion of this chapter, one should be able to

1. Identify methods for extracting oil from various biomass resources.
2. Describe the various refining processes for crude vegetable oil and animal fats and oils and calculate the efficiencies of these processes.
3. Describe the transesterification process for oils and be able to calculate conversion efficiencies.
4. Differentiate the various ASTM properties and standards for biodiesel and describe how the standards are performed.
5. Describe how biofuels are tested in internal combustion engines for engine performance and emissions.

5.1 INTRODUCTION

The trajectory of the US biodiesel industry—its growth, decline, and resurgence—has been quite remarkable. Biodiesel became a very important biofuel when the price of refined vegetable oil was relatively inexpensive and competitive with the price of commercial diesel fuel per unit volume. The US government also provided initial subsidies on the oil feedstock. For the first few years when biodiesel was quite popular, beginning in 2005, several commercial production facilities emerged, bringing the production to record highs each year. Figure 5.1 shows the growth of the biodiesel industry in recent years in the United States (USEIA, 2022). Maximum production was first achieved in 2008 with about 2,650 million liters (700 million gallons) of biodiesel produced, after which the price of the feedstock slowly increased and subsidies for the oil feedstock for biodiesel in the United States were removed. When the price of vegetable oil significantly exceeded the price of diesel, some of these plants started to shift to other cheaper feedstocks such as chicken fat or other animal fats, including waste oil from fast-food chains. Still, the price of chicken fat and other alternatives continued to increase, resulting in a substantial drop in production and the number of biodiesel plants that have closed around 2010. There was, however, a rebound of the biodiesel industry in 2011, with a record production of close to 4,000 million liters (~1,000 million gallons). Over 5 billion liters (1.3 billion gallons) of biodiesel has been produced since 2013. The primary reason for the increase in production is the implementation of the Renewable Fuels Standards by the US government.

Biodiesel is a significant additive to diesel fuel because of its lubricating property. Sulfur, a material that contributed to smog, was previously used. Biodiesel is a natural lubricant. The overall biodiesel production and testing flow processes is shown in Figure 5.2.

DOI: 10.1201/9781003294306-5

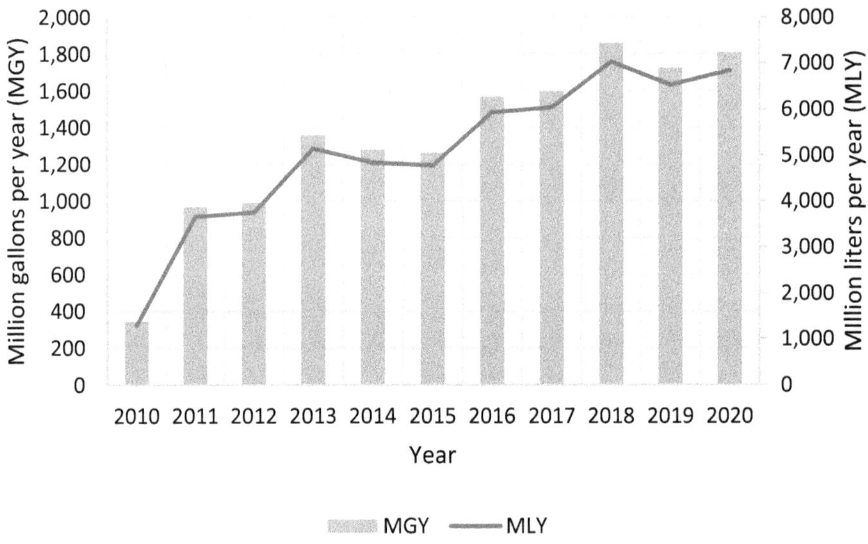

FIGURE 5.1 Growth of the biodiesel industry in the United States by volume of production. (*Source*: USEIA. 2022. US Biodiesel plant production capacity. U.S. Energy Information Administration, Washington, DC. Available at: https://afdc.energy.gov/data/10325. Accessed March 12, 2022)

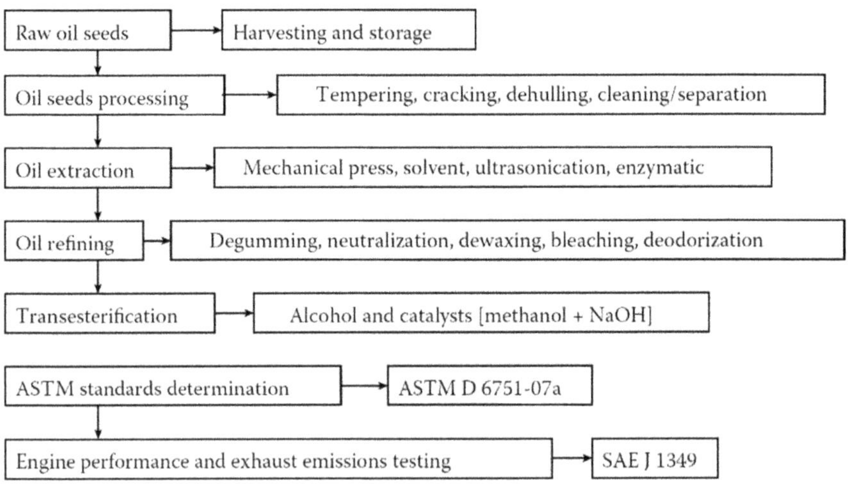

FIGURE 5.2 Overall biodiesel production and testing flow processes chart.

The overall biodiesel conversion process begins with raw oil seeds that are harvested, cleaned, and stored. This is followed by oil seed processing that includes tempering, cracking, dehulling, cleaning, and separation of the meat from the hulls. Oil is extracted through various processes and refined. After refining, the transesterification process is used to produce the esters of oil. ASTM standards characterization is next, followed by engine testing. Exhaust emissions may also be evaluated and measured during engine testing. There were reports that the NO_x emissions of engines fueled by biodiesel may have increased because of more efficient combustion, thereby increasing the combustion temperature that leads to higher NO_x levels. In some states, such as Texas, the local environmental air quality administrations are quite strict in air quality regulations, ensuring that the use of biodiesel will not significantly increase emissions of NO_x compared with diesel. Some additives may be required at very high levels of the biodiesel mixture. However, at low levels, the increase may not be significant.

Example 5.1: World Energy Consumption in Other Units

Convert the US primary biodiesel consumption of 1,864 million gallons in 2020, in units of barrels and compare this with 3,800 million barrels of diesel (percentage of use). Comment on this ratio of biodiesel use with diesel, barrel-wise.

Solution:

a. The number of barrels of biodiesel used in 2005 is calculated as follows:

$$\text{Barrels} = 1{,}864 \times 10^6 \,\text{gal} \times \frac{\text{barrels}}{42 \text{ gal}} = 44{,}380{,}952 \text{ barrels}$$

b. The number of barrels of diesel used in 2020 was around 3,800 million barrels, and thus, this percentage is calculated as follows:

$$\%\text{ Biodiesel used} = \frac{44{,}380{,}952 \text{ barrels}}{3{,}800 \times 10^6 \text{ barrels}} \times 100\% = 1.17\%$$

c. Clearly, the biodiesel usage is still a small fraction of diesel used in units of barrels.

5.2 VEGETABLE OIL AND ANIMAL FATS PRODUCTION IN THE UNITED STATES

In Chapter 1, various sources of oil from crops were discussed. In recent years, because of the high price of oil from crops, the supply of oil feedstocks for biodiesel production has expanded to include animal fats and oils, spent oil from some commercial facilities, and other new sources like algae. Vegetable oil and animal fats production in the United States in the year 2020 is presented in Table 5.1. Data was directly taken from the yearly National Agricultural Statistics Service (NASS, 2022) report. Note that soybean oil is the highest oil-producing crop as well as the largest contributor to biodiesel production, followed by corn oil. Edible tallow is the third largest oil-producing feedstock, followed by canola oil and lard. Most of these crops may be expanded to meet the requirements of the growing biodiesel industry. Edible tallow and lard are also potential sources of biodiesel; with canola oil, they rank in the top five production feedstocks.

TABLE 5.1
Vegetable Oil and Animal Fat Production in the United States in 2020

Oil Type	Metric Tonnes	English Tons
Soybean oil	11,650,117	12,815,000
Corn oil	2,693,209	2,962,500
Edible tallow	895,464	985,000
Canola oil	878,191	966,000
Lard	484,096	532,500
Cottonseed oil	206,820	227,500
Sunflower oil	228,639	251,500
Peanut oil	118,183	130,000
Safflower	23,637	26,000
Total	17,194,263	18,913,500

Source: NASS, Oil Crops Yearbook. Available at: https://www.ers.usda.gov/data-products/oil-crops-yearbook/oil-crops-yearbook/#Vegetable%20Oils%20and%20Animal%20Fats. National Agricultural Statistics Service, NASS, Washington, DC, 2022.

The energy ratios (defined as biodiesel energy divided by the energy of fossil fuel used for conversion) vary for various feedstocks and may range from a low of 2.5 for rapeseed oil to a high of 9.0 for palm oil (Rosillo-Calle et al., 2009). Soybean oil has an energy ratio of 3, indicating there is three times more energy in the biodiesel produced than all the energy used for its conversion, whereas recycled vegetable oil has an energy ratio of 5.5 from the same source.

5.3 VEGETABLE OIL AND ANIMAL FAT CHARACTERISTICS

Raw oil, which is derived from oil seeds or animals and fats, cannot be directly converted into biodiesel. Before the refining process, the raw oil must be thoroughly evaluated for its exact composition. The biodiesel production process begins by refining the raw oil to remove unwanted impurities. There are excellent publications on oil characteristics because of the use of edible oil as food. A complete publication by the American Oil Chemists Society (Firestone, 2006) shows the important parameters of vegetable oils and fats. The foremost parameter is fatty acid composition, that is, the degree of saturation of the fatty acids as well as other physical parameters such as specific gravity, density, and energy content. In chemistry, oils and fats are called lipids. Oils are substances that are liquid at room temperature, whereas fats are solids at room temperature. The chemical structure of oil includes a glycerol backbone with fatty acids attached to this backbone (shown in Figure 5.3), hence the term triglyceride (or triacid glycerol [TAG]) to describe oils and fats chemically. In Figure 5.3, the three fatty acids could be of a similar type (or length) or of various fatty acid types, and they are all bonded by the ester linkage. Thus, to make biodiesel, it is necessary to break this bond that separates the glycerol (or glycerin) from the esters of the oil, hence the name soybean methyl ester (SME), to describe the biodiesel from soybean

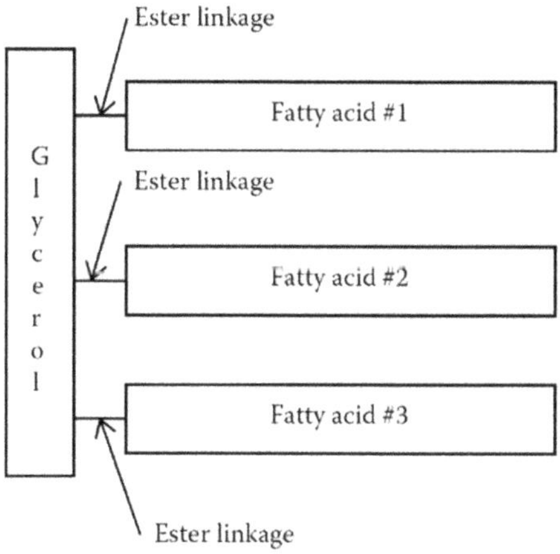

FIGURE 5.3 Basic chemical configuration of triglyceride.

oil. About 95%–98% of crude seed oil is TAG, and all the rest includes diglycerides (having only two fatty acids instead of three) or monoglycerides (having only one fatty acid attached to the glycerol backbone). The other constituents include free fatty acids (FFAs), phosphatides, sterols, triterpenes, alcohols, tocopherols, and fat-soluble vitamins. Crude oil must be refined to remove these constituents.

5.4 FATTY ACID COMPOSITION

The biggest variation in the various vegetable oil types (canola, coconut, or soybean oil) is the percentage of the various fatty acids present. Fatty acids are either saturated (when there are no double bonds on the carbon) or unsaturated (when the carbon atoms have at least one double bond). Table 5.2 lists the most common fatty acids found in vegetable oils and fats. For each acid listed, a common name, the structure in terms of the number of carbons present, and a source are shown. For example, the description C14:0 means there are 14 carbon atoms; the zero means there are no double bonds present. In addition, unsaturated fatty acids may be classified as monounsaturated fatty acids (meaning there is only one double bond). They may also be classified as polyunsaturated (meaning there are two or more double bonds). Furthermore, fatty acids could also be classified as short-chain (between 4 and 10 carbons), medium-chain (between 12 and 14 carbons), or longer chain (more than 16 carbons).

Some vegetable oils and fats have more than five fatty acids present in various amounts. The particular vegetable oils and fats may also be described according to the majority of fatty acids present. For example, coconut oil contains a large fraction of lauric oil. Consequently, it is called lauric oil by the oil industry. The equipment used to measure the fatty acid composition of vegetable oil is a gas chromatograph (GC). The chromatograph must be equipped with a flame ionization detector. To begin the analysis process, a fatty acid standard from representative fatty acids enumerated in Table 5.2 must be used. Fatty acid standards may be purchased from certified chemical manufacturers. The fatty acid composition of some edible fats and oils is shown in Table 5.3.

TABLE 5.2
Systematic, Common Name, and Structure of Most Common Fatty Acids

Systematic Name	Common Name	Structure	Source
Saturated fatty acids			
Tetranoic acid	Butyric acid	C4:0	Butterfat
Pentanoic acid	Valeric acid	C5:0	Valerian plant
Hexanoic acid	Caproic acid	C6:0	Butterfat
Octanoic acid	Caprylic acid	C8:0	Coconut oil
Decanoic acid	Capric acid	C10:0	7% In goat butterfat
Dodecanoic acid	Lauric	C12:0	47% In coconut oil
Tetradecanoic acid	Myristic acid	C14:0	18% In coconut oil
Hexadecanoic acid	Palmitic acid	C16:0	45% In palm oil
Octadecanoic acid	Stearic acid	C18:0	38% In cocoa butter
Eicosanoic acid	Arachidic acid	C20:0	3% In some corn oil
Monounsaturated FA			
Hexadecenoic acid	Palmitoleic acid	C16:1	17% In macadamia oil
Octadecenoic acid	Oleic acid	C18:1	71% In olive oil
Docosenoic acid	Erucic acid	C20:1	42% (Mustard seed)
Polyunsaturated FA			
Octadecadienoic	Linoleic acid	C18:2	76% In safflower oil
Octadecatrienoic	Linolenic acid	C18:3	53% In flaxseed oil
Eicosatetraenoic	Arachidonic acid	C20:4	

TABLE 5.3
Fatty Acid Composition of Some Common Edible Fats and Oils

	Saturated Fats					Mono	Poly	
Oil or Fat	**C10**	**C12**	**C14**	**C16**	**C18**	**C:18:1**	**C:18:2**	**C:18:3**
Almond oil				7	2	69	17	
Beef tallow			3	24	19	43[a]	3	1
Butterfat (cow)	3	3	11	27	12	29[a]	2	1
Butterfat (goat)	7	3	9	25	12	27[a]	3	1
Butterfat (man)	2	5	8	25	8	35[a]	9	1
Canola oil				4	2	62[a]	22	10
Cocoa butter				25	38[a]	32	3	
Cod liver oil			8	17		22[a]	5	
Coconut oil	6	47[a]	18	9	3	6	2	
Corn oil				11	2	28	58[a]	1
Cottonseed oil			1	22	3	19	54[a]	1
Flaxseed oil				3	7	21	16	53[a]
Grape seed oil				8	4	15	73[a]	
Lard (pork fat)			2	26	14	44[a]	10	

(Continued)

TABLE 5.3 (*Continued*)
Fatty Acid Composition of Some Common Edible Fats and Oils

	Saturated Fats					Mono	Poly	
Oil or Fat	**C10**	**C12**	**C14**	**C16**	**C18**	**C:18:1**	**C:18:2**	**C:18:3**
Olive oil				13	3	71[a]	10	1
Palm oil			1	45[a]	4	40	10	
Palm olein			1	37	4	46[a]	11	
Palm kernel oil	4	48[a]	16	8	3	15	2	
Peanut oil				11	2	48[a]	32	
Safflower oil				7	2	13	78[a]	
Sesame oil				9	4	41	45[a]	
Shea nut		1		4	39	44[a]	5	
Soybean oil				11	4	24	54[a]	7
Sunflower oil				7	5	19	68[a]	1
Walnut oil				11	5	28	51[a]	5

Source: Zamora, A., *Fatty Acid Composition of Some Common Edible Fats and Oils*. Scientific Psychic Publications, Bethesda, MD, 2010.

[a] The majority of fatty acids that dominate a certain oil type. In some cases, more than five fatty acid types are present in one oil crop.

5.5 OTHER BASIC OIL PROPERTIES

Some important oil properties include heating value, density, and viscosity. The heating value is important for calculating conversion efficiencies, whereas density and viscosity are important for conveying the oil from one point to another. Plant oils have slightly less energy content than diesel fuel (between 5% and 13% lower). Note also that oils have high viscosity values compared with diesel fuel. Perhaps, viscosity is the one property of biodiesel (the ester of oil) that is significantly reduced when oil is converted into its ester form. The density of oil is also higher than diesel fuel, and this is important for transport purposes. Table 5.4 summarizes some recent data on some basic properties of oils and diesel fuel.

5.6 OIL EXTRACTION PROCESSES

There are various ways of extracting oils from oil seed crop. The most common extraction method is mechanical oil separation using an oil seed press or extruder.

The extruder has provisions for heating the material while it is being pressed. It has several apertures to accommodate different sizes of seeds. During extraction, seeds are normally heated to around 150°C using an electronic controlled band heater. Note that a higher temperature will lead to discoloration. Consequently, the oil will be much darker than usual. Example 5.2 shows how to calculate the percent oil yield. This example is from Jatropha seed samples.

TABLE 5.4
Common Properties of Oil and Diesel

Oil/Fuel	Density (kg/L at °C)		Kinematic Viscosity (cSt or mm²/s)	High Heating Value (MJ/kg)	(MJ/L)
Diesel fuel (Std#2)	0.829	20	2.45	45.79	37.96
Canola oil	0.922	24	34.8	39.57	36.48
Soybean oil	0.926	24	33.2	39.28	36.37
Corn oil	0.927	24	34.9	39.36	36.49
Corn + canola	0.930	24	35.9	39.43	36.67
Olive oil	0.908	22	39.2	39.55	35.91
Safflower oil	0.911	22	38.2	39.48	35.97
Sunflower oil	0.916	22	32.3	39.29	35.99
Grape seed oil	0.918	22	34.0	39.38	36.15
Coconut oil	0.917	30	27.9	37.65	34.53
Peanut oil	0.909	22	38.6	37.51	34.10

Example 5.2: Oil Yield Calculations

Determine the amount of oil extracted from Jatropha seeds (with the hull already removed). The initial amountsof seed used was 136 g, and the amount obtained at the oil collection bottle was 57.9 g. The pressed meat weighed 74.4 g. Determine the percent losses from this experiment.

Solution:

a. The percent oil yield is calculated from the ratio of the amount of oil to the initial amounts of seed used:

$$\%\text{ Oil yield} = \frac{57.9}{136} \times 100\% = 42.6\%$$

b. The percent pressed meat is calculated similarly as follows:

$$\%\text{ Pressed meat } = \frac{74.4}{136} \times 100\% = 54.7\%$$

c. Thus, the losses are calculated as follows:

$$\text{Losses } = 100\% - 42.6\% - 54.7\% = 2.7\%$$

A series of experiments to determine the effect of aperture size on the amount of oil extracted from Jatropha seeds is summarized in Table 5.5. From the data, it was shown that the best aperture size is #8, with an oil yield of around 44% and the lowest losses of around 2%. Aperture size #10 provided slightly lower oil yield (53.4%) but higher losses (3.1%).

TABLE 5.5
Oil Extraction Yield Data from Jatropha Seeds

	Aperture Size #4			Aperture Size #8			Aperture Size #10		
Trial	%Oil	%Meat	Loss	%Oil	%Meat	Loss	%Oil	%Meat	Loss
1	31.6	50.0	18.4	44.9	52.2	2.9	42.6	54.7	2.7
2	40.4	50.0	9.6	44.1	53.7	2.2	44.5	52.3	3.2
3	39.7	52.2	8.1	42.4	56.0	1.6	43.5	53.2	3.3
Mean	37.2	50.7	12.0	43.8	54.0	2.2	43.5	53.4	3.1
SD	4.9	1.3		1.3	1.9		0.95	1.2	

The oil from seeds can also be obtained through solvent extraction. The most popular solvent is hexane, while a cheaper alternative is petroleum ether. In the laboratory, this process uses a solvent extraction setup. A certain amount of seed is placed in the thimble, while a known amount of solvent is placed at the bottom flask. The extraction takes at least 8 h. The oil and solvent mixture (called miscella) is then collected and the solvent recovered or evaporated using the Soxhlet facility to completely separate the oil. This procedure recovers nearly 100% of the oil from the seed.

The product of this process is crude oil. The crude oil then undergoes various refining steps to remove the waxes, phosphorus (degumming), FFAs (by neutralization), and other impurities. Other oil extraction methods include enzymatic processing with the use of inert gases such as CO_2. In some instances, such as extracting oil from microalgae species, where the oil is bound within the cell wall of the algae, some lysing or physical maceration is first done using ultrasound. As such, the use of an ultrasonication facility has become a common oil extraction step. Other strategies involve infusion, where a certain chemical is used to soak the ground oil seed. Then, the infused chemicals are also removed. Nevertheless, the crude oil will have to be refined, bleached, and deodorized before producing the biodiesel.

5.7 OIL REFINING PROCESSES

Refining crude oil that has been extracted from crops requires a series of steps to remove unwanted components in the oil. In the United States, it is customary to begin the transesterification process with what industry people would call "RBD" oil. This means that the oil must be refined, bleached, and deodorized. Refining includes the following substeps: degumming, neutralization, and dewaxing. The degumming step removes unwanted phosphorus content, while neutralization removes some of the FFAs. If FFAs are not removed, they will be converted into soap. The FFAs will also use up some of the catalysts, thereby reducing the conversion efficiency to produce biodiesel. The steps during the refining process will be described in the next section (Sections 5.7.1–5.7.6).

5.7.1 Degumming

Dry gums are unwanted phosphorous-related compounds in crude oil. Dry gums are removed during the transesterification process. Lecithin has the highest concentration in the gum. Lecithin is soluble in crude oil but becomes insoluble upon hydration or in the presence of appreciable amounts of water. The usual procedure to remove these gums is to add phosphoric acid or citric acid to the water. This also enhances the rehydration rate. After mixing and with enough contact time to hydrate the gums, the gums are separated with the use of a centrifuge or through a simple filtering process. There are various filter presses that may be used for this purpose.

The ideal degumming process is one that effectively controls the amount of water and acid while maintaining the appropriate temperature. The length of contact time is also important to achieve the maximum extent of degumming while minimizing losses and improving product quality. Water degumming is perhaps the simplest degumming process. Normally, hot water between 60°C and 80°C is used with about 1%–3% of water used by volume of oil. However, some organizations use an acid degumming process that includes a large fraction of acids. Examples of acid used for acid degumming process include malic acid or EDTA, aside from phosphoric and citric acid. The degumming efficiency is simply defined as the weight of oil after the degumming process divided by the initial oil weight used before the degumming process.

Other refining processes may use enzymes for degumming. Usually, the crude oil is treated with lipases to solubilize the triglycerides. Less waste is generated when using enzymes for degumming. However, there is additional cost when using an enzyme-based process.

Example 5.3: Degumming Efficiency Calculations

Water and phosphoric acid were used to degum crude sunflower oil. The following is the data gathered for this experiment. Determine the degumming efficiency by weight basis and the percent gum recovery by weight as well.

- Weight of oil used = 914.8 g
- Weight of oil after degumming process = 872.8 g
- Weight of gum recovered = 42 g

Solution:

a. The degumming efficiency is the ratio of oil recovered without gum divided by the amount of crude oil used at the start of the degumming process as shown below:

$$\%\text{ Degumming efficiency}(\%) = \frac{\text{Amt of degummed oil}}{\text{Initial amt of oil used}} \times 100\% = \frac{872.8}{914.8} \times 100\% = 95.41\%$$

b. The amount of gum extracted was 42 g, and the initial weight of the crude oil was 914.8 g. Thus, the percent gum recovered will be the ratio of these two values as shown below:

$$\%\text{ Gum Recovered} = \frac{\text{Amt of gum (g)}}{\text{Amt of oil (g)}} \times 100\% = \frac{42}{914.8} \times 100\% = 4.59\%$$

5.7.2 Neutralization

Crude vegetable oils contain FFAs from the natural hydrolysis of triglycerides. These FFAs are normally removed by the addition of an aqueous sodium hydroxide solution. This solution will convert the fatty acids into water-soluble salts. The soap is then removed by draining or centrifugation. After the solution is removed, the oil is washed once or twice with water to ensure the removal of all traces of soap then dried.

Used oil from restaurants, as feedstock, has the highest amount of FFAs. Its high FFAs are because of the oil undergoing processing at elevated temperatures. The fatty acids are broken down into its fragments and become "free" from the glycerol bond. It is a common practice to first determine the FFA concentration when using oils from restaurants.

As mentioned earlier, these FFAs may interfere in esterification by reacting with the alkaline catalysts before the reaction can proceed. Thus, to determine how much lye (or sodium hydroxide) is required to neutralize the FFAs (beyond that required for the transesterification process), a simple acid value determination is made. The experiment is designed to determine the amount (in milligrams of sodium or potassium hydroxide) that is required to neutralize the FFA present in 1 g of sample. The laboratory experimental process is a simple base-to-acid titration; the acid value is calculated from Equation 5.1 if KOH is used.

$$\text{Acid value} = \frac{V \times M \times 56.1}{S} \tag{5.1}$$

where

56.1 = molar mass of titrant (KOH)

V = volume of KOH solution required for the titration, mL

M = molarity of KOH solutions used

S = mass of sample

The percent FFA is then calculated from Equation 5.2.

$$\%\text{ FFA} = \frac{V \times M \times M_A \times 100}{1000 \times S} \tag{5.2}$$

where

M_A = molar mass of the acid chosen for expression of the result

In this exercise, a solvent mixture is also required. This is made by mixing equal volumes of ethanol and diethyl ether and neutralizing before use by adding KOH solution in the presence of a phenolphthalein indicator. The end point of the titration is the appearance of a permanent pink color. Example 5.4 shows simple calculations for determining percent FFA in oils.

Example 5.4: Determining Percent FFA in Oils

About 10 g of crude oil with a high percentage of oleic acid ($C_{18}H_{34}O_2$: MW = 282) was used to measure percent FFA. The following data was gathered during the test:

- Volume of KOH used during titration = 17.7 mL
- Molarity of KOH used = 0.1

Determine the percent FFA for this sample and the acid value.

Solution:

a. The percent FFA is calculated using Equation 5.2:

$$\%\text{FFA} = \frac{17.7 \times 0.1 \times 282 \times 100}{1000 \times 10} = 5.0$$

b. The acid number is calculated from Equation 5.1:

$$\text{Acid value} = \frac{17.7 \times 0.1 \times 56.1}{10} = 9.9$$

c. Because this crude oil has other fatty acids present (other than oleic acid), there will be minor but acceptable errors in calculations. The true value is when all FFAs are identified and calculations are made for each FFA component.

5.7.3 Dewaxing

Among the oil crops, sunflower and corn oils contain relatively large amounts of waxes that must be removed so that the oil does not cloud when brought to lower temperatures. The oil is normally cooled to 5°C, mixed with water containing a surfactant, and allowed to stand for several hours. When neutralization is done at this temperature, the soaps wet the wax crystals and bring them into suspension in the aqueous phase.

Alternatively, sodium lauryl sulfate is added to the water to serve the same purpose. The aqueous wax suspension is removed by centrifugation or filtration. In the biodiesel industry, a small amount of acid is commonly used (citric or ascorbic) and fed in the hydration tank. After a given time of retention, some caustic material is added to further neutralize the oil and the soap is removed. The neutral oil is then subjected to the chiller for a few hours to crystallize the wax, along with whatever soap that remains. All wax is then separated from the oil. The next step is bleaching. Dewaxing efficiency is defined as weight of dewaxed oil versus the initial oil weight before dewaxing. Example 5.5 shows the calculation for dewaxing efficiency.

Example 5.5: Efficiency of Dewaxing Calculations

The degummed and neutralized crude sunflower oil was used to determine the dewaxing efficiency and the percent wax recovered by placing the oil in a freezer at −5°C after treatment with sodium hydroxide and water. The following was the data gathered for this experiment. Determine the dewaxing efficiency by weight basis and the percent wax recovery by weight.

- Weight of oil used for dewaxing process = 908.3 g
- Weight of oil after dewaxing process = 785.6 g
- Weight of wax = 122 g

Solution:

a. The dewaxing efficiency is the ratio of oil recovered without wax divided by the amount of crude oil used at the start of the dewaxing process as shown below:

$$\% \text{ Dewaxing efficiency } (\%) = \frac{\text{Amt of dewaxed oil}}{\text{Initial amt of oil used}} \times 100\%$$

$$= \frac{785.6}{908.3} \times 100\% = 86.49\%$$

b. The amount of wax recovered was 122 g, and the initial weight of the crude oil was 908.3. g. Thus, the percent wax recovered will be the ratio of these two values as shown below:

$$\% \text{ Wax Recovered} = \frac{\text{Amt of wax } (\text{g})}{\text{Amt of oil } (\text{g})} \times 100\% = \frac{122}{908.3} \times 100\% = 13.43\%$$

5.7.4 Bleaching

Bleaching removes colored substances and oxidation products. The materials used for bleaching oils are natural clays (bentonites) treated with acids to improve their absorptivityand filterability. Proper bleaching with acid-activated earth is one of the most critical steps in oil processing. When done correctly, the oil emerging from the bleaching press is nearly colorless, with a peroxide value of zero.

The bleaching earth must be completely removed because clay acts as a very strong oxidation agent that can foul the downstream processing equipment. Bleaching earth must be accomplished by a good filtration facility. The recovered bleached oil should also be protected from thermal and oxidative abuses, as the oil in this condition is in its least stable state. Inert and inexpensive gases such as nitrogen are normally used to fill the headspace of tanks of recently bleached oils to exclude oxygen from the system.

The purpose of bleaching is not only to remove the coloring compounds (carotene, chlorophyll) by adsorption to provide a lighter color but also to remove peroxides and secondary oxidation products from the oil.

There are several types of processes for the bleaching, which are as follows:

- Batch atmospheric: In the batch atmospheric process, a temperature of about 104°C (or 220°F) is used. The bleaching earth is added from the top of the tank with the agitator running. The temperature is then raised to the appropriate bleaching temperature. Finally, the oil is circulated through a filter press.
- Batch vacuum: In the batch vacuum process, a temperature of about 104°C (or 220°F) is commonly used. First, slurry is made in a slurry tank and transferred to the vacuum bleacher. After the prescribed time at the bleaching temperature under vacuum, the oil is cooled, the vacuum is broken, and the oil is filtered.
- Continuous vacuum: In the continuous vacuum process, bleaching clay is continuously fed into a stream of oil at 104°C (or 220°F). This mixture is sprayed into a vacuum chamber to remove both water and air from the clay and the oil. After the temperature is raised, the oil is sprayed into another chamber for bleaching. Finally, the oil is cooled and the vacuum is broken.

5.7.5 Deodorization

Deodorization is usually the last step in the edible oil refining process. Basically, it is a vacuum–steam distillation process at an elevated temperature during which FFAs and volatile components are removed to obtain bland and odorless oil.

To remove excess FFAs, the deodorizing temperature is usually raised 250°C (482°F). However, this action results in higher production of trans fatty acids (~1%–3%) and a higher loss of tocopherols. Deodorization also removes odor- and flavor-causing compounds. These compounds, which are relatively volatile, can be removed by bubbling live steam through the hot oil (200–275°C) under high vacuum (3–10 mm/Hg). Deodorization does not have any significant effect on the fatty acid composition of most fats and oils. This process is not a necessary step in biodiesel production. However, numerous companies deodorize their oils to improve their shelf life.

Deodorization also removes FFAs, however, the live steam also causes some hydrolysis. The optimal conditions are as follows: 0.02–0.04 FFAs and 0.3%–0.5% monoglycerides. The oil at this step is referred to as refined, bleached, and deodorized (RBD) oil. Figure 5.4 shows the different stages of an oil refining process for various oil crops such as (a) safflower, (b) peanut oil, and (c) sunflower oil. The yield from oil refining varies with oil crops. Refined oil yields may range from 50% to 70%, depending on the variety and seed type. Some oils, such as peanut, are noted for their high wax content.

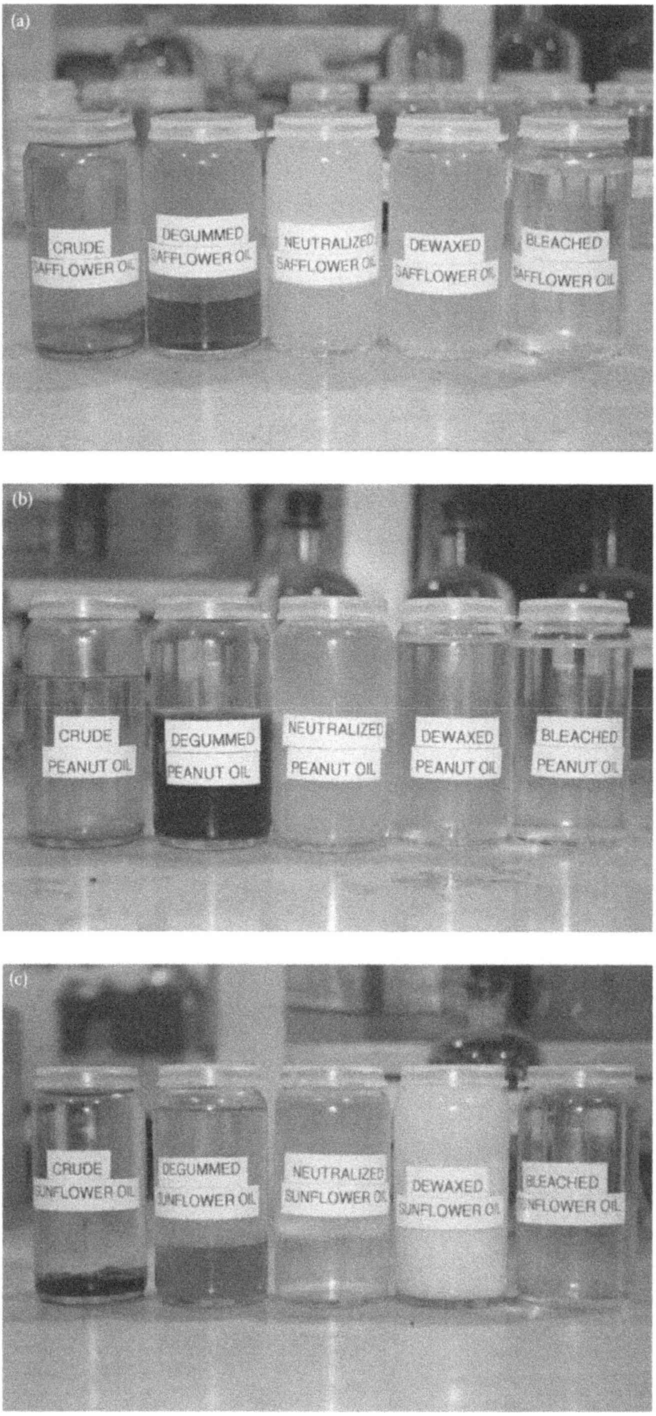

FIGURE 5.4 Refining processes for some oils showing stages of refining.

Various biodiesel producers are developing criteria for refining and have recommended pretreatment criteria. One set of criteria was recommended by Desmet Ballestra (Willits, 2007):

1. FFA less than 0.10% (1,000 ppm). Note that 1 mol of FFA destroys 1 mol of catalyst and creates 1 mol of soap.
2. H_2O less than 0.05% (500 ppm). Also note that 1 mol of H_2O creates 1 mol of FFA, which destroys 1 mol of catalyst and creates 1 mol of soap.
3. Unsaponifiables less than 1%. These include cholesterol, sterols, polymers, etc.
4. Soap less than 0.05% (500 ppm). Too much soap increases glycerin salt content.
5. Phosphorus less than 20 ppm. Two-thirds of phosphorus partitions to the glycerin stream and one-third remains in the biodiesel stream.
6. Sulfur less than 15 ppm.
7. Impurities (sediment) less than 0.10% (1,000 ppm).
8. Breakdown "fatty acid–sterol–glucose" molecules.

5.7.6 Vegetable Oil Refining Calculations

Various types of fresh oil seeds were processed at the BioEnergy Testing and Analysis Laboratory at Texas A&M University, and the results of the refining processes are summarized in Table 5.6. This table will be used to estimate refining efficiencies as illustrated in the following Example 5.6. Safflower (*Carthamus tinctorius* L), sunflower (*Helianthus annus*), and peanut (*Arachis hypogaea*) oil seeds were processed to extract the oil and then refined to arrive at an RBD oil. Around 1,000 g of freshly extracted oils were used in each case. The process started with degumming, neutralization, dewaxing, and bleaching (removal of oxidation products) to arrive at the various efficiencies as shown in the example. The final weight of the resulting product was then measured, and the removal efficiencies were calculated as the ratio of the final output divided by the original weight of the product that was used for the test, then subtracted from 100%. The refined sunflower oil yield was around 68%. For other oil seeds, their yield may vary from a low of 50% to over 70%. Figure 5.5 shows the results of the laboratory experiments for the oil refining processes.

TABLE 5.6
Results of Refining Studies for Three Types of Oil Seeds Showing Initial and Final Weights after Each Refining Step

	Sunflower Oil		Safflower Oil		Peanut Oil	
Refining Process	Initial (g)	Final (g)	Initial (g)	Final (g)	Initial (g)	Final (g)
Degumming	1,000	958.2	1,000	972.5	1,000	981.9
Neutralization	1,000	937.1	1,000	902.7	1,000	873.5
Dewaxing	1,000	867.0	1,000	896.8	1,000	818.7
Bleaching	1,000	916.2	1,000	930.3	1,000	868.8

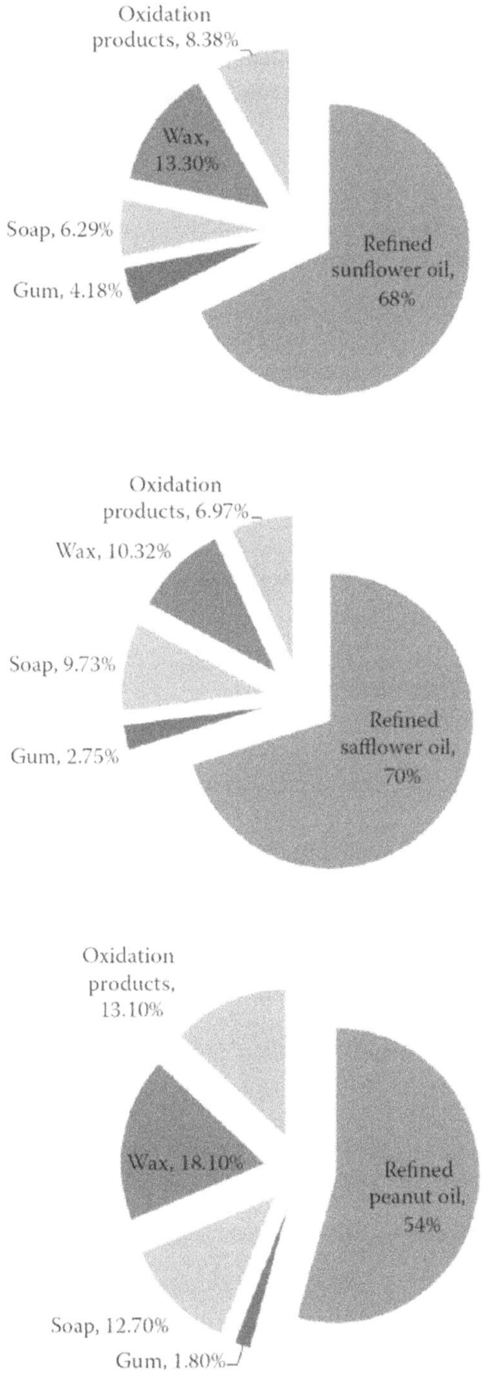

FIGURE 5.5 Various refining process efficiencies from three types of oil seed.

Example 5.6: Refining Efficiency Calculations

Use Table 5.6 to estimate the overall refining efficiency for sunflower oil.

Plot your results to show the refining efficiency for this type of oil.

Solution:

a. The degumming efficiency is calculated as the ratio of oil recovered without wax divided by the amount of crude oil used at the start of the dewaxing process as previously shown in Example 5.3 and reiterated here:

$$\%\ \text{Degumming efficiency} = \frac{958.2}{1000} \times 100\% = 95.82\%$$

$$\%\ \text{Gum Recovered} = \left(\frac{1000 - 958.2}{1000}\right) \times 100\% = \frac{41.8}{1000} \times 100\% = 4.18\%$$

b. The amount of soap is measured during the neutralization step and calculated as follows:

$$\%\ \text{Soap Recovered} = 100\% - \frac{937.1}{1000} \times 100\% = 6.29\%$$

c. The dewaxing efficiency is calculated accordingly as follows:

$$\%\ \text{Dewaxing Efficiency} = \frac{867}{1000} \times 100\% = 86.7\%$$

d. Finally, the amount of oxidation products are removed after the bleaching step and calculated as follows:

$$\%\ \text{Oxidation Products Recovered} = 100\% - \frac{916.2}{1000} \times 100\% = 8.38\%$$

e. Thus, the amount of refined oil recovered is found adding all those components that are removed; then, subtracting from 100% as follows:

$$\begin{aligned} \text{Refined oil} &= 100\% - (\text{degumming}\ \% - \text{neutralization}\ \% \\ &\quad -\text{dewaxing}\ \% - \text{bleaching}\ \%) = 100 \\ &\quad - (4.18 - 6.29 - 13.3 – 8.38) = 67.9\% \end{aligned}$$

f. When the calculations are done for all the other oils, one will be able to come up with those values shown in Figure 5.9. There are slight changes in the values because of rounding-off errors.

5.8 TRANSESTERIFICATION

In very simplistic terms, the transesterification process is illustrated by the schematic shown in Figure 5.6. Many commercial facilities begin the transesterification process with an RBD. Many experts in the field recommend pretreatment of the oil before making biodiesel rather than making biodiesel and then removing the unwanted products. Some commercial systems split the transesterification process into several stages. Thus, the methanol and catalysts are not used up in one processing stage. Some biodiesel facilities use a two-step process by splitting the methanol and catalysts into 60:40 parts or into 50:50 parts. Those producers believe that this approach improves the process efficiency.

The simple biodiesel process balance is as follows (see Figure 5.7): for every 100 kg of RBD oil used, about 10 kg of methanol is used together with the catalyst, resulting in the production of 100 kg of biodiesel and 10 kg of glycerin. This mass balance has been used by biodiesel manufacturers and is achievable within a commercial facility. The volumetric balance is slightly different because the density of the refined vegetable oil is slightly higher than the density of the biodiesel product (see Example 5.7).

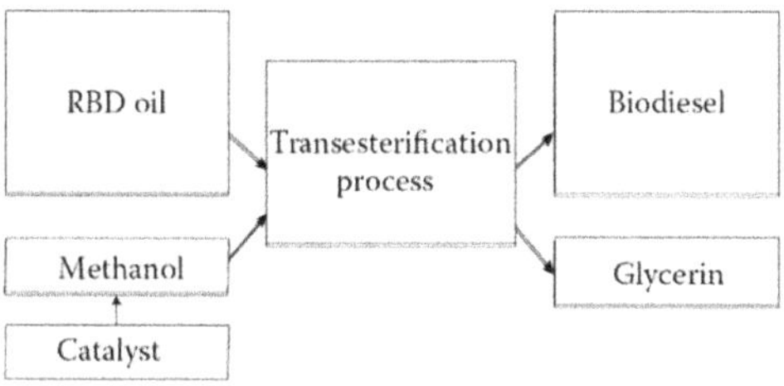

FIGURE 5.6 Schematic of a simple oil transesterification process.

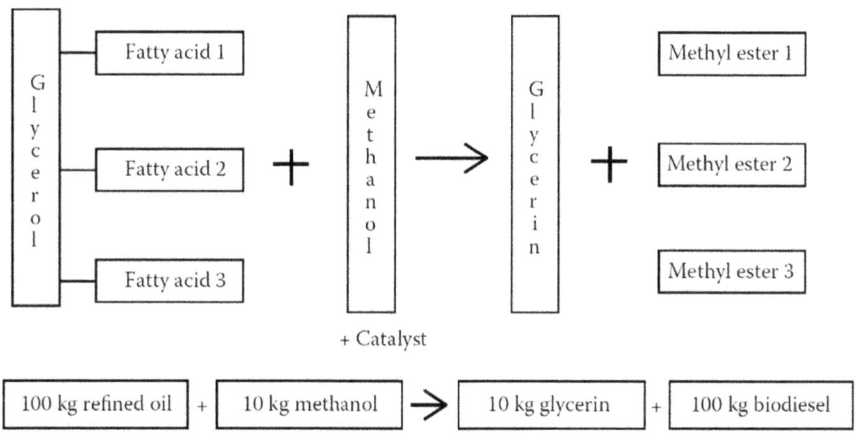

FIGURE 5.7 Mass balance in transesterification process.

Example 5.7: Transesterification Process Mass Balance

How much glycerin (in tonnes and tons and in million liters or gallons) per year will be produced by a biodiesel plant whose output is 11.355 million liters per year (3 MGY) of biodiesel? Assume the density of glycerin is 560 kg/m³ (4.66 lb/gal) and that of biodiesel is 877 kg/m³ (7.3 lb/gal). Use Figure 5.12 for mass balance. Convert the units to the English system.

Solution:

a. The yield of glycerin is 10% by weight; thus, the 11.355 ML/yr (3 MGY) facility would yield the following:

$$\text{Biodiesel (tonnes)} = \frac{11.355\times10^6\ \text{L}}{\text{yr}}\times\frac{877\ \text{kg}}{\text{m}^3}$$

$$\times\frac{\text{m}^3}{1{,}000\ \text{L}}\times\frac{\text{tonne}}{1{,}000\ \text{kg}} = 9{,}958\ \text{tonnes}$$

$$\text{Glycerin (L)} = 995.8\ \text{tonnes}\times\frac{\text{m}^3}{560\ \text{kg}}\times\frac{1{,}000\ \text{kg}}{\text{tonne}}\times\frac{1{,}000\ \text{L}}{1\ \text{m}^3} = 1.8\times10^6\ \text{L}$$

b. Converting the units to English system is shown next.

$$\text{Biodiesel (tons)} = \frac{3\times10^6\ \text{gal}}{\text{yr}}\times\frac{7.3\ \text{lb}}{\text{gal}}\times\frac{\text{ton}}{2{,}000\ \text{lb}} = 10{,}950\ \text{tons}$$

$$\text{Glycerin (gal)} = 10{,}950\ \text{tons}\times\frac{\text{gal}}{4.66\ \text{lb}}\times\frac{2{,}000\ \text{lb}}{\text{ton}} = 4.7\times10^6\ \text{gal}$$

5.9 ASTM CHARACTERIZATION

Biodiesel that is sold commercially or that is to be blended with diesel must conform to the ASTM D 6751–07 (Standard Specification for Biodiesel Fuel Blend Stock [B100] for Middle Distillate Fuels). This is an important fuel standard for biodiesel manufacturers and is summarized in Table 5.7. The standard is a collection of individual standards that must be met before selling or blending the product. There are more than 20 parameters listed; each parameter has its own ASTM standard analysis method. In many biodiesel plants, not all parameters are analyzed. Depending on the feedstock used, analysis is done for only those that may affect the engine usage and for those that may vary quickly. In the next section (Sections 5.9.1–Section 5.9.6), the most common analyses will be discussed. For detailed information about test methods, it is suggested that the complete standards, either ASTM or European Union (EN), be consulted.

TABLE 5.7
ASTM D6751–07: Standards Specification for Biodiesel Fuel

Property	Test Method	Grade 15 Limits	Grade S500 Limits
Calcium and magnesium	EN 14538	5 ppm (max)	5 ppm (max)
Flash point (closed cup)	D 93	130°C (min)	130°C (min)
Water and sediment	D 2709	0.050% (max)	0.050% (max)
Kinematic viscosity, 40°C	D 445	1.9–6.0 mm^2/s	1.9–6.0 mm^2/s
Sulfated ash	D 874	0.20% (max)	0.20% (max)
Sulfur (max)	D 5453	0.0015 ppm	0.05 ppm
Copper strip corrosion	D 130	No. 3 (max)	No. 3 (max)
Cetane number	D 613	47 (min)	47 (min)
Cloud point	D 2500	Report in °C	Report in °C
Carbon residue	D 4530	0.50% (max)	0.50% (max)
Acid number (max)	D 664	0.50 mg KOH/g	0.50 mg KOH/g
Free glycerin	D 6584	0.020% mass	0.020% mass
Total glycerin	D 6584	0.240% (max)	0.240% (max)
Phosphorus content	D 4951	0.001% (max)	0.001% (max)
Distillation temperature	D 1160	360°C (max)	360°C (max)
Sodium and potassium	EN 14538	5 ppm (max)	5 ppm (max)
Oxidation stability	EN 14112	3 h (min)	3 h (min)

5.9.1 Flash Point

The ASTM procedure for measuring the flash point of biodiesel is ASTM D 93 (Standard Test Method for Flash Point by Pensky–Martens Closed Cup Tester), which is also equivalent to the European standard EN ISO 3679. The flash point parameter is used to indicate the residual methanol content in the biodiesel. However, one must be careful that the water content in the biodiesel does not mask a low flash point value. This parameter is also important during the storage, handling, and transport of biodiesel, as well as defining transportation flammable rating.

There are various manufacturers that develop flash point analyzers. It is not the intent of this textbook to recommend a certain manufacturer for specific equipment to use. Although prices vary, one of the most popular units is the Pensky–Martens Flash Point Analyzer, manufactured by Koehler Instrument Company (Bohemia, NY).

A sample of biodiesel is normally placed in a brass test cup of specified dimensions, filled to the inside mark with test specimen, and fitted with a cover of specified dimensions. The sample is heated while the specimen is stirred at specific times and rates. An ignition source is directed into the test cup at regular intervals with a simultaneous interruption of stirring, until a flash is detected. Flash point is defined as the lowest temperature at which application of an ignition source causes the vapors of a specimen sample to ignite, corrected to a barometric pressure of 101.3 kPa (760 mm/Hg or 1 atm) and under the specified conditions of the test. This temperature is then reported as the flash point of the fuel. Several trials must be made to obtain an accurate flash point value.

The experiment must be conducted in a room with no draft or air movement so that the reading is not affected. Light must also be dimmed to see the flash clearly. The ASTM standard for flash point is 130°C minimum. There must also be proper room ventilation to carry the exhaust fumes from the heater to outside of the laboratory premises. The flash point temperature is one measure of the tendency of the test specimen to form a flammable mixture with air under controlled laboratory conditions. Flash point is only one property that must be considered in assessing the overall flammability hazard of a material. The US Department of Transportation and US Department of Labor (OSHA) have established that liquids with a flash point under 37.8°C (100°F) are flammable, as determined by these test methods.

5.9.2 Kinematic Viscosity

The ASTM procedure for measuring the kinematic viscosity of biodiesel is ASTM D 445 (Standard Test Method for Kinematic Viscosity of Transparent and Opaque Liquids [and Calculation of Dynamic Viscosity]), also equivalent to the European standard EN ISO 3104. The viscosity parameter is used to indicate how well the biodiesel would flow in conveying equipment, as well as in engine delivery lines. The parameter is determined by measuring the time required for a volume of liquid to flow under gravity through a calibrated glass capillary viscometer. Low viscosity causes injector leakage and loss of power to the injector pump. High viscosity leads to insufficient fuel flow through the injector. Viscosity also affects the behavior of injector lubrication and fuel atomization. This parameter is also used as an indicator of potential cold-flow properties. There are various distributors for kinematic viscosity facilities. There are numerous instruments available to measure viscosity of liquids. Manufacturers should provide certification that their instrument conforms to the above standards. The unit must be calibrated for various types of glassware used for the test.

One must only use calibrated viscometers of the glass capillary type that are capable of being used to determine the kinematic viscosity and are within the limits of precision provided in the standards. These viscometers must be mounted in the constant temperature bath. The bath temperature must not vary by more than ±0.02°C. An accurate amount of sample is placed in the glass capillary tube and then placed in the middle of the bath. The suction tube on top of the analyzer is used to bring the top level (meniscus) of sample in the start line indicator in the glass capillary and released while simultaneously activating the timer. The time stops when the oil level reaches the marked end line. The test should be performed at least twice and the results averaged. The timing device must be capable of taking readings with a discrimination of 0.1 s or better and should have accuracy within ±0.07% of the reading when tested over the minimum and maximum intervals expected flow times. Calculate each of the determined kinematic viscosity values, ν_1 and ν_2, from the measured flow times, t_1 and t_2, and the viscosity constant, C, using Equation 5.3.

$$\nu_{1,2} = C \times t_{1,2} \tag{5.3}$$

where

$\nu_{1,2}$ = determined kinematic viscosity values for ν_1 and ν_2, respectively, mm^2/s

C = calibration constant of the viscometer, mm^2/s^2

$t_{1,2}$ = measured flow times for t_1 and t_2, respectively, s

Calculate the kinematic viscosity result, ν, as an average of ν_1 and ν_2. Calculate the dynamic viscosity, η, from the calculated kinematic viscosity, ν, and the density, ρ, by means of Equation 5.4.

$$\eta = \nu \times \rho \times 10^{-3} \tag{5.4}$$

where

η = dynamic viscosity, mPa-s

ρ = density, kg/m^3, at the same temperature used for the determination of kinematic viscosity

ν = kinematic viscosity, mm^2/s

The density of the sample can be determined at the test temperature of the kinematic viscosity. The determination is made by an appropriate method, such as Test Method D 1217 (Standard Test Method for Density and Relative Density [Specific Gravity] of Liquids by Bingham Pycnometer). The kinematic or dynamic viscosity report must be given in four significant figures together with the test temperature. It will also be important to note the type of equipment or model used for the testing, date of testing, and the name and address of the test laboratory as recommended in the standards.

Example 5.8: Kinematic Viscosity Calculations

A biodiesel sample was tested for kinematic viscosity following ASTM D445 (Standard Test Method for Kinematic Viscosity of Transparent and Opaque Liquids); the measured flow times t_1 and t_2 for a liquid (density 895 kg/m^3) are 803 and 799 s, respectively. Assume the calibration constant for a viscometer at 40°C is 0.008198 mm^2/s^2. (a) What is the kinematic viscosity of the liquid? (b) What is its dynamic viscosity? (c) Will it pass the ASTM 6751 standard?

Solution:

a. Calculate each of the determined kinematic viscosity values, $\nu_1$and ν_2, from the measured flow times, t_1 and t_2, and the viscometer constant, C, by means of the following equations:

$$\nu_{1,2} = C \times t_{1,2}$$

$$\nu_1 = \frac{0.008189\,mm^2}{s^2} \times 803\,s = 6.576\,cSt\,or\,mm^2/s$$

$$\nu_2 = \frac{0.008189\ mm^2}{s^2} \times 799\ s = 6.543\ cSt\ or\ mm^2/s$$

$$v = \frac{6.576 + 6.543}{2} = 6.560\,\text{cSt or mm}^2/\text{s}$$

b. Calculate the dynamic viscosity, η, from the calculated kinematic viscosity, ν, and the density, ρ, by means of the following equations:

$$\eta = v \times \rho \times 10^{-3}$$

$$\eta = 6.56 \frac{\text{mm}^2}{\text{s}} \times 895 \frac{\text{kg}}{\text{m}^3} \times 10^{-3} = 5.871\,\text{mPa} - \text{s} = 5.871\,\text{cSt}$$

c. This biodiesel sample will not pass the ASTM 6751 standard. The value should be within 1.9–6.0 cSt to be categorized as Grade S15 liquid fuel.

5.9.3 Cloud Point

Cloud point measurements should conform to ASTM D 2500 (Standard Test Method for Cloud Point of Petroleum Products). This test method covers onlypetroleum products and biodiesel fuels that are transparent in layers 40 mm in thickness and with a cloud point below 49°C. Cloud point is defined as the temperature of a liquid specimen when the smallest observable cluster of hydrocarbon crystals (first) occurs upon cooling under prescribed conditions. The cloud point parameter indicates the temperature of crystal formation. To many observers, the cluster of wax crystals looks like a patch of whitish or milky cloud, hence the name of the test method. The cloud appears when the temperature of the specimen is low enough to cause wax crystals to form.

The biodiesel specimen is simply put in a test tube vial and placed in a cooled bath as shown. Dry ice or ice from water may be used. However, water has a limit of freezing temperature while most cloud points are well below freezing, which is why dry ice is necessary. The temperature at which a cloud is first observed at the bottom of the test jar is recorded as the cloud point. Several trials should be done and the results averaged. There is no limit for the cloud point in biodiesel; the standard calls for simply reporting the value of the cloud point to the nearest 1°C.

For petroleum products and biodiesel fuels, cloud point is an index of the lowest temperature of their utility for certain applications. Some states in northern locales have low ambient temperatures during winter times of the year; the fuel should not solidify during these low-temperature conditions.

5.9.4 Water and Sediment

Water and sediment are important biodiesel parameters for biodiesel storage. Note that biodiesel is a solvent. Thus, when biodiesel is placed in a container with some dirt or sludge, it will remove the dirt or sludge or allow them to settle at the bottom, resulting in sediment formation. In time, because of deterioration, some water may also form. The limit for water and sediment is below 0.05% for either type of biodiesel (Grade 15 and Grade S500). This is a very low limit and an important parameter to measure routinely. The standard procedure for measuring water and sediment in biodiesel is the ASTM standard D2709 (Standard Test Method for Water and Sediment in Middle Distillate Fuels by Centrifuge). The test requires a 100 mL sample of freshly made biodiesel. The sample is put in a centrifuge tube that is readable to 0.005 mL and measurable to 0.01 mL. The tube is then placed in a centrifuge using a relative centrifugal force of 800 rpm for 10 min at 21°C–32°C (70°F–90°F). After centrifugation, the volume of water and sediment that has settled into the tip of the centrifuge is read to the nearest 0.005 mL and then reported as the volumetric percentage of water and sediment by centrifuge.

5.9.5 Total and Free Glycerin

The procedure for reporting total and free glycerin in a biodiesel sample is covered by ASTM D 6584 (Standard Test Method for Determination of Free and Total Glycerin in B100 Biodiesel Methyl Ester by Gas Chromatography). This test method covers the quantitative determination of free and total glycerin in B100 methyl ester sample by using a GC. The GC is a unit dedicated solely to the measurement of glycerin in a biodiesel sample and is configured to conform to the ASTM D 6584 standard (residual and bound glycerin in biodiesel). The analysis takes about 25 min and is simple enough for an untrained operator to perform in the laboratory or in the field. There are portable GCs (SRI GC, Torrance, CA) that costs less than $10,000. The unit includes software (PeakSimple Chromatography Data Systems) that displays the calibrated results directly without offline calculations.The operating conditions for the GC are summarized in Table 5.8. This table describes the injector type, the column oven temperature program to use, the type of detector, and the carrier gas flow rate.

TABLE 5.8
Operating Conditions of GC Used to Measure Total and Free Glycerin Samples in Biodiesel

Injector		
Cool on column injection		
Sample size	1 μL	
	Column Temperature Program	
Initial temperature	50°C	Hold 1 min
Rate 1	15°C/min to 180°C	
Rate 2	7°C/min to 230°C	
Rate 3	30°C/min 380°C	Hold 10 min
	Detector	
Type	Flame ionization detector	
Temperature	380°C	
	Carrier Gas	
Type	Hydrogen or helium	Measured at 50°C
Flow rate	3 mL/min	

A freshly made biodiesel sample is analyzed by the GC after silylating with *N*-methyl-*N*-trimethylsilytrifluoracetamide. Calibration must first be made using two internal standards and four reference materials. Mono-, di-, and triglycerides are determined by comparing to monooelin, diolein, and triolein standards, respectively. Average conversion factors are applied to the mono-, di-, and triglycerides to calculate the bonded glycerin content of the sample. Please refer to the standard methods for details of the analysis.

Free and bonded glycerin reflects the quality of biodiesel. A high content of glycerin may cause problems during storage, or in the fuel system, because of separation of the glycerin. High total glycerin content can lead to injector fouling and may also contribute to the formation of deposits at injection nozzles, pistons, and valves.

5.9.6 ASTM Analysis of Various Oil Seed Methyl Esters

Following standard methods discussed in this chapter, the complete ASTM analysis of various biodiesel samples processed at the BETA Lab is summarized in Table 5.9. Note that some of the biodiesel samples will not pass the ASTM standards, particularly kinematic viscosity for sunflower and peanut biodiesel. Thus, better refining and transesterification procedures must be conducted on these samples. Most of the other parameters are within the standards. Note the variations in the cloud point for various samples from below freezing to above freezing. The other samples not discussed in this chapter include SME, canola, or rapeseed methyl ester, and chicken fat methyl ester (CFME).

TABLE 5.9
Results of ASTM Analysis of Various Biodiesel Samples at BETA Lab

Property	Diesel	SME[a]	RME	SFME	SaffME	PME	CFME[b]
Flash point (°C)	128	199	171	192	197	190	131
Water and sediment (%vol)	<0.01	<0.01	<0.01	<0.01	<0.01	<0.05	<0.05
Kinematic viscosity,40°C (mm²/s)	2.26	4.69	4.79	6.33	5.54	7.0	4.69
Sulfur (ppm)	–	4	–	–	–	–	<0.001
Cetane number	–	55	–	–	–	–	48.2
Cloud point (°C)	35	−6	−5	2	2	15	7
Carbon residue (% mass)	–	0.009	–	–	–	–	0.02
Acid number (mg KOH/g)	0.04	0.19	0.20	-	0.68	0.13	0.14
Distillation temperature (°C)		329					352
Oxidation stability (h)	–	7.167	–	–	–	–	3.3
Gross heating value (Btu/lb)	18,994	17,242	17,144	16,970	16,962	17,426	–
Gross heating value (MJ/kg)	44.1	40.0	39.8	39.4	39.4	40.4	

[a] Analyzed at AAA Chemicals, Shoeacres, TX, except for gross heating value, which was analyzed at the BETA Lab.

[b] Analyzed at Precision Labs Inc., Houston, TX.

5.10 ENGINE PERFORMANCE AND EXHAUST EMISSIONS

The engine test facility used for testing biodiesel fuels will be the same as engines using conventional gasoline and diesel fuel. A common practice for engine testing is the use of a Society for Automotive Engineers (SAE) standard for testing, particularly the SAE J 1349 standard entitled "Engine Power Test Code—Spark Ignition and Compression Ignition—Net Power Rating." This standard is intended to provide the industry with a repeatable means of determining the as-installed net power and torque of either a gasoline or diesel engine, including specific fuel consumption.

Although the engine is being tested for performance, exhaust emissions may be recorded simultaneously. The important gaseous compounds in the exhaust are as follows: carbon monoxide (CO), oxides of nitrogen (NO_x), sulfur dioxide (SO_2), total hydrocarbon (THC), carbon soot, and particulate matter. An example of an emissions analyzer is an ENERAC™ Model 700 Emissions Analyzer (Holbrook, NY), capable of measuring CO and SO_2 concentrations of up to 4,500 ppm, NO_2 up to 1,500 ppm, NO up to 3,000 ppm with an accuracy of +1%. In addition, the analyzer also measures the ambient temperature, stack temperature, stack velocity, O_2 concentration in the test room, combustibles (up to 5%), hydrocarbons (up to 30,000 ppm), and CO_2 (0%–20%).

NO_x emissions using SME biodiesel are, on the average, higher than reference diesel, as shown in Figure 5.8. However, this is not true for all rpm and load conditions. An analysis of variance for NO_x concentrations was performed for B100 SME

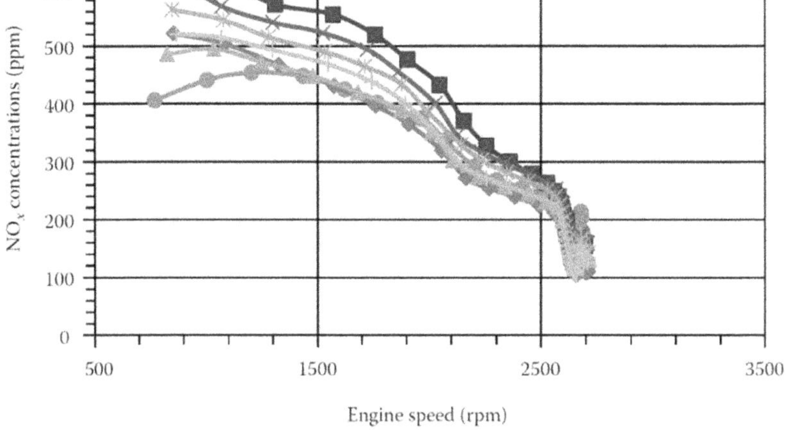

FIGURE 5.8 Typical NO_x emissions for a John Deere engine fueled by soybean oil biodiesel, blends, and reference diesel fuel.

and reference diesel. The results indicated that there are significant differences in NO_x concentrations between the fuels. In other exhaust gases, such as CO, the results indicated that as the amount of biodiesel in the blend was increased and the CO concentration levels decreased from 64% to 33% (Santos, 2009). The SO_2 emission level was significantly below the 15 ppm EPA limit. The THC levels for B50 and B100 were higher than for diesel fuel.

5.11 DESIGN OF BIODIESEL PLANTS IN THE UNITED STATES

5.11.1 General Transesterification Processes

Commercial biodiesel facilities are obviously quite different from laboratory-scale and small-scale biodiesel production plants. Designs vary on how transesterification processes are achieved. In laboratory-scale plants, a common practice is to use all refined oils in one batch as well as methanol and catalysts. In most commercial facilities, transesterification steps are carried out at various stages of a continuous process. Each stage of the continuous process will usually have multiple steps. The full volume of oil normally goes through the first stage, where only a fraction of methanol and catalysts are used. The glycerin produced in this stage is then removed either by settling or by centrifugation. Various other commercial biodiesel facilities break down the transesterification process into two or three stages. Plants will vary in how the glycerin is removed. More expensive plants would utilize centrifugation, whereas the less expensive plants would simply use gravity settling.

The following sections describe various basic designs of commercial plants. Most commercial plants are highly automated, with minimal manual intervention. Sampling of the products is conducted at various points to keep track of purity and completion of the conversion process.

5.11.2 Crown Iron Works Design

The Crown Iron Works design is perhaps one of the simplest, where refined oil is converted into biodiesel in two major stages. A heat exchanger is normally used to bring the conversion reaction temperature to a desired value by using steam at the first transesterification reactor. A decanter follows to remove the glycerin formed. The oil that is not reacted goes to the second stage; additional methanol and catalysts are added to complete the transesterification process. Some acids are then added to the biodiesel before it is placed in the second decanter and mixer and where the final glycerin formed is collected. Water washing begins after this stage, followed by removal of solids and recycling of water. The final stages remove excess methanol. Then, the temperature is lowered to remove any solids (normally some wax leftover). After being pumped into the final filtering system, the biodiesel is ultimately transferred to a storage facility. Figure 5.9 shows the simple schematic diagram of refined oil flow in the Crown Iron Works plant design.

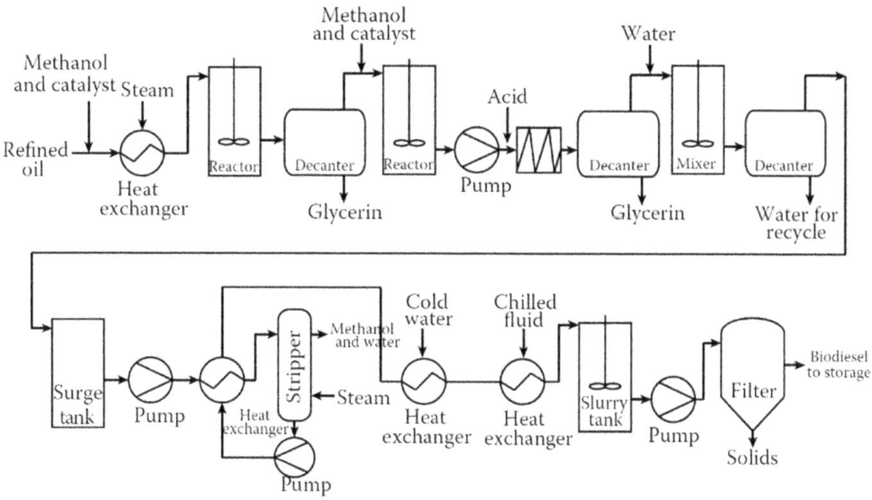

FIGURE 5.9 Crown Iron Works Inc. biodiesel plant design.

5.11.3 Westfalia Biodiesel Plant Design

The schematic of the Westfalia biodiesel plant design is shown in Figure 5.10. In this design, the transesterification reaction is also divided into two major stages. Glycerin is removed by centrifugation. Following two stages of water washing, the moisture is removed by a vacuum decanter system. There are several high shear mixers used (oval contraptions) to improve the transesterification reactions. The unit is fully automated with excellent control of the mass flow rate through the system. Steam is used to bring the temperature to the proper reaction level.

5.11.4 Hydrotreated Vegetable Oil

Hydrotreated vegetable oil (HVO) is a diesel-like fuel that can be produced without fossil resources by processing renewable waste lipids. HVO is sometimes referred to as green diesel. Examples of commercial facility manufacturing HVO include Neste NExBTL,Topsoe Hydroflex Technology, Axens Vegan Technology, H-Bio, ConocoPhilips process, and the UOP/Eni Ecorefining Process. These renewable diesel fuels derived from vegetable oil lower greenhouse gas emissions by 40%–90%, have higher energy content than petroleum diesel, no aromatics, and excellent cold-flow properties. They can be used 100% in diesel engines without modification. They comply with ASTM D7566-11a specifications for aviation turbine fuel.

HVO is produced by hydrogenation (addition of hydrogen) and hydrocracking of vegetable oils and animal fats using hydrogen and catalysts at high temperatures and pressures. In this hydrotreating process, oxygen is removed from the triglycerides and fatty acids. The resulting products consists of straight-chained hydrocarbons (paraffins) with varying properties and molecular size depending on the feedstock

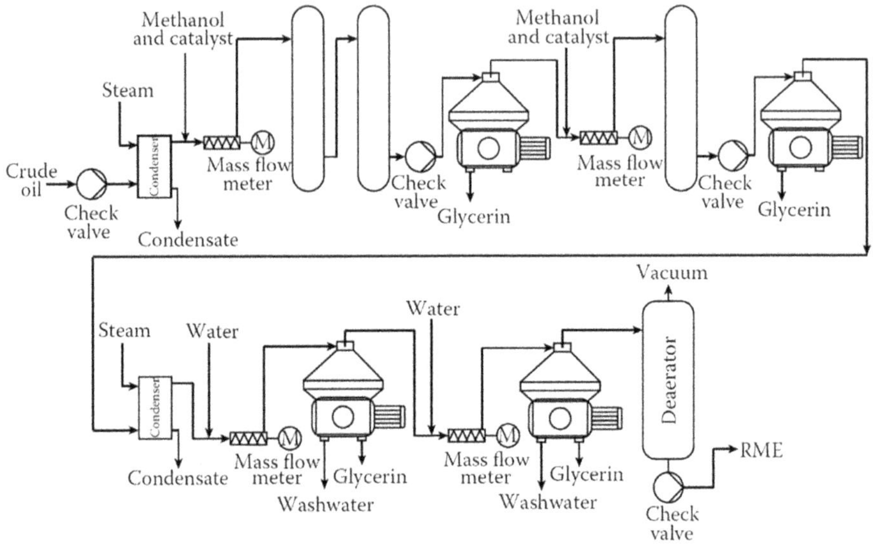

FIGURE 5.10 Wesfalia biodiesel plant design. (Under license of Archer Daniels Midland Company, ADM, Decatur, IL.)

characteristics and process conditions (ETIP Bioenergy, 2020). Hence, it is possible to generate lights, gasoline, aviation fuel, and diesel-like fuels. The hydrotreatment typically takes place between 300°C and 390°C. For treatment of triglycerides, propane is a typical product as shown in Equation 5.5.

$$C_3H_5(RCOO)_3 + 12H_2 \rightarrow C_3H_8 + 3RCH_3 + 6H_2O \tag{5.5}$$

Hydrogen is first added to the double binds in the renewable feedstock. Thereafter, more hydrogen is added to remove propane by cleaving the triglycerides to fatty acids. Lastly, the fatty acids are converted into hydrocarbons by hydro-deoxygenation (removing oxygen as water) and/or decarboxylation (removing oxygen as carbon dioxide). Thereafter, the hydrocarbons are converted into a quality that meets the end-user criteria. HVO can also be upgraded into sustainable aviation fuel. Compared with diesel, HVO has lower density at 20°C (0.78 vs 0.83 kg/L); higher heating value (44.4 vs 43.1 MJ/kg); higher cetane number (>80 vs 50); and lower GHG emissions (11.5–50 vs 88.7 gCO_{2eq}/MJ) (ETIP Bioenergy, 2020).

5.11.5 Economics of Biodiesel Facility

The feasibility of a bioenergy project may be evaluated using common economic parameters, such as net present value (NPV), benefit cost ratio (BCR), payback period (PBP), breakeven point (BEP) analysis, and internal rate of return (IRR). The analyses must note the relationship between time and the value of money. The basic equations for estimating the present and future value of investments are as follows:

$$\text{Present Value} = \text{PV} = \text{FV} \times \frac{1}{(1 + r)^n} \tag{5.6}$$

$$\text{Future Value} = \text{FV} = \text{PV} \times (1 + r)^n \tag{5.7}$$

where

r = rate of return or discount rate (decimal)
n = number of periods (unitless)
PV = present value of cash flow
FV = future value of cash flow

The IRR is a discounted rate that makes the NPV of all cash flows from a particular project equal to zero. The higher the IRR, the more economically desirable the project. The NPV (Equation 5.5–5.8) is the difference between the present value of cash inflows and the present value of cash outflows. A positive NPV means that the project earnings exceed anticipated costs. The BCR (Equation 5.5–5.9) is the ratio between the project benefits and costs. Values greater than 1 are desirable. The PBP (Equation 5.5–5.10) is the length of time required to recover the cost of investment.

$$\text{Net Present Value} = \text{NPV} = \sum_{n=1}^{N} \frac{\text{cash inflow}}{(1 + i)^n} - \text{cash outflowsw} \tag{5.8}$$

$$\text{Benefit Cost Ratio} = \text{BCR} = \frac{\text{project benefits}}{\text{project costs}} \tag{5.9}$$

$$\text{Payback Period} = \text{PBP (years)} = \frac{\text{project costs}}{\text{annual cash inflows}} \tag{5.10}$$

When estimating the fixed cost of a project, the major cost components are depreciation and interest on borrowed money. There are many ways to estimate the depreciation of a facility. The two most common and simple methods are straight-line depreciation (Equation 5.5–5.11), and the sum-of-years' digit depreciation method (SYD) (Equation 5.5–5.12).

$$\text{Straight Line Depreciation (\$)} = \frac{\text{Principal} - \text{Salvage Value}}{\text{LIfe of Unit}} \tag{5.11}$$

$$\text{SYD Depreciation (\$)} = \text{Depreciation Base} \times \frac{\text{Remaining Useful Lfe}}{\text{Sum of Years' Digits}} \tag{5.12}$$

In Equations 5.5–5.12, the depreciation base is the difference between the initial capital cost ($) and the salvage value of the asset ($). The sum of the years' digits is the sum series: 1, 2, 3, up to n, where n is the useful life of the asset in years, as shown in Equations 5.5–5.12:

$$\text{Sum of Years' Digits} = \text{SYD} = \frac{n(n+1)}{2} \tag{5.12}$$

The other large portion of capital cost is interest on borrowed money. This is usually the percentage (interest rate) charged by a bank based on the amount of the loan. The governing equation without including the salvage value (Equation 5.5–5.13) is like the amortization calculation for a loan amount:

$$\text{Annuity} = A = P \times \left(\frac{r \times (1+r)^n}{(1+r)^n - 1} \right) \tag{5.13}$$

where

A = annuity or payment amount per period ($)
P = initial principal or amount of loan ($)
r = interest rate per period (%)
n = total number of payments or period (unitless)

The above equations may be used in the case studies given at the end of this chapter.

5.12 CONCLUSION

In the United States and around the world, biodiesel comprises an increasing percentage of biofuels produced from biomass. Widespread research on the development of newer feedstock for oil production, such as algae, will pave the way for the resurgence of biofuels. Non-feedstock sources of oil will dominate the biodiesel feedstock in the future. Developing countries have embarked on production of *Jatropha* and *Pongamia* tropical plants. The biodiesel produced from these oil crops is highly compatiblewith conventional diesel fuel, a fuel that is very popular in developing countries.

The chemistry for making biodiesel is simple and quite easy to implement. However, steps before biodiesel production, as well as satisfaction of the ASTM standards for biodiesel, may affect the economics of this biofuel. For example, feedstock cost is a large percentage of the fuel cost—normally around 85%. Thus, if the price of crude oil is the same as diesel fuel, it will be very difficult for biodiesel manufacturers to achieve satisfactory profit margins.

Obtaining biodiesel from oil seed crops is a complex process of harvesting, drying, dehulling, and oil extraction. The resultant crude oil must then undergo refining to remove the phosphorus (degumming), neutralization (remove FFA), dewaxing, bleaching, and deodorization before one can begin making biodiesel. Methanol, even though it would require only 10% of the oil equivalent, may be scarce in some countries. These countries may resort to a much cheaper alcohol, such as ethanol. However, the water content must be virtually zero for this alcohol to be effective. Otherwise, the water in the ethanol may react with catalysts to form soap.

ASTM standards for biodiesel impose a 15-parameter gauntlet that must be traversed before biodiesel may be blended with diesel. A facility compliant with the complete range of ASTM requirements is both costly and difficult to maintain. Thus, commercial biodiesel facilities would simply send their samples for analysis to certified laboratories.

HVO is an emerging renewable fuel that may replace fossil-fuel-based diesel. The production of HVO is now well known and well-developed at industrial scale. The investment cost for HVO facilities are generally higher than that for biodiesel production plants discussed in this chapter. The hydrogen used to produce this fuel primarily comes from fossil fuel. Combustion of HVO diesel has lower NO_x emissions and reduced issues of poor cold properties (e.g., flow properties), storage stability, and fuel aging unlike biodiesel (ETIP Bioenergy, 2020).

The combustion of biodiesel in engines may also have some emission issues, such as NO_x. For biodiesel to achieve widespread use, these issues must be addressed. Finally, one should also consider the fact that biodiesel has slightly lower energy content in both mass and volume than diesel.

5.13 PROBLEMS AND DISCUSSION ISSUES

5.13.1 Various Types of Fatty Acids

P5.1 Differentiate the following fatty acids and provide an example for each: saturated, unsaturated, and polyunsaturated.

5.13.2 Refined, Bleached and Deodorized (RBD) Oil

P5.2 What is the meaning of RBD oil? Do you think that what the letters B and D stand for are necessary for technical performance of an internal combustion engine?

5.13.3 Transesterification Process

P5.3 Why do the fatty acid esters (biodiesel) and glycerol products separate into two layers after the reaction is completed?

5.13.4 Oil Refining Processes

P5.4 In 2010, the United States produced 190.11 million pounds of crude peanut oil (NASS Crops Branch, (2012) 720–2127; ERS; and Bureau of the Census). Based on the pie chart given in Figure 5.5, how much refined peanut oil, in metric tonnes, can be produced after the RBD process? What amounts of by-products (i.e., wax, gum, soap, and oxidation products, also in metric tonnes) can be formed?

5.13.5 Acreage to Establish 100 MGY Biodiesel Plant

P5.5 Calculate the area (acres) required to produce oil feedstock that will satisfy the requirements of a 100 million gallon per year (100 MGY) biodiesel facility based on peanuts. It was reported that peanuts could generate 151.4 lbs of peanut oil per acre. Assume: for every 100 lbs of refined oil and 10 lbs of methanol and catalyst, about 100 lbs of biodiesel and 10 lbs of glycerin are produced. The density of crude peanut oil is 7.66 lb/gal and that of biodiesel is 7.418 lbs/gal. The biodiesel recovery is 54% by weight from raw peanut oil.

5.13.6 Refining Processes

P5.6 How important is bleaching in the vegetable oil refining process? Please list at least five important reasons for bleaching vegetable oils.

5.13.7 Refining Efficiency Calculations

P5.7 Use Table 5.6 to estimate the overall refining efficiency for safflower oil. Plot your results to show the refining efficiency for this type of oil.

5.13.8 ASTM Fuel Properties

P5.8 By following the procedure in ASTM D445 (Standard Test Method for Kinematic Viscosity of Transparent and Opaque Liquids), the measured flow times t_1 and t_2 for a liquid (density 880 kg/m^3) are 564 and 560 s, respectively. Assuming the calibration constant for a viscometer at 40°C is 0.008198 mm^2/s^2, (a) What is the kinematic viscosity of the liquid in units of centistokes (cST) or mm^2/s? (b) What is its dynamic viscosity in units of mPa-s? (c) Will it meet the ASTM 6751 standard?

5.13.9 Energy Density of Biodiesel

P5.9 Determine the energy density (MJ/L) of soybean oil biodiesel fuel (specific gravity 0.85) if it has a calorific value of 37.3 MJ/kg.

5.13.10 Biodiesel Production Processes

P5.10 Given the following conditions:

Weight of starting material (safflower oil)	1,000 g
Reaction time	1 h
Weight of catalyst (KOH)	0.4% wt. of initial weight of oil
Volume of methanol	10% vol. of oil
Reaction temperature	50°C

How much catalyst is required for biodiesel reaction to proceed? Assume: RBD oil has a total acid number of 2 mg KOH/g oil.

5.13.11 Major Fatty Acid in Vegetable Oil Types

P5.11 What is the major fatty acid in coconut oil?
a. lauric acid b. palmitic acid c. oleic acid d. erucic acid

5.13.12 Density of Diesel Fuel Versus Vegetable Oil

P5.12 Diesel fuel is heavier than vegetable oils per unit volume.
a. True b. False c. No idea d. It depends

5.13.13 Biodiesel with Lowest Flash Point

P5.13 Which of the following biodiesel has the lowest flash point?
a. CFME b. SME c. SaffME d. RME

5.13.14 Acreage for Biodiesel Feedstock

P5.14 Determine the acreage (hectares) required for a 3.785 million liters of biodiesel from mustard seed. Enumerate all assumptions made including yields of vegetable oil from mechanical extraction and transesterification processes.

5.13.15 Glycerin Production and Marketing

P5.15 Design a system to valorize glycerin co-product of biodiesel production from a 3.785 million liters biodiesel facility from mustard seed. Enumerate all assumptions made including target use of glycerin in pharmaceuticals. Propose a specific purification process to process glycerin at the highest purity level required by the pharmaceutical industry.

5.14 CASE STUDY 5.1. BIODIESEL ECONOMICS AND SUSTAINABILITY

Evaluate a project to produce 3.785 million liters per year (MLY) of biodiesel from soybean oil (soybean methyl ester, SME). The biodiesel and glycerin will be sold on a commercial scale. Overall revenue comes from the sale of biodiesel, glycerin, and tax credit. Develop a spreadsheet to perform simple economic and sensitivity analyses. Fixed costs include depreciation, interest on investment, and taxes and insurance. Variable costs include refined oil cost, processing cost, utilities cost, labor cost, and management cost. Data for the project is summarized in the following table:

Refined Soybean Oil Price	\$0.90/kg	Facility Cost	\$264,201/L
Processing Cost	\$0.06/L SME	Biodiesel Selling Price	\$1.0568/L
Yearly Operation	350 days	Glycerin Selling Price	\$0.60/kg
Financial Parameters		Other Variable Costs	
Project Life	20 years	Labor Hours	10 h/day
Interest on Investment	8%	# of personnel	6 full-time
Tax on Capital	2%	# of Manager	1 full-time
Discount Rate	2.36%	Labor Wage Rate	\$15/h
Tax Credit	30%	Manager Salary	\$80,000/pax/yr
Product Yields		R&M Cost	5% of capital
Ideal Biodiesel Yield	449 L/ha	Utilities Cost	10% of capital
Conversion Efficiency	76.4%	Biodiesel Conversion	0.89 kg/L
Glycerin Yield (Wt%)	10% of soybean oil	Soybean Oil Density	0.917 kg/L
Yield of Oil from Seed	20%	SME Density	0.89 kg/L
Oil Refining Efficiency	68%	Soybean Seed Yield	2245 kg/ha

Required:

a. Determine the acreage required for the operation assuming an ideal biodiesel yield of 449 L/ha with biodiesel conversion efficiency of 76.45.
b. What is the total weight of biodiesel required by weight per year (tonnes/yr), weight of refined oil required (tonnes/yr), and weight of glycerin produced assuming a mass balance given in textbook (i.e. 100 kg refined oil + 10 kg catalyst → 100 kg biodiesel + 10 kg glycerin?
c. Report also the total weight of raw soybean oil required (tonnes/yr), weight of soybean seeds required (tonnes/yr).
d. Develop a simple spreadsheet to estimate economic parameters such as internal rate of return (IRR), payback period (PBP), benefit cost ratio (BCR), net present value (NPV) and report on the viability of this project given the data.
e. Perform sensitivity analyses to determine the impact of changing input parameters on the economic indicators. A graph for relationship would be beneficial visually with predictive equations and with R^2 value in the chart. The following are suggested changes:
   i. Acreage on NPV (3.785 MLY to 9.4625 MLY)
   ii. Biodiesel selling price versus IRR and PBP (\$1.0568/L to \$1.255/L)
   iii. Glycerin selling price versus IRR and PBP (\$0.21/kg to \$1.0/kg)
   iv. Tax credit on IRR and PBP (20% to 40%)
   v. Capital cost on IRR and PBP (\$132,100/MLY to \$660,502/MLY)
   vi. Interest on investment on IRR and PBP (5% to 25%)
   vii. Feedstock cost on IRR and PBP (\$0.75/kg to \$1.0/kg)
   viii. Processing cost on IRR and PBP (\$0.026/L to \$0.159/L)

REFERENCES

European Technology and Innovation Platform (ETIP) Bioenergy. 2020. Hydrogenated vegetable oil (HVO) Fact Sheet. Brussels-Capital Region, Belgium. Available at: https://etip-bioenergy.eu/images/ETIP_B_Factsheet_HVO_feb2020.pdf. Accessed 27 September 2022.

Firestone, D. 2006. *Physical and Chemical Characteristics of Oils, Fats and Waxes*. American Oil Chemist Society (AOCS) Press, Urbana, IL.

NASS. 2022. Oil Crops Yearbook. National Agricultural Statistics Service, NASS, Washington, DC. Available at: https://www.ers.usda.gov/data-products/oil-crops-yearbook/oil-crops-yearbook/#Vegetable%20Oils%20and%20Animal%20Fats.

Rosillo-Calle, F., L. Pelkmans and A. Walter. 2009. A Global overview of vegetable oils, with reference to biodiesel. A Report for the IEA Bioenergy Task 40. IEA Bioenergy, International Energy Agency, Paris, France.

Santos, B. S. 2009. Engine Performance and exhaust emissions of a diesel engine from various biodiesel feedstock. MS Thesis. Department of Biological and Agricultural Engineering, Texas A&M University, College Station, TX.

USEIA. 2022. US Biodiesel plant production capacity. U.S. Energy Information Administration, Washington, DC. Available at: https://afdc.energy.gov/data/10325. Accessed March 12, 2022.

Willits, J. 2007. Desmet Ballestra biodiesel production technology. Lecture presented at the 4th Practical Short Course on Biodiesel and Industrial Applications of Vegetable Oils. August 19-22, 2007, at Texas A&M University Organized by the Food Protein Research and Development Center (FPRDC), Texas Engineering Experiment Station, College Station, TX.

ASSESS YOUR KNOWLEDGE

1. Where does the US renewable jet fuel come from or what feedstock was used to make renewable jet fuel in the United States?
   a. Crude oil
   b. Natural gas
   c. Switchgrass
   d. Sugar and algae oil

11. True or False. The United States will be able to meet its goal of reducing greenhouse gas emissions to satisfy the Paris Agreement and have a targeted reduction of GHG emissions by 7.6% each year from 2020 to 2031 through its implementation of Renewable Fuels Standards.
    a. True
    b. No Idea
    c. False
    d. It depends

2. How much percentage of the biodiesel cost per gallon comes from the feedstock cost?
   a. 15%
   b. 85%
   c. 50%
   d. 75%

3. Between the engine speed of 1,000 and 2,000 rpm, which among the fuels below had the highest NO_x concentration in engine exhaust?
   a. B100 Biodiesel
   b. Reference Diesel
   c. B5
   d. B20

4. In the year 2019, the United States consumed 43 million barrels of biomass-based diesel fuel, nearly all as biodiesel blends with petroleum diesel. The world's biodiesel consumption that year was 9.3 billion gallons. How much percentage of US biodiesel production comprised the world's total biodiesel consumption?
   a. 0.46
   b. 1.94
   c. 19.4
   d. 10.8

12. This is an American federal program that requires transportation fuel sold in the United States to contain a minimum volume of renewable fuels. Name this program.
    a. Renewable fuel standard
    b. Renewable identification number
    c. Energy Policy Act of 2006
    d. Energy Independence Act of 2007

13. Name the most popular catalysts used to make biodiesel in the United States from vegetable oils and fats?
    a. Methanol
    b. Sodium
    c. Amylase
    d. Sodium methoxide

14. To guarantee biofuels production, sale, and blending, the United States issues this sort of currency of the US Renewable Fuels Standards (RFS). They are **credits** that the US EPA uses to track and enforce compliance with the renewable fuel mandates set by the RFS in the United States. These credits are essentially record of individual batches of renewable fuel being blended into the US gasoline and diesel pools. Name this credit tool.
    a. Renewable fuel standard
    b. Renewable identification number
    c. Energy Policy Act of 2006
    d. Energy Independence Act of 2007

5. Name the most common co-product of biodiesel transesterification process.
   a. Glycerin
   b. Soap
   c. Methanol
   d. Gum
6. This is a vegetable oil refining process that removes lecithin from crude vegetable oils. Usually phosphoric acid or citric acid is used to separate this from the crude oil.
   a. Dewaxing
   b. Bleaching
   c. Degumming
   d. Neutralization
7. This is a new additional ASTM requirement for biodiesel and added due to filter clogging during winter months (ASTM D7501–09 Standard Test Method for Determination of Fuel Filter Blocking Potential of Biodiesel (B100)), Name this new test method.
   a. Flash point
   b. Oxidation stability
   c. Cold soak test
   d. Cloud point
15. What is the primary oil extraction process used for microalgae samples with very hard cell wall? The cell wall must be broken down requiring cell lysis and this process is usually implemented.
   a. Acid washing
   b. Ultrasonication
   c. mechanical press
   d. solvent extraction
16. Name the European Standards for biodiesel.
   a. ASTM D 6751–08
   b. D975–08a
   c. D396–08b
   d. EN 14214
17. True or False. Biodiesel has lower specific fuel consumption than commercial diesel fuel.
   a. False
   b. True
   c. It depends
   d. No idea

8. Peanut oil was used to produce biodiesel. During lab analysis, the kinematic viscosity was found to be 6.9 mm^2/s. Will this biodiesel be blended with diesel or sold to commercial refineries for blending with commercial diesel?
   a. Yes
   b. No
   c. I do not know
   d. It depends

9. What is the single most important property of biodiesel as a blending material for diesel fuel?
   a. Energy content
   b. High oxidation
   c. High viscosity
   d. Lubricity

10. True or False. Biodiesel has higher energy per unit of volume than diesel fuel.
    a. True
    b. No idea
    c. False
    d. It depends

18. Name the US Standards for biodiesel.
    a. ASTM D 6751–08
    b. D975–08a
    c. D396–08b
    d. EN 14214

19. Biodiesel comes from organic vegetable oil and hence vulnerable to oxidation. Biodiesel storage and the presence of air and moisture will affect its stability and hence this test method is usually implemented and based from a European standard. Name this standard as a basis for this biodiesel oxidation stability property.
    a. Cold soak test
    b. ASTM D 6751–08a
    c. Viscosity
    d. Rancimat EN 14112

20. True or False. Biodiesel has lower NO_x emissions than diesel fuel, especially at higher engine rpm.
    a. True
    b. False
    c. It depends
    d. No Idea

Answers: 1d, 2b, 3a, 4c, 5a, 6c, 7c, 8b, 9d, 10c, 11c, 12a, 13d, 14b, 15b, 16d, 17a, 18a, 19d, 20b

ADDITIONAL LEARNING RESOURCES

- Access the Lecture Notes at the Instructor and Student Resources: https://routledgetextbooks.com/textbooks/instructor_downloads/
- The Case Study Videos can be found in the ebook+ version of this book.

6 Bioethanol Production

LEARNING OBJECTIVES

Upon completion of this chapter, one should be able to

1. Describe the various methods for producing ethanol from biomass resources.
2. Discuss the different microorganism types responsible for the conversion of sugars, starch, or lignocellulosic materials into ethanol.
3. Describe the various biomass pretreatment processes and enumerate advantages and disadvantages of various processes.
4. Estimate the ethanol yields from various biomass feedstocks.
5. Design simple ethanol distillation facilities, including description of full-scale ethanol plants.

6.1 INTRODUCTION

Ethanol (chemical formula C_2H_5OH) can be produced from biomass from its sugar, starch, and cellulosic components. The most common bioconversion process occurs through the action of certain specific yeasts or enzymes produced by microbes that act on the sugar, starch, or cellulosic components. Ethanol production from sugars is the most common process and makes use of yeast for conversion into ethanol, a one-step fermentation process. The most popular yeast is *Saccharomyces cerevisiae*, which has now been genetically modified to improve its ability to produce high percentages of ethanol while being able to withstand the high sugar concentration of some biomass. Ethanol from sugar or starchy crops is called "first-generation" ethanol, as opposed to "second-generation" ethanol that comes from lignocellulosic biomass. Plant cellulose is considered the most abundant renewable resource in the world.

Conversion of starch and cellulosic biomass into ethanol requires an additional step and additional sets of microbes to break down the starch components first into sugars and then subsequently into ethanol through fermentation by yeast. The most common microbes used for starch conversion into sugars are those that produce enzymes to break down starch into sugars. These liquefying enzymes, or α-amylases, are used to break down the starch into sugars such as maltose. These enzymes may be isolated from *Aspergillus niger*, a certain type of mold, or *Aspergillus awamori*.

The conversion of cellulosic biomass into ethanol is a little bit more complex and requires other microbial populations to first hydrolyze the rigid cellulosic structure of biomass. The more popular microbial species are certain strains of *Trichoderma reesi*, which can break down the cellulosic matrix into simple sugars through the

DOI: 10.1201/9781003294306-6

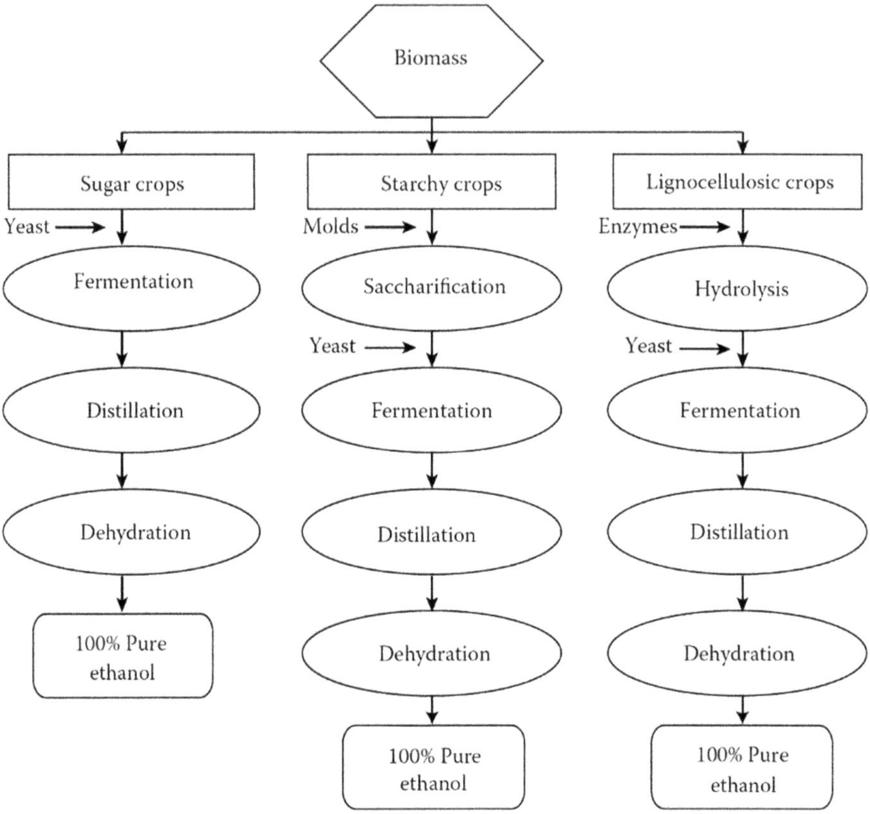

FIGURE 6.1 Schematic of ethanol production from various biomass components.

action of enzymes produced by these microbes. *Clostridium thermocellum* is another type of microbe that breaks down cellulose and synthesizes alcohol but produces some other by-products such as acetate and lactate, which may reduce overall conversion efficiencies. Some fungi species are now used for the same purpose (e.g., white rot fungi).

Figure 6.1 shows the various pathways to produce ethanol from sugary, starchy, and lignocellulosic biomass as described above. The ethanol concentration during fermentation is around 10%–18% by volume and will have to be distilled to generate high-purity ethanol and dehydrated to obtain 100% pure ethanol. Simple distillation of the beer will not yield 100% ethanol because of its nature to bind with water and form what we call an azeotrope, a mixture that cannot be changed by simple distillation. These compound mixtures are also called constant boiling mixtures because when the ethanol–water mixture is boiled, the ratio of constituents remains the same. For a water–ethanol mixture, the concentration is around 95.63% ethanol and 4.37% water by weight. The mixture will boil at 78.2°C (Fair, 1987), which is lower than the boiling temperatures of the individual components. Because of this, it is virtually impossible to separate the components by distillation.

Many researchers consider bioethanol production from cellulosic biomass to be the "holy grail" of ethanol production. These lignocellulosic biomass components, aside from being the most abundant, are non-food-competing sources. As such, they would be an ideal resource for biofuel production. However, the cost of genetically modified organisms that produce the required enzymes to break down the cellulose matrix is still quite prohibitive. More research is required to lower the cost of enzyme production.

Ethanol is very hygroscopic and an azeotrope with water. An additional step is required to completely remove water, such as the use of a specially sized molecular sieve. The theoretical ethanol production from pure sugar (e.g., pure glucose) is illustrated in Equation 6.1. In this equation, 1 mol of sugar (represented by glucose), with the action of yeast, will produce 2 mol of ethanol and 2 mol of carbon dioxide. The density of pure ethanol is about 0.789 g/cm^3. Thus, the mass balance shows that 180 g (or kg) of pure glucose is required to produce 92 g (or kg) of pure ethanol, while producing 88 g (or kg) of carbon dioxide. These numbers can also be recalculated to show that 1.54 kg of pure sugar is required to generate a liter of ethanol (or around 13 lb/gal), as illustrated in Example 6.1. These numbers are guidelines to use for the conversion of pure sugars into ethanol.

$$\underset{(180\ \text{kg})}{C_6H_{12}O_6} + \text{yeast} \rightarrow \underset{(92\ \text{kg})}{2C_2H_5OH} + \underset{(88\ \text{kg})}{2CO_2} + \text{heat} \tag{6.1}$$

Example 6.1: Ethanol Production from Glucose

Determine the amount of pure sugar (in kg) required to generate a liter of pure ethanol following Equation 6.1. Convert these units into the number of pounds of sugar required per gallon of ethanol.

Solution:

a. Amount of pure glucose required to generate a liter of ethanol

$$\frac{\text{kg glucose}}{\text{L ethanol}} = \frac{180\ \text{kg glucose}}{92\ \text{kg ethanol}} \times \frac{0.789\ \text{kg ethanol}}{\text{L ethanol}} = 1.54\frac{\text{kg}}{\text{L}}$$

b. Amount of sugar required to produce a gallon of ethanol

$$\frac{\text{lb sugar}}{\text{gal ethanol}} = \frac{1.54\ \text{kg glucose}}{\text{L ethanol}} \times \frac{2.2\text{lb}}{1\ \text{kg}} \times \frac{3.785\ \text{L}}{\text{gal}} = 12.8\frac{\text{lb}}{\text{gal}}$$

6.2 SUGAR CROPS

Sugars are crystalline carbohydrates and are required for human growth. Sugars can be classified as monosaccharides (glucose, fructose, or galactose), disaccharides (sucrose, lactose, or maltose), or polysaccharides (dextrose). The general formula for sugars is $C_nH_{2n}O_n$, where $3 < n < 7$. Glucose, a simple monosaccharide sugar, in the above example, has the formula $C_6H_{12}O_6$. From Equation 6.1, one can also show that for every kilogram of ethanol produced, almost 1 kg (92/88 = 1.04) of carbon dioxide is produced during the process.

Various sugar crops that are potential feedstock for ethanol production are described in Chapter 1. Sugarcane is grown in the states of Texas, Louisiana, Florida, and Hawaii; the yields vary because of different climates. Sugar beets are grown in 19 states in the United States. Sweet sorghum can be grown in the central states. Jerusalem artichokes are well adapted to northern climates and are not feasible to be grown in the south of the United States. Fruit crops are grown almost everywhere, and yields depend on the varieties grown.

The main advantage of using sugar crops as feedstocks for ethanol production is that minimal preparation is needed. Most sugar crops have high yields per acre. The coproducts of fermentation have value as feed or fuel. They can also be used for livestock feed and soil amendment. The main disadvantage is that sugar decomposition is very fast. Prolonged storage may result in sugar loss. It must be processed immediately after harvesting.

One additional disadvantage (true for any microbiological conversion process) is the requirement for sterility of the conversion process. Most feedstocks are sterilized. The conditions are adjusted so that only the target microbial population will remain in the fermentation broth. The yeast *S. cerevisiae* can withstand a lower pH than other microbial populations. As such, the fermentation broth is almost always brought to a lower pH (between 4 and 5) to ensure that only this group of microbes will remain active in the fermentor. Other types of microbes that have been proven to convert sugary materials into ethanol include the following: *Saccharomyces uvarum* (yeast), *Zymomonas mobilis* (bacteria), *Escherichia coli* (bacteria), *Candida shehatae* (yeast), and *Candida acidothermophilum*. *Candida shehatae* converts xylose sugars into ethanol.

6.2.1 Production of Ethanol from Sweet Sorghum

Sweet sorghum is an ideal feedstock for ethanol production. The process flow chart for converting sorghum into ethanol is shown in Figure 6.2. Sweet sorghum grows within 3–4 months; after harvest, the stalks are cleaned and the leaves are removed. Once the stalks are cleaned, the juice is extracted in a simple roller press.

The juice formed is dirty with leaves and some soil. The crushed stalks are placed in a pile. The juice is then filtered. Clean water is normally added (or removed by evaporation or heating) to adjust the sugar content appropriately (normally around 20%). The sugary broth is then sterilized, normally by heating slightly or adjusting the pH to around 5. The yeast is prepared separately, added to the broth, and fermented. Fermentation takes about 3–5 days, depending on the environment and temperature for the growth and propagation of yeast. In village-level systems, fermentation is done in 100 L containers (25 gal).

After the fermentation process, the beer (normally around 10%–15% alcohol) is distilled to recover a high percentage of ethanol. Additional energy is used in this process to boil the beer in order to separate the water from the alcohol. Ethanol

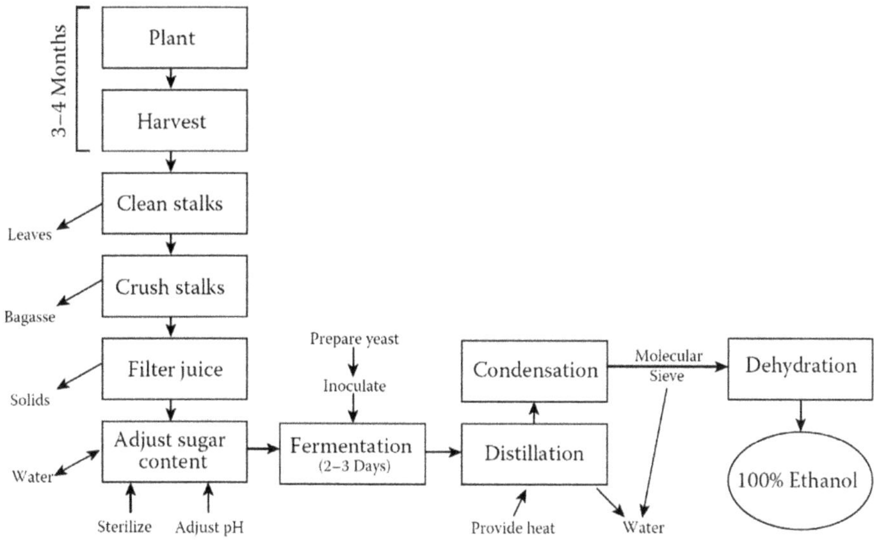

FIGURE 6.2 Process flow chart for conversion of sorghum into ethanol.

boils at around 80°C (the actual boiling point of pure ethanol at 1 atm is 78.3°C). With proper design of a distilling column, ethanol can be separated effectively. As mentioned earlier, not all water can be removed by distillation. Further separation must be conducted using an appropriately sized molecular sieve to dehydrate the condensed liquid into 100% ethanol (or 200 proof).

Example 6.2: Sweet Sorghum Conversion into Ethanol

Sweet sorghum stalks were pressed to obtain 1 L of the juice (mass, 1 kg). The total sugar concentration was found to be 160 g/L. The juice was further fermented using yeast and produced an ethanol concentration of 55 g/L. Calculate the ethanol yield per weight of juice and that of sugar as the substrate. Compare this with the theoretical glucose-to-ethanol yield assuming the density of ethanol is 0.8 g/mL.

Solution:

a. The ethanol yield per juice weight is calculated as follows:

$$\text{Yield (g ethanol/g juice)} = \frac{1\text{ L}}{1{,}000\text{ g}} \times \frac{55\text{ g}}{\text{L}} = 0.055\text{ g/g}$$

b. Thus, the yield of ethanol yield based on juice is about 55 g/kg of juice.
c. The yield based on sugar is calculated as follows:

$$\text{Yield (g ethanol/g sugar)} = \frac{55\text{ g}}{\text{L}} \times \frac{1\text{ L}}{160\text{ g sugar}} = 0.34\text{ g/g}$$

d. Thus, the yield of ethanol per gram of sugar is around 340 g/kg sugar.
e. In the calculation above, the 340 g of ethanol may be converted into the volume basis if the density is known as follows:

$$\text{Yield (L ethanol/kg sugar)} = \frac{340\,\text{g}}{\text{kg}} \times \frac{1\,\text{mL}}{0.8\,\text{kg}} \times \frac{\text{L}}{1{,}000\,\text{mL}} = 0.425\,\text{L/kg}$$

f. In Example 6.1, 1.54 kg sugar is required to generate a liter of ethanol, and this is also equivalent to 0.65 L/kg.
g. The theoretical yield is about 35% higher than some actual experiments using real juice from sorghum.

6.2.2 Ethanol Dehydration Methods

There are various methods of separating alcohol from water, beer, or fermentation mash. Coble et al. (1985) presented an excellent discussion of the methods for ethanol dehydration. The procedures are summarized as follows:

1. Tertiary azeotropic distillation
2. Vacuum distillation
3. Solvent extraction
4. Simple distillation
5. Physical separation using molecular sieves
6. Use of desiccants

Tertiary azeotropic distillation makes use of a third component such as benzene or pentane to separate ethanol from water. Benzene or pentane is added to the ethanol–water solution, and this mixture forms a minimum boiling point azeotrope, which can then be distilled to leave pure ethanol in a separate container called a reboiler. The benzene can then be easily separated with water using another reboiler system. Benzene must be recovered to make the process economical. The energy consumption of this process to bring 190 proof ethanol into a 200-proof pure compound is reported to be around 9,400 kJ/L (Eakin et al., 1981).

Vacuum distillation uses pressure below 11.7 kPa such that at this condition, the ethanol–water azeotrope does not form. The anhydrous ethanol is separated by simple distillation. Note that this reduced pressure would lower the boiling point, thereby lowering the distillation temperatures for the separation process. However, the reported total energy consumption for this process is higher than the conventional methods using simple distillation and tertiary distillation (10,300 kJ/L).

Solvent extraction is also used to separate ethanol from water using a suitable solvent. Liquid carbon dioxide at 6,210–6,900 kPa is an example of a solvent; at this condition, ethanol is more soluble in CO_2 than water and the ethanol may be separated from the water molecules. Once this is achieved, the pressure is reduced (say to

4,800 kPa) to flash the CO_2, leaving behind pure ethanol. Other solvents have been tested to provide much lower operating pressures.

Ethanol and water can also be separated by a series of distillation processes called multi-effect distillation or vapor recompression distillation. This separation system uses various distillation columns or "effects," and the distillate vapor from one column is used to heat the liquid in the other column or reboiler. For effectiveness, the pressure must be lower in each successive column.

The most common separation system separates ethanol from water by using size-selective molecular sieves. Molecular sieves are nothing more than "filters" with accurately manufactured pore sizes. The sieves are made from crystalline metal aluminosilicates consisting of a network of interconnecting pores. The separation is based on the selective adsorption of molecules onto the pore walls of the sieve. A sieve with a 3 Å effective pore diameter is used. The diameter of water molecules is less than this size; consequently, water is adsorbed by the sieve. Ethanol molecules are larger than the pore diameter, cannot be adsorbed, and, thus, are separated. The molecular sieves are reusable after simple drying. The reported energy requirement for this process is quite low and may range from 120 to 160 kJ/L (Coble et al., 1985).

Finally, the ethanol–water mixture may also be separated with the use of special desiccants that can either adsorb only water or only ethanol. Calcium oxide is the most common material used; cornstarch, microcrystalline cellulose, $CaSO_4$, and NaOH have also been used.

The energy balance for the complete combustion of pure and dry ethanol in air-using engines may be estimated by developing the stoichiometric combustion equation as shown in Equation 6.2. In this balanced equation, for every mole of ethanol, 3 mol of air are used and 2 mol of carbon dioxide and 3 mol of water are produced. Example 6.2 shows the air-to-fuel ratio for the complete combustion of ethanol in air.

$$C_2H_6O + 3(O_2 + 3.76\ N_2) \rightarrow 2CO_2 + 3H_2O + 11.28\ N_2 \qquad (6.2)$$

Example 6.3: Combustion of Pure Ethanol

Determine the air-to-fuel ratio for the complete combustion of ethanol (C_2H_6O) gas in air. Note that air contains 79% nitrogen and 21% oxygen by volume. The main products are water and carbon dioxide. Also, estimate the amount of carbon dioxide produced per weight of ethanol combusted. Compare this with the amount of CO_2 produced during the fermentation process from glucose.

Solution:

a. Balance the combustion equation with air to yield CO_2 and H_2O:

$$C_2H_6O + 3(O_2 + 3.76\ N_2) \rightarrow 2CO_2 + 3H_2O + 11.28\ N_2$$

b. With the molecular weights of each compound known, calculate the air-to-fuel ratio using the coefficients from the balanced equation above:

$$\frac{\text{Air}}{\text{Fuel}} = \frac{3\times(O_2 + 3.76\ N_2)}{C_2H_6O} = \frac{3\times\left(16\times 2 + 3.76\times(14\times 2)\right)}{\left((2\times 12)+(6\times 1)+(1\times 16)\right)} = \frac{411.84}{46} = 8.95$$

c. Thus, one would require about 8.95 kg of air for every kilogram of ethanol combusted in air (or 8.95 lb/lb).
d. The amount of CO_2 produced per weight of ethanol is calculated similarly as follows:

$$\frac{2CO_2}{C_2H_6O} = \frac{2\times\left(12+(16\times 2)\right)}{\left((2\times 12)+(6\times 1)+(1\times 16)\right)} = \frac{88}{46} = 1.9\frac{\text{kg}}{\text{kg}}$$

e. About 0.96 kg of CO_2 is produced per kilogram of ethanol combusted.
f. About 0.48 kg of CO_2 is produced for every kilogram of glucose used to make ethanol.

6.3 STARCHY CROPS

Most of the ethanol produced in the United States comes from corn, the main component of which is starch. This starch component will have to be converted into simple sugars for subsequent action of fermentation yeasts. The reported ethanol yield from corn in the United States is around 354 L/ha (44 gal/acre), which is lower than the reported yield from cassava, from a study in Nigeria of around 410 L/ha (44 gal/acre) (Brown, 2006). Table 1.7 presents some ethanol yields from sugar and starchy crops that may be used for estimating bioethanol yields from starch. Starch conversion to ethanol is slightly different from sugar crops. Starch molecules should be broken down into simple sugars with the use of enzymes before the sugars are converted into ethanol. Enzymes are actually protein molecules, and they act as catalysts to speed up the conversion process. One type of enzyme is called amylase, and this specifically breaks down starch into sucrose. Sucrose is a disaccharide having two molecules of glucose as shown in Equation 6.3.

$$2C_6H_{10}O_5 + H_2O + \text{ amylase } \rightarrow C_{12}H_{22}O_{11} \tag{6.3}$$

The other type of enzyme is called invertase, also called saccharifying enzyme, and breaks down sucrose into glucose sugar (Equation 6.4).

$$C_{12}H_{22}O_{11} + H_2O + \text{ invertase } \rightarrow 2C_6H_{12}O_6 \tag{6.4}$$

Finally, the yeast acts on this glucose sugar to generate ethanol as shown in Equation 6.4.

Example 6.4 shows the conversion factor to produce ethanol from starch. Theoretically, 0.757 kg of starch is required to generate 1 kg of ethanol or 0.566 kg/L by volume. This is also equivalent to 4.71 lb pure starch per gallon of ethanol produced. The current state-of-the art ethanol production from corn in the United States is about 2.8 gal/bushel. A bushel of corn is around 56 lb, equivalent to around 20 lb of cornstarch to a gallon of ethanol. As such, there is a 4-fold factor when using actual corn grain. The other agricultural crops that are potential sources of ethanol (from their starch components) include cassava, highly digestible grain sorghum, and soybean. The primary advantage of starch crops, unlike sugar crops, is that they can be processed into starch grain and stored for longer periods before conversion into ethanol. The cultivation of starch crops is well-established and well-understood, and its coproducts are suitable for animal feed, as they are high in protein content. Disadvantages include the need for additional equipment for preparation, additional sets of enzymes, additional labor, and additional energy costs for processing.

Example 6.4: Conversion Factor for Starch Conversion into Ethanol

Determine the amount of starch required (in kg) to generate 1 kg or 1 L of ethanol. Convert these into equivalent English units.

Solution:

a. The mass balance for Equation 6.2 is as follows:

$$\underset{324}{2C_6H_{10}O_5} + \underset{18\ \text{kg}}{H_2O}\ \text{amylase} \rightarrow \underset{342\ \text{kg}}{C_{12}H_{22}O_{11}}$$

b. The mass balance for Equation 6.3 is as follows:

$$\underset{342}{C_{12}H_{22}O_{11}} + \underset{18\ \text{kg}}{H_2O} + \text{invertase} \rightarrow \underset{452\ \text{kg}}{2C_6H_{12}O_6}$$

c. For every 342 kg of starch, we produce 452 kg of ethanol or 0.757 kg starch/kg ethanol.
d. Likewise, if the density of pure ethanol is 0.789 kg/L, the conversion from weight to volume is calculated as follows:

$$\frac{324\ \text{kg starch}}{452\ \text{kg ethanol}} \times \frac{0.789\ \text{kg ethanol}}{\text{L ethanol}} = 0.566\frac{\text{kg starch}}{\text{L ethanol}}$$

e. In English units, the equivalent is as follows:

$$\frac{0.566\ \text{kg starch}}{\text{L ethanol}} \times \frac{2.2\ \text{lb}}{1\ \text{kg}} \times \frac{3.785\ \text{L}}{\text{gal}} = 4.71\frac{\text{lb starch}}{\text{gal}}$$

Commercial ethanol plants in the United States convert corn into ethanol in two ways:

1. Wet milling
2. Dry milling

The two processes are shown schematically in Figure 6.3.

In the wet milling process, kernels are separated into components and only the starch portion is fermented; in the dry grind process, whole kernels are ground into flour and fed directly into the fermentation process. Although wet milling produces more coproducts such as oil, gluten meal, and gluten feed, it requires higher capital and operating costs than the dry grind process (Nichols et al., 2008). Thus, most of the expansion of the ethanol industry in the United States is in dry grind facilities. About 82% of the US ethanol facilities were dry grind plants and 18% were wet milling plants (RFA, 2023). Dry grind facilities produce around 401.5–416.4 L/tonne of corn (2.7–2.8 gal/bushel) (1 bushel corn = 56 lb) along with distillers' grain (DDG).

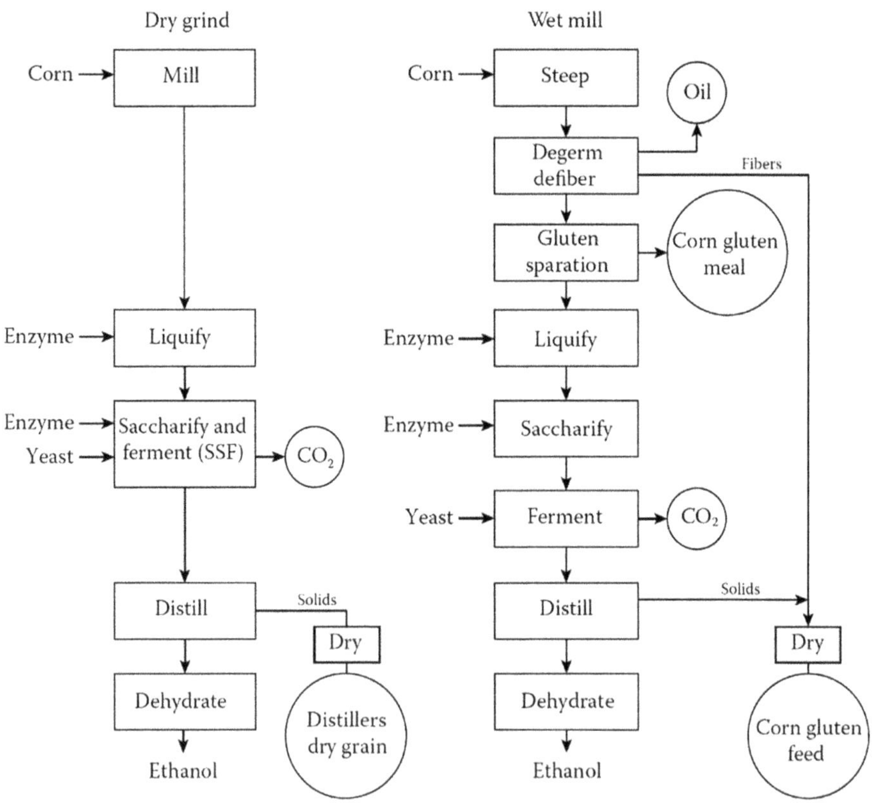

FIGURE 6.3 Dry and wet milling processes for conversion of starch into ethanol. (Adapted from Nichols, N. N., B. S. Dien, R. J. Bothast and M. A. Cotta, The corn ethanol industry. In: *Alcoholic Fuels* (S. Minteer, ed.). CRC Press, Boca Raton, FL. pp. 59–78, 2006.)

DDG is used as animal feed and comprises mainly yeast and nonfermentable parts of corn such as germ, fiber, and protein (Nichols et al., 2008).

The basic dry grind process involves grinding, cooking, liquefaction, saccharification, fermentation, distillation, and coproduct recovery. Corn is ground and mixed with water to produce a slurry. This slurry is then cooked to break down the crystalline structure of the starch granules; the resulting mash is further liquefied and saccharified by amylase to reduce the viscosity and produce the sugars. The fermentable sugars are then converted into ethanol by yeast during fermentation (Naidu et al., 2007; Singh et al., 2006).

The wet milling process is only different at the beginning of the process when the corn is first cleaned to ensure it is free of dust and dirt. As shown in Figure 6.3, this is followed by soaking the corn in water (called steep) at around 50°C for several hours (usually 24–48 h). During this time, it doubles in size. Sulfur dioxide (or sulfurous acid) is added to separate the grain into its component parts (and oil in particular). After steeping, the corn slurry is processed through a series of grinders to separate the corn germ. The corn oil from this germ is extracted onsite or sold to crushers who extract the oil. The remaining fiber, gluten, and starch slurry components are separated using centrifugal, screen, and hydro-cyclonic separators. The fibers are first separated by simple screens, allowing the gluten and the starch to pass through. The starch–gluten suspension passes through a centrifuge where the gluten, which is less dense than the starch, is easily separated. The starch component is then converted into ethanol as in the dry grind process. The gluten is dried and sold as animal feed along with the fiber. New feedstocks, such as hybrid sorghum grain, are now being developed for more efficient conversion of grain starch into ethanol. One such study (Hernandez et al., 2011) showed that highly digestible grain starch from hybrid sorghum has the potential to generate cost-effective ethanol using the simultaneous saccharification and fermentation process.

Example 6.5: Conversion of Sorghum Grain into Ethanol

Determine the amount of sorghum grain (called milo in some literature) required to generate 3.785 million liters of ethanol per year (1 MGY). Assume that the current ethanol yield from sorghum grain is 416 L/tonne (2.8 gal/bushel). Note that 25.45 kg of sorghum grain (56 lb) is equivalent to a bushel in the English unit system. Grain sorghum yield is around 6287 kg/ha (100 bushels/acre). Also, determine the number of hectares (acres) required to satisfy this ethanol yield. Repeat the calculations using the English system of units.

Solution:

a. The required sorghum grain is calculated as follows:

$$\text{Grain}(\text{tonnes}) = 3.785 \times 10^6\,\text{L} \times \frac{\text{tonne}}{416\ \text{L}} = 9{,}099\,\text{tonnes}\ (10{,}008\,\text{tons})$$

b. The area required is calculated as follows:

$$\text{Area}(\text{ha}) = 9{,}099\,\text{tonnes} \times \frac{1{,}000\ \text{kg}}{1\ \text{tonne}} \times \frac{\text{ha}}{6{,}287\ \text{kg}} = 1{,}447\,\text{ha}\,(3{,}575\,\text{acres})$$

c. In the English system of units, the following is the calculation procedure:

$$\text{Grain}(\text{tons}) = 1 \times 10^6\,\text{gal} \times \frac{\text{bu}}{2.8\,\text{gal}} \times \frac{56\,\text{lb}}{1\,\text{bu}} \times \frac{\text{ton}}{2{,}000\,\text{lb}} = 10{,}000\,\text{tons}$$

d. The area required to satisfy this requirement is calculated as follows:

$$\text{Area}(\text{acres}) = 10{,}000\,\text{tons} \times \frac{\text{acre}}{100\,\text{bu}} \times \frac{1\,\text{bu}}{56\,\text{lb}} \times \frac{2{,}000\,\text{lb}}{\text{ton}} = 3571\,\text{acres}$$

6.4 CELLULOSIC BIOMASS

The main problem with using sugar and starchy crops for ethanol production is that they are primary sources of nutrients and food for the human population. When fuel ethanol was flourishing in the United States, the price of corn increased dramatically, because of competition by corn ethanol refineries. The use of cellulosic material, on the other hand, will not compete with human food sources and can be ideal for ethanol production. Cellulose is the backbone of most sugary and starchy crops. Potential crop residues that have very high lignocellulosic content include corn stover, rice and wheat straw, cotton gin trash, rice hulls, sorghum, sugarcane biomass, bagasse, and many others. Forage grasses—switchgrass, alfalfa, Napier grass, forage sorghum, and Sudan grass—are also potential sources. Animal manure also contains a large portion of lignocellulosic biomass and could be converted into ethanol.

These materials do not compete with food crops and are widely available. However, hydrolysis of the cellulose chain is still quite expensive, and no commercial cost-effective process has yet been established.

Biomass is composed of three major components: cellulose, hemicellulose, and lignin. The hemicellulose components include carbohydrates and sugars (both the five- and six-carbon sugars) and are about 20%–40% of biomass weight. The cellulosic component is around 40%–60%. The lignin component is around 10%–25%. The lignin components are not broken down by microbes or by their enzymes. Thus, the conversion of lignocellulosic biomass into ethanol requires some pretreatment processes.

Example 6.6: Cellulosic Biomass Conversion to Ethanol

One pathway of lignocellulosic biomass to ethanol production reported 345.675 L/dry tonne (83.8 gal/dry ton) of lignocellulosic biomass (Dutta et al., 2011). How many liters per year (gallons per year) will be produced by 405 ha (1,000 acres) of woody biomass that has a yield of 44.9 tonnes/ha/yr (20 tons/acre/yr). The biomass has 35% moisture after harvest.

Solution:

a. The first step is to determine the amount of biomass that can be harvested in a year as follows:

$$\text{Biomass}(\text{dry tonnes}) = 405\,\text{ha} \times \frac{44.9\ \text{tonnes}}{\text{ha}} \times 0.65 = 11{,}820\,\text{dry tonnes}$$

$$\text{Biomass}(\text{dry tons}) = 1{,}000\,\text{acres} \times \frac{20\ \text{tons}}{\text{acre}} \times 0.65 = 13{,}000\,\text{dry tons}$$

b. Then, the amount of ethanol can be directly computed as follows:

$$\text{Ethanol}(\text{L}) = 11{,}820\,\text{dry tonnes} \times \frac{345.675\,\text{L}}{\text{dry tonne}} = 4{,}085{,}879\,\text{L}\ (4\,\text{MLY})$$

$$\text{Ethanol}(\text{gal}) = 13{,}000\,\text{tons} \times \frac{83.8\,\text{gal}}{\text{dry ton}} = 1{,}089{,}400\,\text{gal}(1.1\,\text{MGY})$$

6.5 BIOMASS PRETREATMENT PROCESSES

Various pretreatment procedures have been developed to improve the conversion of lignocellulosics into ethanol. They can be classified into physical, chemical, and enzymatic processes.

6.5.1 Physical Pretreatment of Biomass

Physical pretreatment of biomass entails the breakdown of cellulosic components into the smallest sizes, which is possible by the following processes:

1. Pulverization or grinding of biomass
2. Ultrasonication
3. Steam explosion

The simplest physical pretreatment is the reduction of the physical size of the biomass components. This exposes or separates the cellulosic and lignin components. When biomass is first harvested, the materials are either baled or formed into modules. The biomass is first chopped or shred into small particles, then ground or milled into smaller particles that are a few millimeters (mm) in size. A chipper shredder is typically used. A hammer mill is used for further processing. A much finer mill (e.g., a medium-sized Wiley Mill) reduces the particles even more. The particle size following this stage is about 1 mm. The submillimeter-sized particles may be further broken down into micron sizes using ultrasound. Depending on the length of time in the ultrasonicator, as well as the power level used, materials can be broken down to the micron level. If even further breakdown of particles is required, the material may be placed in a pressure reactor. There, in the presence of steam, they will undergo what is called the steam explosion process. This reactor subjects the ground biomass to pressures and temperatures up to 5,000 psi and 1,200°C. It is noted that high energy costs limit the use of high levels of pressure and temperature.

6.5.2 Chemical Pretreatment of Biomass

The cellulose, hemicellulose, and lignin components may also be separated by chemical means. Cellulose and lignin are difficult to separate. The NREL compositional analysis protocol discussed in Chapter 4 showed some simple acid treatment processes used to separate the acid-soluble and acid-insoluble lignin. Similar processes are commercially done, for the same purpose, to separate the lignin from the cellulose, because microbes cannot act or convert the lignin fractions.

One other problem in this process is the formation of by-products that may inhibit the production of ethanol. These include the production of phenolic compounds and the formation of furfurals. As a practical matter, it is not economically feasible to use large amounts of chemicals. The most common approach is to use concentrated or mild acids at elevated temperatures. This pretreatment step also alters the structure of the cellulose matrix and makes it more accessible to the enzymes used during the hydrolysis process. Pretreatment has been recognized as one of the most expensive processing steps in cellulose-to-ethanol conversion. Various review articles have been published concerning this issue (Carvalheiro and Duarte, 2008; Hendriks and Zeeman, 2008).

A partial list of other popular chemical pretreatment process is as follows:

1. Liquid hot water
2. Weak acid hydrolysis
3. Strong acid hydrolysis
4. Alkaline hydrolysis
5. Organosolv
6. Oxidative delignification
7. Room temperature ionic liquids

A review by Harmsen et al. (2010) on the above processes will be summarized here.

The liquid hot water treatment uses water at high temperature (200°C–230°C) and pressure for several minutes (e.g., 15 min). This process is also called "hydrothermolysis" or solvolysis. It was reported that between 40% and 60% of the total biomass is dissolved by the process, with 4%–22% of the cellulose, 35%–60% of the lignin, and all hemicellulose removed. More than 90% of the hemicellulose sugars are recovered with the use of mild acid to hydrolyze the resulting liquid. This is when some of the furfurals are reportedly formed. These furfurals will inhibit the fermentation process and must be further separated.

A weak acid treatment is normally carried out at either a low temperature (<160°C) with acid concentrations between 10% and 40% or at temperatures higher than 160°C with acid concentrations between 5% and 10%. The most common acid used is sulfuric acid. It is mixed with the biomass and held at the desired temperature for a few minutes. The use of other organic acids, such as malic and fumaric acids, has also been reported.

Very strong concentrations of sulfuric and hydrochloric acids may also be used in the pretreatment process. The high concentration improves hydrolysis, and subsequent enzymatic pretreatments may not be necessary. However, as one would expect, acids are very corrosive to metals in the process vessels. Acids also need to be recovered to reduce overall pretreatment costs. At least with strong acid treatment, there is no need to raise the temperature of the process.

The use of alkaline chemicals, such as lime (calcium hydroxide) or sodium hydroxide, to remove lignin from the biomass, is also popular. This pretreatment, initially discussed in Chapter 2, was popularized by the MixAlco™ Process of Mark Holtzapple at Texas A&M University (TAMU). Reaction conditions are mild. However, reaction times may be quite long. An advantage of this process is the minimal production of furfural compounds, which may inhibit subsequent fermentation.

Aqueous ammonia is also popularly used, with the mixture brought to higher temperatures to reduce the lignin content, remove the hemicellulose, and decrystallize the cellulose portions. There are several techniques; most common are ammonia fiber explosion (AFEX), ammonia recycle percolation (ARP), and soaking in aqueous ammonia (SAA).

The ARP technique is normally conducted at higher pressures (about 2.3 MPa); the process recovers ammonia, and lignin and other sugars are separated. The SAA technique is usually conducted at low temperatures. Typical conditions for ARP would be 15%–30% aqueous ammonia at 30°C–37°C for 12 h to several days with minimal agitation and a solid to liquid ratio of 1:12. The solids are recovered after this long soaking process by filtration. The solids are then washed and analyzed. As reported, lignin solubilization of 66% and saccharification yields of 83% for glucan and 63% for xylan, if 15% ammonia is used at 75°C for 48 h. As in all the other chemical pretreatment processes, the cost of ammonia and the degree of recovery affect pretreatment expenses.

The use of organic solvent (organosolv) is also becoming popular. Common solvents used include acetone, methanol, ethanol, or ethylene glycol. The temperature used may be as high as 200°C. Note that some of the solvents may inhibit the fermentation process and should be completely removed after the pretreatment. However, high-quality lignin is reportedly produced. In addition, enzyme costs are reduced when effective lignin separation is achieved.

Oxidizing agents such as hydrogen peroxide and ozone have reported high delignification efficiencies because of their reactivity in breaking the complex aromatic rings of the lignin matrix and converting them to simple carboxylic acids. One disadvantage is that these resulting chemicals will inhibit the fermentation process. Consequently, they must be completely removed. It was also reported that some hemicellulose portions were degraded, thereby limiting their use in the subsequent fermentation process. A reported yield of 95% during the subsequent enzymatic hydrolysis is possible with this method with a solution of only 2% hydrogen peroxide at 30°C. The lignin

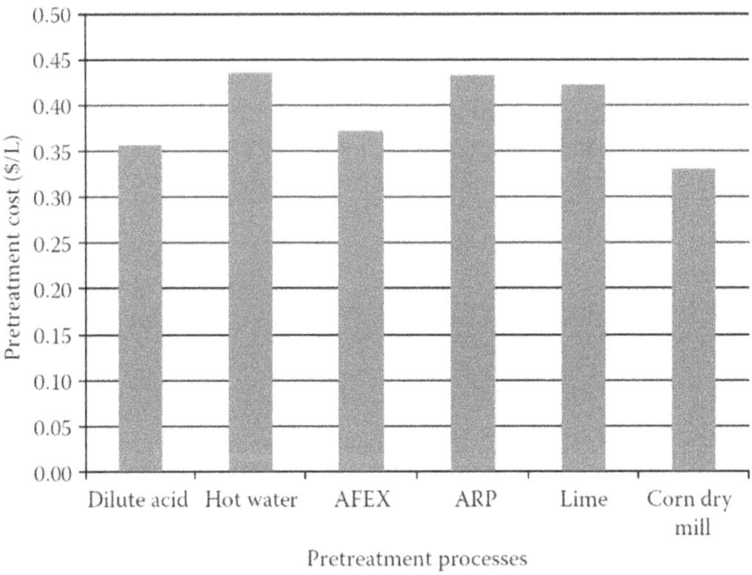

FIGURE 6.4 Pretreatment economic analysis for various chemical pretreatment processes. (From Dale, B., E. Newton, M. Laser, S. Kim, L. Laureano-Perez, H. Alizadeh, H. Jin and L. Lynd, *Recent Improvements for the Ammonia Fiber Explosion (AFEX) Process and Resulting Reductions in Projected Ethanol Prices*. Great Lakes Bioenergy Research Center Report, Michigan State University, Ann Arbor, MI, 2009.)

dissolution was around 50%. The use of ozone is as effective as hydrogen peroxide, with the advantage of reduced hemicellulose decomposition. This process was tried using wheat straw, bagasse, pine, peanut, cotton straw, and poplar sawdust.

Finally, ionic liquids (inorganic anion or organic cation) are also used for chemical pretreatment processes. These compounds are salts at normal temperatures but will enter the liquid phase at low temperatures when they are utilized for the pretreatment process. However, limited reports have been published, and these chemicals will have the same issues of toxicity, recoverability, and cost. Figure 6.4 shows the different pretreatment processes described and compares their relative costs. The dilute acid and corn dry mill are still the most cost-effective pretreatment processes per gallon of ethanol produced. However, the results are comparable to those of other pretreatment processes.

6.5.3 Enzymatic Pretreatment of Biomass

There is also a growing trend toward using microorganisms for the pretreatment of lignocellulosic biomass. The most common organisms used to degrade lignin and hemicelluloses are brown and white rot fungi. A main advantage is the low cost of these microbes. Major disadvantages include the long retention times to complete the process and the low conversion efficiencies. Research on the use of genetically modified organisms is also underway. Process improvements await the identification of high-performing microbes.

6.5.4 Combination of Various Pretreatment Methods

One may use the combination of physical, chemical, and enzymatic pretreatment processes to achieve a goal of high conversion efficiency at the lowest cost. For example, the simple conversion of cotton gin trash into ethanol may follow the exhaustive route shown in Figure 4.3. A study was conducted using most of these pretreatment processes to convert cotton gin trash into ethanol. In this process, the initial step was the use of mechanical pretreatment, such as grinding and reduction of size mechanically, with the final mechanical pretreatment involving ultrasonication. The resulting slurry was then brought to a mild steam explosion process (mild temperature and pressure) followed by using enzymes for the final hydrolysis of the cellulose fractions. Initial results revealed improved conversion efficiencies with less energy and power. However, more studies are required to establish the overall economics.

6.6 PILOT-SCALE ETHANOL PRODUCTION SYSTEMS

The Biological and Agricultural Engineering Department at TAMU has a pilot-scale conventional ethanol production facility designed since the 1980s (Egg et al., 1985). The plant has four 1,300-L batch cooker/fermenters and a continuous distillation column with a capacity of 30 L of ethanol per hour (see Figure 6.5 for detailed system components). Specific areas within this facility have been outlined in the figure and include the following important components:

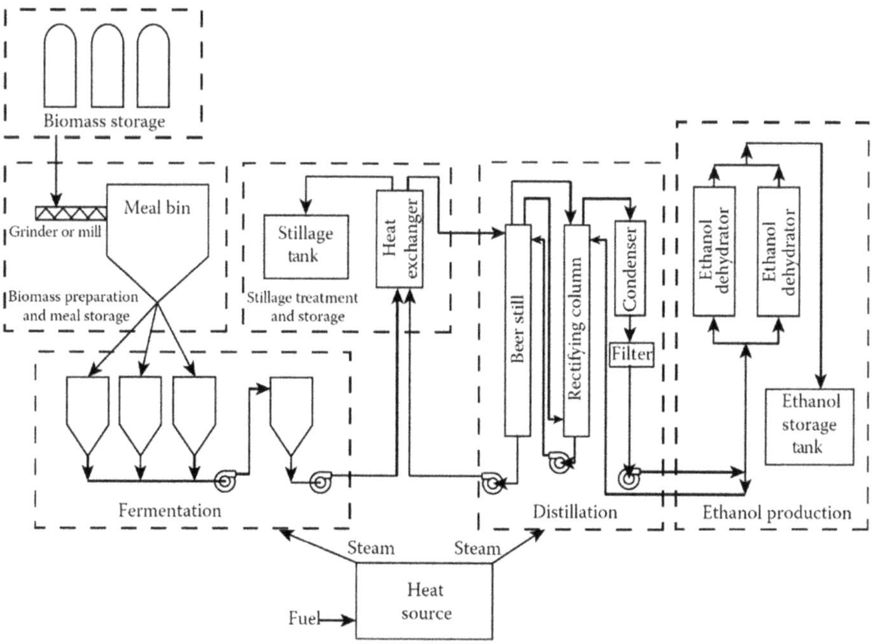

FIGURE 6.5 Schematic of 1,300 L (350 gal) ethanol production facility for grain sorghum and corn.

1. Biomass storage facility (in this case, corn or sorghum grain storage)
2. Biomass preparation and meal storage facility (grinder and meal bin)
3. Fermentation tanks (which also act as the boiler during distillation)
4. Heat source to generate steam during distillation (use of natural gas)
5. Beer still, rectifying column, and condenser
6. Stillage treatment and storage (stillage tank)
7. Ethanol production facility (to generate 100% ethanol product)

The unit can be operated daily to provide a continuous production of 200 proof ethanol from grain sorghum or corn.

The facility uses 1.98 m^3 (70 ft^3) of natural gas for every 3.785 L (gal) of ethanol processed. Of this, 0.71 m^3 (25 ft^3) is used for cooking and the other 1.27 m^3 (45 ft^3) is used for the distillation process. The total water usage is around 155.5 L of water per 3.785 L of ethanol (41 gal of water per gallon of ethanol), where 129 L (34 gal) is used for cooking and another 26.5 L (7 gal) is used for the distillation process per 3.795 L of ethanol produced (1 gal ethanol). The electricity usage is around 0.185 kWh/L (0.7 kWh/gal). The yields established through several pilot runs are as follows:

- 386.6 L/tonne (2.6 gal/bushel) for corn at 182–184 proof
- 371.7 L/tonne (2.5 gal/bushel) for grain sorghum at 182–184 proof
- 356.9 L/tonne (2.4 gal/bushel) corn at 100% ethanol
- 342 L/tonne (2.3 gal/bushel) grain sorghum at 100% ethanol

The current industry standard for ethanol production from grain is around 416.4–431.2 L/tonne (2.8–2.9 gal/bushel). The TAMU facility fermenting log for the grain-to-ethanol production process is shown in Figure 6.6. The process starts by heating the grain mash. In about 30 min, or around 50°C (120°F), the liquefying enzyme

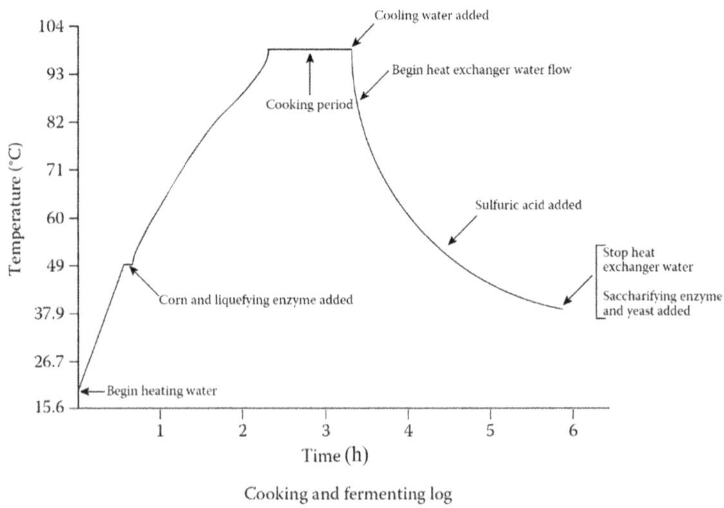

FIGURE 6.6 Cooking and fermenting log for grain-to-ethanol production process.

is added. The heating continues for about 2 h until the temperature reaches about 93°C (200°F). This temperature is then maintained for about an hour. Thus, after about 3.5 h, the heating is stopped. Cooling water is added, and at this stage, the heat exchanger water flow is initiated. Sulfuric acid is added at about 4 h when the broth temperature is around 50°C (120°F). After 6 h of cooling, the heat exchanger is stopped and the saccharifying enzyme and yeast are added. The temperature is around 38°C (100°F) at this point. The distillation log is also quite simple. The still is also heated, and in around 30 min, the beer is fed into the distillation column. In less than an hour, alcohol flows and is directly delivered to the ethanol storage facility. The distillation for a 950 L (250 gal) beer boiler will end in around 4 h, and this is also the time that the ethanol will cease to flow out of the condenser. The still is shut down in around 4.5 h. The process is repeated each day until feedstocks are available. Example 6.7 shows the calculations of energy, electricity, and water usage in an ethanol facility.

Example 6.7: Energy Calculations in Ethanol Plants

Estimate the natural gas, electricity, and water usage in a 3.785-million-liter (1 MGY) ethanol facility based on data previously presented for grain sorghum above and summarized below. Make an estimate of the cost of energy and electricity. Assume the cost of natural gas is $4/MMBtu and electricity cost is $0.12/kWh. Assume the heat content of natural gas is around 1,025 Btu/ft^3.

- Natural gas = 1.98 m^3/3.785 L (70 ft^3/gal)
- Electricity = 0.185/kWh/L (0.7 kWh/gal)
- Water usage = 155.5 L/3.785 L ethanol (41-gal H_2O/gal ethanol)
- Note 1 MMBtu = 1 million Btu

Solution:

a. The natural gas usage is calculated as follows:

$$\text{Natural gas}\left(\text{m}^3\right) = 3.785\times10^6\ \text{L}\times\frac{1.98\,\text{m}^3}{3.785\text{L}} = 1.98\times10^6\ \text{m}^3$$

$$\text{Natural gas}\left(\text{ft}^3\right) = 1\times10^6\ \text{gal}\times\frac{70\,\text{ft}^3}{\text{gal}} = 70\times10^6\ \text{ft}^3$$

b. The electricity usage is calculated as follows:

$$\text{Electricity}(\text{kWh}) = \frac{0.185\,\text{kWh}}{\text{L}}\times3.785\times10^6\ \text{L} = 700{,}000\,\text{kWh}$$

c. The water usage calculations are as follows:

$$\text{Water}(\text{L}) = \frac{155.5\,\text{L water}}{3.785\ \text{L ethanol}} \times 3.785 \times 10^6\ \text{L ethanol} = 155.5 \times 10^6\ \text{L}$$

$$\text{Water}(\text{gal}) = 1 \times 10^6\ \text{gal} \times \frac{41\,\text{gal}}{\text{gal ethanol}} = 41 \times 10^6\ \text{gal}$$

d. The cost calculations are as follows:

$$\text{Natgas cost (\$)} = \frac{\$4}{1 \times 10^6\ \text{Btu}} \times \frac{1025\,\text{Btu}}{\text{ft}^3} \times 70 \times 10^6\ \text{ft}^3 = \$287{,}000$$

$$\text{Electricity cost (\$)} = 700{,}000\,\text{kWh} \times \frac{\$0.12}{\text{kWh}} = \$84{,}000$$

In Example 6.7, the natural gas cost exceeds the electricity cost ethanol production by more than three times. Thus, one must be aware that separation of ethanol from the beer and from the azeotrope will entail a lot of heat energy cost in the form of natural gas. In developing countries where fuelwood is cheap, the fuelwood is normally used to heat the boiler as well as for the distillation process.

6.7 CELLULOSIC ETHANOL STUDIES

A study was conducted at the Bioenergy Testing and Analysis Laboratory at TAMU (https://betalab.tamu.edu) to convert sweet sorghum plant into ethanol (Imam and Capareda, 2011). The sweet sorghum juice was extracted from the plant and was subsequently converted into ethanol using dry alcohol yeast (*S. cerevisiae*) (Ethanol Red) provided by Fermentis (Lesaffre Yeast Corporation, Milwaukee, WI). The yeast was stored in a refrigerator and activated immediately before fermentation. Dry yeast was activated by adding 0.5 g of dry yeast (Ethanol Red) to 10 mL of preculture broth. The 10 mL of preculture broth contained 0.2 g glucose, 0.05 g peptone, 0.03 g yeast extract, 0.01 g KH_2PO_4, and 0.005 g $MgSO_4 \times 7\ H_2O$. The preculture broth was shaken at 200 rpm in an incubator shaker at 38°C for 25–30 min.

The concentration of the inoculated cells was about 1×10^6 per milliliter, determined by plate count agar (PCA). Microbial cell cultures were serially diluted using peptone saline diluents (1 g/L peptone and 8.5 g/L NaCl) and were counted on a PCA that consisted of glucose (1 g/L), yeast extract (2.5 g/L), tryptone (5 g/L), and agar (15 g/L). Two varieties of sorghum (V-1 and V-2) were obtained from the Sorghum Breeding Program of the Soils and Crop Science Department at TAMU. The juice yield from pressing the plants was 40%–50%. Variety V-1 contained 64% sucrose, 22% glucose, and 14% fructose. Variety V-2 contained 56% sucrose, 30% glucose, and 14% fructose, as determined by the high-performance liquid chromatography (HPLC) method. The HPLC (Waters 9,600 with Consta Metric 3,200 solvent delivery system from LCD Analytical) is equipped with an autosampler, Shodex SP 810 packed column, and a refractive index detector. The column temperature was maintained at 60°C. Each sample was run for 25 min at a flow rate of 0.7 mL/min by using water as the eluent. Both the sugar composition and ethanol produced from experiments were measured using this HPLC unit.

The sorghum juice obtained from pressing was filtered by using 25 mm Whatman 1005325 Grade 5 qualitative filter paper. Fermentation efficiency was tested for an autoclaved sample (30 min at 60°C) and a non-autoclaved sample (frozen), with sugar contents of 25% and 30%. The first two samples evaluated the benefit of sterilization, and the second two samples evaluated whether higher sugar concentration is feasible. One liter of juice was supplemented with 3 g of yeast extract in a 1.5 L Erlenmeyer flask. The pH value of the juice was adjusted with extract to approximately 4.2–4.3 by using 2N hydrochloric acid. The juice was then inoculated with 10 mL of freshly activated dry yeast and run in a 3 L fermenter for a period of 72 h at 32°C and 750 rpm for ethanol production. All experiments were conducted in triplicate to determine ethanol production.

Results of this experiment are shown in Figure 6.7 for fermentation kinetics and in Figure 6.8 for fermentation efficiency. A few of the things garnered from these experiments are as follows:

1. There is very little advantage in sterilizing the juice once it is processed into ethanol, so long as it comes fresh from the plants.
2. Higher sugar concentration, lower fermentation efficiency, and sugar concentration below 25% are ideal.
3. Residual sugars are present for higher concentrations above 25%.
4. In this study, variety V-1 had a lower ethanol yield compared with variety V-2. However, rates of sugar consumption and ethanol production were

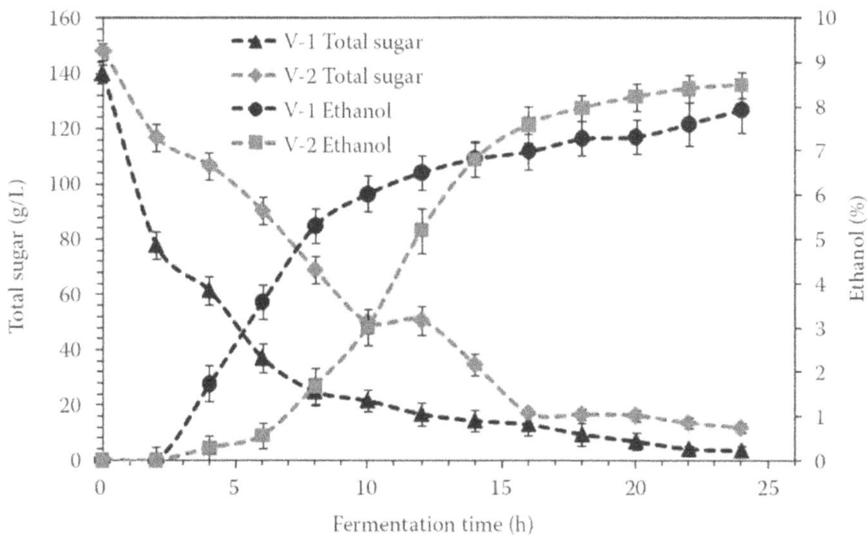

FIGURE 6.7 Glucose consumption and ethanol production during fermentation of sweet sorghum juice.

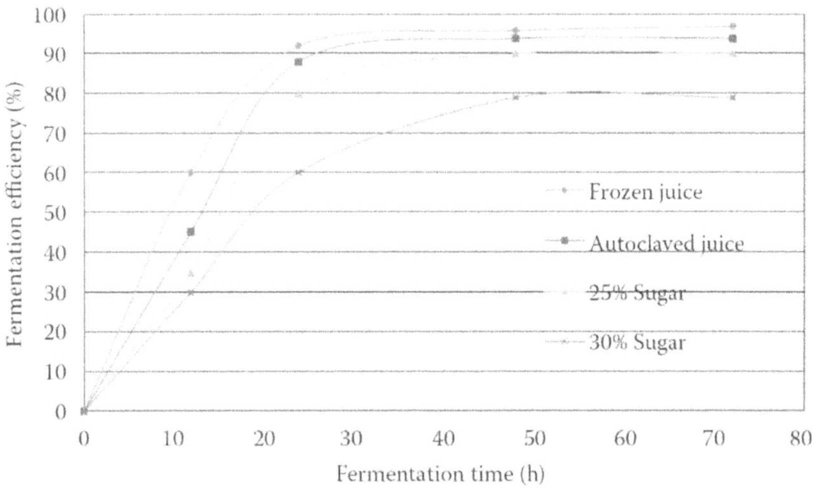

FIGURE 6.8 Plot of fermentation efficiency against fermentation time for ethanol production from sweet sorghum juice.

higher for V-1 because of its lower initial sugar concentration, as depicted in Figure 6.7. This illustrates the importance of conducting kinetic studies rather than basing the results on ethanol yield alone.

The maximum ethanol concentration in the final fermentation broth was 8.3% for variety V-1 and 9.2% for V-2. In terms of energy efficiency, variety V-1 may be a better option in the long run because of its higher ethanol production. It was also discovered that variety V-1 had a shorter maturation period than variety V-2. In terms of ethanol yield, variety V-2 may be the choice.

The sorghum biomass remaining after juice extraction was also used to produce ethanol using the lignocellulosic option discussed in this chapter. The dry milled sample of sweet sorghum biomass underwent further ethanol production using a genetically modified strain of *T. reesi*. The following enzymes were used for this study:

1. Accellerase 1500—ensures complete conversion of cellobiose to glucose
2. Accellerase XY—an accessory enzyme from *T. reesi*, used to increase specific xylanase activity of glucan and enhance various polysaccharide conversions
3. Accellerase XC—an accessory enzyme from a strain of *Penicillium funiculosum* and used to improve both xylan and glucan conversion
4. Accellerase BG—β-glucosidase from modified *T. reesi*, effective in converting cellubiose into monosaccharides and is also an inhibitor of cellulase enzyme

Before the enzymatic hydrolysis and fermentation process was implemented, an attempt was made to develop a pretreatment process to improve the succeeding fermentation process as follows:

1. The dry milled sorghum biomass sample was ground further in a Wiley Mill to obtain a particle size of about 1 mm in diameter and with a moisture content of 8% as determined by ASTM D3173.
2. Two different pretreatment procedures were conducted and compared on the 1 mm ground sample. First, the ground biomass was homogenized for 25 min using an ultrasonicator (Hielscher Ultrasonics, Ringwood, NJ). Second, the ultrasonicated sample was pretreated in a Parr pressure reactor (Parr Instrument Company, Moline, IL) in a 2 L reaction vessel. The ground and homogenized biomass were mixed with water in a ratio of 1:9 wt/wt of biomass and water. The biomass and water slurries were loaded in the reactor and treated at 150°C for 30 min. The pretreated biomass was then washed with hot distilled water and centrifuged for 20 min at 12,000 rpm to remove the dissolved sugar from the pretreatment step.
3. The supernatant from the pretreatment step was collected and analyzed using HPLC for the sugar released during this step. Most sugars released at this point are soluble sugars and some glucose could have been produced from the cellulosic part of the biomass.
4. The remaining portion was washed with water to ensure that soluble sugars did not influence the final glucose or ethanol yields.
5. This pretreated biomass then underwent enzymatic hydrolysis using the enzymes enumerated earlier.
6. The pretreated biomass was enzymatically hydrolyzed in 50 mM sodium acetate buffer at pH 4.8 for 96 8. Hydrolysis was conducted at 50°C and 125 rpm using an incubator/shaker (Innova, New Brunswick Scientific, NJ).

There were four sets of enzymatic treatment experiments that included the following enzyme loadings: (a) 0.15 mL/g of Accellerase 1500, (b) 0.15 mL/g of Accellerase 1500 + 0.07 mL/g of Accellerase XC, (c) 0.15 mL/g of Accellerase 1500 + 0.03 mL/g of Accellerase XY, and (d) 0.15 mL/g of Accellerase 1500 + 0.15 mL/g of Accellerase BG. Samples of hydrolysis slurries were collected at 0, 3, 6, 8, 24, 48, 72, and 96 h after the enzyme addition for soluble carbohydrate analysis. The sample slurries collected were centrifuged at 13,000 rpm for 20 min in a 1.5 mL autosampler vial and the supernatants were filtered using 0.5 pm hydrophilic polytetrafluoroethylene syringe filters (Millipore, Billerica, MA). These filtered samples were then analyzed on HPLC for sugar release during the enzyme hydrolysis procedure. Cellulose and hemicellulose conversion efficiency was presented by the percentage of lignocellulose enzymatically converted to glucose, pentose, and other hexose sugars. It was calculated by comparing the hexose or pentose yield (g) after enzymatic hydrolysis with the initial hexose or pentose content in the untreated biomass using Equation 6.5.

$$\% \text{ Enzymatic conversion of lignocellulose } = \left[\frac{c \times V}{m}\right] \times 100\% \quad (6.5)$$

where

c = concentration of glucose/xylose in the sample hydrolyzed, g/L

V = total volume hydrolyzed, L

m = weight of glucose or xylose before enzymatic hydrolysis determined through compositional analysis, g

The results are summarized as follows:

1. Cellulose and hemicellulose concentration increased by 35% and 15% for the hot water treatment and 49% and 25% for the combination of the mechanical and hot water treatment, respectively, increasing exposure of cellulose and hemicellulose to enzymatic hydrolysis after the pretreatment.
2. Lignin concentration was decreased by 52% after the ultrasonic treatment in combination with hot water pretreatment in a pressure reactor, while lignin increased by 1% when treated with hot water alone, which did not go through further homogenizing.
3. The cellulose conversion to glucose was the highest when Accellerase 1500 was used in combination with Accellerase XC enzyme (89%), followed by Accellerase 1500+BG (84%), and Accellerase 1500+XY (83%), and Accellerase 1500 (82%) (see Figure 6.9).
4. The hemicellulose conversion to xylose and arabinose was the highest when Accellerase 1500 was used in combination with Accellerase XC enzyme (48%) followed by Accellerase 1500+XY (40%).
5. On the basis of cellulosic conversion only, ethanol yield varied from 3.2 to 4.2 g ethanol per 100 g of sweet sorghum biomass in this study.

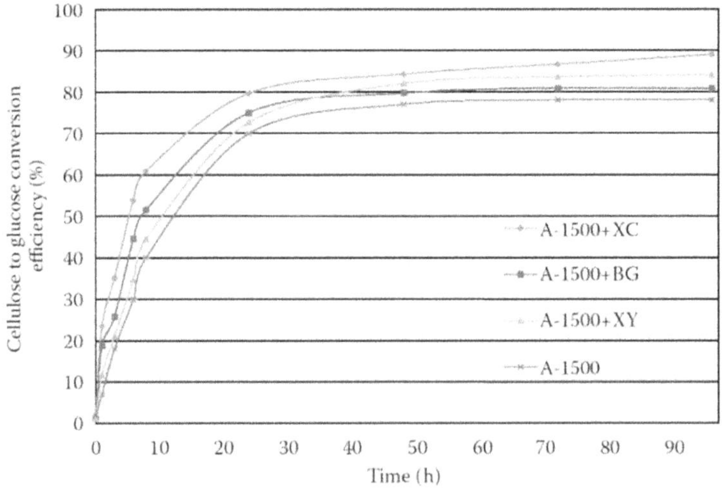

FIGURE 6.9 Lignocellulosic conversion into glucose using sweet sorghum biomass.

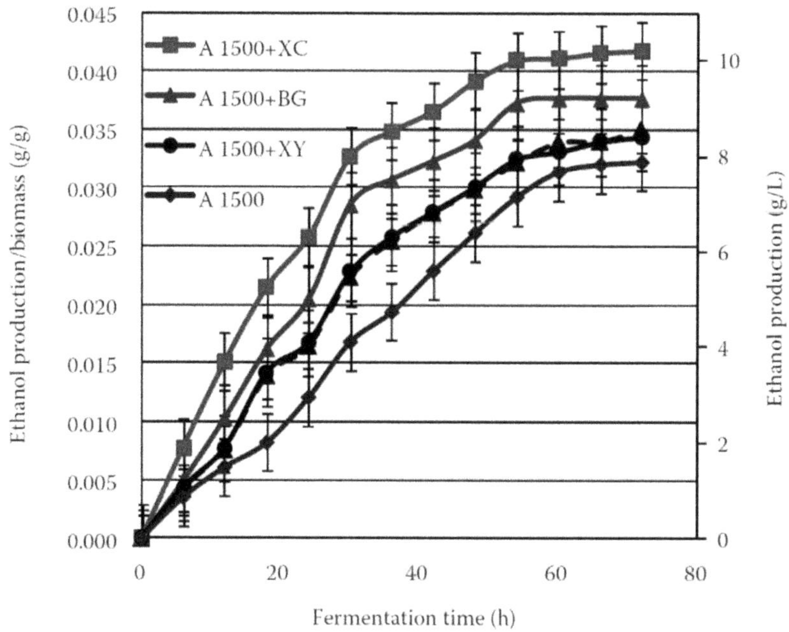

FIGURE 6.10 Effect of various enzymes used in this study on ethanol production.

6. The ethanol yields of Accellerase 1500+XC was 10%, 24%, and 31% higher than Accellerase 1500+BG, Accellerase 1500+XY, and Accellerase 1500, respectively (see Figure 6.10).
7. Using a combination of Accellerase 1500 and Accellerase XC enzyme in combination with mechanical and hot water pretreatment can be a potential process for lignocellulosic conversion to glucan and xylan, which in turn can result in increased ethanol production. The ethanol production can further be increased if a pentose-specific microbe can be used during fermentation from pentose sugars to ethanol.

In summary, producing ethanol from the sugar, starch, or cellulosic components of the plants is technically possible to provide higher ethanol yields; the cellulosic components are not wasted. Ethanol contribution from the juice will be added to the ethanol coming from the cellulose components. If there is residual biomass remaining after the fermentation process, it is possible that it may be further converted into ethanol by thermal means. Thus, the ethanol pathways are quite remarkable if the price of enzymes is significantly reduced. Example 6.8 shows how the efficiency of conversion is calculated.

Example 6.8: Enzymatic Conversion of Lignocellulosic Biomass

From a biomass compositional analysis, a sample of switchgrass was found to contain 6.4 g of glucose. After enzymatic hydrolysis, the product was analyzed using

HPLC and had a concentration of 36 g/L. The volume ofthe hydrolysis sample was 150 mL. What is the enzymatic hydrolysis efficiency for the saccharification of cellulose to glucose?

Solution:

a. The hydrolysis efficiency is calculated by comparing the hexose or pentose yield (g) after enzymatic hydrolysis with the initial hexose or pentose content in the untreated biomass as follows:

$$\%\text{Enzymatic conversion of lignocellulose} = \left[\frac{c \times V}{m}\right] \times 100\%$$

where c =36 g/L = 0.036 g/mL, V = 1450 mL, and m = 6.4 g

$$\%\text{Enzymatic conversion of lignocellulose} = \left[\frac{0.036 \times 150}{6.4}\right] \times 100\% = 84.4\%$$

b. Thus, the enzymatic conversion efficiency is 84.4%.

6.8 COMMERCIAL CELLULOSIC ETHANOL PLANTS IN THE UNITED STATES

There were numerous cellulosic ethanol plants constructed in the United States during the biofuel bloom in early 2005. Most plants were supported by subsidies and grants. Iogen, POET, Abengoa, DuPont, Diversa, and Novozymes and Dyadic built refineries that can process biomass and turn it into bioethanol using either conventional processes or their unique enzymes. However, most of these facilities were either canceled or closed because of technical obstacles and costs.

Many of the companies developing cellulosic ethanol technology or demonstrations in the United States are members of the Advanced Ethanol Council, which provides extensive information and data on the ethanol industry. INEOS New Planet BioEnergy commenced commercial production at its 30.3M liters-per-year cellulosic ethanol biorefinery at Vero Beach, Florida. The feedstock for the process includes citrus and agricultural wastes and wood wastes. Poet-DSM Advanced Biofuels, LLC also commenced production of its 75.7 M liters per year of cellulosic ethanol from corn stover and cob.

6.9 CONCLUSION

Considerable research has been accomplished in the conversion of biomass feedstock into ethanol. During the energy crisis in the 1970s and 1980s, Brazil embarked on a massive project of the conversion of sugarcane into ethanol, while other countries embarked on their own indigenous feedstock sources. As early as that time, the question of a positive net energy balance emerged, with sugary feedstock having the edge over other sources.

It was shown in this chapter that sugar, starch, and lignocellulosic biomass may be converted into ethanol. Sugar feedstocks such as sugarcane and sweet sorghum, called the first-generation biofuels, have been the most popular source of inputs for direct ethanol production. They require only the action of yeast, with *S. cerevisiae* being the most popular and still the best yeast being used thus far. Biomass high in starch requires additional enzymes to convert the starch into sugars before adding yeast for the final fermentation of the sugars into ethanol. Majority of ethanol produced in the United States comes from the starch material in corn. However, both biomass feedstocks (sugar and starch) are primarily food sources, and this creates a food-to-fuel issue. The price of corn in the United States has always been affected by the proliferation of ethanol plants in the corn-belt areas, and there has been a move to seek alternative sources of ethanol. The most logical option would be the conversion of the most abundant biomass resource on Earth, lignocellulosic biomass, into ethanol. Currently, the only remaining issue for the conversion of cellulose into sugars and ethanol is the high cost of cellulosic enzymes, as well as the high cost of pretreatment processes outlined in this chapter.

Numerous enzyme companies such as Genencor International (Palo Alto, CA), Iogen Corporation (Ottawa, Ontario, Canada), and Dyadic International (Jupiter, FL) have been conducting research to improve the efficiencies of enzymatic conversion of lignocellulosics into ethanol as well as reduce costs. Genencor International has been genetically modifying *T. reesi* to achieve high ethanol conversion yields. Some results are presented in this chapter. With continued government support, these companies can lower the cost of enzymes and thereby make the production of cellulosic ethanol economically feasible in the future. Other emerging routes for ethanol production include the conversion of synthesis gas directly into ethanol by identifying microbes that are very efficient in this conversion pathway. *Clostridium ljungdahlii* bacteria is said to ingest carbon monoxide and hydrogen and produce ethanol effectively. CO and H_2 may be produced from the pyrolysis or gasification of lignocellulosic biomass as well. Developing countries should investigate local sources of similar enzymes for the same purpose. Reports have been published on the use of certain fungi (e.g., white rot fungi) that hydrolyze cellulosic biomass into simple sugars. In all of these undertakings, the support of local governments will be instrumental in the success of these projects, including high amounts of research dollars and some form of subsidy for either the feedstock or the enzyme costs. Although cellulose is the most abundant biomass resource on Earth, researchers have yet to identify the most cost-effective enzymes produced by indigenous microbes. There are hundreds of candidate microbes for the purpose, and the search will continue until that "super bug" is found.

6.10 PROBLEMS AND DISCUSSION ISSUES

6.10.1 Rules of Thumb in Ethanol Production

P6.1 Determine the amount of pure sucrose sugar (in kg) to generate a liter of ethanol using the idealized equation shown below. The density of ethanol is 0.8 kg/L. Convert this unit into its English equivalent (lb/gal).

$$\underset{\text{(sugar)}}{C_{12}H_{22}O_{11}} + H_2O + \text{yeast} \rightarrow \underset{\text{(ethanol)}}{4C_2H_5OH} + 4CO_2 + \text{heat}$$

6.10.2 First- and Second-Generation Biofuels

P6.2 Differentiate between the first- and second-generation biofuels in the bioethanol industry in terms of biomass source type, processing technology, and biomolecules used.

6.10.3 Sugars and Ethanol Yield from Sweet Sorghum

P6.3 Sweet sorghum stalks were pressed to obtain 1 L of juice (mass, 1 kg). The juice was then filtered and analyzed on HPLC for its total sugar concentration of 140 g/L. It was further fermented using yeast and produced an ethanol concentration of 65 g/L, as analyzed on HPLC. Determine the following: (a) the ethanol yield per weight of juice (g/g and g/100 g juice) and (b) the ethanol yield per weight of sugar as substrate (g/g and g/100 g).

6.10.4 Fermentation Efficiency and Ethanol Yields per Unit of Area

P6.4 Ethanol yield is calculated to be 46 g per 100 g of glucose from a fermentation experiment. What is the efficiency of this fermentation process? Assume the theoretical ethanol production of 51.5 g ethanol/100 g glucose. Sweet sorghum production is about 44.91 tonnes/ha (20 dry tons of sorghum/acre), and 50% juice by weight is derived from pressing the stalks. Calculate the ethanol yield per hectare (acre) of land. Assume density of ethanol is 0.789 kg/L. The actual yield of ethanol from the juice was found to be 46 g ethanol/liter of juice. You may also assume a liter of juice is approximately 1 kg.

6.10.5 Pretreatment and Ethanol Production Process

P6.5 What are the steps for lignocellulosic biomass conversion to bioethanol? Briefly explain what happens at each step. List a few pretreatment platforms for lignocellulosic conversion.

6.10.6 Differences in First- and Second-Generation Ethanol

P6.6 State the differences between first-generation (starch) and second-generation (lignocellulose) biomass structures and how these differences influence the breaking of the structure during pretreatment.

6.10.7 Enzyme and Its Role in Ethanol Production

P6.7 What is an enzyme? Describe the enzyme system used for first-generation (starch) versus second-generation (lignocellulose) biomass because of the structural differences in the second-generation biomass.

6.10.8 Enzymatic Conversion of Lignocellulosics into Ethanol

P6.8 An experiment was carried out using sweet sorghum biomass using enzyme A and enzyme B for a 96-h period. These enzymes were cellulase enzymes that converted the cellulose in the biomass to glucose. The volume of biomass hydrolyzed was 150 mL and the biomass contained 6.4 g glucose before hydrolysis. Table P6.1 presents the data over the 96 h period from HPLC. Using the data, calculate the percent enzymatic conversion of lignocellulose to glucose. Plot the data.

TABLE P6.1
Experimental Data for Glucose-to-Ethanol Conversion

Time (h)	Enzyme A Glucose (g/L)	Enzyme B Glucose (g/L)
0	0.5	0.8
1	10	3
3	15	7.5
6	23	13
8	26	17
24	34	30
48	36	33
72	37	33.1
96	38	33.2

6.10.9 Advantages and Disadvantages of Ethanol Production

P6.9 If you are to recommend the research and development of a company between the first and second generations of bioethanol production, which one would you recommend and why?

6.10.10 Lignin Conversion and Uses

P6.10 Most biological conversions use cellulose and hemicellulose to produce bioethanol: the lignin remains unused for biofuel production. Can we make any bioproducts from lignin?

6.10.11 Improving Biofuel Production

P6.11 List some of the technologies that can be improved for second-generation biofuel production.

6.10.12 Feedstock with the Highest Bioethanol Production Potential

P6.12 Name the feedstock that will potentially yield the most bioethanol per unit of area.

a. Jerusalem artichoke b. grapes c. corn d. sorghum

6.10.13 Feedstock with the Most Carbohydrates

P6.13 Identify the biomass feedstock with a reported highest carbohydrates percentage by weight.

a. wheat b. rice c. corn d. sugarcane

6.10.14 Crop with the Highest Residue to Crop Ratio

P6.14 Identify the agricultural crop having the highest published residue to crop ratio.

a. rice b. bagasse c. corn d. barley

6.10.15 Acreage Required for a Bioethanol Refinery

P6.15 Determine the acreage required for an18.925 [5 MGY] bioethanol facility from dry dates. List all assumptions made.

6.10.16 Bioethanol from Whole Plant Biomass

P6.16 Determine the potential bioethanol produced from oat and oat straw. List all assumptions made. Assume you have 405 ha [1,000] acres of land for this project.

6.11 CASE STUDIES

Propose a project to produce 3.785 million liters per year (MLY) of bioethanol derived from corn using dry milling process. The bioethanol and dry distiller's grain (DDG) will be sold on a commercial scale while taking advantage of tax credit from the government. Develop a spreadsheet to perform simple economic and sensitivity analyses. Data is summarized in the table below.

Corn Price	$0.154/kg dry corn	Facility Cost	$264,201/L
Processing Cost	$0.132/L ethanol	Ethanol Selling Price	$0.5284/L
Yearly Operation	350 days	DDG Selling Price	$0.198/kg
Financial Parameters		Other Variable Costs	
Project Life	20 years	Labor Hours	10 h/day
Interest on Investment	8%	# of personnel	6 full timc
Tax on Capital	2%	# of Manager	1 full-time
Discount Rate	2.36%	Labor Wage Rate	$15/h
Tax Credit	30%	Manager Salary	$80,000/pax/yr
Product Yields		R&M Cost	5% of capital
Bioethanol	2,001 L/ha	Utilities Cost	10% of capital
DDG Yield	30% of corn usage	Corn Conversion	2.52 kg/L

Required:

a. Determine the acreage required for the operation assuming 2,001 L of ethanol is produced per hectare of land.
b. What is the total plant cost?
c. Report on the yearly corn feedstock tonnage requirement (tonnes/yr).
d. Develop a simple spreadsheet to estimate economic parameters, such as internal rate of return (IRR), payback period (PBP), benefit cost ratio (BCR), and net present value (NPV), and report on the viability of this project given the data.
e. Perform sensitivity analyses to determine the impact of changing input parameters on the economic indicators. A graph for relationships would be beneficial visually, with predictive equations and with R^2 value n the chart. The following are suggested changes:
   i. Acreage on NPV (3.785 MLY to 9.4625 MLY)
   ii. Ethanol selling price versus IRR and PBP ($0.500/Lto $0.9247/L)
   iii. DDG selling price versus IRR and PBP ($0.20/kg to $0.55/kg)
   iv. Tax credit on IRR and PBP (20% to 40%)
   v. Capital cost on IRR and PBP ($132,100/MLY to $660,502/MLY)
   vi. Interest on investment on IRR and PBP (5% to 20%)
   vii. Feedstock cost on IRR and PBP ($0.055/kg to $0.165/kg)
   viii. Processing cost on IRR and PBP ($0.026/L to $0.159/L)

REFERENCES

Brown, L. 2006. *Plan B 2.0. Rescuing a Planet Under Stress and a Civilization in Trouble.* Earth Policy Institute, W. W. Norton and Co. Publishing, New York.

Carvalheiro, F. and L. C. Duarte. 2008. Hemicellulose biorefineries: A review on biomass pretreatment. *Journal of Scientific and Industrial Research* 67 (11): 849–864.

Coble, C. G., J. M. Sweeten, R. P. Egg, E. J. Soltes, W. H. Aldred and D. A. Givens. 1985. Biological conversion and fuel utilization: Fermentation for ethanol production. In: *Biomass Energy: A Monograph* (E. A. Hiler and B. A.S tout, eds.). Texas A&M University Press, College Station, TX. pp. xx–xx.

Dale, B., E. Newton, M. Laser, S. Kim, L. Laureano-Perez, H. Alizadeh, H. Jin and L. Lynd. 2009. *Recent Improvements for the Ammonia Fiber Explosion (AFEX) Process and Resulting Reductions in Projected Ethanol Prices*. Great Lakes Bioenergy Research Center Report, Michigan State University, Ann Arbor, MI.

Dutta, A., M. Talmadge, J. Hensley, M. Worley, D. Dudgeon, D. Barton, P Groenendijk, D. Ferrari, B. Stears, E. M. Searcy, C. T. Wright and J. R. Hess. 2011. Process design economics for conversion of lignocellulosic biomass to ethanol; thermochemical pathway by indirect gasification and mixed alcohol synthesis. Technical Report NREL/TP-5100-51400 for DOE Task No. BB07-3710 under contract No. DE-AC36-08GO28308. USDOE, Golden, CO.

Eakin, D. E., J. M. Donovan, G. R. Cysewski, S. E. Petty and J. V. Maxham. 1981. Preliminary evaluation of alternative ethanol/water separation processes. Pacific Northwest national Laboratory, Battelle Memorial Institute, Report No. PNL-3823. Prepared for the US Department of Energy under contract DE-AC06-76RLO 1830.

Egg, R. P., J. M. Sweeten and C. G. Coble. 1985. Grain sorghum stillage recycling: Effect on ethanol yield and stillage quality. *Biotechnology and Bioengineering* XXVII: 1735–1738.

Fair, J. R. 1987. Distillation. In: *Handbook of Separation Process Technology* (R. W. Rousseau, ed.). John Wiley & Sons, Ltd., New York. pp. 229–339.

Harmsen, P F. H., W. J. J. Huijgen, L. M. Bermudez Lopez and R. R. C. Bakker. 2010. Literature review of physical and chemical pretreatment processes for lignocel-lulosic biomass. ECN Report No. ECN-E-10-013. Energy Research Center of the Netherlands (ECN), Wageningen, Netherlands.

Hendriks, A. T. W. M and G. Zeeman. 2008. Pre-treatments to enhance the digestibility of lignocellulosic biomass. *Bioresource Technology* 100 (1): 10–18.

Hernandez, J. R., S. C. Capareda, O. Portillo, D. B. Hays and W. L. Rooney. 2011. Simultaneous saccharification and fermentation (SSF) of high digestible grain sorghum for ethanol production. *Biological Engineering Transactions* 4 (1): 3–15.

Imam, T. and S. C. Capareda. 2011. Fermentation kinetics and ethanol production from different sweet sorghum varieties. *International Journal of Agricultural and Biological Engineering* 4 (3): 33–40.

Naidu, K., V. Singh, D. B. Johnston, K. D. Rausch and M. E. Tumbleson. 2007. Effects of ground corn particle size on ethanol yield and thin stillage. *Cereal Chemistry* 84 (1): 6–9.

Nichols, N. N., B. S. Dien, R. J. Bothast and M. A. Cotta. 2006. The corn ethanol industry. In: *Alcoholic Fuels* (S. Minteer, ed.). CRC Press, Boca Raton, FL. pp. 59–78.

Nichols, N. N., D. A. Monceaux, B. S. Dien and R. J. Bothast. 2008. *Production of Ethanol from Corn and Sugarcane. Bioenergy*. ASM Press, Washington, DC. pp. 3–15.

Renewable Fuels Association (RFA). 2023. *How Ethanol is Made*. Ellisville, MO. Available at: https://ethanolrfa.org/ethanol-101/how-is-ethanol-made or How Ethanol is Made (ethanolrfa.org) (accessed July 12, 2023).

Singh, V., C. J. Batie, G.W. Aux, K. D. Rausch and C. Miller. 2006. Dry-grind processing of corn with endogenous liquefaction enzymes. *Cereal Chemistry* 83 (4): 317–320.

ASSESS YOUR KNOWLEDGE

1. The US average annual consumption of bioethanol in 2020 was 800,000 barrels per day. The gasoline usage for that year was around 123.49 billion gallons. Determine the % contribution of bioethanol in the overall gasoline market.
   a. 2.72
   b. 27.3
   c. 0.293
   d. 9.96

2. Sweet sorghum stalks were pressed to obtain a gallon of juice (mass = 8.2 lbs). The total sugar concentration was found to be 1.2 lb/gallon. The juice was further fermented using yeast and produced an ethanol concentration of 0.45 lb/gal. Calculate the amount of sugar required in lbs to generate a gallon of ethanol. Assume the density of ethanol is 6.7 lb/gal.
   a. 10
   b. 18
   c. 20
   d. 25

3. Determine the air-to-fuel ratio by volume (i.e., L/L) for the complete combustion of pure propanol (C_3H_8O) in air. Note that air contains 79% nitrogen and 21% oxygen by volume. The main products are water and carbon dioxide.
   a. 12.06
   b. 2.20
   c. 10.3
   d. 4.5

11. Sweet sorghum stalks were pressed to obtain a gallon of juice (mass = 8.2 lbs). The total sugar concentration was found to be 1.2 lb/gallon. The juice was further fermented using yeast and produced an ethanol concentration of 0.45 lb/gal. Calculate the amount of ethanol produced per lb of juice. Assume the density of ethanol is 6.7 lb/gal.
    a. 10
    b. 18.2
    c. 0.0549
    d. 0.45

12. Determine the amount of carbon dioxide (in lbs) produced when a gallon of pure ethanol is made based on the ideal equation shown below using sugar as substrate. Assume the density of ethanol is 6.57 lbs/gallon.

    $C_6H_{12}O_6$ + yeast à $2C_2H_5OH$ + $2CO_2$ + heat

    a. 6.28
    b. 3.14
    c. 12.9
    d. 1.96

13. Determine the acreage required per year to produce 5 million gallons of ethanol from corn starch. Assume corn productivity is 4.45 tons per acre per year. Use ethanol yield of 2.8 gallons per bushel of corn produced.
    a. 6,742
    b. 50,000
    c. 11,236
    d. 25,000

4. Determine the amount of sorghum grain or milo required (in tons) to generate 3 million gallons (MGY) per year of bioethanol. Assume the current ethanol yield from this grain to around 2.8 gallon per bushel. Milo weighs 56 lbs to a bushel. The grain sorghum yield is around 100 bushels per acre.
   a. 30,000
   b. 10,714
   c. 21,430
   d. 60,000

5. A farmer wanted to produce bioethanol from corn stover. The corn stover yield was reported at 8.8 tons/acre/year (wet basis). It was reported in literature that one can generate about 85 gallons of bioethanol per dry ton of the material. The average moisture of harvested corn stover is 10%. He owns 100 acres of corn field from where he will acquire the corn stover for this project after corn is harvested. How much bioethanol (gallons) will he generate each year?
   a. 134,640
   b. 269,280
   c. 67,320
   d. 33,660

14. Determine the acreage required (in acres) to produce sorghum grain or milo to generate 3 million gallons (MGY) per year of bioethanol. Assume the current ethanol yield from this grain to be around 2.8 gallon per bushel. Milo weighs 56 lbs to a bushel. The grain sorghum yield is around 100 bushels per acre.
   a. 30,000
   b. 10,714
   c. 21,430
   d. 60,000

15. A farmer would like to produce bioethanol from corn stover. The corn stover yield was reported at 8.8 tons/acre/year (wet basis). It was reported in literature that one can generate about 85 gallons of bioethanol per dry ton of the material. The average moisture of harvested corn stover is 10%. He owns 100 acres of corn field where he will acquire the corn stover for this project after corn is harvested. How much gross revenue (in $) will he generate each year if the selling price of his bioethanol yield is $1.50/gallon?
   a. 302,940
   b. 201,960
   c. 50,490
   d. 100,980

6. In the pretreatment of lignocellulosic materials for making ethanol, a popular pretreatment process is called AFEX. What is the meaning of this acronym or what does AFEX stands for in ethanol production process?
   a. Associated Fiber Expressions Process
   b. Ammonia Fiber Expansion Process
   c. Acid Flowthrough Expansion Process
   d. *Aspergillus flavus* Extract Process

7. This is a pretreatment process for the decomposition of cellulose using mild to string chemical compounds with very low pH.
   a. Enzymatic hydrolysis
   b. Acid hydrolysis
   c. AFEX
   d. Steam explosion

8. These are several varieties of enzymes produced chiefly by fungi, bacteria, and protozoans that catalyze cellulose or decompose cellulose and of some related polysaccharides, usually for the production of simple sugars. What do you call those enzymes that decompose cellulose?
   a. Fuelzyme
   b. Amylases
   c. Sugarase
   d. Cellulase

9. This is a pretreatment process for the decomposition of cellulose using the combination of high temperature and pressure steam.
   a. Steam explosion
   b. Acid hydrolysis
   c. Enzymatic hydrolysis
   d. Lysis

16. Name the most popular commercial dry alcohol yeast (common trade name) used for making ethanol from sugar crops and distributed by Fermentis Group.
   a. Leavening Yeast
   b. Baker's Yeast
   c. Fleischmann's Yeast
   d. Ethanol Red

17. Approximately how much water is used (gallons) to generate a gallon of ethanol?
   a. 21
   b. 41
   c. 82
   d. 160

18. What is the approximate natural gas usage (ft^3) when a gallon of ethanol is produced?
   a. 70
   b. 35
   c. 7
   d. 140

19. In the pretreatment economics for making ethanol, which process had the lowest pretreatment cost in units of $/gallon?
   a. Dilute Acid
   b. AFEX
   c. Lime
   d. Hot water

10. Most ethanol produced in the United States comes from which general pathway?
    a. Dry Grind
    b. Wet milling
    c. Lignocellulosic
    d. Sugar

20. This is a cellulose biomass pretreatment process using the following compound O_3.
    a. Tri-oxi process
    b. AFEX
    c. Organosolv
    d. Ozonolysis

Answers: 1d, 2b, 3d, 4a, 5c, 6b, 7b, 8d, 9a, 10a, 11c, 12a, 13c, 14b, 15d, 16d, 17b, 18a, 19a, 20d

ADDITIONAL LEARNING RESOURCES

- Access the Lecture Notes at the Instructor and Student Resources: https://routledgetextbooks.com/textbooks/instructor_downloads/
- The Case Study Videos can be found in the ebook+ version of this book.

7 Biogas Production

LEARNING OBJECTIVES

Upon completion of this chapter, one should be able to

1. Describe the process of producing methane from various biomass resources, particularly animal manure.
2. Enumerate the different types of microorganisms that are responsible for the conversion of lignocellulosic materials into biogas (methane and carbon dioxide).
3. Estimate the amount of biogas produced for the different animal manure types and their conversion efficiencies.
4. Enumerate the advantages and disadvantages of anaerobic digestion systems.
5. Design a simple biogas digester and differentiate between first- and second-generation anaerobic digesters.
6. Discuss the various uses of biogas.

7.1 INTRODUCTION

Biogas production is one of the oldest and most widely used conversion processes for most wastewater treatment plants (Turovskiy and Mathai, 2006), including animal manure (NAS, 1976). The technical term for the process is anaerobic digestion. It is a biochemical conversion process carried out by various microorganisms in which the biomass organic material is converted into organic acids, which, in turn, are converted into methane gas (with carbon dioxide as the other by-product). The process takes place under very strict anaerobic conditions (no oxygen). The entire process takes place in four clearly defined steps, as shown in Figure 7.1, and includes:

1. Hydrolysis—converting insoluble complex biomass into soluble compounds
2. Acidogenesis—converting higher organics into simpler organics
3. Acetogenesis—converting higher organics and organic acids into acetic acid
4. Methanogenesis—converting organic acid and hydrogen into methane

7.1.1 Hydrolysis

Hydrolysis is the process of converting insoluble complex biomass into soluble compounds by the action of extracellular enzymes of hydrolytic bacteria. These bacteria are the first set of microorganisms that breaks down the biomass into simpler

DOI: 10.1201/9781003294306-7

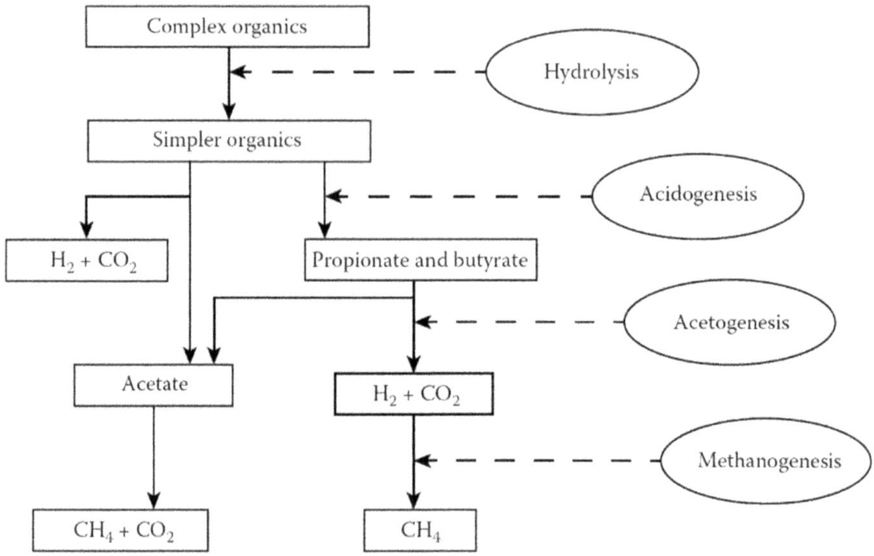

FIGURE 7.1 Schematic of anaerobic digestion process.

compounds. Carbohydrates, fats, and proteins are converted into sugars, fatty acids, and amino acids. The bulk of microorganisms are fermentative in nature (i.e., organisms that digest cellulose, hemicellulose, starch, proteins, and sugars, converting them into some other forms such as fatty acids and alcohols). In most anaerobic digesters, this stage can be a rate-limiting step if the digester is not properly inoculated with fermentative microbes. Thus, in most newly operated anaerobic digesters, the first few days of operation could be challenging if the microbial population in the reactor is too small to tackle this initial step. At the end of hydrolysis, the complex organics are converted into simpler organics, mostly volatile fatty acids.

7.1.2 Acidogenesis

The second step (acidogenesis) is the conversion of the soluble organics from hydrolysis into short-chain fatty acids and alcohols. Most acids are converted into acetic acid using the acetogenesis step (conversion into acetic acid). Other organic acids are also produced, such as propionic, butyric, valeric, and others. At this stage, hydrogen is produced by the fermentation of glucose. Because of the significant production of acidic compounds, the pH of the system goes down. The microbes responsible for this process are called "acid formers" and may be composed of homoacetogenic bacteria and some facultative bacteria. The doubling time of these organisms is normally 14h. The hydrogen gases formed are sometimes produced by photosynthetic bacteria. In this case, acetic acids and water are converted into carbon dioxide and hydrogen.

7.1.3 Acetogenesis

The acetogenesis step is the widespread production of acetic acid (CH_3COOH) by hydrogenation, as shown in Equation 7.1. In this relationship, 2 mol of carbon dioxide and 4 mol of hydrogen gas are combined to generate 1 mol of acetic acid and 2 mol of water.

$$2CO_2 + 4H_2 \rightarrow CH_3COOH + 2H_2O \tag{7.1}$$

Acetic acid may also be formed by the dehydrogenation process, as shown in Equation 7.2. This representative equation is the breakdown of propionic acid into acetic acid with the coproduction of hydrogen and more carbon dioxide. The weight balance for the reactants and products is also shown.

$$\underset{(74\ \text{kg})}{C_3H_6O_2} + \underset{(54\ \text{kg})}{3H_2O} \rightarrow \underset{(60\ \text{kg})}{CH_3COOH} + \underset{(6\ \text{kg})}{3H_2} + \underset{(44\ \text{kg})}{CO_2} + \underset{(18\ \text{kg})}{H_2O} \tag{7.2}$$

During the anaerobic digestion process, acetic acids are the most abundant organic acids produced, among many other fatty acids. A list of the most common fatty acids produced during the digestion process is summarized in Table 7.1. Volatile acids are classified as short-chain groups if the carbon atom is less than six. These acids are measured accurately using a gas chromatograph, although other methods such as direct titration, steam distillation, and conventional distillation are also used.

7.1.4 Methanogenesis

The final step in the anaerobic digestion process is the conversion of organic acids into methane by the action of methane-producing microbes (methanogenesis). The methanogenesis process is broken down into two parts: (a) the conversion of acetic acid into methane (called acetotrophic), which comprises about 70% of the process reactions, and (b) the conversion of hydrogen into methane utilizing the carbon dioxide produced during the process (also called hydrogenotropic), which comprises the remaining 30% of the process reactions. These reactions are shown in Equations 7.3

TABLE 7.1
Volatile Fatty Acids in Anaerobic Digestion Processes

Common Name	Chemical Formula	Chemical Name	MW
Formic acid	HCOOH	Methanoic acid	46
Acetic acid	CH_3COOH	Ethanoic acid	60
Propionic acid	C_2H_5COOH	Propanoic acid	74
Butyric acid	C_3H_7COOH	Butanoic acid	88
Valeric acid	C_4H_9COOH	Pentanoic acid	102
Caproic acid	$C_5H_{11}COOH$	Hexanoic acid	116

and 7.4, respectively, including the estimated amounts of products formed by volume (mole basis) and by weight (kg).

$$\underset{(60\ \text{kg})}{CH_3COOH} \underset{\rightarrow}{\rightarrow} \underset{(16\ \text{kg})}{CH_4} + \underset{(44\ \text{kg})}{CO_2} \quad (7.3)$$

$$\underset{(8\ \text{kg})}{4H_2} + \underset{(44\ \text{kg})}{CO_2} \underset{\rightarrow}{\rightarrow} \underset{(16\ \text{kg})}{CH_2} + \underset{(36\ \text{kg})}{2H_2O} \quad (7.4)$$

These reactions are very sensitive to pH, and normally should occur in a pH range between 6.8 and 9.1. In addition, these reactions are very slow, and the doubling times of microbes are normally from 12 h to as long as 9 days. The reactions are also very sensitive to environmental changes, such as temperature and the concentrations of other compounds such as ammonia (NH_3) and hydrogen sulfide (H_2S). Note that by weight, it would seem as though the methane production is small relative to the amount of acetic acid used. However, as will be shown in Example 7.1, a large volume of methane is produced per unit weight of acetic acid. The theoretical production is around 370 L (98 gal or 13 ft^3) of methane per 1 kg (2.2 lb) of acetic acid or 390 L (103 gal or 13.8 ft^3) of methane per 1 L (0.26 gal) of acetic acid. The reason for this is the low density of methane gas compared with the liquid acetic acid.

Example 7.1: Amount of Methane Formed from Acetic Acid

Equation 7.3 shows that for every mole of acetic acid, 1 mol of methane is formed. Estimate the theoretical pure methane production (in L or ft^3) for every amount of acetic acid in units used (L or kg). Assume the density of components as follows:

- Density of pure acetic acid at STP = 1.049 kg/L
- Density of pure methane at STP = 0.717 kg/m^3

Solution:

a. The methane production per kilogram or liter of acetic acid is estimated as follows:

$$\text{Methane}\left(\frac{m^3}{kg}\right) = \frac{16\ \text{kg methane}}{60\ \text{kg acetic acid}} \times \frac{m^3}{0.717\ \text{kg}} = 0.372\frac{m^3}{kg}$$

$$\text{Methane}\left(\frac{L}{kg}\right) = 0.372\frac{m^3}{kg} \times \frac{1{,}000\ L}{1\ m^3} = 372\frac{L}{kg}$$

b. The methane production on a volume basis is calculated as follows:

$$\text{Methane}\left(\frac{m^3}{L}\right) = \frac{16\ \text{kg}\ CH_4}{60\ \text{kg}\ C_2H_4O_2} \times \frac{1\ m^3}{0.717\ \text{kg}} \times \frac{1.049\,\text{kg}}{L} = 0.39\frac{m^3}{L}$$

$$\text{Methane}\left(\frac{L}{L}\right) = \frac{0.39\,m^3}{L} \times \frac{1{,}000\,L}{m^3} = 390\,\frac{\text{L methane}}{\text{L acetic acid}}$$

Temperature effects this conversion process because different microbes thrive under different zones of comfort. It is also known that methane conversion processes are faster when high-temperature-tolerant organisms (called thermophilic organisms) are used in the reactor. The thermophilic temperature range is usually from 45°C to 55°C. In cold climates, this would mean heating the reactor to favor the growth and conversion of these thermophilic microbes. Mesophilic microbes are those that thrive from 20°C to 42°C. Psychrophilic microbes are those that thrive below 20°C. Note the decrease in hydraulic retention times (HRTs) as thermophilic microbes are utilized.

HRT is defined as the volume of sludge in the digester (m^3) divided by the volume of digester sludge withdrawn daily (m^3/day). HRT is also the time that a given volume of sludge remains in the reactor. Another parameter used is the solids retention time (SRT) to identify the amount (kg) of solids in the digester divided by the mass of solids withdrawn daily (kg/day). Both the HRT and SRT are easily established by conducting a simple experiment on a batch process utilizing the feedstock in question, as well as the use of proper microorganisms for the anaerobic digestion process. In a simple laboratory, this is achieved by developing a batch reactor system shown schematically in Figure 7.2, with an accompanying photo of the setup shown in Figure 7.3. Two 3.785 L (1 gal) glass bottles are used with a graduated cylinder to measure the volume of biogas produced. The first bottle on the left is where the biomass and animal manure mixture are placed and inoculated with microbial population (from an existing operational digester). A mixture of some biomass with manure is better to improve or raise the carbon-to-nitrogen ratio (C/N). The first bottle is completely sealed, and the outlet is directed to the second bottle that is filled with clean water. The second bottle also has another tube immersed at the bottom. The other end of this tube is directed to the graduated cylinder. Note that the tube coming from the digester is short (and should not touch the top water level) and the tube going

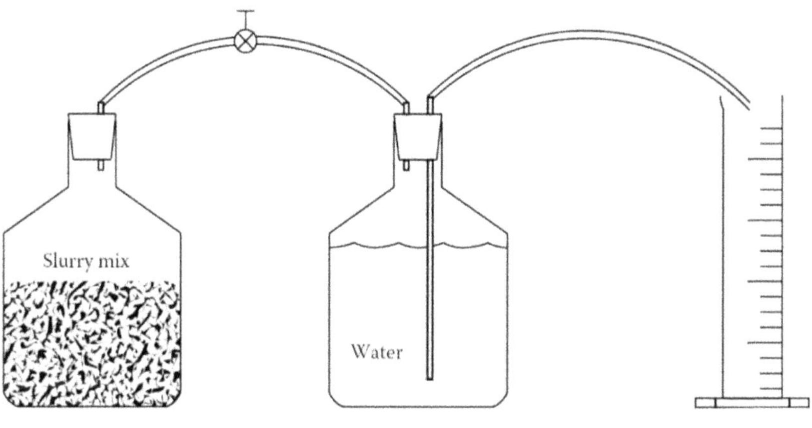

FIGURE 7.2 Schematic of laboratory batch digestion system used to establish HRT or SRT.

FIGURE 7.3 Photo of simple laboratory setup to determine retention times in the digester.

to the graduated cylinder is immersed up to the bottom of this water displacement bottle. Each day, the reactor is shaken to mix the microbes and biomass.

The amount of water displaced on the graduated cylinder indicates the amount of biogas produced. One should observe typical results similar to a graph shown in Figure 7.4, which records the actual data taken from the digestion of a mixture of manure and ground biomass (e.g., alligator grass) and pure manure (dairy manure). Note that there will be a point where the rate of gas production is diminished (in the graph, around 60 days) until the amount of gas produced becomes marginal. This

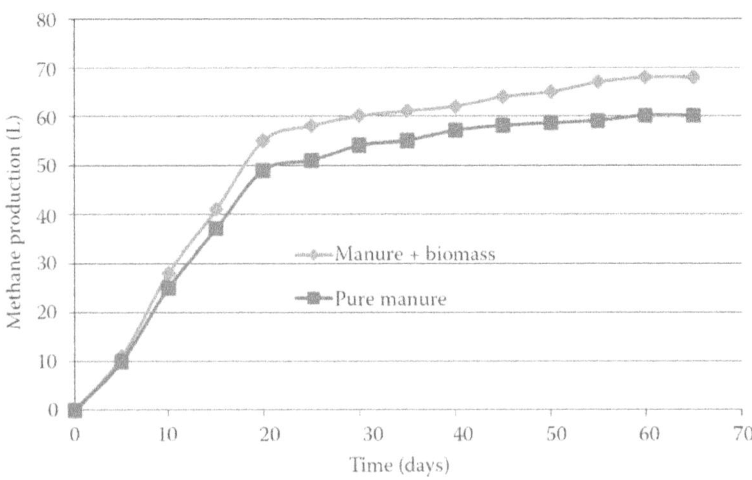

FIGURE 7.4 Typical biogas production from anaerobic digestion of biomass and pure manure.

peak denotes the HRT of this manure and biomass combination. The HRT would then be used to design a scaled-up system. When the biogas production becomes marginal, it would be wise to replenish the system with fresh new samples. The basis for the sizing of the reactor is the product of the daily manure input multiplied by the HRT, as shown in Equation 7.5.

$$\text{Digester size}\left(\text{m}^3\right) = \text{HRT(days)} \times \left(\frac{\text{L or m}^3}{\text{day}}\right) \tag{7.5}$$

Example 7.2: Use of Laboratory Data to Estimate Digester Size

An experiment was conducted to measure the methane produced by mixing alligator grass with dairy manure in equal amounts on a weight basis and then adding equal amounts of water. The following data was established:

- Maximum methane production occurred after 60 days.
- The total amount of methane produced per kilogram of mixed material was 25 L.

If one were to design a digester for daily conversion of 100 kg of animal manure plus 100 kg of alligator grass and 200 kg of water, what would be the minimum size of the digester and the gas holder? Assume the density of combined mixture slurry is 1.2 kg/L. Design the gas holder to accommodate 3 days of biogas output.

Solution:

a. Every day, there will be 400 kg of slurry that will be generated, and thus, if the HRT is 60 days, the minimum size of the digester is calculated as follows:

$$\text{Size}\left(\text{m}^3\right) = 60\ \text{days} \times \frac{400\ \text{kg slurry}}{\text{day}} \times \frac{\text{L}}{1.2\ \text{kg}} \times \frac{\text{m}^3}{1{,}000\ \text{L}} = 20\ \text{m}^3$$

b. The minimum volume of the gas holder is estimated as follows:

$$\text{Size}\left(\text{m}^3\right) = 3\ \text{days} \times \frac{25\ \text{L}}{\text{kg slurry}} \times \frac{400\ \text{kg slurry}}{\text{day}} \times \frac{\text{m}^3}{1{,}000\ \text{L}} = 30\ \text{m}^3$$

The typical composition of biogas is methane (50%–75%), carbon dioxide (25%–50%), nitrogen (up to 10%), hydrogen sulfide (<3%), and hydrogen (<1%). The rule of thumb commonly used for biogas is about 65% CH_4, 35% CO_2, with traces of H_2S. The methanogenic bacteria generally require a pH of between 6.4 and 7.2. However, the acid-producing bacteria would tend to lower the pH when they are more active than the slow-growing methane producers. The low pH is detrimental to the methane-producing microbes. The retention times of most biogas digestion processes are quite long, in order to maintain an ecological balance between acid formers and methane converters. In addition, the methane formers are more temperature-sensitive than the acid formers. At very low temperatures, the methane formers have difficulty keeping up with the acid formation. Consequently, there is a tendency for the pH to go down further (more acidic), and this eventually leads to reactor failure.

Animal manure is the most common feedstock for biogas production due to the presence of inherent anaerobic microbes in the stomach of ruminants. The common biogas production rate is approximately 283–708 L (10–25 ft^3) of biogas (containing between 50% and 70% methane) per 454 g (1 lb) of volatile solids added to the digester reactor (0.6–1.6 L/g) when the organic matter is highly biodegradable (Gerardi, 2003). Example 7.3 shows a simple way of estimating biogas production from animal manure. The energy content of biogas ranges from 22 to 26 MJ/m^3 (500–700 Btu/ft^3), while its methane content ranges from 60% to 70% (NAS, 1976).

Example 7.3: Amount of Biogas Produced in Dairy Facility

Suppose you have 500 head of dairy cows producing about 26,500 kg of wet manure each day. Determine the amount of biogas produced each day (minimum and maximum) if the volatile solids are about 10%. Assume further that 0.6–1.6 L of biogas is produced per gram of volatile solids.

Solution:

a. Minimum biogas production per day is calculated as follows:

$$\text{Biogas}\left(\frac{m^3}{day}\right)=\frac{0.6\,L}{gVS}\times\frac{26{,}500\,kg\,manure}{day}\times\frac{0.1\,kgVS}{kg\,manure}$$

$$\times\frac{1{,}000\,g\,VS}{kg\,VS}\times\frac{1\,m^3}{1{,}000\,L}=1590\frac{m^3}{day}$$

$$\text{Biogas}\left(\frac{ft^3}{day}\right)=1{,}590\frac{m^3}{day}\times\left(\frac{3.28\,ft}{1\,m}\right)^3=56{,}107\frac{ft^3}{day}$$

b. Maximum biogas production per day is calculated as follows:

$$\text{Biogas}\left(\frac{m^3}{day}\right)=\frac{1.6\,L}{gVS}\times\frac{26{,}500\,kg\,manure}{day}\,\frac{0.1\,kgVS}{kg\,manure}\,\frac{1{,}000\,gVS}{kgVS}$$

$$\times\frac{1\,m^3}{1{,}000\,L}=4{,}240\frac{m^3}{day}$$

$$\text{Biogas}\left(\frac{m^3}{day}\right)=4{,}240\frac{m^3}{day}\times\left(\frac{3.28\,ft}{m}\right)^3=149{,}619\frac{ft^3}{day}$$

The amount of methane produced by the anaerobic digestion process is dependent on the nature of raw materials, temperature, and type of reactor used. The methane production reported in the literature varies. The most common practice is to report methane production based on the amounts of volatile solids present and converted. Volatile solids, by definition, are those portions of the solid manure material that are volatilized at 550°C ± 50°C (NAS, 1976). This is normally done by determining the difference between the total solids (i.e., when the manure is completely dried) and the ash remaining after ignition at the above temperature. Unfortunately, many laboratories do not have facilities to measure volatile solids. Consequently, these laboratories correlate methane production with just the weight of fresh manure.

7.2 BIOMASS/WASTE PARAMETERS IMPORTANT IN ANAEROBIC DIGESTION

Before using anaerobic digestion as a conversion process, one must be familiar with the common parameters that are used to evaluate the biomass and other wastes. Common parameters include:

1. Biochemical oxygen demand (BOD)
2. Chemical oxygen demand (COD)
3. Total organic carbon (TOC)
4. Total volatile solids (TVS) or volatile suspended solids (VSS)
5. C/N ratio
6. Inhibitory substances

This list will be expanded when dealing with second-generation anaerobic digesters.

7.2.1 Biochemical Oxygen Demand

BOD is perhaps the most widely used parameter in wastewater treatment facilities. This quantity is the measure of the amount of oxygen used by microorganisms in the biological oxidation of organic matter. Note that the BOD varies as a function of time, and thus, most research results on BOD are based on a given timeframe, such as a 5-day test to report, which is called the 5-day BOD (or BOD_5). The procedure for calculating BOD is as follows: measure the dissolved oxygen before and after the test; take the difference; then, divide this difference by the volumetric fraction of the sample used (see Equation 7.6). Example 7.4 demonstrates this simple calculation. Note that the higher the BOD, the more treatment is required, and, in relation, the more organic matter is present. Of course, it may be used for further conversion. The typical BOD of untreated domestic wastewater treatment plants ranges from 110 to 400 mg/L (Tchobanoglous and Burton, 1991).

$$\text{BOD}_5,\ \text{mg/L} = \frac{D_1 - D_2}{f} \tag{7.6}$$

where

D_1 = dissolved oxygen of diluted sample immediately after preparation, mg/L
D_2 = dissolved oxygen of diluted sample after 5 days incubation, mg/L
f = volumetric fraction of sample used (decimal)

Example 7.4: Calculation of BOD

Suppose you have a 500 mL wastewater sample that was used to measure BOD_5. The initial concentration of the sample immediately after preparation was 8.5 mg/L. The 5-day dissolved oxygen content was measured to be 4.3 mg/L. Only 5 mL of the sample was used for the test. Determine the BOD_5 in units of mg/L. Comment on this value.

Solution:

a. Equation 7.6 may be used directly to calculate BOD_5 as follows:

$$BOD_5, \text{mg/L} = \frac{8.5-4.3}{5/500} = 420 \text{ mg/L}$$

b. This is a lower BOD_5 value.

7.2.2 Chemical Oxygen Demand

COD is a measure of the amount of oxygen required to oxidize the organic matter present in the sample. The test would normally use a strong chemical reagent (usually potassium dichromate) in an acidic medium and performed at an elevated temperature. Sometimes, a catalyst (e.g., silver sulfate or mercuric sulfate) is required to aid the oxidation of certain classes of organic compounds (Tchobanoglous and Burton, 1991). The COD is usually higher than the BOD because there are additional chemical compounds that cannot be degraded biologically by microorganisms; these compounds would require strong chemicals or the use of higher temperatures for complete oxidation. Likewise, the higher the COD of the biomass sample, the more organic matter is present for further conversion into energy and fuels. The theoretical COD of a sample may also be calculated if the chemical formula of the organic sample is known (i.e., true for pure compounds). This is done by completely oxidizing the material, as shown by the combustion equation for glucose ($C_6H_{12}O_6$) (Equation 7.7).

$$\underset{(180 \text{ kg})+(192 \text{ kg}) \rightarrow}{C_6H_{12}O_6 + 6O_2 \rightarrow} \underset{(264 \text{ kg})+(108 \text{ kg})}{6CO_2 + 6H_2O} \tag{7.7}$$

The COD of glucose is simply the ratio of the amount of oxygen required for every amount of glucose converted as follows:

$$\frac{\text{Oxygen}}{\text{Glucose}} = \frac{6\times(16\times 2)}{\left[(6\times 12)+(12)\times(1)+(6\times 16)\right]} = \frac{192}{180} = 1.1 \text{ gO}_2/\text{gC}_6\text{H}_{12}\text{O}_6$$

Note that many feedstocks for anaerobic digestion are in slurry form. Thus, it is convenient to use units of volume (mg/L) rather than weight. The ultimate analysis of the sample must be determined (as discussed in Chapter 3) to calculate the theoretical

COD if the chemical formula of the biomass is not known. More recently, COD has been easily measured using online thermal oxidation process equipment to provide real-time COD fluctuations in wastewater. The typical COD of domestic wastewater ranges from 250 to 1,000 mg/L (Tchobanoglous and Burton, 1991).

Example 7.5: Calculation of COD

Estimate the theoretical COD for maltose sugar in units of amount of oxygen per weight of maltose (g/g). The chemical formula for maltose is $C_{12}H_{22}O_{11}$. If 1 g of maltose sugar is placed in 1 L of water, what would be the COD in units of mg/L?

Solution:

a. The first step is to balance the oxidation equation for maltose and calculate the relative weights as follows:

$$\underset{(342\ \text{kg})}{C_{12}H_{22}O_{11}} + \underset{(384\ \text{kg})}{12O_2} \rightarrow \underset{(528\ \text{kg})}{12CO_2} + \underset{(198\ \text{kg})}{11H_2O}$$

b. Thus, the amount of oxygen used per weight of maltose is as follows:

$$\frac{\text{Oxygen}}{\text{Maltose}} = \frac{384}{342} = 1.12\ gO_2/gC_{12}H_{22}O_{11}$$

c. If 1 g of maltose is placed in 1 L of water, then the COD expressed in mg/L is calculated as follows:

$$\frac{\text{Oxygen}}{\text{Glucose}} = \frac{1.12\ gO_2}{gC_6H_{10}O_5} \times \frac{1\ gC_6H_{10}O_5}{\text{L}} \times \frac{1{,}000\ \text{mg}}{1\ \text{g}} = 1{,}120\ \text{mg/L}$$

7.2.3 Total Organic Carbon

TOC is another means of measuring the organic matter content of biomass that is used for conversion when using anaerobic digestion processes. The sample is injected into a high-temperature furnace or a suitable chemically oxidizing reactor, where the TOC is measured indirectly by the amount of carbon dioxide that is formed during the process. The presence of inorganic carbon may cause an error in measurement. This is corrected by acidification and aeration of the sample appropriately before analysis. This test is performed in many state-of-the-art wastewater treatment plants on a routine basis and done very rapidly. The typical TOC concentration in domestic wastewater treatment plants ranges from 80 to 290 mg/L (Tchobanoglous and Burton, 1991).

7.2.4 Total Solids and Volatile Suspended Solids

The amount of potential methane that may be produced during the anaerobic digestion process is accurately estimated when based on the amounts of volatile solids concentrated in the sample. Total solids (TS) are the measure of all the particles in slurry sample. This is broken down into total dissolved solids (TDS), volatile suspended solids (VSS), and ash. TDS and VSS are further differentiated into their fixed and volatile components. The overall balance is shown in Equation 7.8.

$$TS = TDS + VSS + ash \tag{7.8}$$

TDS are solids that pass through a filter with a pore size of around 1.2 microns or smaller. After the filtration process, the known liquid (of known amount) is dried and the remaining residue is measured. The ratio of the dried material (mg) to the volume of liquid (L) provides the concentration of TDS (mg/L). VSS becomes the amount of suspended solids that are volatilized at 550°C. Note that the sum of VSS and ash is termed total suspended solids (TSS) in some textbooks. The ash is the residue after the solids are exposed to the higher volatilization temperature of 550°C.

Example 7.6: Calculation of TS, VS, SS, and VSS

A 50 mL wastewater sample was used to measure the TS, VS, SS, and VSS. This sample was dried at 105°C until the material's weight became constant (at least 24 h), resulting in a final weight of 40 mg dried sample.This sample was then placed in a furnace at 550°C, and the amount of material loss was 36.1 mg. A replicated 50 mL sample was also filtered using a 1.2-micron pore size filter, and the amount of material lost during this step was measured to be 15 mg. This material was then placed in a furnace at 550°C, and the amount of material lost was around 3.0 mg. Determine the TS, VS, SS, and VSS.

Solution:

a. The TS is simply the difference in the weight of the sample after drying as follows:

$$TS = \frac{40\text{ mg}}{50\text{ mL}} \times \frac{1{,}000\text{ mL}}{\text{L}} = 800\text{ mg/L}$$

b. The VS is the amount of material volatilized after placing the dried sample in the furnace as follows:

$$VS = \frac{36.1\text{ mg}}{50\text{ mL}} \times \frac{1{,}000\text{ mL}}{\text{L}} = 722\text{ mg/L}$$

c. The SS is calculated from the sample after filtration as follows:

$$SS = \frac{15\text{ mg}}{50\text{ mL}} \times \frac{1{,}000\text{ mL}}{\text{L}} = 300\text{ mg/L}$$

d. The VSS is calculated after the filtered sample is exposed to 550°C in the furnace as follows:

$$VSS = \frac{3.0\ \text{mg}}{50\,\text{mL}} \times \frac{1{,}000\,\text{mL}}{\text{L}} = 60\,\text{mg/L}$$

7.2.5 Typical Septic Tanks or Municipal Solid Wastes Effluent Data

Most feedstocks for anaerobic digestion are quite high in BOD, COD, and VSs compared with wastewater. Table 7.2 summarizes the typical ranges of sludge concentrations from septic tanks. These ranges are much higher than for domestic wastewater. Note that the TS ranges from a low of 5,000 mg/L to as high as 100,000 mg/L with a typical average value of about 40,000 mg/L. The ratio of COD to 5-day BOD is about 5, while nitrogen content could be about 400 mg/L.

Wastewater treatment plants are primary sources of feedstock for biogas production due to the presence of high COD and BOD and the presence of microbes already familiar with conversion of organics into other simpler compounds. In some treatment plants, some wastes are diverted to an anaerobic digestion system, whereas others undergo aerobic treatments. The cost for building anaerobic digesters in many municipal sewage plants is quite high, and only a fraction is used to generate methane in most cases.

7.3 ACID- AND METHANE-FORMING MICROBES

Biogas production is made possible by the action of various microbes whose performance characteristics are quite diverse. More than 300 different species of bacteria are found in the feces of a single individual (Gerardi, 2003) or an animal, and most of these bacteria are strict anaerobes. Most oxygen-tolerant bacteria such as *Desulfovibrio* sp. and *Desulfomarculum* sp. produce H_2S from the reduction of sulfur-related compounds (sulfates), whereas oxygen-intolerant bacteria such as *Methanobacterium formicium* and *Methanobacterium propionicium* produce methane.

TABLE 7.2
Typical Characteristics of Septage Effluent

Parameter	Range (mg/L)	Typical (mg/L)
Total solids (TS)	5,000–100,000	40,000
Suspended solids (SS)	4,000–100,000	15,000
Volatile suspended solids (VSS)	1,200–14,000	7,000
5-day BOD, 20°C BOD5	2,000–30,000	6,000
Chemical oxygen demand (COD)	5,000–80,000	30,000
Total Kjedhal nitrogen (TKN as N)	100–1,600	700
Ammonia, NH_3 as N	100–800	400
Total phosphorus as P	50–800	250
Heavy metals	100–1,000	300

Source: Tchobanoglous, G. and F. L. Burton, *Wastewater Engineering: Treatment, Disposal and Reuse.* 3rd Ed. McGraw Hill Book Co., New York, 1991.

Bacteria degrade substrates using enzymes. Facultative anaerobic bacteria, such as *Enterobacter* spp., produce a variety of acids, alcohols, carbon dioxide, and hydrogen from carbohydrates, lipids, and proteins. Some organisms, such as *Escherichia coli*, produce odorous compounds such as indole and skatole (Gerardi, 2003).

The intermediate products from the anaerobic digestion process have already been shown schematically in Figure 2.4. In stage 1 (hydrolysis and fermentation), around 20% organic acids are readily produced, composed primarily of acetic acid. Some 4% hydrogen is already produced at this stage. Around 76% of the complex organics are converted into higher organic acids. Of this amount, some 52% are later converted into acetic acid through acetogenesis and dehydrogenation, while 24% are converted into hydrogen gas. This is considered stage 2 of the process. In the final stage of the process, the acetic acids are converted into methane by the methane-producing microorganisms, around 28% of which comes from conversion of hydrogen into methane (Capareda, 2011).

Acidogenesis is carried out primarily by bacteria, while acetic acid and hydrogen are converted into methane by another group of microorganisms called archaea. Two groups are normally involved in this process, the aceticlastic methanogens, which convert acetic acid into methane and carbon dioxide, and the hydrogen-oxidizing methanogens, which reduce carbon dioxide (Whiticar, 2020). The two most popular methanogens are the *Methanosarcina* and the *Methanothrix* (*Methanosarcina barkeri, Methanosarcina acetivorans, Methanosarcina mazei, Methanothrix sochngenii*, or *Methanobacterium bryantii*). Another diagram of the anaerobic digestion system is shown in Figure 7.5, depicting other feedstock that may be used, products and coproducts, and applications.

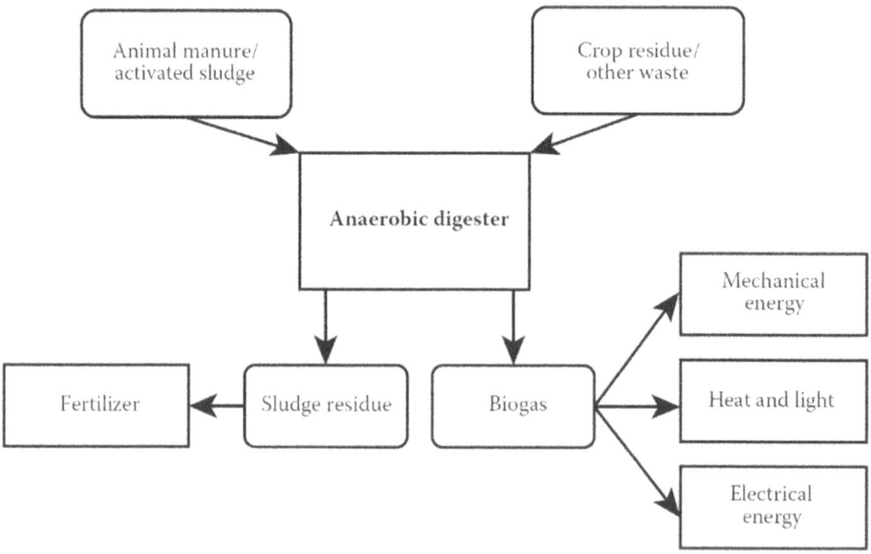

FIGURE 7.5 Schematic diagram of an anaerobic digestion system showing feedstock that may be used, products, and applications.

7.4 ADVANTAGES AND DISADVANTAGES OF ANAEROBIC DIGESTION PROCESSES

There are various advantages and disadvantages in the use of anaerobic digestion processes for the conversion of biomass resources into methane gas. They are summarized as follows.

7.4.1 Advantages of Anaerobic Digestion Processes

1. Renewable combustible gas is produced.
2. The digestion of solids removes objectionable odors after the conversion process.
3. A highly valued sludge is coproduced, which is considered an excellent source of nutrients and has fertilizer value.
4. There is reduction in the amounts of wastes to handle at the end of the conversion process.
5. The waste resources are contained in an enclosed vessel, preventing spoilage and generation of odor.

7.4.2 Disadvantages of Anaerobic Digestion Processes

1. The overall system cost is generally high because of long retention times.
2. The microbes involved are very sensitive to changes in the conditions within the reactor and may be subject to shock loading or toxicity.
3. The daily effluent requires further treatment before disposal in waterways.
4. The volume reduction after the conversion is not very significant, leaving a large amount of sludge to handle.

Example 7.7: Determining Total Weight of Sludge Left after Digestion

The total weight of slurry added to a certain digester was calculated to be 2,800 kg (6,160 lb) comprising 50% manure and 50% water. During the digestion process, 10% of the total slurry weight was volatilized. Determine the net weight of the slurry withdrawn per day and the amount of sludge left, after drying to about 10% moisture.

Solution:

a. The net weight of slurry withdrawn per day is calculated as follows:

$$\text{Slurry removed} = 2{,}800\,\text{kg} \times 0.90 = 2{,}520\,\text{kg}\left(5{,}544\ \text{lb}\right)$$

b. If all the moisture is assumed retained and the sludge is brought to 10% moisture content, the weight of sludge left can be calculated as follows (recall the moisture equation from Chapter 1):

$$10\% = \frac{\text{Wet weight} - 1{,}400\ \text{kg}}{\text{Wet weight}} \times 100\%$$

$$\text{Sludge wet weight} = 1{,}555.6\,\text{kg}$$

7.5 BIOGAS CONVERSION PROCESS AND DIGESTER DESIGNS

Ruminant animals have been converting cellulosic grasses into energy for their own metabolic uses. There are numerous anaerobic microbes in the stomach of ruminants responsible for conversion. Human beings have not been able to simulate the activities in the animal's stomach and build a refinery like how the animal generates its energy. The best that can be done is to collect the manure samples daily right after excretion, placing those samples in a reactor, and making sure that the entry of any oxidant is avoided.

Villages in China and India have been utilizing methane gas from animal manure for many years (NAS, 1976). A few head of animals are sufficient to meet the household's cooking and lighting needs. The amount of biogas produced from animal manure is approximately 150–200 L/kg of dry matter (or 2.4–3.2 ft^3/lb) of manure (Stout et al., 1979). The combustion of about 30 L of biogas (1 ft^3) will release the amount of energy equivalent to lighting a 25 W bulb for about 6 h (NAS, 1976). This is equivalent to about 0.54 MJ of energy (512 Btu). Example 7.8 shows a simple way of determining the volume of biogas required to satisfy a household's electrical requirement.

These reactors developed in China and in India are what we may refer to as first-generation digesters, as opposed to newer designs that are considered advanced or high rate. Improvements have been made primarily because of better understanding of the microbiology of anaerobic digestion. First-generation digesters are characterized by high HRTs, large reactors, and usually a plug-flow system. Second-generation digesters are characterized by short HRTs as well as different reactors for differentiating some of the digestion processes.

In the past, the assumption was that there was only one set of microorganisms converting these organic wastes into methane. However, it took too long for these microorganisms to process the organic wastes into methane (longer retention times). Further understanding of the microbiology of anaerobic digestion revealed that there are two distinct processes involved and two different types of microbes. The first stage is the conversion of organic material into organic acids by the acid-producing microbes. The second stage is the conversion of the first-stage acids into methane by methane-producing microbes. The first-stage acid-producing microbes thrive on low pH, primarily because that is their product. However, the second-stage methane producers are very sensitive to low pH and their population will diminish if the pH becomes too low.

If the pH imbalance is not corrected, all the methane formers will be eliminated, thereby bringing the reaction to stop and making the reaction "sour." This will lead to complete stoppage of the reaction, and no methane will be produced. The first-generation system allows for the ecological balance among the microorganisms at the expense of lengthening the time it takes to convert the organic wastes completely. These types of digesters will be discussed in the next section (Section 7.6)

Example 7.8: Volume of Biogas Required for Electrical Power Production

A household in India requires 500 kWh of electricity per day. Biogas will be used to fuel an engine-generator to produce the needed electricalpower. The thermal efficiency of the engine is 30%, and the electrical to mechanical efficiency is 90%. Thirty liters (1 ft^3) of biogas with a methane content of 65% will provide 0.63 MJ (600 Btu) of energy. Determine the volume of biogas required (in cubic meters and cubic feet per day) for this purpose.

Solution:

a. From the efficiency equation, we have the energy required from the biogas:

$$\%\,\text{Efficiency} = \frac{\text{output}}{\text{input}}$$

$$\text{Input} = \frac{500\,\text{kWh}}{(0.30)\times(0.90)} = 1{,}852\,\text{kWh}$$

b. The biogas required can be calculated easily as follows:

$$\frac{1{,}825\ \text{kWh}}{\text{day}} \times \frac{3.6\ \text{MJ}}{1\ \text{kWh}} \times \frac{30\ \text{L}}{0.63\ \text{MJ}} \times \frac{\text{m}^3}{1{,}000\ \text{L}} = 317.5\frac{\text{m}^3}{\text{day}}\left(11.203\frac{\text{ft}^3}{\text{day}}\right)$$

7.6 FIRST- AND SECOND-GENERATION BIOGAS DIGESTERS

Anaerobic digesters originated when humans disposed of their daily human waste in a controlled enclosed tank, or what we now call the septic tank. Influent is introduced daily into an enclosed tank on one side. Effluent is removed daily on the opposite side. The dimensions of the unit are based on the size of the household (or the volume of wastes generated each day) and the number of days the wastes are held in the tank.

For example, if one wanted to have a 60-day retention time for the wastewater to stay in the enclosed reactor, and the household is utilizing 100 L (26.4 gal) of wastewater each day, the size of the septic tank will be around 6,000 L or 6 cm^3 (1,585 gal or 212 ft^3). There have been several improvements for these waste tanks, and they have been called Travis or Imhoff tanks. When communities began to handle large volumes of waste, the design became more sophisticated. Currently, many municipal wastewater treatment plants rely on sludge digesters to treat municipal sewage.

7.6.1 Chinese Dome Type

A very common digester design in rural China is called the dome-type digester. A dome digester is usually made of bricks and located slightly below ground level. A feed of animal manure is introduced each day. The fresh manure is mixed with water, usually in a 1:1 ratio of solids to water and fed into the digester. Because the digester is made of bricks and is well sealed, the biogas produced is slightly pressurized. Consequently, the pressurized biogas is available for cooking or lighting without the need for a biogas pump or compressor blower. There is usually an agitator to mix the slurry well with the aged materials. There should also be an overflow area where the spent slurry is discharged. There should also be an access point to the reactor for cleaning or for removal of sludge. The sludge accumulates in the digester and, at some point, must be removed. Sludge removal is normally done once a year. In the design, leaks must be avoided both in the reactor and in the gas-conveying systems. Some settled sludges will accumulate in the overflow tank. Settled sludge can be used for soil amendment or for plant fertilizer.

7.6.2 Indian Gobar Gas Plant

The gobar gas plant is popular in India. This design has a floating gas holder, usually made of metal with proper water seal and a cylindrical tank as the digester is usually below the gas holder. An inlet mixing tank is placed on one end and used to mix the manure with the appropriate amount of water. The biomass is also fed each day. The reactor is baffled so that the incoming feed will not immediately exit. The floating steel gas holder has sufficient weight such that the biogas formed is slightly pressurized. Consequently, the biogas can be easily moved to its point of use. An outlet pipe must also be in place to remove the digested slurry at the end of its conversion stage. The volume of overflow is usually equal to the volume of incoming feed. As in the Chinese dome-type design, sludge accumulates at the bottom of the reactor and should be removed during long-term maintenance procedures. The reactors are usually made of concrete blocks and properly sealed by a nonpermeable sealant to ensure strict anaerobic conditions. A variation on the design described is to have a separate gas holder that is not directly above the reactor.

7.6.3 Taiwanese Rectangular Digester Design

The Taiwanese Rectangular Digester Design is popular in developing countries for large-scale biogas digestion systems. This type of digester is used where there are many confined animals. The design has two rectangular boxes (digester and effluent discharge) that contain animal manure. The walls and flooring are made of concrete blocks with the mortar made of impermeable materials to prevent the seepage and entry of water or air. Mixing is done daily to ensure good contact between the microbes and the manure that is loaded each day. This is a common plug-flow reactor design where the influent feed moves from one side to the opposite side and into the second reactor after the HRT is achieved. There is a separate gas holder. There is also a separate outlet box for the discharge of the already digested manure.

One successful swine company in the Philippines (Maya Farms, Antipolo, Rizal) built a digester like the Taiwanese rectangular design and produced biogas enough to satisfy 100% of the company's requirement for electrical power. The facility had a population of 10,000 head swine where biogas was produced and used for processing meat and other by-products (Maramba, 1978). In the early 1980s, this facility was completely independent of local utilities for both its heat and electricity needs. The company generated power from biogas using various commercial four-cylinder gasoline engines that had been converted to run on biogas. Each engine was connected to a generator that produced electricity used by the facility.

Engine modification included removing the carburetor and replacing it with a simple pipe for inlet air and another pipe for the biogas. The ratio of the pipe sizes was based on the ratio between the air required to complete the combustion of the biogas. Old Toyota 4K-type engines were used. In India, however, it is more common to use diesel-type engines because of their higher compression ratios and higher efficiencies.

7.6.4 Cylindrical Above-Ground Digesters

There are various other innovative digester designs. One unique design is the above-ground anaerobic digester made using cylindrical tanks. These unique designs are available at the Center for Pig Husbandry at Lipa City, Philippines. The advantage of this design is that, for ease of maintenance, sludge can be easily removed each day. System leaks can be easily checked. Sunlight is used to heat the reactor, thereby increasing the efficiency. A ladder can be mounted to aid in inspecting the unit from an accessible top cover. This means of access also assists in the task of mixing the daily slurry. A major disadvantage of this system is that it requires a pump to bring the manure slurry to the reactor. These pumps normally corrode easily, because of the acidic nature of the slurry. Another disadvantage is degradation from sunlight. Periodically, the unit must be repainted to eliminate rust. The unit is also expensive when durable materials are used. Another variation of the system uses concrete culverts instead of metal cylinders. However, a separate gas holder will be required. If the container is not 100% filled, that void space above the slurry level will contain the biogas, thereby creating some pressure in the reactor. Consequently, biogas may be used without the need for a gas pump. Monitoring the system is quite convenient by accessing several exit ports from both ends of the reactor and the mid-section.

All reactors described thus far are first-generation digesters. They are single-reaction systems characterized by long retention times. These retention times normally range from 30 to 60 days, depending on the type of manure used as well as additional biomass that is mixed with the manure. The primary advantage of first-generation digesters is that they are not affected much by shock loading and changes in environmental conditions. These systems have very low maintenance costs with minimal attention required during the operation.

7.6.5 High-Rate Biogas Digesters (Second-Generation Biogas Digesters)

Second-generation digesters are those characterized by short retention times. In addition, they have a smaller footprint and are more compact in design. The emergence of high-rate biogas digesters is because of a better understanding of the microbiology of methane production. The only major setback to these digesters is the tedious monitoring and constant adjustments during operation. For example, these systems are very prone to shock loading. As with all advanced technologies, more checks and balances are required during the operation of these high-rate digesters. Considering what is important to first-generation digesters, one identifies daily mixing and daily feed. However, high-rate digesters are monitored for an extensive set of operational criteria, including pH, alkalinity, VSS loadings, and the presence of generated inhibitory substances. Table 7.3 summarizes a list of these parameters and ranges for optimal operation.

Foremost of the many factors affecting biogas production is the nature of feedstocks and their substrates. This has to do with COD, or BOD, and their ratios. The digester design is important as to whether it is batch, continuous, or multistage. Before start-up operations, the microbes to be used are crucial. The inoculant seeding must be made to ensure that proper microbial populations are used with a good balance between acid- and methane-producing microbes. In most digester operations, the pH must be maintained near neutral. Because of this, in many large-scale biogas digestion plants, especially the high-rate systems, there will be a separate acid and base tank. The contents of these tanks can be used to bring the reactor pH to neutral. The temperature must also be appropriate for the type of microbes used. The different thermophilic, mesophilic, and psychrophilic organisms are important. For high-rate reactors, the loading rates are also important (SRT, HRT, etc.). The C/N ratio must be maintained at around 30:1. Other factors that may affect the operation of a biogas digester include the concentrations of volatile fatty acid (VFA), the nutrient and micronutrients present, the VSS, and the TSS. Finally, the buffering capacity, or the resistance of a system to change in pH, must be excellent. Recall that resistance to pH change is perhaps the important characteristic of first-generation digesters. Buffering capacity is measured as the concentration of calcium carbonate that would have the equivalent capacity to neutralize acids.

TABLE 7.3
Optimum Conditions for a Single-Stage Anaerobic Digestion Process

Parameter	Optimum	Maximum
BOD/N/P	100:1.1:0.2	–
pH	6.8–7.4	6.4–7.8
VFA (volatile fatty acid)	50–500	2,000
Alkalinity	1,500–1,000	3,000
$NH_4^+ - N$	550–1,000	3,000
Na	100–200	3,500–5,500
K	200–400	2,500–4,500
Ca	100–200	2,500–4,500
Mg	7.5–150	1,000–1,500
SO_4^{-2}	–	500
S-	0.1–1.0	100
CL-	–	15,000
Temperature	37°C	50–55
VFA/alkalinity	<0.4 but not more than 0.8	

Source: KMIT, Training course on biogas reactor design and development lecture notes, May–June 1987. King Mongkut's Institute of Technology Continuing Education Center, Bangkok, Thailand, 1987.

A good anaerobic digestion facility must have equipment to measure all these parameters and to keep the values within optimal conditions, and never near the maximum values listed in Table 7.3. The facility must have a gas chromatograph or similar devices to measure VFA and additional analyzers to measure parameters such as the ultimate analysis of the slurry and to estimate the C/N ratio. Both VSS and TSS should be measured, and thus filtering and drying facilities must be in place. Other requirements include an online pH monitoring system, good temperature control, and heating capabilities.

The system pH results from the equilibrium of acid-producing and acid-consuming organisms. If the acid-producing organisms dominate, the system could be buffered with calcium carbonate. Nutrient availability must also be checked. These are necessary to build up the microbial biomass. The most important elements are N, P, K, and trace elements (Ni, Co, Mo, and Te). Animal manure is a good supplier of these trace elements. However, there are other nutrients that may be toxic when in excess amounts, such as NH_4^+, Na, Cu, Ni, Zn, and some antibiotics that are fed to animals. The toxicity level for NH_3 is usually in the range from 1,500 to 3,000 mg/L. There will be no adverse effect from 200 to 1,000 mg/L, and the beneficial range is from 50 to 200 mg/L. For H_2S, the toxicity level is around 200 mg/L at neutral pH. There are unwanted heavy metals at higher concentrations, such as Co, Cu, Fe, Ni, and Zn (Turovskiy and Mathai, 2006).

The high-rate digesters were originally designed for low-strength wastes; however, recent units can now handle high-strength wastes with high quantities of suspended solids, such as livestock manure. Callander and Barford (1983) made an extensive review of the development of the anaerobic digester technology. The improvements of such digesters can be largely attributed to the fact that researchers have been able to understand the microbiology of the production of methane gas from waste streams.

7.6.6 Anaerobic Filters

Anaerobic filters were the first high-rate digesters developed (Young and McCarthy, 1969). These digesters are characterized using filter media as part of biomass support particles. These are in the form of sand, Raschig rings, or stationary fixed film/nets. Conversion rates are minimal. However, the pressure drop through the system is high due to the presence of filter media. The systems were improved to consider different flows (upflow or downflow) as well as expanded and fluidized bed systems, still with filter media present.

The design that attracted widespread attention is the upflow anaerobic sludge blanket (UASB) by Prof. Letingga from the Netherlands (Letingga et al., 1980). This design eliminated the need for filter media by creating a blanket of sludge as the feed inlet of wastewater. This sludge blanket is filled with a high population of acid- and methane-producing microbes. The removal of filter media and the replacement of a blanket filled with microbes improved the conversion rates and lowered the pressure drop through the system. The reactor size was reduced. The secret behind this innovation, which was not immediately revealed, was the use of special coagulants to enhance the growth of microbes and to build the sludge blanket as the influent feed was introduced. The technology has gained popularity worldwide and has also been the technology of choice in many developing nations, especially in Europe. Nowadays, one will observe various innovations and combinations of these systems. The latest high-rate reactors are those that have separate reactors for the acid-producing microbes and methane-producing groups. This way, one system is not affected by the performance of the other. There will be an NaOH tank in between the reactors because the acid-producing microbes always have more acid output that is detrimental to the methane producers.

UASB systems have been marketed by various commercial entities, one of which is Biotim, a company purchased recently by Waterleau (Leuven, Belgium).

7.7 DESIGN OF BIOGAS DIGESTER

In the design of biogas digesters, one must first rely on published data about manure production from various animals. Table 7.4 is an excellent reference that can be used to estimate manure production, given the type of animal. Note that manure production varies with animal weight, and this table should be based on specific animal weight. Simple digester sizing may be easily calculated once this table is available and validated. Example 7.9 illustrates the use of Table 7.4. Table 7.5 may be further used to design the size of the gas holder, including associated heating values of the biomass from various animal sources. The biogas production data table also shows the approximate equivalents of the gas with gasoline, diesel, natural gas, and propane. The heating values used for conversion are as follows: gasoline, 33.4 MJ/L; diesel, 37.1 MJ/L; natural gas, 37.3 L; and propane, 48.7 MJ/L. Example 7.10 shows the calculation of equivalents. This example also assumes that the biogas contains 65% methane with a heating value of 22.4 MJ/m^3 (600 Btu/ft^3). If biogas from plant sources is to be used, Table 7.6 summarizes some yields of methane gas from various species of plants. Note the highest yield, from hybrid poplar, is 410 L/kg.

TABLE 7.4
Daily Manure Production from Various Animals

	Total Solids	Volatile Solids	Total Manure		Moisture
Animal Type	**(kg)**	**(kg)**	**(kg/[d-a])**	**(L/[d-a])**	**(%w.b.)**
Beef—growing calf	2.7	2.3	22	22	88
Dairy—lactating cow	8.9	7.5	68	68	87
Dairy—heifer, 440 kg	3.7	3.2	22	22	83
Horse—sedentary, 500 kg	3.8	3.0	25	25	85
Layer	0.022	0.016	0.088	0.088	75
Swine—gestating, 200 kg	0.50	0.45	5.0	5.0	90
Swine—lactating, 192 kg	1.2	1.0	12	12	90
Swine—boar, 200 kg	0.38	0.34	3.8	3.8	90

Source: ASABE D384.2 MAR2005 (R2010), Manure Production and Characteristics Standards. Standard Revised. ASABE, St. Joseph, MI, 2010.

TABLE 7.5
Biogas Production Heating Values and Approximate Equivalents to Gasoline, Diesel, Natural Gas, and Propane

Livestock (454 kg Weight)	Biogas Production (m^3/day)	HV MJ	Approximate Equivalents Gasoline L	Diesel L	Natural Gas m^3	Propane Kg
Beef	0.85	19.0	0.57	0.53	0.51	0.4
Dairy	1.27	28.5	0.87	0.76	0.76	0.6
Broilers	2.61	58.2	1.74	1.59	1.56	1.2
Layers	2.04	45.6	1.36	1.21	1.22	1.0
Swine	0.82	18.4	0.57	0.49	0.48	0.4

Source: Stout, B. A., *Energy Use and Management in Agriculture.* Breton Publishers, Belmont, CA, 1984; Converse, J. C. and R. E. Graves, Facts on methane production from animal manure. Fact Sheet A-2636. University of Wisconsin, Madison, WI, 1974.

TABLE 7.6
Methane Yield of Various Woody Biomass Types

Common Name	Scientific Name	Methane Yield L/kg	Methane Yield ft^3/lb
Cellulose powder		390	6.26
Pussy willow	*Salix eriocephala*	291	4.66
Shining willow	*Salix lucida*	270	4.33
Coyote willow	*Salix exigua*	280	4.49
Common osier	*Salix viminalis*	340	5.46
White willow	*Salix alba*	301	4.82
Purple osier willow	*Salix purpurea* (SP3)	230	3.69
Peach leaf willow	*Salix amygdaloides* (SAM5)	210	3.37
Halberd willow	*Salix hastata* (SH2)	230	3.69
Hybrid poplar	*Platanus nigra* × *P. maximowiczii*	390	6.26
Hybrid poplar	*Platanus nigra* × *P. euramericana*	410	6.58
Eastern cottonwood	*Platanue deltoids*	301	4.82
American sycamore	*Platanus occidentalis*	380	6.10
Black locust	*Robinia pseudoacacia*	301	4.82
Sweet gum	*Liquidambar styraciflua*	260	4.17

Source: Badger, P. and P. Pullammanappallil, Biomass conversion to energy and fuels. In: *Wood to Energy: Using Southern Interface Fuels for Bioenergy* (C. Staudhammer, A. L. Hermansen-Baez, D. R. Carter and E. A. Macie, eds.). USDA Forest Service General Technical Report GTR-132, 2011.

Example 7.9: Minimum Size of Digester

Determine the minimum size of a digester from manure of 100 heads of gestating swine weighing 91 kg (200 lbs) each. Determine the total amount of wet manure produced per day (in kg) and the volume (in m^3). If the retention time is 30 days, what is the minimum size of the anaerobic digestion reactor? Use Table 7.4 for your calculations.

Solution:

a. The daily manure production is calculated from the equation as follows:

$$\text{Daily production (kg)} = 100\,\text{hds} \times \frac{5\text{ kg}}{\text{hd} - \text{day}} = 500\text{ kg}$$

b. The daily volume of manure is calculated as follows:

$$\text{Daily volume (L)} = 100\,\text{hds} \times \frac{5\text{ L}}{\text{hd} - \text{day}} = 500\text{ L}$$

c. The minimum size of reactor is the daily production multiplied by the retention time:

$$\text{Reactor size}\left(\text{m}^3\right) = \frac{500\text{ L}}{\text{day}} \times 30\text{ days} \times \frac{1\text{ m}^3}{1{,}000\text{ L}} = 15\text{ m}^3$$

The design of a biogas digester depends on various criteria and goals. Usually, if the criterion is to size the digester based on the number of heads of the facility, then the initial step will be to make an inventory of the amount of manure produced each day as well as the mixing ratio that will be used for the manure and water that will be introduced into the reactor. Normally, for wet manure such as swine or dairy, the mixing ratio is 1 part of water to 1 part of wet manure. For some dried manure, such as that of poultry litter, the mixing ratio could be 1 part manure to 2 parts water. This will be followed by establishing the retention time for the type of manure used. Sometimes, however, the methane production is required for a given application and the required calculation will be to determine the number of head for an animal facility that will be required to deliver this methane output. Thus, the steps in the design are as follows:

1. Determine the production requirement of methane.
2. Determine the amount of raw material needed.
3. Determine the optimal digestion period.
4. Determine the size of the digester.
5. Identify the type of digester used.
6. Determine the size of the gas holder.

Example 7.10: Digester Design

Suppose 100 kWh/day electrical power is needed by a household. The energy will come from the combustion of biogas in an internal combustion engine with a thermal conversion efficiency of 25% and a mechanical-to-electrical conversion efficiency of 80%. The heating value of the biogas is 22.4 MJ/m³ (600 Btu/ft³). The manure will come from the dairy facility (lactating cow, 440 kg or 968 lbs). The mixing ratio is 1:1 (fresh manure to water). The hydraulic retention time is 21 days. Determine the following:

- The amount of biogas required for this purpose
- The required number of dairy animals
- The digester volume and dimensions
- The gas holder volume and dimensions (assume gas holder volume is half of the daily biogas production)
- The weight of slurry withdrawn each day (assuming 10% volatilization)
- Use Table 7.4 for calculation of manure volume
- Use Table 7.5 for calculating biogas production

Solution:

a. The biogas required per day is calculated as follows:

$$\frac{m^3}{day} = \frac{100\ kWh}{day} \times \frac{3.6\ MJ}{kWh} \times \frac{m^3}{22.4\ MJ} \times \frac{1}{0.25} \times \frac{1}{0.80} = 80\ m^3/day$$

b. The number of animals requires is calculated as follows:

$$\text{No. of cows} = \frac{80\ m^3}{day} \times \frac{hd - day}{1.27\ m^3} \times \frac{1{,}000}{968} = 65 \text{ dairy cows}$$

c. The digester volume and dimensions are estimated as follows:

$$\text{Digester volume}\left(m^3\right) = 65 \text{ cows} \times \frac{68\ L}{hd - day} \times 1 \frac{m^3}{1{,}000\ L} \times 21 \text{days} \times 2$$

d. The gas holder volume and dimensions are calculated as follows:

$$\text{Volume}\left(m^3\right) = \frac{80\,m^3}{day} \times \frac{1}{2} = 40\,m^3/day$$

e. The weight of slurry withdrawn each day is calculated as follows:

$$\text{Weight}\left(kg\right) = 65 \text{cows} \times \frac{68\,kg}{hd - day} \times 2 \times 0.9 = 7{,}956\,kg$$

7.8 BIOGAS UTILIZATION

Biogas may be used in so many ways. It may be used directly for heat generation without pretreatment; however, in some applications, the biogas must be cleaned or properly pretreated before it is used. In many developing countries, the most common uses of biogas include cooking, heating, and lighting. The biogas may also be used to generate electrical power by using internal combustion engines, or other nonconventional systems such as fuel cells. Listed are some major applications and uses of biogas, followed by a summary discussion of some of those applications:

1. Heat and steam generation
2. Drying
3. Power generation
4. Cogeneration
5. Fuel cell applications
6. Liquid fuel production

Biogas is very combustible despite the presence of carbon dioxide and other trace gases. The energy content of the biogas has already been discussed in earlier sections. Designing a burner system is quite easy, by considering that about 35% of the volume is occupied by carbon dioxide, which does not aid in the combustion process. In many developing countries, power generation from biogas is not a new development for use with slightly modified gasoline or diesel engines to generate electrical power. Harmful gases such as H_2S and other corrosive trace elements must be removed.

The simultaneous use of heat and power is also becoming popular (cogeneration). In some applications, electrical power is first produced (called the topping cycle); then, excess heat is recovered for use in other areas within the utilization facility. In other designs, the biogas is first used to generate thermal energy or heat, as for steam generation, and the excess energy is then used to generate electrical power (called the bottoming cycle in cogeneration terminology).

More recently, biogas (with some CO_2 removal) has been used as the base fuel for some fuel cells. Fuel cells are devices that convert the chemical energy of a fuel into electricity through the reaction of the fuel with oxygen or air. The methane gas from biogas is an excellent fuel for use in fuel cells. However, the current system cost for fuel cells is still quite high. Biofuel from biogas must be produced cheaply to be economically feasible.

Biogas can also be converted into liquid biofuel through various catalytic or some noncatalytic conversion pathways. The closest liquid fuel from the methane in biogas is methanol and other higher alcohols. Liquid fuel production will be discussed in later chapters.

Biogas must be purged of H_2S, a foul-smelling compound, before the fuel is used in engines. Techniques for H_2S removal have been highly documented. Most commonly, a mixture of iron fillings and wood shavings is used to remove this "rotten egg" smell. Approximately 35 L (1.24 ft^3) of iron sponge (or one bushel) will remove 3.7 kg (8.2 lb) of sulfur (0.0352 m^3/3.7 kg) (Stout, 1984). The normal H_2S

concentration in biogas is approximately 0.2%, and thus 35 L (1 bushel) of this material can treat 2,465 m^3 (87,000 ft^3) of biogas before being regenerated. The scrubbing reaction is shown in Equation 7.9 as well as the regeneration reaction (Equation 7.10).

Scrubbing reaction

$$Fe_2O_3 + 3H_2S \rightarrow Fe_2S_3 + 3H_2O \tag{7.9}$$

Regeneration reaction

$$2Fe_2S_3 + 3O_2 \rightarrow Fe_2O_3 + 3\ S_2 \tag{7.10}$$

An alternative material is zinc oxide. However, this material is quite expensive. Iron filings may be sourced from any machine shop that does lathe work. Metal filings can be easily mixed with sawdust or wood shavings.

7.9 CONCLUSION

Methane may be easily produced from organic biomass wastes through anaerobic digestion processes. Manure from ruminant animals is the primary source of anaerobic microbes for the conversion process. The ideal C/N ratio must be kept to about 30. The use of other biomass residues, such as grasses and other lignocellulosic biomass, may be necessary. There are generally two sets of microbes responsible for biogas production—acid- and methane-producing microbes. The methane-producing microbes are very sensitive to reactor pH conditions.

Numerous digesters have been designed through the years. The first-generation digesters include those developed in China, India, and Taiwan. Because of recent knowledge about microbial activities within these digesters, numerous other high-rate biogas digesters have been developed. These include the use of anaerobic filter reactors developed in Canada and the UASB developed in the Netherlands.

A primary advantage of the anaerobic digestion process is its compatibility with biomass wastes and residues that are high in moisture. The biomass must be kept in slurry form such that the microbes can effectively perform their conversion processes. The main disadvantage of the anaerobic digester is the fact that a large volume of sludge must be handled after the conversion process and disposed of properly. The system costs vary, with the high-rate reactor having increased operational costs due to necessary monitoring of many parameters for optimal operation. However, these costs are offset by the smaller digester footprint and lower initial capital cost.

The application and use of biogas has been quite remarkable due to the availability of engines that now operate appropriately on this methane and carbon dioxide mixture. Low Btu engines are now available for landfill biogas and digester gases for electrical power generation. One should be aware that the raw biogas needs to be properly cleaned before its use in engines. There are also other gaseous materials that must be removed to address odor issues with the presence of hydrogen sulfide and other trace gases that may be corrosive.

7.10 PROBLEMS AND DISCUSSION ISSUES

7.10.1 Acid- and Methane-Producing Microbes

P7.1 Differentiate between acid- and methane-producing microbes.

7.10.2 Theoretical Acetic Acid Production

P7.2 Equation 7.2 showed that for every mole of propionic acid, 1 mole of acetic acid is formed. Estimate the theoretical acetic acid production (in L or gal) for every amount of proprionic acid used in units of (L or kg or lb). Assume the density of component as follows:

- Density of pure propionic acid at STP = 0.990 kg/L
- Density of pure acetic acid at STP = 1.049 kg/L

7.10.3 Calculation of BOD

P7.3 Suppose you have 1,000 mL wastewater sample that was used to measure BOD_5. The initial concentration of the sample immediately after preparation was 15 mg/L. The 5-day dissolved oxygen content was measured to be 8.5 mg/L. Only 5 mL of the sample was used for the test. Determine the BOD_5 in units of mg/L. Comment on the strength of the BOD value.

7.10.4 Calculation of COD

P7.4 Estimate the theoretical COD for fructose sugar in units of the amount of oxygen per weight of fructose (g/g). The chemical formula for sucrose is as follows: $C_6H_{12}O_6$. If half a gram of fructose sugar is placed in a liter of water, what would be the COD in units of mg/L?

7.10.5 Use of Laboratory Data to Estimate Digester Size

P7.5 An experiment was conducted to measure the methane produced by mixing alligator grass with dairy manure using equal amounts on a weight basis. Then, equal amounts of water were added. The following data was established:

- Maximum methane production occurred after 90 days.
- The total amount of methane produced per kilogram of mixed material was 40 L.

If one were to design a digester for daily conversion of 50 kg or animal manure plus 50 kg of alligator grass and 100 kg of water, what would be the minimum size of the digester and the gas holder? Assume the density of combined mixture slurry is 1.2 kg/L. Design the gas holder to accommodate 3 days of biogas output.

7.10.6 Energy Produced from Biogas Daily

P7.6 Suppose you have 1,000 head of dairy cows producing about 53,000 kg of wet manure each day. Determine the amount of energy produced each day derived from biogas (minimum and maximum) if the volatile solids are about 12%. Use a heating value of gas at 22 MJ/m^3. Assume further that 0.6–1.6 m^3 of biogas is produced per kilogram of volatile solids.

7.10.7 Volume of Biogas Required for Electrical Power Production

P7.7 A household in India requires 500 kWh of electricity per day. Biogas will be used to fuel an engine-generator to produce the required electrical power. The thermal efficiency of the engine is 30%, and the electrical to mechanical efficiency is 90%. Thirty liters (1 ft^3) of biogas with a methane content of 55% will provide 0.58 MJ (550 Btu) of energy. Determine the volume of biogas required (in cubic meters and cubic feet per day) for this purpose.

7.10.8 Amount of Sludge Produced in an Anaerobic Digester

P7.8 Determine the amount of sludge taken out of a digester if the daily input is 5,300 kg of wet manure. Only 10% of this is converted into biogas by weight. Also, determine the amount of moisture to be removed to bring the slurry to bone dry assuming the slurry has 70% of water at the start and there is no moisture loss during the conversion process.

7.10.9 Theoretical COD from Biomass

P7.9 Estimate the theoretical COD for starch in units of amount of oxygen per weight of starch (g/g). The chemical formula for starch is as follows: $C_6H_{10}O_5$. If 1 g of starch is suspended in 1 L of water, what would be the COD in units of mg/L?

7.10.10 Estimating Minimum Size of Digester

P7.10 Determine the minimum size of a digester from manure of 500 heads of lactating dairy cow. Determine the total amount of wet manure produced per day (in kg) and the volume (in m^3). If the retention time is 45 days, what is the minimum size of the anaerobic digestion reactor? Use Table 7.4 for your calculations.

7.10.11 Feedstock with the Highest Methane Yield

P7.11 Identify the feedstock with the highest methane yield.
a. sweet gum b. hybrid poplar c. coyote willow d. black locust

7.10.12 Manure Type with the Highest Potential Methane Yield

P7.12 Which among the manure type will yield the most methane assuming equal animal weights?

a. broilers b. layers c. dairy d. swine

7.10.13 Manure with the Highest Volatile Solids

P7.13 Identify the manure type with the highest volatile solids.

a. dairy b. beef c. swine d. layer

7.10.14 Sizing of Anaerobic Digester

P7.14 Design the size of the anaerobic digester (rectangular shape with length twice the width and a maximum of 5 m deep) including gasholder (3-day biogas storage) for a 40,000-head chicken layer facility. Enumerate all assumptions made.

7.10.15 Power Production from Layer Biogas

P7.15 Estimate the power output (kW) of an anaerobic digester that will handle the manure of 40,000-head chicken layer facility. Assume year-round operation. Enumerate all assumptions made.

7.11 CASE STUDIES

A dairy farmer has 300 heads of milking cows. He plans to collect manure each day and design an anaerobic digester to generate biogas. The biogas will be cleaned of impurities and used to run an engine coupled with a generator to produce electrical power (overall conversion efficiency of 25%). This power will be used to operate his aerator for his wastewater treatment system. The established biogas production each day was 1.28 m^3/head/day [45 ft^3/hd/day]. You are to specify the design parameters for his project. The retention time for his digester is 45 days. You are to specify the minimum size of digester in cubic meters, the size of gas holder assuming 3-day storage supply. The actual digester size is 20% more than the minimum size (as freeboard). Specify the electrical power output (kW) that could be generated for this facility to operate the genset for 10 hours per day. Each day, each animal generates around 26 kg [58 lbs/hd/day] of fresh manure with a slurry density of around 962 kg/m^3 [60 lbs/ft^3]. He uses an equal measure of water (1 kg/L or 62.4 lbs/ft^3) by weight for his slurry to feed into the digester. The daily biogas production is approximately 1.28 m^3/day/hd [1.27 m^3/day/hd]. Determine the amount of iron sponges required to clean up the biogas assuming 0.1% sulfur content. This sponge would remove 3.7 kg of sulfur/bushel [8.2 lbs of sulfur per bushel] and a bushel of iron sponge has a volume of 0.03514 m^3/bu [1.24 ft^3/bu]. The density of pure sulfur is 7.2 lbs/gal [0.86466 kg/m^3]. Assume the heating value of biogas as 24.20 MJ/m^3

[650 Btu/ft^3]. A bushel is around 35.24L. The revenue for this project includes the sale of electricity, sale of sludge as fertilizer, and tax credit. The table summarizes some data that may be used for the project.

Additional Data for the Project

Interest on Investment	8% of capital	Fertilizer Yield	20% of dry sludge
TSI	2% of capital	Recoverable Fertilizer	90% of dry sludge
Sale of Electricity	\$0.12/kWh	Tax Credit	30% of capital
Operation/year	350 days	Genset Cost	\$325/kW
Genset operation	10 h/day	R&M Cost	2% of capital
Fertilizer Cost	\$82.50/tonne	Biogas Cleanup Cost	\$0.05/L
Miscellaneous Costs	5% of capital	Cleanup System Cost	10% Gasholder Cost
Anaerobic Digester (AD) Cost	\$52.93/m^3	Gasholder Cost	\$35.29/m^3

Required:

a. Daily manure production (kg/day)
b. Daily manure volume (m^3/day)
c. Water requirement (L/day)
d. Slurry volume (L/day)
e. Minimum and actual digester size (m^3)
f. Daily biogas production (m^3/day)
g. Gasholder size (in m^3) and digester diameter (in m) assuming heigh and diameter are the same
h. The daily biogas energy (MJ/day)
i. The electrical power output (kW)
j. Sulfur generation rate (m^3/day and kg/day)
k. H_2S cleaning sponge required (m^3/day and L/day)
l. Setup the economic analysis for this project to report on internal rate of return (IRR), payback period (PBP), benefit cost ratio (BCR), and net present value (NPV)
m. Perform sensitivity analyses by changing values of some parameters and report on variations in IRR and PBP. Use the following items as guide:
   i. Population versus NPV (300–5,000 heads facility)
   ii. Soil amendment price (\$82.50–\$550/tonne)
   iii. Net metering cost (\$0.05–\$0.20/kWh)
   iv. Tax credit (5%–10%)
   v. Biogas heating value (18.61–27.92 MJ/m^3)
   vi. Anaerobic digester (AD) capital cost (\$17.64–\$176.44/m^3)
   vii. Gasholder capital cost (\$17.64–\$176.44/m^3)
   viii. Interest on Investment (3%–25%)
   ix. Gas cleanup cost (\$0.05–\$0.75/L)

REFERENCES

ASABE D384.2 MAR2005 (R2010). 2010. *Manure Production and Characteristics Standards*. Standard Revised. ASABE, St. Joseph, MI.

Badger, P. and P. Pullammanappallil. 2011. Biomass conversion to energy and fuels. In: *Wood to Energy: Using Southern Interface Fuels for Bioenergy* (C. Staudhammer, A. L. Hermansen-Baez, D. R. Carter and E. A. Macie, eds.). USDA Forest Service, Southern Research Station, General Technical Report GTR-132. Asheville, NC. pp. 33–48. Available at: https://www.srs.fs.usda.gov/pubs/gtr/gtr_srs132.pdf or gtr_srs132.pdf (usda.gov) (accessed July 12, 2023).

Callander, I. J. and J. P. Barford. 1983. Recent advances in anaerobic digestion technology. *Process Biochemistry* 18(4): 24–30, 37.

Capareda, S. 2011. Biomass energy conversion. In: *Sustainable Growth and Applications in Renewable Energy Sources* (M. Nayeripour and M. Kheshti, eds.). InTech Publishing (InTechWeb.org), Rijeka, Croatia. pp. 209–226.

Converse, J. C. and R. E. Graves. 1974. *Facts on Methane Production from Animal Manure*. Fact Sheet A-2636. University of Wisconsin, Madison, WI.

Gerardi, M. H. 2003. *The Microbiology of Anaerobic Digester*. John Wiley & Sons, Ltd., Hoboken, NJ.

King Mongkut's Institute of Technology (KMIT). 1987. Training course on biogas reactor design and development lecture notes, May–June 1987. Continuing Education Center, Bangkok, Thailand.

Letingga, G., A. F M. Van Velsen, S. W. Hobma, W. de Zeeuw and A. Klapwijk. 1980. Use of the upflow sludge blanket (USB) reactor concept for biological wastewater treatment, especially for anaerobic treatment. *Biotechnology and Bioengineering* 22: 699–734.

Maramba, F. D., Sr. 1978. *Biogas and Waste Recycling: The Philippine Experience*. Regal Printing Company, Manila, Philippines.

National Academy of Sciences (NAS). 1976. *Methane Generation from Human, Animal and Agricultural Wastes, Energy for Rural Development Report: Renewable Resources and Alternative Technologies for Developing Countries*. National Research Council, Washington, DC.

Stout, B. A. 1984. *Energy Use and Management in Agriculture*. Breton Publishers, Belmont, CA.

Stout, B. A., C. A. Myers, A. Hurand and L. W. Faidley. 1979. *Energy for World Agriculture*. FAO Agriculture Series. Food and Agriculture Organization of the United Nations, Rome, Italy.

Tchobanoglous, G. and F. L. Burton. 1991. *Wastewater Engineering: Treatment, Disposal and Reuse*.3rd Ed. McGraw Hill Book Co., New York.

Turovskiy, Izrail S. and P. K. Mathai. 2006. *Wastewater Sludge Processing*. John Wiley and Sons, Ltd.

Whiticar, M.J. 2020. The biogeochemical methane cycle. In: *Hydrocarbons, Oils and Lipids: Diversity, Origin, Chemistry and Fate. Handbook of Hydrocarbon and Lipid Microbiology* (H. Wilkes, eds.). Springer, Cham. pp. 1–38. https://doi.org/10.1007/978-3-319-54529-5_5-1.

Young, J. C. and P. L. McCarthy. 1969. The anaerobic filter for waste treatment. *Journal of Water Pollution Control Federation* 41: 160–169.

ASSESS YOUR KNOWLEDGE

1. These are types of microorganisms that thrive under very low environmental temperature of below 20°C. They produce the most biogas with less retention times in cold regions.
   a. Thermophilic
   b. Psychrophilic
   c. Mesophilic
   d. Halophilic

2. These are types of microorganisms that thrive under medium environmental temperature of around 20°C–42°C. They produce the most biogas with less retention times in semi-arid regions with mild temperature ranges.
   a. Thermophilic
   b. Psychrophilic
   c. Mesophilic
   d. Halophilic

11. For every mole of acetic acid, 1 mole of methane and carbon dioxide is formed (see equation below). Estimate the theoretical amount of carbon dioxide formed (ft^3) for every lb of acetic acid used. Assume the density of components as follows:density of pure acetic acid at STP = 8.74 lb/gal; density of pure methane at STP = 0.045 lb/ft^3; density of pure carbon dioxide at STP = 0.114 lb/ft^3.
    $CH_3COOH \rightarrow CH_4 + CO_2$
    a. 5.9
    b. 6.43
    c. 51.6
    d. 56.2

12. Suppose you have 100 heads of dairy heifers producing a total of about 2,200 of wet manure each day that is fed into a plug-flow anaerobic digester. Determine the total amount of biogas produced each day (ft^3) if the biogas production is 45 ft^3 per animal per day.
    a. 254
    b. 4,500
    c. 190.5
    d. 381

3. This parameter is defined as the mass of solids in the digester divided by the mass of solids withdrawn daily in the digester in an anaerobic digestion process.
   a. Hydraulic retention time
   b. Total volatiles
   c. Slurry density
   d. Solids retention time

4. This parameter is defined as the volume of sludge in the digester divided by the volume of digested sludge withdrawn each day in an anaerobic digestion process.
   a. Hydraulic retention time
   b. Solids retention time
   c. Total volatiles
   d. Slurry density

13. Suppose you have 100 heads of dairy heifers producing a total of about 2,200 of wet manure each day that is fed into a plug-flow anaerobic digester. Determine the total amount of energy in units of Btu/day produced if the biogas production is 45 ft^3 per animal per day. The heating value of this gas is 600 Btu/ft^3.
    a. 13,000
    b. 27,000
    c. 2,700,000
    d. 156,000

14. For every mole of acetic acid, 1 mol of methane and carbon dioxide is formed (see equation below). Estimate the theoretical amount of carbon dioxide formed (ft^3) for every gallon of acetic acid used to produce 1 mol of methane. Assume the density of components as follows: density of pure acetic acid at STP = 8.74 lb/gal; density of pure methane at STP = 0.045 lb/ft^3; density of pure carbon dioxide at STP = 0.114 lb/ft^3.
    $CH_3COOH \rightarrow CH_4 + CO_2$
    a. 56.2
    b. 15.6
    c. 6.43
    d. 5.9

5. This process is the conversion of complex biomass into methane by the action of anaerobic microbes (complete absence of oxygen).
   a. Anaerobic digestion
   b. Aerobic digestion
   c. Methanogenesis
   d. Acidogenesis

6. In typical anaerobic digestion process, methane is produced most from what compound?
   a. Hydrogen gas
   b. Complex organics
   c. Acetic acid
   d. Higher organics

15. The biogas from a 500-head dairy facility will be used to run an engine and generate electrical power. Assuming conversion efficiencies of 35% thermal and 90% mechanical, what size of generator in kW may be used for this application if the biogas will be used continuously throughout the day? Use the following basic information from this animal type: gas production per animal = 45 ft^3/head/day; and heating value of biogas = 550 Btu/ft^3.
    a. 476
    b. 4.8
    c. 33
    d. 47.6

16. A farmer had 5,000 heads of dairy cattle. The manure produced was used to generate biogas to replace natural gas for heating purposes. Each animal had a biogas yield of 45 ft^3/day with a heating value of 550 Btu/ft^3. In 2015, the average heat content of natural gas was about 1,000 Btu/ft^3. The price of natural gas in that year had an average price of \$5/1,000 ft^3 for heating purposes. Determine the equivalent energy savings each day (\$) if natural gas usage was replaced by the equivalent energy derived from biogas.
    a. \$160
    b. \$800
    c. \$618.75
    d. \$200

7. This anaerobic digestion stage or process is characterized mainly by the production of acetate or acetic acid by hydrogenation or dehydrogenation. In this stage, acetic acid is mainly produced. Name this AD stage or process.
   a. Acidogenesis.
   b. Acetogenesis.
   c. Methanogenesis
   d. Hydrolysis

8. This anaerobic digestion stage or process is characterized by the conversion of carbohydrates, fats, and proteins into sugars, fatty acids, and amino acids. The process is conducted by exo-enzymes of fermentative micro-organisms. This can be a rate limiting for anaerobic digestion process. Name this AD stage or process.
   a. Acidogenesis
   b. Acetogenesis
   c. Methanogenesis
   d. Hydrolysis

17. For every mole of acetic acid, 1 mol of methane and carbon dioxide is formed (see equation below). Estimate the theoretical amount of carbon dioxide formed (lb) for every lb of acetic acid used. Assume the density of components as follows: density of pure acetic acid at STP = 8.74 lb/gal; density of pure methane at STP = 0.045 lb/ft^3; density of pure carbon dioxide at STP = 0.114 lb/ft^3.
    $CH_3COOH \rightarrow CH_4 + CO_2$
    a. 6.43
    b. 51.6
    c. 0.733
    d. 56.2

18. Determine the amount of iron sponge (ft^3) that is required to remove the H_2S in a biogas for a 500-head dairy heifer facility each day. Assume 45 ft^3 of biogas is produced per animal per day. Further assume that the biogas has 0.15% sulfur. The performance of the iron sponge is that for every 1.5 ft^3 of sponge about 10 lbs of sulfur is removed. Assume further the sulfur density of 7.2 lbs/gal.
    a. 167
    b. 78.8
    c. 125
    d. 250

9. This anaerobic digestion stage or process is characterized by the conversion of carbohydrates, fats, and proteins into sugars, fatty acids, and amino acids. The process is conducted by exo-enzymes of fermentative micro-organisms. This can be a rate limiting for anaerobic digestion process. Name this AD stage or process.
   a. Acidogenesis
   b. Acetogenesis
   c. Methanogenesis
   d. Hydrolysis

10. What is considered the optimum pH condition or range for an ideal anaerobic digestion process?
   a. 6.8–7.4
   b. 4–8
   c. 0.1–1.0
   d. 8–12

19. About 4 mol of hydrogen is required for every mole of methane produced during anaerobic digestion process using the ideal equation below. The density of methane is 0.045 lbs/ft^3 and the density of hydrogen is 0.0056 lb/ft^3. Determine the amount of methane produced (in ft^3) per lb of hydrogen used using the ideal equation shown.
   $4H_2 + CO_2 \rightarrow CH_4 + 2H_2O$
   a. 0.124
   b. 8.95
   c. 1.25
   d. 44.44

20. About 4 mol of hydrogen is required for every mole of methane produced during anaerobic digestion process using the ideal equation below. The density of methane is 0.045 lbs/ft^3 and the density of hydrogen is 0.0056 lb/ft^3. Determine the amount of methane produced (in ft^3) per cubic feet of hydrogen used using the ideal equation shown.
   $4H_2 + CO_2 \rightarrow CH_4 + 2H_2O$
   a. 0.249
   b. 22.2
   c. 0.124
   d. 44.44

Answers: 1b, 2c, 3d, 4a, 5a, 6c, 7b, 8d, 9d, 10a, 11b, 12b, 13c, 14a, 15d, 16c, 17c, 18b, 19d, 20a

ADDITIONAL LEARNING RESOURCES

- Access the Lecture Notes at the Instructor and Student Resources: https://routledgetextbooks.com/textbooks/instructor_downloads/
- The Case Study Videos can be found in the ebook+ version of this book.

8 Torrefaction

LEARNING OBJECTIVES

Upon completion of this chapter, one should be able to

1. Discuss the importance of torrefaction and biomass pretreatment processes for thermal conversion.
2. Describe the various ways of improving the property of biomass resources by thermal means.
3. Describe the various torrefaction processes and enumerate the advantages and disadvantages to those processes.
4. Estimate the energy and mass balances and yields during torrefaction processes.
5. Identify the various equipment used for torrefaction and pelleting.
6. Describe the various properties of torrefied biomass.

8.1 INTRODUCTION

Torrefaction is a process of modifying the properties of biomass using mild thermal conversion. The process is normally conducted under mild temperatures ranging from 200°C to 320°C. Biomass physical properties, as well as some transport characteristics, are improved. The most common process is slow heating in an inert atmosphere. Inert gas is used during the process to limit the conversion of carbon into other compounds and volatiles in the biomass. Nitrogen is the most common inert gas used. The goal of the process is to retain the energy content and only remove unwanted compounds, particularly moisture (water).

The two most common properties of torrefied biomass are low moisture content and high energy density. Other beneficial properties include hydrophobicity, less smoke when burned, and less volatility when stored at elevated temperatures. Figure 8.1 shows a typical fluidized bed torrefaction facility. The reactor is very similar to the fluidized pyrolysis facility, with an electrical heater to control the torrefaction temperature. The solid, gaseous, and liquid condensible products are carefully analyzed. The energy densification is normally in the range from 1.1 to 1.5 and is defined as the ratio of percent of energy conserved divided by the percentage of mass retained.

The energy densification may be as high as 30% in some instances. Example 8.1 shows the calculation for energy densification via a simple torrefaction process.

Example 8.1: Energy Densification for Torrefied Product

One kilogram of biomass (2.2 lb) with an energy content of 15.5 MJ (14,692 Btu) was torrefied at 300°C, and the resulting 0.7 kg of biomass (1.54 lb) that was left had an energy content of 19.9 MJ/kg (8,573 Btu/lb). Determine the energy densification for this torrefaction process.

DOI: 10.1201/9781003294306-8

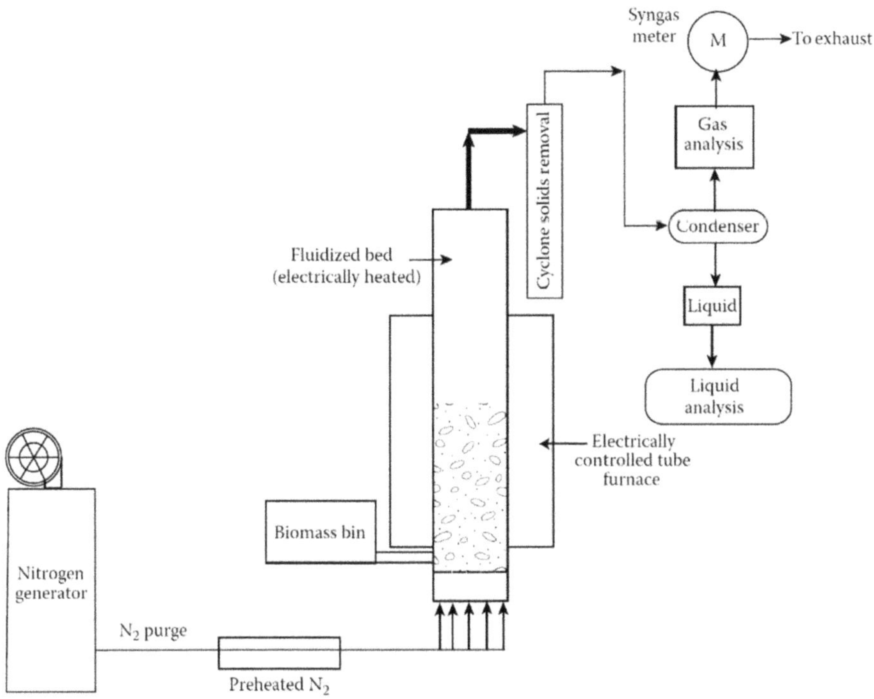

FIGURE 8.1 Schematic of torrefaction facility.

Solution:

a. The percentage of energy conserved on a unit mass basis is calculated as follows:

$$\frac{19.9\,\text{MJ}}{\text{kg}} \times 0.7\,\text{kg} = 13.93\,\text{MJ}$$

b. The amount of energy lost will be the difference between the original 15.5 and 13.93 MJ = 1.57 MJ or about 10% of the energy. Thus, the percentage of energy conserved was 0.90 or 90%.

c. The energy densification is calculated by taking the ratio of 90% (or 0.90) to 70% (or 0.70) (i.e., the percentage of mass retained) shown as follows:

$$\text{Energy densification}\left(\frac{\%\text{Energy}}{\%\text{Mass}}\right) = \frac{90}{70} = 1.29$$

8.2 BIO-PHYSICO-CHEMICAL CHANGES IN BIOMASS DURING TORREFACTION

The bio-physico-chemical changes that occur during biomass torrefaction are best exhibited by Figure 8.2 (Bergman and Kiel, 2005). Biomass is generally composed of hemicellulose, lignin, and cellulose. Each of these major components reacts differently as the temperature is increased. The rightmost side of the figure shows the

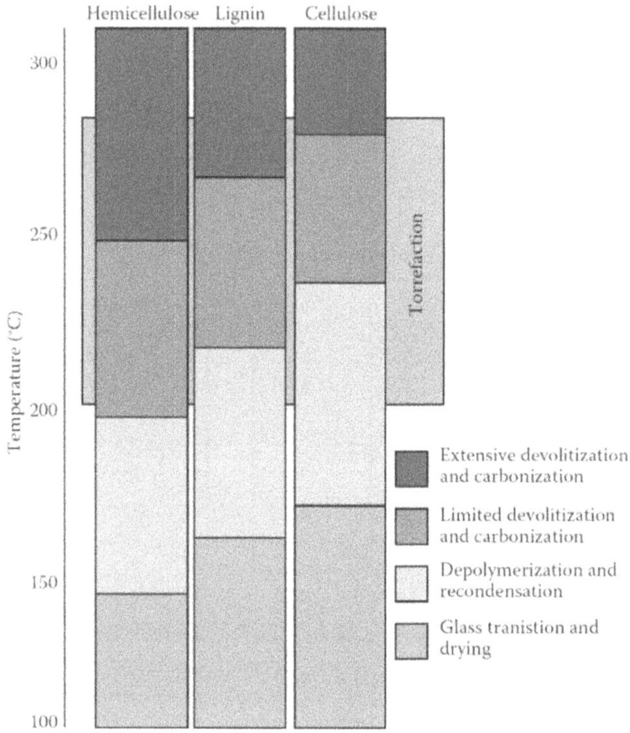

FIGURE 8.2 Physico-chemical changes in biomass during torrefaction. (Modified from Bergman, P. C. A., A. R. Boersma, R. W. H. Zwart and J. H. A. Kiel, Torrefaction for biomass cofiring in existing coal-fired power plants. Energy Research Center of the Netherlands (ECN), ECN Biomass, Petten, The Netherlands. ECN Report No. ECN-C-05–013, 2005.)

color change as biomass is heated from 100°C to 300°C. From the original light brown material, the biomass turns dark brown, and into a black material as it nears 300°C. The following events occur: (a) drying, (b) glass transition/softening, (c) depolymerization and recondensation, (d) limited devolatilization and carbonization, and (e) extensive devolatilization and carbonization. Drying occurs slightly above 100°C for all components (hemicellulose, lignin, and cellulose), while the cellulose components are the last to dry at a temperature nearing 200°C. At around 150°C, the hemicellulose components start to depolymerize and recondensate. The primary reason for this event is the presence of some sugars in the hemicellulose portion. Lignin depolymerizes at a much higher temperature, followed by the cellulose component. At 200°C, most of the moisture is driven away for all components with a little bit of moisture left in the cellulose components. As temperature is further increased beyond 200°C, limited devolatilization occurs on the hemicellulose portion, along with some carbonization. The lignin portion devolatilizes at a much higher temperature, followed by the cellulose components. At 250°C, most of the hemicellulose portions are devolatized and extensive devolatilization occurs as the temperature is increased further. The lignin portion will undergo extensive

TABLE 8.1
Compositional Analysis of Some Selected Biomass and Wastes

	Hemicellulose (%Weight)	Lignin (%Weight)	Cellulose (%Weight)
Sugarcane bagasse	27	20	38
Sugarcane leaves	21	16	36
Napier grass	20	9	32
Sweet sorghum	16	10	36
Eucalyptus saligna	12	25	45
Municipal solid waste	9	17	33
Newspaper	16	21	62

Source: Pandey, A., *Handbook of Plant-Based Biofuels*, CRC Press, Taylor & Francis Group, Boca Raton, FL, 2009.

devolatilization near 275°C, while the cellulose portion will not undergo extensive devolatization until slightly above 300°C (Bergman and Kiel, 2005). As such, contrary to the common perception that the lignin component is the recalcitrant component in the biomass, the cellulose components are those degraded last during thermal conversion.

Table 8.1 summarizes the different hemicellulose, lignin, and cellulose components of some plant materials, including other biomass waste resources such as municipal solid wastes and newspapers. Hemicellulose ranges from a low of 9% for MSW to as high as 27% for sugarcane bagasse. The cellulose component ranges from 32% to 62%, the highest being that from newspapers. The lignin content ranges from 9% to 25%, the highest being that from *Eucalyptus saligna*. Note the lignin content of biomass is never converted by biochemical means and remains undigested.

8.3 TORREFACTION PRODUCTS

Torrefaction of biomass produces three primary products: (a) solid product of a dark brown color; (b) condensible liquids, including moisture, acetic acids, and some oxygenates; and (c) noncondensible gases that are composed mainly of CO, H_2, and small amounts of other hydrocarbons such as methane. A detailed graphic illustration of these products is shown in Figure 8.3.

The mass yields during torrefaction of willow at 280°C for 17.5 min showed that about 87.7% of the yield is composed of solid products. The remaining components include lipids, organics, gases, and water comprising 4.5% by weight of the total biomass. A large portion of the biomass that was removed during the torrefaction process was moisture (8%). The lipids comprised terpenes, phenols, fatty acids, waxes, and tannins, and accounted for 1.4% of the total weight of the biomass. The organics include sugars, acids, alcohols, furans, and ketones (1.7%), while the gaseous products include hydrogen (H_2), carbon monoxide (CO), methane (CH_4), toluene (C_7H_8), and benzene (C_6H_6) and about 1.4% of the converted

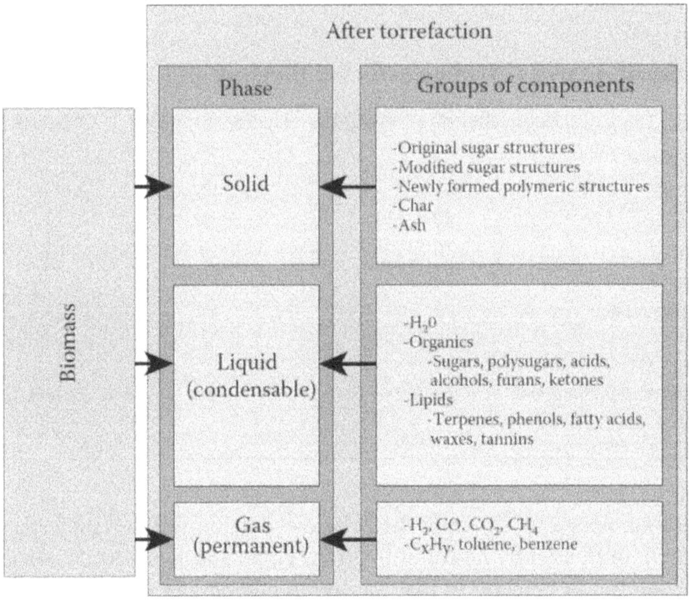

FIGURE 8.3 Products formed during torrefaction of biomass. (Modified from Bergman, P. C. A., A. R. Boersma, R. W. H. Zwart and J. H. A. Kiel, Torrefaction for biomass co-firing in existing coal-fired power plants. Energy Research Center of the Netherlands (ECN), ECN Biomass, Petten, The Netherlands. ECN Report No. ECN-C-05–013, 2005.)

biomass weight (Bergman and Kiel, 2005). In this torrefaction process, about 94.5% of the energy is retained in the torrefied biomass, while 3.4% is contained in the lipids and 1.6% in the organics. Surprisingly, only 0.1% of the energy is contained in the gaseous product.

8.3.1 Solid Products

Solid products from the torrefaction process include the original and modified sugar structures. There will be newly formed polymeric structures due to the action of mild heat, as well as the conserved carbon in the form of char, and the inorganic ash portion of the biomass. The biomass conserves most of the cellulose and lignin components, where some highly volatile organic compounds are converted into a gaseous or liquid phase. Most of the weight reduction of the biomass due to torrefaction results from the evaporation of moisture. However, because most biomass has high volatile organic matter content, some organic carbon is converted into gaseous or liquid components.

8.3.2 Liquid Condensibles

The liquid portion of the condensibles also includes (aside from water) organics such as volatilized sugars, polysugar acids, alcohols, and very minimal amounts of furans and ketones. The lipids that are volatilized include terpenes, phenols, fatty acids, waxes, and tannins. The reduction in the energy content of the biomass is due to the conversion of the volatiles into liquid condensibles. Unlike water, these compounds have energy values leading to reduction in the energy content of the solid biomass that remains after the process. The amount of the condensible is low, and only a minimal amount of energy is lost during the process.

8.3.3 Gaseous Products

The gaseous (noncondensible) components released during torrefaction are made up of hydrogen gas, carbon monoxide, carbon dioxide, methane, and some hydrocarbons such as benzene and toluene. These, of course, are also in very small amounts and are not beneficial for recovery. The release of these gaseous compounds, together with moisture, also improves the overall density of the material and makes the resulting solid product denser. Example 8.2 shows the calculation to determine the energy and mass balance during torrefaction process.

Example 8.2: Determining Energy and Mass Balances in Torrefaction

One kilogram of biomass with a heating value of 19 MJ/kg torrefied at 290°C resulted in various solid, liquid, and gaseous components. The remaining solid component was 0.85 kg with a heating value of 20 MJ/kg. About 0.10 kg of liquid was collected with about 15% moisture. The organic component of this liquid portion had a heating value of 21 MJ/kg. Estimate the energy content of the gaseous component assuming no losses. Determine also the percentage of energy contained in each component.

Solution:

a. The energy contained in the solid portion is estimated as follows:

$$\text{Solid (MJ)} = 0.85\ \text{kg} \times \frac{19\ \text{MJ}}{\text{kg}} = 16.15\ \text{MJ}$$

b. The liquid component has 85% organic portion and has the following energy:

$$\text{Liquid (MJ)} = 0.10\ \text{kg} \times 0.85 \times \frac{21\ \text{MJ}}{\text{kg}} = 1.785\ \text{MJ}$$

c. Thus, the gaseous component, assuming no energy losses, is calculated by taking the difference as follows:

$$19\ \text{MJ} = 16.15\ \text{MJ} + 1.785\ \text{MJ} + \text{energy in the gas}$$

$$\text{Energy in the gas} = 1.065\ \text{MJ}$$

d. The energy balance is as follows: solid = 85.0%, liquid = 9.39%, and gas = 5.61%.

8.4 PHYSICAL PROPERTIES OF TORREFIED BIOMASS

8.4.1 Bulk Density

The mass loss in the form of solids, liquids, and gases during the torrefaction process causes the biomass to become more porous. This results in a significant reduction in volumetric density and increase in the bulk density. Table 8.2 summarizes the changes in bulk density for *Eucalyptus grandis* wood at three different treatments (Rodrigues and Rousset, 2009). The temperature is varied from 220°C to 280°C. As the temperature is increased, the bulk density is reduced, with a corresponding large percentage loss compared with the control. Example 8.3 shows the process for calculating the bulk density of torrefied biomass.

Example 8.3: Bulk Density of Torrefied Biomass

Switchgrass was chopped after harvest and found to have a bulk density of 108 kg/m^3 (6.73 lb/ft^3). A sample of switchgrass was placed in a 305 mm×305 mm×305 mm box (1 ft×1 ft×1 ft). The sample weighed 3.06 kg (6.73 lb). This sample was torrefied at 300°C for 1 h, and there was a 30% reduction in weight. The resulting volume of the product was also measured and found to be 0.00714 m^3 (0.252 ft^3). Calculate the new bulk density of the torrefied material. Convert the units to the English system.

Solution:

a. The new bulk density is calculated from the new weight and the new volume occupied by the sample as follows:

$$\text{New bulk density}\left(\frac{\text{kg}}{\text{m}^3}\right)=\frac{3.06\times 0.7\,\text{kg}}{0.00714\,\text{m}^3}=\frac{2.142\,\text{kg}}{0.00714\,\text{m}^3}=300\frac{\text{kg}}{\text{m}^3}$$

b. In the English system, the units are as follows:

$$\text{New bulk density}\left(\frac{\text{lb}}{\text{ft}^3}\right)=\frac{300\,\text{kg}}{\text{m}^3}\times\frac{2.2\,\text{lb}}{1\,\text{kg}}\times\left(\frac{1\,\text{m}}{3.28\,\text{ft}}\right)^3=18.7\frac{\text{lb}}{\text{ft}^3}$$

c. There is an almost a 3-fold increase in the bulk density of the material.

TABLE 8.2
Bulk Density (Dry Basis) of *Eucalyptus grandis* as Affected by Torrefaction Temperature

Treatment	Bulk Density (kg/m^3)	Percentage Loss (%)
Control	850[a]	–
T2: 220°C	830[a]	2.35
T3: 250°C	790[b]	7.06
T4: 280°C	730[c]	14.12

Source: Rodrigues, T. O. and P. L. A. Rousset, *Cerne*, 15, 446, 2009.

Note: Means followed by the same letter in superscript are statistically similar at the 5% probability level. The average moisture content of the control was 15%.

The bulk density of the biomass changes as it goes through the drying and torrefaction conditions in the reactor. Example 8.4 shows changes in the bulk density as a function of torrefaction temperature.

Example 8.4: Bulk Density of Biomass as a Function of Torrefaction Temperature

Eucalyptus grandis was placed in a gravimetric torrefaction facility and exposed to various torrefaction temperatures. The weight and final volume of the sample at each temperature setting was measured as summarized in the table. The original weight of the sample was 850 g in a 1,000 mL filled graduated cylinder. Data shows the final weights and volumes of the sample at each temperature setting. Calculate the bulk density of the original sample and at the highest temperature of exposure. Convert all units in the English system. Comment on the change in bulk density as torrefaction temperature was increased.

Torrefaction Temperature (°C)	Initial Weight (g)	Final Weight (g)	Final Volume (mL)
220	850	819.2	987
250	850	750.5	950
280	850	681.1	933

Solution:

a. The original bulk density of the biomass can be calculated by taking the ratio of its weight to its volume as follows:

$$\text{Bulk Density}\left(\frac{\text{kg}}{\text{m}^3}\right)=\frac{850\,\text{g}}{1{,}000\,\text{mL}}\times\frac{\text{kg}}{1{,}000\,\text{k}}\times\frac{1\,\text{mL}}{1\,\text{cm}^3}\times\frac{1\times10^{-6}\,\text{cm}}{\text{m}^3}=850\frac{\text{kg}}{\text{m}^3}$$

b. The new bulk density of the biomass at the highest exposure temperature of 280°C is calculated as follows:

$$\text{Bulk Density}\left(\frac{\text{kg}}{\text{m}^3}\right)=\frac{681.1\,\text{g}}{933\,\text{mL}}\times\frac{\text{kg}}{1{,}000\,\text{k}}\times\frac{1\,\text{mL}}{1\,\text{cm}^3}\times\frac{1\times10^{-6}\,\text{cm}}{\text{m}^3}=730\frac{\text{kg}}{\text{m}^3}$$

c. Converting the units in the English system is as follows:

$$\text{Bulk Density}\left(\frac{\text{lb}}{\text{ft}^3}\right)=850\frac{\text{kg}}{\text{m}^3}\times\frac{2.2\,\text{lb}}{1\,\text{kg}}\left(\frac{1\,\text{m}}{3.28\,\text{ft}}\right)^3=53\frac{\text{lb}}{\text{ft}^3}$$

$$\text{Bulk Density}\left(\frac{\text{lb}}{\text{ft}^3}\right)=730\frac{\text{kg}}{\text{m}^3}\times\frac{2.2\,\text{lb}}{1\,\text{kg}}\left(\frac{1\,\text{m}}{3.28\,\text{ft}}\right)^3=45.5\frac{\text{lb}}{\text{ft}^3}$$

d. The bulk density was slightly lowered.

8.4.2 Grindability

Biomass in its raw form is difficult to grind. The components, most especially the cellulose and lignin, are very fibrous and quite hard to break during mechanical processing. Torrefaction makes the biomass lose its fibrous nature, breaking down the hemicellulose matrix and the depolymerizing the cellulose components. Studies conducted by Bergman and Kiel (2005) indicated that the energy requirement for grinding torrefied biomass is reduced by about 70%–90% and the mill capacity increases by a factor of 7.5–15. The grinding characteristics of torrefied biomass make it like coal. The grinding energy is decreased as torrefaction temperature is increased. For example, as the temperature is increased from 220°C to 260°C, the grinding energy is significantly decreased from around 300 to 150 kWh/ton at a mere 5 min of treatment. As the torrefaction period is increased from 5 to 60 min, the grinding energy is likewise reduced. To measure the energy required for grinding the biomass, simple measurement of amperage and the voltage source of the grinding unit (in case of electric motors) is made. For engine prime movers, the power may be estimated by the amount of fuel used during the grinding process and the efficiency of conversion from fuel energy into mechanical energy of the unit. Otherwise, data should be used, if available, on the engine concerning torque and brake horsepower as well as specific consumption. Example 8.5 shows how this is estimated, using a grinding mill powered by an electric motor.

Example 8.5: Grinding Energy Usage

A hammer mill was used to grind corn stover (12% moisture content) with a 3.2 mm screen opening. The unit is powered by a 1.5 kW electric motor (115 V). The power drawn by the hammer mill motor was measured using a wattmeter. Around 1 kg (2.2 lb) of biomass was used during the 6-min test and the measured amperage was 1 A. Estimate the specific energy (in kWh/tonne) or (Btu/lb) required to accomplish this task.

Solution:

a. The power required to accomplish the task is simply the product of voltage and amperage used as follows:

$$\text{Power (W)} = \text{volt} \times \text{amp} = 115\,\text{V} \times 1\,\text{A} = 115\,\text{W}$$

b. The energy required is the product of power and time, and is calculated as follows:

$$\text{Energy (W)} = 115\,\text{W} \times 6\,\text{min} \times \frac{\text{h}}{60\,\text{min}} = 11.5\,\text{Wh}$$

c. Thus, the energy required per weight of biomass is the ratio of energy and weight, and is calculated as follows:

$$\text{Specific Energy}\left(\frac{\text{kWh}}{\text{tonne}}\right) = \frac{11.5\,\text{W}-\text{h}}{1\,\text{kg}} \times \frac{1{,}000\,\text{kg}}{\text{tonne}} \times \frac{\text{kW}}{1{,}000\,\text{W}} = 11.5\frac{\text{kWh}}{\text{tonne}}$$

$$\text{Specific Energy}\left(\frac{\text{Btu}}{\text{lb}}\right) = \frac{11.5\,\text{kWh}}{\text{tonne}} \times \frac{1\,\text{tonne}}{2{,}200\,\text{lb}} \times \frac{3{,}412\,\text{Btu}}{1\,\text{kWhr}} = 17.8\frac{\text{Btu}}{\text{lb}}$$

8.4.3 Particle Size Distribution, Sphericity, and Surface Area

Particle size distribution, sphericity, and surface area are important parameters used for understanding the ability of the biomass to flow, as well as the ability to combust during co-firing with a conventional fuel such as coal. Many researchers observed that ground and torrefied biomass produced more uniform particle sizes compared with untreated biomass due to its brittle nature, more like coal. Phanphanich and Mani (2011) found that torrefied pine chips and logging residues produced smaller particle sizes compared with untreated biomass. Furthermore, the results also indicated that sphericity and particle surface area increase as the torrefaction temperature is increased up to 300°C. Other studies also made use of a scanning electron microscope (SEM) to describe the changes in the particle behavior after torrefaction, showing structural breakage as temperature is increased (Pilon and Lavoie, 2011), resulting in a powdered material that is almost spherical instead of linear in shape. The surface area is also increased when micropore development is enhanced. The pore development may not be observed in the SEM photos. However, using the Brunauer, Emmet, and Teller (BET) surface area and pore volume analyzer, the development may be indirectly deducted from the surface area analysis. A study was also made on the torrefaction of bamboo, banyan, and willow biomass in terms of effect on particle size distribution (Chen et al., 2011). Results revealed that the raw biomass was characterized by agglomeration in the regime of smaller particle size, but as the materials underwent torrefaction, the dispersion of powder was improved. These characteristics of torrefied biomass enhance its application as a co-firing fuel in boilers and blast furnaces, especially during the injection of the powdered biomass. SEM images were also made to show the tubular structures before the torrefaction and a more uniform powdery structure after torrefaction.

8.4.4 Moisture Content

The moisture content of torrefied biomass, as expected, decreases as temperature increases and with exposure to higher temperature. The trend is to be expected, as in many drying operations. The relationship shows that the rate of moisture removal varies with the initial moisture content of the biomass. In other words, the higher the moisture content, the higher the rate of decrease in moisture; the lower the moisture content, the lower the rate of decrease in moisture. This is evident if the moisture content of the material is significantly higher. If the moisture content is significantly low, say 2%, there is also a marginal decrease in the moisture content. Example 8.6 illustrates the changes in weights of the biomass as water is removed during torrefaction.

Example 8.6: Moisture Content of Torrefied Biomass

A tonne (2,200 lb) of switchgrass having 8% moisture was torrefied and the moisture content of the final product was 1.5%. How much moisture was removed from this process? Assume that only the moisture was removed. Calculate the moisture removed per ton of wet material.

Solution:

a. The amount of moisture from 1 tonne (1,000 kg) material is calculated as follows:

$$\text{Moisture (kg)} = 1{,}000\ \text{kg} \times 0.08 = 80\ \text{kg}$$

b. Thus, the total amount of moisture in the material is 80 kg, and the dry weight is 920 kg (1,000–80), which is conserved.
c. The final weight of the resulting 1.5% moisture torrefied product is as follows (recall the moisture content equation from Chapter 1):

$$1.5\% = \frac{\text{Wet weight}}{\text{Wet weight} + \text{Dry weight}} \times 100\% = \frac{\text{Wet weight}}{\text{Wet weight} + 920\ \text{kg}} \times 100\%$$

$$\text{Wet weight (kg)} = 14\ \text{kg}$$

d. The amount of moisture to be removed will be 80 kg −14 kg = 66 kg. This is also the amount of moisture to be removed per tonne of material.

8.4.5 Storage Properties

Storage issues for biomass, such as off-gassing and self-heating, may be avoided by using the torrefaction processes. This is primarily due to loss in solid, liquid, and gaseous products that are chemically and microbiologically active. Studies on off-gassing of torrefied wood chips at the University of British Columbia, Canada, indicated that CO and CO_2 emissions were very low; nearly one-third of the emissions from regular wood chips at room temperature (Tumuluru et al., 2008). Storage properties are enhanced when the elements for subsequent biological degradation are removed.

Microorganisms thrive in moist conditions as well as in the presence of nutrients for their growth. In addition, these degrading elements are also affected by inhibitory elements produced during the torrefaction process, such as polycyclic aromatic hydrocarbons. During storage of raw biomass, gases are normally emitted even at low temperatures. For example, at 5°C, the maximum CO_2 and CO concentrations observed were 1,000 and 500 ppm, respectively, but at 40°C, the CO and CO_2 concentrations increased to 15,989 and 25,456 ppm, respectively, for a 23-day storage period (Tumuluru et al., 2009). The oxygen depletion was also quite significant during this period; it was reduced from 19.5% to 2.1%. Note that these off-gases may have indoor environmental consequences. For example, the safe level of CO concentration is 35 ppm for an average 8-h exposure (Cooper and Alley, 1994). The 500 ppm concentration measured in the above-cited study is above this threshold. One should be familiar with this indoor safety regulation.

8.4.6 Hydrophobicity

The hydrophobicity of torrefied biomass can be due to loss of the OH group, which reduces the ability to form hydrogen bonds. This makes the material less reactive to the storage environment. The chemical rearrangementreaction results in the formation of non-polar unsaturated structures that preserve the biomass for a long period without biological degradation, such as coal. In a study by Bergman (2005), it was indicated that raw pellets swelled rapidly and then disintegrated into their original particles. However, torrefied biomass pellets that were produced under optimal conditions did not disintegrate and showed little water uptake (only about 7%–20% on a mass basis).

Tumuluru et al. (2011a,b) compared the moisture uptake of torrefied biomass to untreated biomass and found that there was about a 25% decrease in the water uptake when compared with the control.

8.4.7 Pelletability

Some researchers have investigated the feasibility of densifying torrefied biomass by turning it into even more dense pellets. The flow diagram to produce torrefied wood pellets is shown in Figure 8.4. The biomass is first dried then undergoes torrefaction. The torrefied material is then ground to reduce the particle size and pelleted using conventional pellet mills. A binder may be required for some processes. If a binder is not used, higher pressure is normally required.

The advantages of densification of torrefied biomass are as follows:

1. The resulting product has high bulk density.
2. The material has higher calorific value.
3. The material is said to be more durable.
4. The material is even more hydrophobic.

The mechanical strength is enhanced. It was shown that the mechanical strength is increased by a factor of 1.5–2.0 over conventional pellets (i.e., those that are pelleted directly from raw biomass).

A measure of durability is presented by Theerarattananoon et al. (2011) and is given by Equation 8.2. Example 8.7 shows the relative durability of biomass pellets. The procedure for durability testing is contained in the ASABE Standard S269.4 (R2007), entitled "Cubes, Pellets, and Crumbles—Definitions and Methods for Determining Density, Durability, and Moisture Content." (ASABE, 1991). The process is conducted

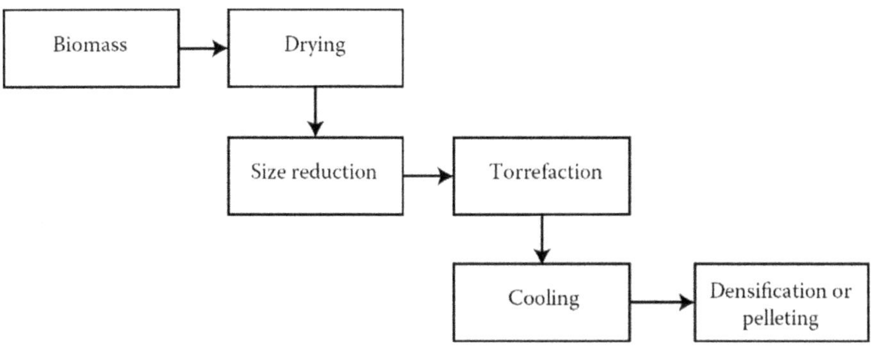

FIGURE 8.4 Flow diagram for converting torrefied biomass into densified material or pellets.

by fabricating a tumbler box where the pellets are tumbled at 50 rpm for about 10 min (in this dust-tight enclosure). After tumbling, the sample is removed and sieved, and the percentage of whole pellets is calculated using Equation 8.1. Note that fines must be removed before the test. The pellets should be sieved (before the test) using a No. 6 US sieve having an aperture size of 3.36 mm.

$$\text{Durability}\,(\%) = \frac{\text{Mass of pellets after tumbling}}{\text{Mass of pellets before tumbling}} \times 100\% \tag{8.1}$$

Example 8.7: Durability of Biomass Pellets

A clean and previously sieved 500 g (1.1 lb) sample of sorghum stalk pellets was placed in a tumbling box to determine the durability. After tumbling for 10 min, the pellets were removed from the box and sieved again. The weight of the material was measured and found to be 461 g. A similar test was made using wheat straw, and the final weight of the tumbled pellets was 487 g. Determine the percent durability of these pellets and compare the durability.

Solution:

a. The durability equation is used as shown below:

$$\text{Sorghum durability}\,(\%) = \frac{461\,\text{g}}{500\,\text{g}} \times 100\% = 92.2\%$$

$$\text{Wheat durability}\,(\%) = \frac{487\,\text{g}}{500\,\text{g}} \times 100\% = 97.4\%$$

b. In this test, the wheat straw pellets were found to be more durable.

8.5 COMPARISON BETWEEN TORREFIED BIOMASS VERSUS PELLETED BIOMASS

There is a marked difference between a torrefied biomass and a biomass that has been pelleted. The first effect is a decrease in the moisture content as the pelleted biomass is further pelleted. Pelleting is an additional energy input to the biomass to make it denser and improve the energy density. The heating value increases consequentially. More importantly, the bulk density will also increase significantly.

In a study performed by Boerrigter et al. (2006), the moisture content was reduced from 3% to 1%, whereas the energy content (dry basis) was improved from 20.4 to 22.7 MJ/kg. In addition to the above changes, the strength was improved while dust formation was minimized during handling. In both cases, further biological degradation was minimized, thereby improving its storage behavior. Both samples were described to be hydrophobic, but the pelleted sample has poor swelling abilities, and a high-moisture environment will affect the pelleted product more than a torrefied product. Both materials, however, have poor response to seasonal conditions, which must be avoided during storage. Example 8.8 illustrates the increase in the bulk density of pelleted biomass. In the above study, the bulk density was improved from 230 kg/m^3 of torrefied biomass to 850 kg/m^3 of the pelleted biomass. The energy density increased from 4.6 to 18.4 GJ/m^3.

Example 8.8: Bulk Density of Torrefied Biomass

Torrefied switchgrass sample processed as in Example 8.3 was pelleted and placed in the bulk density box measuring 305 mm × 305 mm × 305 mm and reweighed. The final weight of the pelleted material in this box was 24 kg. Calculate the new bulk density of the torrefied and pelleted material. Convert the units to the English system.

Solution:

a. The new bulk density is calculated from the new weight and the new volume occupied by the sample as follows:

$$\text{Bulk density}\left(\frac{\text{kg}}{\text{m}^3}\right) = \frac{24\text{ kg}}{305\,\text{mm}^3} \times \left(\frac{1{,}000\,\text{mm}}{\text{m}}\right)^3$$

$$\text{Bulk density}\left(\frac{\text{kg}}{\text{m}^3}\right) = 846$$

b. In the English system, the units are as follows:

$$\text{New bulk density}\left(\frac{\text{lb}}{\text{ft}^3}\right) = \frac{846\,\text{kg}}{\text{m}^3} \times \frac{2.2\,\text{lb}}{1\,\text{kg}} \times \left(\frac{1\,\text{m}}{3.28\,\text{ft}}\right)^3 = 52.7\frac{\text{lb}}{\text{ft}^3}$$

8.6 THERMAL GRAVIMETRIC ANALYSIS STUDIES OF BIOMASS

A common procedure for evaluating the behavior of biomass at elevated temperatures is to make use of a thermal gravimetric analyzer (TGA). In this procedure, a known quantity of the biomass is placed in a receptacle and as either the temperature or residence time is increased, the weight of the sample is reported and recorded. If there are provisions to capture the resulting gaseous and liquid condensibles, these materials are further analyzed in separate facilities. Note that at lower temperatures, there is a linear decrease in the percent of mass; however, at temperatures above 230°C, there will be a marked decrease in the mass as residence time is increased. As such,when the highest temperature of 300°C is used, as much as 65% of the mass may be lost if the residence time is brought to over 100 min. Example 8.9 shows how the energy density is increased in this process. However, this example assumes that only the moisture is driven away during the process. As we will observe in the pyrolysis process at elevated temperatures, there is a decrease in the heating value of the solid products when the temperature is further increased, indicating a reduction in the overall energy contained or what remains in the solid biomass. The energy content of the torrefied biomass may have increased. However, because of the weight loss, the available energy is reduced.

Example 8.9: Energy Density Changes in Torrefied Biomass

One kilogram of a biomass sample had an initial heating value of 19 MJ/kg. This material was placed in a thermal gravimetric analyzer and the loss in weight is summarized in the table at various residence times. Complete the table for the estimated heating value of this biomass, assuming that only moisture is removed and that the energy is conserved. What is the energy content of this material on a dry basis?

Loss in Weight (%)	Residence Time (min)	Energy Content (MJ/kg)
0	0	19
1.0	20	
2.0	40	
5.0	60	
6.0	80	
7.0	100	

Solution:

a. Note that the heating value unit is in MJ/kg, and thus, the denominator is the total weight of the material. The reduction in weight, when attributed to energy loss, will be the new denominator to base the calculation for the new HV as follows:

$$\text{New HV}\left(\frac{\text{MJ}}{\text{kg}}\right)=\frac{19\,\text{MJ}}{0.99\,\text{kg}}=19.19\,\text{MJ/kg}$$

b. The second case is as follows:

$$\text{New HV}\left(\frac{\text{MJ}}{\text{kg}}\right)=\frac{19\,\text{MJ}}{0.98\,\text{kg}}=19.39\,\text{MJ/kg.}$$

	Residence Time (min)	Energy Content (MJ/kg)
0	0	19
1.0	20	19.19
2.0	40	19.39
5.0	60	20.0
6.0	80	20.21
7.0	100	20.43

8.7 CHEMICAL COMPOSITION CHANGES DURING TORREFACTION OF BIOMASS

A classic figure to present changes in the chemical composition of torrefied biomass is presented by Van Krevelen and reported by Bridgeman et al. (2010). The Van Krevelen diagram may be plotted based on the complete ultimate analysis of the biomass. The *y*-axis in this diagram is usually the ratio of atomic hydrogen to carbon (H/C). The *x*-axis is usually the ratio of atomic oxygen to carbon (O/C). Thus, if the ultimate analysis of the biomass is known, the Van Krevelen plot can easily be made and presented (see Example 8.10). The H/C ratios for torrefied samples are almost always above 1.0 but are never above 0.25 in the raw biomass. This indicates that if the process calls for feedstock with a high H/C ratio, a torrefied product is ideal. At lower temperatures and lower torrefaction times, the higher H/C ratios will also be quite low.

In the report by Bridgeman et al. (2010), the lignite and miscanthus samples have similar H/C and O/C ratios when torrefied at 290°C for 60 min. One issue in this process is the energy required to maintain the torrefaction temperature and the exposure time.

In general, as torrefaction temperature is increased, the carbon content will increase while hydrogen and oxygen contents would decrease. This may be due to the evolution of carbon monoxide and carbon dioxide and some moisture. If ratios are calculated, the H/C ratio as well as the O/C ratios will decrease with increasing torrefaction temperature (Tumuluru et al., 2011a). If a torrefied product is used for thermal conversion, such as combustion or gasification, there will be less smoke and vapor release during conversion and energy will be conserved.

Example 8.10: Van Krevelen Plot of Raw Biomass

Use the ultimate analysis data in Table 3.6 to make a Van Krevelen plot (i.e., plot of atomic H/C ratio on *y*-axis versus atomic O/C ratio on *x*-axis) for the raw biomass. Discuss the results and identify where those points are in the quadrant (i.e., upper right, upper left, lower right, or lower left corner of the graph).

Solution:

a. The ratio of H/C and O/C from the table is calculated and some values are presented in the table and plotted as shown:

Lignite:

$$\text{No. Moles Oxygen} = \frac{19.2}{16} = 1.200$$

$$\text{No. Moles Carbon} = \frac{64}{12} = 5.333$$

The mole ratio is also the atomic ratio as follows:

$$\text{Atomic O : C Ratio} = \frac{1.200}{5.333} = 0.2250$$

$$\text{No. Moles Hydrogen} = \frac{4.2}{1} = 4.200$$

The mole ratio is also the atomic ratio as follows:

$$\text{Atomic H : C Ratio} = \frac{4.200}{5.333} = 0.7875$$

The rest of calculations are summarized in the table below:

Biomass	**H/C Ratio**	**O/C Ratio**
Lignite	0.7875	0.2250
Charcoal	0.4633	0.1214
Douglas fir	1.4455	0.5808
Douglas fir bark	1.2598	0.4898
Pine bark	1.3308	0.5564
Rice hulls	1.7766	0.7753
Rice straw	1.5612	0.6849

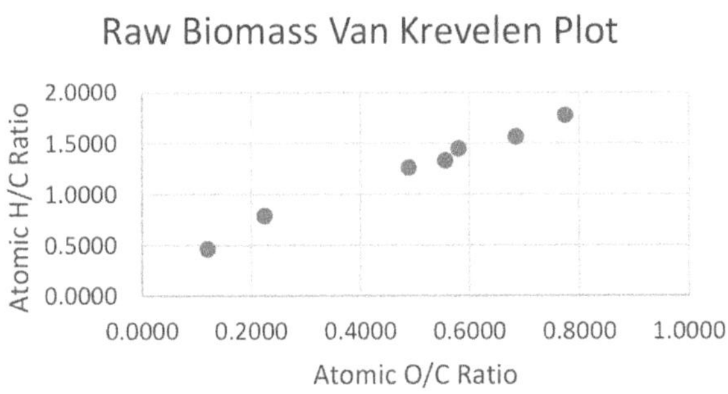

b. Note that the region plotted is at the lower left corner of the Van Krevelen plot. Charcoal and lignite are considered the best materials. Charcoal is considered torrefied material.

8.7.1 Carbon Content

The carbon content of raw biomass, as summarized in Table 3.5, ranges from a low of 27.6% (poultry manure) to a high of 49.5% (peat). The carbon content for solids and biofuels, as summarized in Table 3.6, ranges from a low of 33.0% (feedlot manure) to a high of 77.9% (for Utah coal). If these biomass materials were torrefied, the carbon content would be increased significantly. The carbon content of a biomass sample may be as low as 47%; however, a combination of a 340°C temperature and a 120 min torrefaction time would increase the carbon content to more than 56%, an increase of almost 10% (Tumuluru et al., 2011a). Note that these percentages are relative values since the total weight of the biomass has decreased due to the torrefaction process.

The carbon content of the biomass material is always correlated to the heating value, and one would expect the energy density to increase accordingly. For future conversion of torrefied biomass into liquid fuels, such as hydrocarbon fuels, the improved carbon content is related to the improvement in the liquid hydrocarbon fuels that may be produced in succeeding processes. It should be noted that at higher torrefaction temperatures, volatiles are lost. This also means a loss of carbon because many of those volatiles are carbon-related compounds (e.g., acetic acid, methanol, or furfural).

Some carbon in the volatiles is also lost during the torrefaction process; however, the total bound carbon is still conserved for future conversion processes.

8.7.2 Hydrogen Content

Of the elemental properties of biomass that are important for further conversion into liquid fuel, hydrogen and carbon contents are the most significant. After all, hydrocarbon fuels are mainly composed of hydrogen and carbon. Thus, one would expect that under mild thermal treatment, these elements should be conserved. Unfortunately, elemental hydrogen is also bound in the water molecules (which is one main compound that is removed), and thus, the hydrogen content is expected to have a bigger reduction than the carbon content. Studies by Bridgeman et al. (2008) revealed that the reduction in elemental hydrogen content is up to 10.3% at a torrefaction temperature of 290°C for reed canary grass but only 7.7% reduction in willow at the same temperature. For wheat straw, the elemental hydrogen reduction is about 17.65% at the same temperature. Hydrogen reduction in many biomass samples after torrefaction also varies with the species. The widespread reduction in elemental hydrogen may be detrimental to the future use of the biomass for liquid fuel production. It is essential to use the appropriate torrefaction temperature when upgrading the biomass.

8.7.3 Calorific Value

The decline in the total calorific value as torrefaction temperature is increased may also be significant for some biomass species. Consequently, inappropriate temperatures must be avoided or limited during the torrefaction process. Although the heating value of the product may increase as shown in Figure 8.5, significant loss in energy must be avoided during torrefaction. As shown at the beginning of this chapter, about 90% of the energy may be conserved during the torrefaction processes; however, there are some instances where as much as 65% of the energy may be lost, especially at higher torrefaction temperatures (Medic et al., 2010). The initial moisture content of biomass has a significant effect on the resulting heating value of the final torrefied product since the heating value is based on a unit weight of the product. If the weight is mainly comprised of water, then there

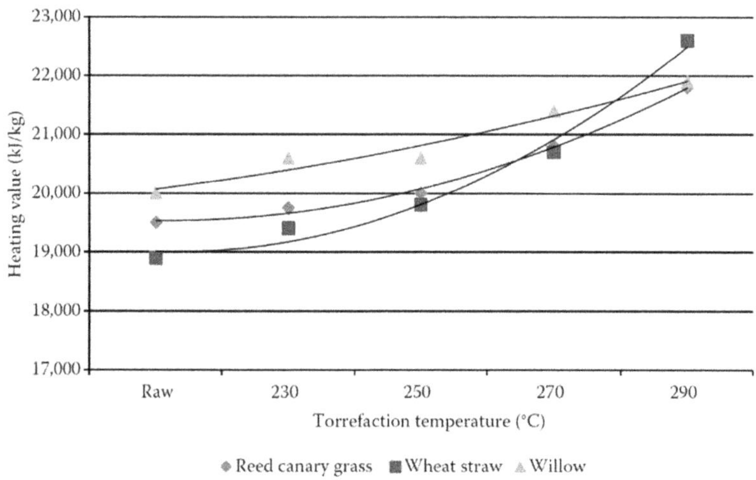

FIGURE 8.5 Changes in heating values of biomass product as torrefaction temperature is increased. (Adapted from data by Bridgeman, T. G., J. M. Jones, I. Shield and P. T. Williams, *Fuel*, 87, 844, 2008.)

will be a significant increase in the energy density. As reported by Medic et al. (2010), temperature has a more significant impact on energy density than moisture, torrefaction time, or exposure period. Biomass torrefied at lower temperatures has the highest energy yield.

8.7.4 Degree of Carbonization or Conservation of Carbon

Torrefaction is said to increase the amount of atomic carbon while decreasing the amount of atomic hydrogen and oxygen. The consequence of this phenomenon is a chemical compositional change in the resulting biomass. Torrefaction decreases the atomic ratio of oxygen to carbon (O/C ratio) and hydrogen to carbon (H/C ratio), compared with the original raw biomass. Medic et al. (2010) reported a decrease in the O/C ratio of about 7% at 200°C, 15% at 250°C, and 45% at 300°C.

Production of hydrocarbon fuels is also synonymous with the production of liquid fuels that are low in oxygen content. Torrefaction is said to remove more oxygen relative to the carbon, thereby improving the quality of the finished product. The loss of oxygen in this case is a better option. The oxygen is bound in the volatiles that are removed during the torrefaction processes. The degree of carbonization is increased as exposure time is increased. The degree of carbonization is further increased as the temperature is also increased.

8.7.5 H/C Ratio

The Van Krevelen diagram is a good way to illustrate the effect of the C/H ratio as the biomass undergoes torrefaction. It was shown that raw biomass has a low H/C ratio and is usually within 1.0. Torrefaction will improve the H/C ratio to more than 1, as well as improving the O/C ratio. At a given increase in torrefaction temperature, there is a significant decrease in the H/C ratio, which is further magnified as temperature continues to increase. The H/C ratio of 0.675, for example, increased to more than 1.4 as the temperature was increased from 250°C to 350°C (Tumuluru et al., 2011a). The opposite is true with the effect of exposure time. The highest H/C ratio is obtained when the exposure time is less. A quick torrefaction period is required if the H/C ratio is to be enhanced.

8.7.6 Nitrogen and Sulfur Content

Nitrogen and sulfur are two of the five most important elements in a biomass. The other three are carbon, hydrogen, and oxygen. One should remember that an ultimate analysis looks at these five elements in addition to the percent ash (the inorganic portion in biomass). Nitrogen is an important plant component and a ubiquitous element found in all biomass. Plants will not grow appreciably without significant amounts of nitrogen in the form of fertilizer. Sulfur is also ever present in the biomass matrix, but it is of lower concentration than nitrogen. The nitrogen and sulfur in the finished torrefied product are functions of the biomass conversion process. Nitrogen and sulfur in the biomass affect the emissions associated with further use and conversion of the biomass. If the carbon is enhanced, the relative concentration of nitrogen and sulfur is also increased. This will also affect any catalytic processes that may be adopted during biomass conversion. Excess sulfur and nitrogen will cause the catalysts to be poisoned, thus limiting their effectiveness. For example, as temperature is raised from 250°C to 350°C and as exposure time is lengthened from 30 to 120 min, the percent increase of nitrogen is raised from 0.58% to 0.80% (Tumuluru et al., 2011b).

The presence of these elements in the biomass, whether torrefied or not, must be considered so that succeeding processes can be optimized. During the thermal conversion of biomass, the emission of NO_x and SO_2 is a strong function of the nitrogen and sulfur content of the biomass that is being converted into fuel. For example, nitrous oxide, a regulated pollutant from the combustion of biomass, comes from either the fuel NO_x or thermal NO_x. The thermal NO_x can be controlled and lowered using a proper operating temperature. However, if the inherent nitrogen in the biomass is quite high, it will be difficult to eliminate.

8.8 ADVANTAGES AND DISADVANTAGES OF TORREFACTION PROCESS

There are various advantages in the use of torrefaction as a pretreatment process for biomass. Foremost is the increased energy density as well as improved storage and transport properties. An exhaustive list of other advantages will be enumerated. There are also disadvantages whenever biomass is preprocessed before being used in a biorefinery. The main disadvantage is the increased cost of the pretreatment process or evaluating whether the additional step is warranted since the cost of biomass as feedstock is already quite high. Disadvantages will also be enumerated in this section.

8.8.1 Advantages of Torrefaction as Pretreatment Process

1. Increase in energy content of the final torrefied product.
2. Increase in energy density per unit of weight.
3. Decrease in moisture content, thereby lowering transport and storage weight.
4. Improvement of bulk density for ease of transport.
5. Improvement of storage characteristics.
6. Repel moisture and inhibits the attack of insects.
7. Biomass can be ground or pelleted easily after torrefaction.
8. Improvement of H/C ratio for better feedstock properties during conversion.
9. Remove some oxygen that is not needed in further conversion process.
10. Overall improvement in the chemical composition of the biomass for latter conversion into biofuels.

8.8.2 Disadvantages of Torrefaction as Pretreatment Process

1. Additional cost for processing.
2. Removal of some important volatiles needed in further conversion.
3. Additional energy and equipment required for processing.
4. If biomass is pelleted, the size will still have to be reduced in some conversion processes, thereby adding to the cost of biofuel production.
5. Ash content is not removed so ash content would likely increase per unit of weight.
6. Some carbonaceous compounds are also removed during the heating process, and thus some energy from the carbon is lost.

In summary, the justification on the use of torrefaction as a pretreatment process should be based on the relative cost of the process. The total cost of the pelleting process, for example, was estimated at around $55/tonne for the complete process, including drying cost (Mani et al., 2006). In some instances, this pelleting cost is equal to the pretreatment and conversion cost. One has to reconsider adding this extra step in the conversion process so that it will not be uneconomical.

8.9 CONCLUSION

Torrefaction is defined as an isothermal pyrolysis process for biomass. Torrefaction occurs in the temperature range of 200°C to 320°C. During the torrefaction process, the biomass loses most of the low-energy content material in the form of gaseous and condensible liquids. The solids, which include the original sugar structures and other newly formed polymeric structures, retain most of the energy while experiencing reduced mass. High mass reduction is primarily due to the loss in zero-calorie water. Common events that occur during torrefaction include drying, depolymerization and recondensation, limited devolatilization and carbonization, and extensive devolatilization and carbonization.

Torrefaction of biomass not only removes moisture but improves some physical properties. Enhancements include improved physical appearance, higher energy density, better grindability, and better pelletability. The ultimate and proximate analysis show a relative increase in the carbon content with a corresponding increase in the calorific value. Torrefaction improves the pelletability of the biomass, apparently because of decomposition of the hemicellulose components and to the softening of the lignin.

If the biomass is torrefied before being densified, it has been shown that the energy consumption during the pelleting process is reduced by a factor of 2 and the throughput is increased, also by a factor of 2.

Increasing the torrefaction temperature and the duration of the torrefaction process significantly decreases the moisture content, hydrogen content, and the H/C ratio. This effect is more pronounced at temperatures above 280°C, which may be due to the devolatilization of the hemicellulose and cellulose components. The calorific value of torrefied biomass is increased as temperature and retention times are increased. A torrefied biomass also experiences minimal off-gassing with a resulting stable temperature during storage. The material becomes hydrophobic, thus limiting the action of microbes for subsequent degradation and results in longer storage times.

Finally, torrefied biomass is said to have a wider range of applications in gasification processes and improves the quality of the fuel in power generation plants as a co-firing fuel.

8.10 PROBLEMS AND DISCUSSION ISSUES

8.10.1 Energy Densification for Torrefied Product

P8.1 Three kilograms of biomass (6.6 lb) with an energy content of 17.2 MJ/kg (7,411 Btu/lb) was torrefied at 300°C, and the resulting 2.25 kg of biomass (4.95 lb) that was left had an energy content of 18.5 MJ/kg (7,971 Btu/lb). Determine the energy densification for this torrefaction process.

8.10.2 Determining Energy and Mass Balances in Torrefaction

P8.2 Three kilogram of biomass with a heating value of 17.2 MJ/kg torrefied at 300°C resulted in solid, liquid, and gaseous components. The remaining solid component was 2.25 kg with a heating value of 18.5 MJ/kg. About 0.25 kg of liquid was collected with about 10% moisture. The organic component of this liquid portion (with water removed) had a heating value of 19 MJ/kg. Estimate the energy content of the gaseous component assuming no losses. Likewise, determine the percentage of energy contained in each component and plot the results.

8.10.3 Bulk Density of Torrefied Biomass

P8.3 Wheat straw was chopped after harvest and found to have a bulk density of 60 kg/m^3 (3.74 lb/ft^3). A sample of wheat straw was placed in a 305 mm × 305 mm × 305 mm box (1 ft × 1 ft × 1 ft). The sample weighed 1.7 kg (3.74 lb). This sample was torrefied at 300°C for 1 h and there was a 10% reduction in weight. The resulting volume of the product was also measured and found to be 0.0102 m^3 (0.36 ft^3). Calculate the new bulk density of the torrefied material. Convert the units in the English system.

8.10.4 Bulk Density of Biomass

P8.4 *Eucalyptus grandis* was placed in a gravimetric torrefaction facility and exposed to various temperatures. The weight and final volume of the sample at each temperature setting was measured as summarized in the table. The original weight of the sample was 850 g in a 1,000 mL filled graduated cylinder. Data shows the final weights and volumes of the sample at each temperature setting. Calculate the bulk densities of all samples at all temperature of exposure and plot the results. Convert all units in the English system. Comment on the change in bulk density as torrefaction temperature was increased.

Torrefaction Temperature (°C)	Initial Weight (g)	Final Weight (g)	Final Volume (mL)
220	850	819.2	987
250	850	750.5	950
280	850	681.1	933

8.10.5 Grinding Energy Usage

P8.5 A chipper/shredder was used to grind wheat straw (15% moisture content) with a 1.6-mm screen opening. The unit is powered by a 770 W electric motor (220 V). The power drawn by the chipper/shredder motor was measured using a wattmeter. Around 3 kg (6.6 lb) of biomass was used during the 15 min test and the measured amperage was 2.3 amps. Estimate the specific energy (in kWh/tonne) or (Btu/ton) required to accomplish this task.

8.10.6 Moisture Content of Torrefied Biomass

P8.6 A tonne (2,200 lb) of wheat straw having 12% moisture was torrefied, and the moisture content of the final product was 2%. How much moisture was removed from this process? Assume only the moisture was removed (not other significant gaseous or liquid condensibles). Calculate the moisture removed per tonne of wet material (kg/tonne). Convert all units in the English system.

8.10.7 Durability of Biomass Pellets

P8.7 A clean and previously sieved 500 g (1.1 lb) sample of big bluestem pellet was placed in a tumbling box to determine the durability. After 10 min tumbling, the pellets were removed from the box, sieved again, and the weight of the material was measured and found to be 484.5 g. A similar test was made using corn stover pellets and the final weight of the tumbled pellets was 491 g. Determine the percent durability of these pellets and compare their durability.

8.10.8 Bulk Density of Torrefied Biomass

P8.8 Torrefied wheat straw sample processed as in Problem 8.3 was pelleted and placed in the bulk density box measuring 305 mm × 305 mm × 305 mm and reweighed. The final weight of the pelleted material in this box was 20.5 kg. Calculate the new bulk density of the torrefied and pelleted material. Convert all the units in the English system.

8.10.9 Energy Density Changes in Torrefied Biomass

P8.9 A biomass sample with 10% moisture has an initial heating value of 19 MJ/kg. This material was placed in a thermal gravimetric analyzer. The moisture reduction is summarized in the table. Complete the table for the estimated heating value of this biomass, assuming that only water removal and the energy is conserved. What is the energy content of this material on a dry basis? You may assume an initial biomass weight of 1 kg (1,000 g) and continue the calculation as weight is decreased due to moisture removal.

Moisture Content (%)	Residence Time (min)	Energy Content (MJ/kg)
10.0	0	19
9.6	20	
9.3	40	
4.7	60	
4.0	80	
3.7	100	

8.10.10 Van Krevelen Plot of Raw Biomass

P8.10 Use the ultimate analysis data in Table 3.5 to make a Van Krevelen plot for the raw biomass. In this plot, make the y-axis as the atomic H/C ratio and the x-axis as the atomic O/C ratio. Discuss the results and identify where these points are relative to the plot in Example 8.10.

8.10.11 Bulk Density Effect After Torrefaction

P8.11 The mass loss in the form of solids, liquids, and gases during the torrefaction process causes the biomass to become more porous. What will happen to its bulk density after this process?

a. reduced b. increased c. no change d. it depends

8.10.12 Grindability of Torrefied Biomass

P8.12 Torrefaction makes the biomass lose its fibrous nature because of breakdown of the hemicellulose matrix and the depolymerization of the cellulose components. What happens to its grindability or its mill capacity when the biomass is torrefied?

a. reduced b. increased c. no change d. it depends

8.10.13 Changes in Chemical Composition of Torrefied Biomass

P8.13 This diagram best describes the difference between lignite coal and biochar based on its chemical compositional analysis. Name this classic compositional analysis plot.

a. Venn diagram b. torrefaction diagram c. kerogen plot d. Van Krevelen

8.10.14 Energy Farming for Torrefied Product

P8.14 You are given 4,000 ha [10,000 acres] of land to grow energy crop to commercialize the sale of biochar from fast growing trees such as camelina. Establish the economics of this process assuming you can sell the biochar for $300/tonne. List all assumptions madefor this project.

8.10.15 Biomass Pellets

P8.15 A project to install a biomass pellet heating system at a remote hospital in British Columbia received $400,000 yearly operational cost funding as part of low-carbon future project. An additional $500,000 was also given to acquire a pellet machine and another $500,000 to acquire a torrefaction facility. You are to develop and establish the economics and sustainability of this project with this initial seed money. Propose on how to best use this funding to achieve this goal of providing heat to this small hospital by using pellet stoves for heating purposes.

REFERENCES

ASAE S269.4 (R2007). 1991. Cubes, Pellets and Crumbles – Definition and Methods for Determining Density, Durability and Moisture Content. American Society of Agricultural and Biological Engineers, St. Joseph, MI.

Bergman, P. C. A. 2005. Combined torrefaction and pelletisation: The TOP Process, Energy Research Center of the Netherlands (ECN), ECN Biomass, Petten, The Netherlands. ECN Report No. ECN-C-05-073. pp. 1–8.

Bergman, P. C. A. and J. H. A. Kiel. 2005. Torrefaction for biomass upgrading. In: *Proceedings of the 14th European Biomass Conference and Exhibition held in Paris, France on October 17-21, 2005.* Jointly organized and published proceedings by the ETA-Florence, Italy and WIP-Munich, Germany. Report also available at the Energy Research Center of the Netherlands (ECN), ECN Biomass, Petten, The Netherlands.

Boerrigter, H., J. Kiel and P. Bergman, 2006. Biomass pre-treatment by torrefaction. 3rd Thermal NET Meeting, April 3–5, 2006, Lille, France. Publication of the Energy Research Center of Netherlands (ECN). ECN Biomass, Petten, The Netherlands.

Bridgeman, T. G., J. M. Jones, I. Shield and P. T. Williams. 2008. Torrefaction of reed canary grass, wheat straw and willow to enhance solid fuel qualities and combustion properties. *Fuel* 87 (6): 844–856.

Bridgeman, T. G., J. M. Jones, A. Williams and D. J. Waldron. 2010. An investigation of the grindability of two torrefied energy crops. *Fuel* 89 (12): 3911–3918.

Chen, W. H., W. Y. Cheng, K. M. Lu and Y. P. Huang. 2011. An evaluation on improvement of pulverized biomass property for solid fuel through torrefaction. *Applied Energy* 88 (2011): 3636–3644.

Cooper, C. C. and G. C. Alley. 1994. *Air Pollution Control: A Design Approach*. Waveland Press Inc., Prospect Heights, IL.

Mani, S., S. Sokhansanj, X. Bi and A. Turhollow. 2006. Economics of producing fuel pellets from biomass. *Applied Engineering in Agriculture* 22 (3): 421–426.

Medic, D., M. Darr, B. Potter and A. Shah. 2010. Effect of torrefaction process parameters on biomass feedstock upgrading. ASABE Technical Paper No. 10-09316. ASABE, St. Joseph, MI.

Pandey, A. 2009. *Handbook of Plant-Based Biofuels*. CRC Press, Taylor & Francis Group, Boca Raton, FL.

Phanphanich, M. and S. Mani. 2011. Impact of torrefaction on the grindability and fuel characteristics of forest biomass. *Bioresource Technology* 102 (2): 1246–1253.

Pilon, G. and J. M. Lavoie. 2011. Characterization of switchgrass char produced in torrefaction and pyrolysis conditions. *BioResources* 6 (4): 4824–4839.

Rodrigues, T. O. and P. L. A. Rousset. 2009. Effects of torrefaction on energy properties of Eucalyptus grandis wood. *Cerne* 15 (4): 446–452.

Theerarattananoon, K., F. Xu, J. Wilson, R. Ballard, L. McKinney, S. Staggenborg, P. Vadlani, Z. J. Pei and D. Wang. 2011. Physical properties of pellets made from sorghum stalk, corn stover, wheat straw and big bluestem. *Industrial Crops and Products* 33: 325–332.

Tumuluru, J. S., X. Kuang, S. Sokhansanj, J. Lim and T. Bi. 2009. Wood pellet properties produced in British Columbia and studies on off-gas emissions at low and high storage temperatures. ASABE Technical Paper No. 09-7265. ASABE, St. Joseph, MI.

Tumuluru, J. S., K. Xingya, S. Sokhansanj, L. C. Jim, T. Bi and S. Melin. 2008. Effects of storage temperature on off-gassing and physical properties of wood pellets. ASABE Technical Paper No. 08-4248. ASABE, St. Joseph, MI.

Tumuluru, J. S., S. Sokhansanj, J. R. Hess, C. T. Wright and R. D. Boardman. 2011a. A review on biomass torrefaction process and product properties for energy applications. *Industrial Biotechnology* 7 (5): 384–401.

Tumuluru, J. S., S. Sokhansanj, C. T. Wright, R. J. Hess, R. Boardman and T. Kremer. 2011b. A review on biomass torrefaction process and product properties. Presented at the S-1041 Symposium on Thermochemical Conversions held on August 1–3, 2011, at Oklahoma State University and Sponsored by the S-1041 Committee of the USDA, Washington, DC.

ASSESS YOUR KNOWLEDGE

1. Which is not a characteristic of a torrefied biomass?
   a. Hydrophobic
   b. Hydrophilic
   c. High energy density
   d. Better grindability

2. Determine the energy densification of a biomass that underwent torrefaction process with the following data: (a) initial biomass weight = 5 kg, (b) final biomass weight = 2.75 kg, (c) initial energy content of biomass = 15 MJ/kg, and (d) final energy content of biomass = 20 MJ/kg.
   a. 1.4
   b. 0.769
   c. 1.11
   d. 0.70

3. What happens to the bulk density of the biomass after it is torrefied or pelleted?
   a. Decreased
   b. Increased
   c. No change
   d. No trend

11. A clean and previously sieved 500 g [1.1 lb] sample of corn stover was placed in a tumbling box to determine the durability. After a 10 min tumbling, the pellets were removed from the box, sieved again, and the weight of the material was measured and found to be 491 g. Determine the percent durability of these corn stover pellets.
   a. 96.9%
   b. 98.2%
   c. 95.3%
   d. 92.8%

12. Three kilograms of biomass [6.6 lbs] with an energy content of 17.2 MJ/kg [7,411 Btu/lb] was torrefied at 300°C and the resulting 2.25 kg of biomass [4.95 lbs] that was left had an energy content of 18.5 MJ/kg [7,971 Btu/lb]. Determine the energy densification for this torrefaction process.
   a. 1.1
   b. 1.3
   c. 0.9
   d. 1.5

13. What is the typical value of energy densification ratio (energy/mass) during torrefaction?
   a. 0.7
   b. 0.9
   c. 1.3
   d. 1.5

4. Three kilograms of biomass with a heating value of 17.2 MJ/kg was torrefied at 300°C resulting in various solid, liquid, and gaseous components. The remaining solid component was 2.25 kg with a heating value of 18.5 MJ/kg. About 0.25 kg of liquid was collected with about 10% moisture. The organic component of this liquid portion had a heating value of 19 MJ/kg. Determine the percent in energy contained in bio-oil.
   a. 8.28%
   b. 80.7%
   c. 11.05%
   d. 50.5%

5. Torrefied switchgrass sample was pelleted and placed in a bulk density box measuring 1 ft^3 and reweighed. The final weight of the pelleted material in this box was 35 lbs. The original bulk density of the raw unpelleted switchgrass was 15 lb/ft^3. Determine the percent increase in bulk density based on this new bulk density of pelleted material.
   a. 57%
   b. 133%
   c. 5.7%
   d. 0.57%

14. Three kilograms of biomass with a heating value of 17.2 MJ/kg was torrefied at 300°C resulting in the various solid, liquid and gaseous components. The remaining solid component was 2.25 kg with a heating value of 18.5 MJ/kg. About 0.25 kg of liquid was collected with about 10% moisture. The organic component of this liquid portion had a heating value of 19 MJ/kg. Estimate the energy content of the gaseous component (MJ) assuming no losses.
   a. 5.7
   b. 41.6
   c. 4.2
   d. 51.6

15. Three kilograms of biomass with a heating value of 17.2 MJ/kg was torrefied at 300°C resulting in the various solid, liquid and gaseous components. The remaining solid component was 2.25 kg with a heating value of 18.5 MJ/kg. About 0.25 kg of liquid was collected with about 10% moisture. The organic component of this liquid portion had a heating value of 19 MJ/kg. Determine also the percent in energy contained in the biochar.
   a. 80.7%
   b. 11.05%
   c. 8.28%
   d. 50.5%

6. A clean and previously sieved biomass pellet weighing 5 lbs was placed in a tumbling box to determine its durability. After tumbling for 10 min, the pellet was removed from the box, sieved again and reweighed. The final weight after this step was 4.95 lbs. Determine the percent durability of this pelleted material.
   a. 95
   b. 97.5
   c. 105
   d. 99

7. In the energy balance for torrefaction, how much energy is usually retained in the biomass after this thermal conditioning process?
   a. 50%
   b. 70%
   c. 90%
   d. 99%

16. Three kilograms of biomass with a heating value of 17.2 MJ/kg was torrefied at 300°C resulting in various solid, liquid, and gaseous components. The remaining solid component was 2.25 kg with a heating value of 18.5 MJ/kg. About 0.25 kg of liquid was collected with about 10% moisture. The organic component of this liquid portion had a heating value of 19 MJ/kg. Determine also the percent in energy contained in syngas.
   a. 11.05
   b. 80.7
   c. 8.28
   d. 50.5

17. Determine the overall conversion efficiency of a torrefaction process with the following data: (a) initial biomass weight = 5 kg, (b) final biomass weight = 4.75 kg, (c) initial energy content of biomass = 15 MJ/kg, and (d) final energy content of biomass = 20 MJ/kg and considered the only output energy of the process.
   a. 70%
   b. 96.5%
   c. 77.75%
   d. 100%

8. Which biomass component would thermally convert the last as temperature is increased in the torrefaction region?
   a. Cellulose
   b. Hemicelluloses
   c. Carbohydrates
   d. Lignin

9. What is the abscissa axis for a Van Krevelen plot?
   a. Atomic oxygen to carbon ratio
   b. Atomic hydrogen to carbon ratio
   c. Oxygen to hydrogen ratio
   d. Hydrogen to carbon ratio

18. Wheat straw was chopped after harvest and found to have a bulk density of 60 kg/m^3 [3.74 lb/ft^3]. A sample of wheat straw was placed in a 305 mm x 305 mm x 305 mm box [1ft x 1 ft x 1 ft]. The sample weighed 1.7 kg [3.74 lbs]. This sample was torrefied at 300°C for 1 h and there was a 10% reduction in weight (i.e. final weight = 1.53 kg). The resulting volume of the product was also measured and found to be 0.0102 m^3 [0.36 ft^3]. Calculate the new bulk density of the torrefied material in units of lb/ft^3.
    a. 9.35
    b. 18.7
    c. 45.3
    d. 150

19. A chopper was used to grind wheat straw (15% moisture content) with a 1.6 mm screen opening. The unit was powered by a 770 W electric motor (220 Volts). The power drawn by the chopper motor was measured using a wattmeter. Around 3 kg [6.6 lbs] of biomass was used during the 15 min test and the measured amperage was 2.3 amps. Estimate the specific energy in units of Btu/lb that is required to accomplish this task.
    a. 65.4
    b. 506
    c. 126.5
    d. 42.2

10. True or False. It is very easy to grind torrefied biomass.
    a. False
    b. True
    c. Can't tell
    d. Don't know

20. Estimate the bulk density of a biomass with the following data in units of lb/ft^3.
    $\rho_w = 1{,}000$ kg/m^3 $X_w = 0.30$
    $\rho_v = 100$ kg/m^3 $X_v = 0.25$
    $\rho_s = 300$ kg/m^3 $X_s = 0.45$
    a. 232.6
    b. 14.5
    c. 45.3
    d. 9.4

Answers: 1b, 2a, 3b, 4a, 5a, 6d, 7c, 8a, 9b, 10b, 11b, 12a, 13c, 14a, 15a, 16a, 17b, 18a, 19a, 20b

ADDITIONAL LEARNING RESOURCES

- Access the Lecture Notes at the Instructor and Student Resources: https://routledgetextbooks.com/textbooks/instructor_downloads/

9 Pyrolysis

LEARNING OBJECTIVES

Upon completion of this chapter, one should be able to

1. Discuss pyrolysis, a biomass thermal conversion process without the use of an oxidant.
2. Identify various types of biomass pyrolyzers according to reaction rates, heating rates, and temperatures.
3. Estimate the energy and mass balances in a pyrolyzer.
4. Distinguish the effect of temperature and heating rates on the composition of products and product yields.
5. Describe the various products of pyrolysis technology and ways to produce various products accordingly.

9.1 INTRODUCTION

Pyrolysis, also called destructive distillation, is an irreversible thermal conversion process for biomass in the complete absence of an oxidant. Pyrolysis of biomass brings about three primary products, in gaseous, liquid, and solid forms. The important components of the gaseous products include hydrogen and carbon monoxide, which are collectively called synthesis gas (syngas) or producer gas, along with other gases such as methane and other hydrocarbons such as ethane and ethylene. When this combustible gas is passed through a condenser, condensible liquids are formed. The liquids are primarily water and tar, with other compounds such as methanol, acetic acid, acetone, phenols, aldehydes, and esters. Because of the wide range of liquid products that may be formed, pyrolysis is possibly the only process that can produce the valuable chemicals and industrial feedstock.

Figure 9.1 shows the block diagram of a simple, auger-based, batch or slow pyrolysis reactor, to evaluate the pyrolysis process for various biomass materials. The important component parts are as follows:

1. An inert gas tank—the source for inert gas (or purge gas) such as nitrogen, the most common inert gas used in a pyrolyzer.
2. Mass flow controller for the inert gas—to measure the inert gas flow rate through the system during evacuation of air through the system.
3. Digitally controlled tube furnace—most tube furnaces can control temperatures of up to 1,200°C.
4. Tar/moisture trap—an initial moisture trap may be placed before the actual condenser to remove water molecules that will condense once a reduction in

DOI: 10.1201/9781003294306-9

temperature is experienced (i.e., lower than 100°C, the condensing temperature of water).

5. Condenser with cooling water—the condenser is used to convert gaseous fuels into liquid form.
6. Temperature readout to monitor temperature in the reactor and the condenser—this is an important component of a pyrolysis system, to be able to monitor reaction temperature.
7. Liquid product (bio-oil) collector—the collector is placed at the bottom of the condenser.
8. Second tar/moisture trap—another moisture trap, if necessary, after the non-condensible gases exit from the condenser.
9. Syngas flow measurement—to conduct complete energy and mass balance of the process, the amount of syngas coming out of the condenser must be measured.
10. Sampling port for syngas—the complete composition and concentration of all gaseous materials formed must be measured and sampled.
11. Real-time gas analyzer—in some laboratories, an online gas analyzer is installed to measure the gaseous materials formed.

Figure 9.1 is a complete setup to analyze energy and mass balance in a pyrolysis system for numerous types of biomass feedstocks. If this system is operated as a batch reactor, a known amount of biomass is placed in the reactor. The tube furnace is set to a certain reactor temperature, and the inert mass flow controller is set to a known purge gas flow rate. The heating rate may also be estimated when

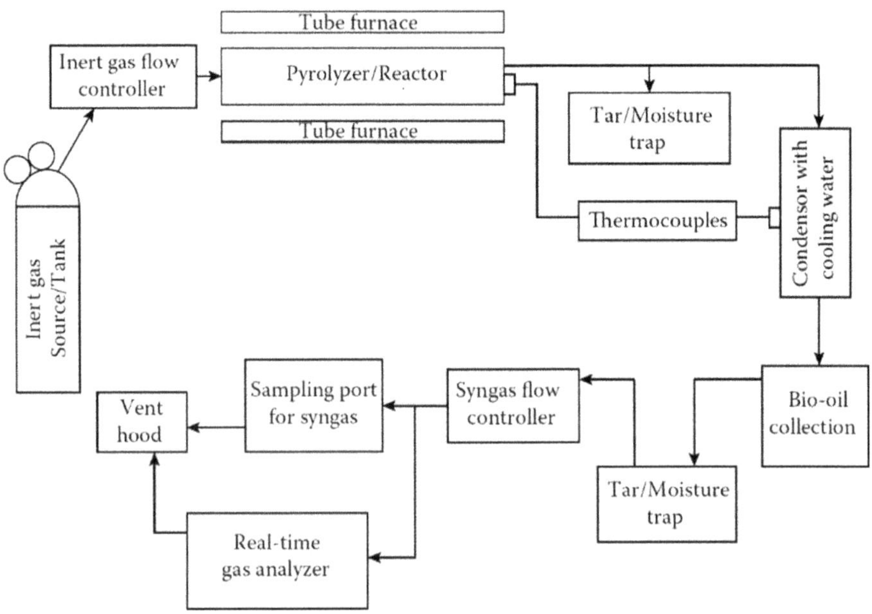

FIGURE 9.1 Schematic of pyrolysis process.

using a digitally controlled furnace. A chiller is normally used to provide cooling water to the condenser. The condenser may be a simple single-pass counterflow heat exchanger. Although an online gas analyzer may be used, it is good practice to also allow some ports where syngas can be sampled during the experiment. A gas flow meter is necessary following the condenser and second tar/moisture trap to determine the exact amount of syngas produced by the process. An additional measuring instrument may be required to measure the energy supplied to the furnace for heat input to the system. If the furnace is electrically controlled, a simple amp probe, or an ammeter, is all that is required. At the end of the process, the amount of char will be measured, including the heating value (HV). Example 9.1 provides a step-by-step analysis of the products and by-products to complete an energy and mass balance in a pyrolysis process. Example 9.2 illustrates how to convert mass yields of bio-oil in a pyrolysis process into volume yields. It is more convenient and intuitive to report bio-oil yields in units of liters per tonne (or gallons per ton) rather than on a mass basis (such as kilograms per tonne). The density of the bio-oil must be known to make this conversion.

Example 9.1: Energy Balance in a Pyrolysis Process

Suppose that 0. 454 kg (1 lb) of biomass with an HV of 17.4 MJ/kg (7,500 Btu/lb) was placed in a batch reactor and the furnace temperature was set at 550°C. About 0.15 kg (0.33 lb) of char was produced with an HV of 20.9 MJ/kg (9,000 Btu/lb). Determine the percentage of energy retained in the char from the original input material.

Solution:

a. The energy available on the input material is calculated as follows:

$$0.454\ \text{kg} \times \frac{17.4\ \text{MJ}}{\text{kg}} = 7.9\ \text{MJ}$$

b. The energy contained in the char is calculated as follows:

$$0.15\ \text{kg} \times \frac{20.9\ \text{MJ}}{\text{kg}} = 3.14\ \text{MJ}$$

c. Thus, the percentage of energy retained in char is calculated as follows:

$$\frac{3.14\ \text{MJ}}{7.9\ \text{MJ}} \times 100\% = 39.7\%$$

d. Thus, 39.7% of the energy is still contained in the char.

The products of a pyrolysis process will depend on how the pyrolyzer is operated in terms of the time the biomass is exposed to a given temperature as well as the

TABLE 9.1
Conditions for Various Thermal Conversion Processes

Mode	Conditions
Fast	~500°C, short hot vapor residence time: ~1 s
Intermediate	~500°C, hot vapor residence time: ~10–30 s
Slow	~400°C, long vapor residence time: approximately hours to days
Gasification	750–900°C
Torrefaction	~290°C, solids residence time: ~10–60 min

Source: Bridgewater, A. V., *Biomass and Bioenergy* 38, 68, 2012.

TABLE 9.2
Yields of Solid, Liquid, and Gaseous Products for Various Thermal Conversion Processes

Mode	Liquid	Solid	Gas
Fast	75%	12% char	13%
Intermediate	50% in two phases	25% char	25%
Slow	30%	35% char	35%
Gasification	5%	10% char	85%
Torrefaction	0% unless condensed, then up to 5%		

Source: Bridgewater, A. V., *Biomass and Bioenergy* 38, 68, 2012.

temperature of the reaction. The most important rule is this: more syngas is formed at higher temperatures, whereas most solid biochar is produced at lower temperatures. Liquid products are produced at moderate temperatures, normally between 400°C and 600°C, and the highest bio-oil yields are those where the biomass is exposed at these temperatures at the fastest rate, also called heating rate. The heating rate is defined as the exposure time of the biomass to a given reactor temperature, or sometimes called the residence time. Table 9.1 is the basis for the definition of slow and fast pyrolyzers as described by the residence time of the biomass in the reactor. Table 9.2 presents the yields of solid, liquid, and gaseous products from various thermal conversion processes. A fast pyrolysis process will yield the highest liquid product yield (as much as 75%), whereas gasification produces the most amounts of syngas (as high as 85%). Slow pyrolysis will produce the most amount of biochar, while gasification will produce the least amounts of liquid products.

Example 9.2: Bio-Oil Yield in Liters per Tonne (Gallons per Ton)

A pyrolysis experiment showed that the bio-oil yield was 75% by weight. If the density of the bio-oil was found to be 1.2 g/mL or 1.2 kg/L (10 lb/gal), calculate the bio-oil yield in liters/tonne and in gallons/ton. Comment on the yield per unit weight of biomass.

Solution:

a. This is a simple conversion to appropriate units as follows:

$$\text{Yield (L/tonne)} = \frac{750 \text{ kg bio-oil}}{\text{tonne}} \times \frac{\text{L}}{1.2 \text{ kg}} = 625 \frac{\text{L}}{\text{tonne}}$$

b. The equivalent English unit is as follows:

$$\text{Yield (gal/ton)} = \frac{75 \text{ lb bio-oil}}{100 \text{ lb biomass}} \times \frac{2{,}000 \text{ lb}}{\text{ton}} \times \frac{\text{gal}}{10 \text{ lb}} = 150 \frac{\text{gal}}{\text{ton}}$$

c. Around 625 L of bio-oil (or 150 gal) may be produced per tonne (or per ton) of biomass, which is quite a lot of liquid product at 75% yield by weight.

The overall pyrolysis process begins with the processing of biomass feedstock and includes drying or particle size reductions. The complete continuous process is schematically shown in Figure 9.2. It is a common practice to grind the biomass below a few millimeters in size (1–3 mm). The biomass must also be dried (<10% is common) because a wet biomass also generates a lot of water in the bio-oil. Heat is normally

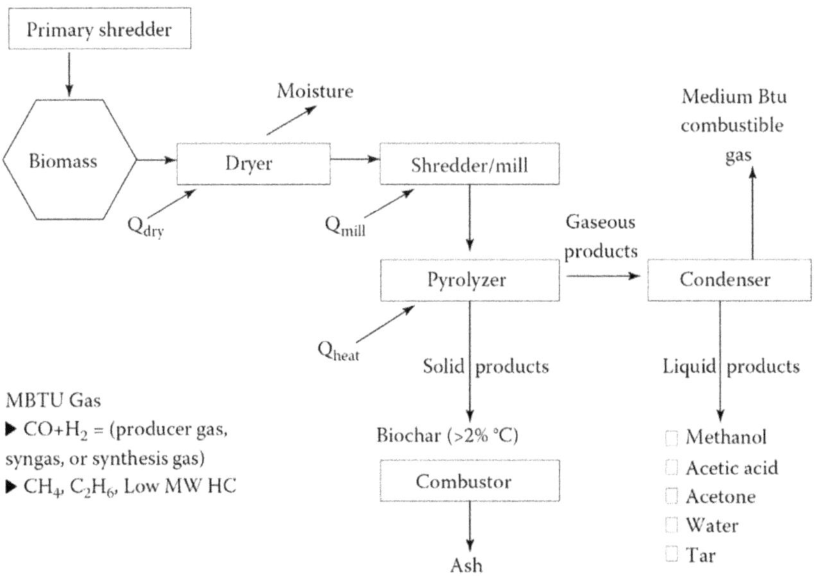

FIGURE 9.2 Process flow in continuous pyrolyzer to generate solid, liquid, and gaseous products.

supplied externally to the pyrolyzer, unless an inert gas that carries the heat for the reaction is used. The syngas produced must be separated quickly from the biochar; otherwise, some reaction between the gas and the char particulate may alter the range of products produced.

In a fluidized bed reactor, the residence time of the biomass is calculated based on the time it takes for the biomass to pass through the reactor. This is simply a function of the volume of the reactor divided by the volumetric flow rate of the biomass and fluidizing air, as shown by Equation 9.1. Example 9.3 illustrates how this calculation is made. Note that in fluidized bed systems, the residence time could be a fraction of a second, or in some cases, 1,000th of a second (milliseconds, ms).

$$\text{Residence time FB (s)} = \frac{\text{FB reactor volume (L)}}{\text{Volumetric flow rate (L/s)}} \tag{9.1}$$

Example 9.3: Estimation of Residence Time in Fluidized Bed Reactor

A fluidized bed pyrolyzer has an internal diameter of 5.2 cm (2 in) and finely ground biomass at a feed rate of 5 g/min (0.011 lb/min) is introduced with an actual flow rate of 70.85 L/min (2.5 cfm). Determine the residence time of the biomass in a 60.96 cm (24 in) tall vertical reactor.

Solution:

a. The reactor volume is calculated as follows:

$$\text{Volume (L)} = \frac{\pi d^2 h}{4} = \frac{\pi \times (0.052\,\text{m})^2 \times 0.6096\,\text{m}}{4} \times \frac{1{,}000\,\text{L}}{1\,\text{m}^3} = 1.3\,\text{L}$$

b. The volumetric flow rate is calculated as follows:

$$\text{Volumetric flow rate (L/s)} = 70.85 \frac{\text{L}}{\text{min}} \times \frac{1\text{ min}}{60\text{ s}} = 1.18\text{ L/s}$$

c. Thus, the residence time in seconds is as follows:

$$\text{Residence time (s)} = \frac{1.3\text{ L}}{1.18\text{ L/s}} = 1.1\,\text{s}$$

9.2 VARIOUS PYROLYSIS PROCESSES BASED ON HEATING RATES

Pyrolyzers may be categorized according to the heating rates provided to the biomass as listed in Table 9.1. Most pyrolyzer units are operated according to the rates with which biomass is introduced into the reactor and the length of time it stays in the active region in the reactor (Del Remedio Hernandez et al., 2006). Thus, we may identify pyrolyzers as either slow, intermediate, or fast. There is still no clear delineation of what is meant by "fast," as some researchers would also use the term "flash." For now, the description in Table 9.1 should be our basis for differentiating the various pyrolyzer types.

9.2.1 Slow/Conventional Pyrolysis

Most auger-based pyrolyzers are considered slow or conventional types. These pyrolyzers have longer residence times, and their main purpose is to produce charcoal or biochar. The unit may provide approximately 1 tonne per 24-h day of biochar output, at around 50% yield by weight of the biomass (i.e., the input biomass is around 2,000 kg). The auger typically runs at 1 revolution per minute (rpm) or slightly faster. The cylindrical reactor has a diameter of around 15.24 cm (6 in) with an auger that moves the biomass from the feed hopper to the char-collecting bin. It is quite difficult to estimate the residence time of the biomass in this reactor. An estimate may be obtained based on the linear flow rate of the biomass as well as the length of the reactor. The linear rate of biomass flow through the reactor may be estimated during the calibration of the unit. The purpose is to determine the biomass feed rate as a function of the auger rpm. Following the calibration procedure, a graph showing the rpm of the auger versus the amount of biomass conveyed will be established. The length of the auger reactor (AR) must be known. Then the reaction time in an AR is easily estimated by dividing the length of the reactor by the biomass moved through the reactor in units of m/s, as shown in Equation 9.2. Example 9.4 shows how this time is calculated. Note that some researchers are using auger-based pyrolyzers as fast pyrolyzer units by increasing the rpm of the auger to reduce the residence time from minutes to a few seconds. Unfortunately, this will also lead to the production of large amounts of unreacted or unconverted biomass, and thus, conversion efficiencies will be quite low.

$$\text{Residence time AR (s)} = \frac{\text{AR reactor length (m)}}{\text{Biomass flow velocity (m/s)}} \tag{9.2}$$

Example 9.4: Estimation of Residence Time in Auger Reactor

A 15.24 cm (6 in) auger pyrolyzer has a length of 0.9146 m (3 ft). Finely ground biomass at a feed rate of 1.4 kg/min (3.1 lb/min) was introduced at 0.1 rpm. Determine the residence time of the biomass if the linear flow velocity of the biomass is 9.2 cm/min (0.3 ft/min). Comment on this residence time.

Solution:

a. The residence time is simply calculated as follows:

$$\text{Residence time AR (s)} = \frac{0.9146\,\text{m}}{0.092\,\text{m/min}} \times 60\,\text{s/min} = 596.5\,\text{s}\left(9.94\,\text{min}\right)$$

b. This is a slow pyrolysis process.

9.2.2 Fast/Flash Pyrolysis Systems

Residence times in fast pyrolyzers were discussed earlier. Various fast pyrolyzers have been used by different researchers. Perhaps the most popular of which is the fluidized bed pyrolyzer. The unit has the following component parts:

1. Enclosed biomass bin hopper
2. An auger-feeding system at the base of the reactor
3. Fluidizing inert gas source and flow controller
4. Electrically heated reactor with electronically controlled temperature
5. Completely insulated series cyclones that will remove the char particulate
6. Condensing train that is cooled by a chiller
7. Syngas measuring instruments, including volumetric flow-measuring instruments
8. Associated temperature gauges, controllers, and data-logging system

Before the start of any test, the feed rate of a particular biomass in question must be established (the calibration curve). The dimensions of the pyrolyzer should be known, including the flow rate to use. The residence time may be computed using the procedure outlined in Example 9.4, so that the biomass will be converted in fast pyrolysis mode. The procedure relating how much biomass is fed into the reactor using the calibration curve will be further illustrated in Example 9.5 using corn stover.

Energy and mass balances may then be calculated after each test to determine the conversion efficiencies, including an estimate of the fractions of energy that are contained in the solid, liquid, and gaseous components.

Example 9.5: Calibration Curve for Augers

The following data was gathered while calibrating the feed rate of corn stover biomass for an auger in a pyrolyzer. Develop an equation relating the rpm of the auger with the feed rate. Estimate the regression coefficients. What would be the rpm so that a feed rate of 40 kg/h would be fed through the reactor?

rpm	Mass Flow Rate (kg/h)
11	5
20	8
30	12
41	18
50	20
60	24
70	27
80	32
85	34

Solution:

a. This problem is best solved using spreadsheet software such as Microsoft Excel to establish the equation relating the rpm with the feed rate. The results are as follows with an R^2 of 0.9953.

$$\text{Feed rate (kg/h)} = 0.4001 \times (\text{rpm})$$

b. Thus, the rpm setting to obtain a feed rate of 40 kg/h is calculated as follows:

$$\text{rpm} = \frac{\text{Feed rate (kg/h)}}{0.4001} = \frac{40}{0.4001} = 100 \text{ rpm}$$

c. The graph is shown as follows:

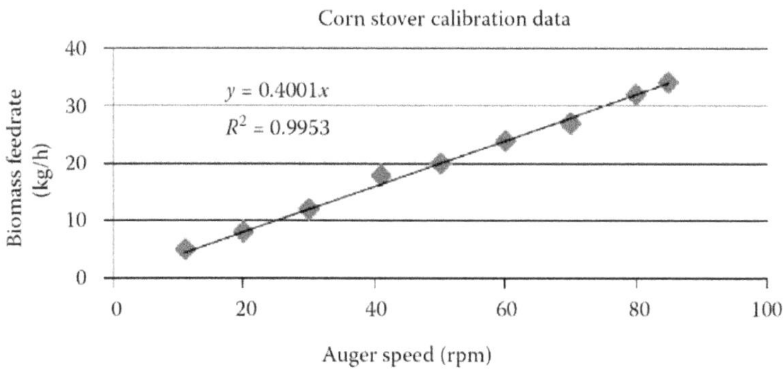

9.2.3 Flash Vacuum Pyrolysis

The flash vacuum pyrolysis process has been used to pyrolyze biomass and is used as a mechanism to generate polycyclic aromatic hydrocarbons (Ammann et al., 1999). Plant sterols were found to yield polycyclic aromatic hydrocarbons under subatmospheric conditions (or in vacuum conditions) with temperatures ranging from 400°C to 700°C. The advantage of using vacuum conditions is the ability to operate at lower temperatures, thereby avoiding the adverse chemical reactions at higher temperatures. Some textbooks refer to this process as flash vacuum thermolysis. The residence time is made so short that secondary reactions are avoided, resulting in simpler, low molecular weight (MW) compounds in the low carbon number group.

Sugarcane bagasse was also used to establish product yields and properties, using a bench- and pilot-scale vacuum pyrolysis process (Garcia-Perez et al., 2002). The process yielded more oil (34.4% vs. 30% by weight) than similar nonvacuum processes and less charcoal (19.4% vs. 25.7% by weight). The resulting bio-oil product also had low ash content, high HV (22.4 MJ/kg), and low viscosity (4.1 cSt at 90°C). Another advantage of this system is that by creating a vacuum in a flash, the gaseous products have very limited contact time with the coproduced char. The process may be on a batch or continuous system. The vacuum pressures used ranged from 3.5 to 10 kPa (Zhang et al., 2008).

These conditions were used for the vacuum pyrolysis of waste tires. Additives were used during the process to further improve the products and coproducts. The effect of additives such as Na_2CO_3 and NaOH has been thoroughly investigated.

9.2.4 Ablative Pyrolysis

The ablative pyrolysis process is quite different from the other methods of fast pyrolysis presented thus far. Note that in the previous methods described, the biomass must be ground to the micron level for the pyrolysis process to be effective. This small size is not necessary for ablative pyrolysis. In this system, the biomass is pressed onto the heated reactor wall to transfer enormous amounts of energy in a short period. The biomass does not need to be ground in this case, as it is moved in a vortex pathway across the reactor. This swift mechanical movement allows the rapid evaporation of vapor for further condensation into a liquid product that generates some form of oil lubricant through the reactor. Bridgewater (2012) described this process as "butter being melted on a very hot frying pan." A strong centrifugal force causes very high pressure contact between the biomass and the heated wall. This action also results in high relative motion between the biomass and the reactor wall for enhanced chemical reactions. Note the wall temperature is not very high and is usually less than 600°C. This design was originally developed at the National Renewable Energy Laboratory of the US Department of Energy (USDOE). This is the only fast pyrolysis system that does not require grinding the biomass; large particles can be used and are easily converted. In addition, the design does not require inert gas. With no need for gas, the entire reactor can be smaller, when compared with other systems. The process is dependent on the surface area of the conical reactor wall and the rotating mechanisms required to bring the contact between the biomass and the heated wall. Thus, scaling up may be an issue, although several units have been demonstrated in Germany (Meier et al., 2005; Peacocke and Bridgewater, 1994).

9.2.5 Microwave Pyrolyzer

Microwave pyrolyzers are used to thermally convert biomass into biofuels, primarily because it is relatively easy to generate this kind of energy and to transfer energy to the biomass, even if it has not been properly ground. One such microwave pyrolyzer was developed at the University of Minnesota (UM) by Dr. Roger Ruan. Magnetron in pyrolysis reactors is used to generate microwave energy. In the UM design, there are two reactors connected in series; the biomass is moved from the hopper to the reactor using a motor-auger mechanism. The syngas is pulled through a vacuum pump and passed through a condensing column. Each reactor has its own condensing system to collect the bio-oil. The non-condensible syngas is then filtered before it is placed in storage tanks for future use. This combustible gas may also be used in a gas turbine to generate electricity that can be used to run all the electric motors and vacuum pumps.

9.3 EFFECTS OF TEMPERATURE ON PRODUCT YIELDS FROM PYROLYSIS OF MICROALGAE

In a pyrolyzer, the energy and product yields may be easily established by completing the energy and mass balances after the pyrolysis process. The first step is the complete characterization of the biomass in question. In this section, the pyrolysis of microalgae will be described using a controlled batch pyrolysis process to complete the energy and mass balance calculations. A suitable facility to conduct this process is a simple batch pyrolyzer. Table 9.3 lists the initial characteristics of the microalgae used to establish the energy and mass balances. This biomass sample was then pyrolyzed at 400°C, 500°C, and 600°C to generate complete energy and mass balance data.

The microalgae sample HV is 24.7 MJ/kg (10,642 Btu/lb), quite high for a biomass resource. This biomass sample was pyrolyzed at various temperatures. Mass and energy balances were established. Note that the HV numbers were reported based on either the "as-received" basis, dry basis, or dry and ash-free basis. Example 9.6 illustrates how these numbers could vary, depending on whether moisture or ash is included.

TABLE 9.3
Initial Biomass Characterization for Microalgae Used for Pyrolysis Studies

Biomass Characteristics	Experimental Value
Proximate Analysis	
MC	6.18 ± 0.14
VCM (db)	81.27 ± 0.98
FC (db)	5.17 ± 0.46
Ash (db)	13.57 ± 0.65
Ultimate Analysis	
C (daf)	48.31 ± 0.26
H (daf)	7.66 ± 0.12
O (daf)	24.85 ± 0.36
N (daf)	4.80 ± 0.02
S (daf)	0.81 ± 0.05
Compositional	
Lignin	4.9 ± 0.46
Extractives	47.35 ± 3.79
Protein	23.95 ± 2.36
Sugars	6.87 ± 0.17
Sugar Analysis	
Glucan	
Xylan	
Galactan	
HV (MJ/kg)	24.7 ± 0.49

Example 9.6: HV on As-Received, Dry, and Ash-Free Basis

The HV of an algae sample (moisture content, 10%) was found to be 24.7 MJ/kg (10,642 Btu/lb). Determine the HV in dry basis and in dry ash-free basis if the ash content was found to be 13.6% (dry basis). Discuss the implications of these units.

Solution:

a. The HV on a dry basis is simply calculated as follows:

$$\text{HV}\left(\text{MJ/kg}\right) = \frac{\text{HV (as received basis)}}{(1-\text{moisture content})} = \frac{24.7}{0.9} = 27.4\,\text{MJ/kg}$$

b. The HV on a dry ash-free basis is further calculated as follows:

$$\text{HV}\left(\text{MJ/kg}\right) = \frac{\text{HV (dry basis)}}{(1-\text{ash content})} = \frac{27.4}{\left(1-0.136\right)} = 31.7\,\text{MJ/kg}$$

c. Note that the denominator in (a) is the dry weight of the material and the denominator in (b) is the dry and ash-free weight of the material. Thus, the relative or reported HV of the material will increase if ash and moisture are removed.

9.3.1 Microalgae Mass Balances

A pyrolysis experiment was conducted in a batch pressure vessel to develop the energy and mass balance of a biomass. The experiment included accurately weighing a known amount of biomass and placing this sample in a reactor where the temperature and heating rate were programmed. At the end of the experiment, the reactor was cooled, and all the solid, liquid, and gaseous components were accurately measured. The mass balance and energy balance were then established from these results. Table 9.4 summarizes the results of one such experiment, including the percentage of weight from each coproduct. Note that the sum of solids, liquids, and gases is not 100% and some losses are observed. It is quite difficult to recover all process coproducts, especially from a continuous process. The range of mass loss was quite significant at around 25%–27%. Figures 9.3 through 9.5 show the mass

TABLE 9.4
Yields of Solid, Liquid, and Gaseous Components of Microalgae Sample by Weight

Temperature (°C)	Syngas Yield (%w/w)	Bio-Oil Yield (%w/w)	Biochar Yield (%w/w)	Losses (%w/w)
400	4.0	19.0	51.8	25.2
500	10.9	27.6	34.4	27.1
600	14.8	23.2	35.8	26.2

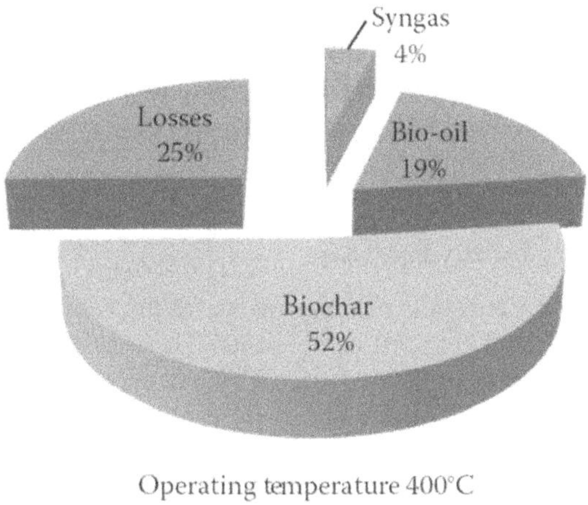

FIGURE 9.3 Mass balance of microalgae pyrolysis at 400°C.

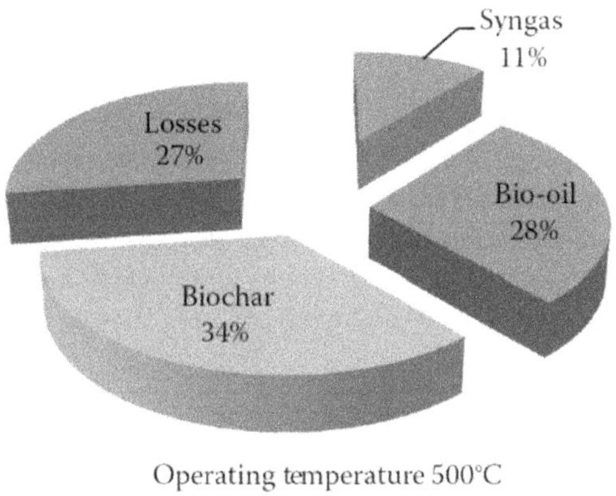

FIGURE 9.4 Mass balance of microalgae pyrolysis at 500°C.

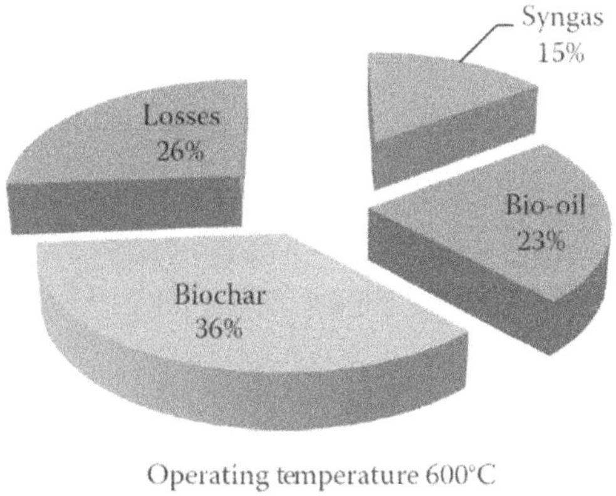

FIGURE 9.5 Mass balance of microalgae pyrolysis at 600°C.

balance results from this experiment at various conversion temperatures. Note that although it is relatively straightforward to measure the weights of all coproducts to establish the mass balance, the energy balance is quite tedious to establish; further analysis of all the coproducts is required. Thus, provisions must be made to appropriately measure the coproducts.

9.3.2 Microalgae Energy Balances

Table 9.5 summarizes the results of experiments for the subsequent analysis of the solid (biochar) and liquid (bio-oil) components. These samples were analyzed in oxygen bomb calorimeters to measure the HVs and reported as shown. The gaseous components may be quite difficult to measure because gravimetric gaseous samplings are hard to accomplish. In pyrolysis experiments such as those described above, the volume of the gaseous component is easily measured at a given temperature and pressure. This component is subsequently cooled to normal temperature (20°C) and pressure (1 atm), or NTP, as opposed to standard atmosphere and pressure (STP), which is 0°C and 1 atm. With the known density of individual components (at NTP), the energy content in units of MJ/kg may be calculated. The gaseous components may be measured using an appropriate gas chromatograph (GC) to determine the volumetric composition of the individual gas present (%v/v). The data for this experiment are summarized in Table 9.6 for the volumetric yields at various temperatures. Table 9.7 summarizes the HVs of the synthesis gas at NTP. Note that the gases are measured at actual temperature and pressure and corrected to NTP usingthe ideal gas law. The other parameters of major syngas components are also summarized in Table 9.7, such as density and MW and were used to estimate the HV of the syngas.

TABLE 9.5
Biochar and Bio-Oil Analysis Results

Temperature (°C)	Biochar HV (MJ/kg)	Biochar Hv (Btu/lb)	Bio-Oil Hv (MJ/kg)	Bio-Oil HV (Btu/lb)
400	26.87	11,577	35.80	15,424
500	20.44	8,806	38.50	16,588
600	20.97	9,035	38.05	16,394

TABLE 9.6
Gaseous Components of Typical Syngas from Pyrolysis Experiments by Volume Basis (%v/v)

Temp (°C)	H_2	CO	CH_4	C_2H_4	C_2H_6	O_2	N_2	CO_2	HV (MJ/m^3)
400	2.4	6.0	5.8	1.1	3.7	0.6	6.4	61.6	6.6
500	15.3	3.6	21.0	3.4	11.6	1.4	6.0	10.3	20.5
600	14.5	4.3	44.6	1.1	11.6	1.3	5.2	6.1	28.5

TABLE 9.7
Parameters of Common Gases Used for Calculating HV of Synthesis Gas

Parameter (at NTP)	H_2	CO	CH_4	C_2H_4	C_2H_6
HV (MJ/kg)	142.4	10.8	58.1	53.5	55.1
HV (MJ/m^3)	12.8	12.6	38.8	63.0	69.7
Density (kg/m^3)	0.0899	1.165	0.668	1.178	1.264
MW (g/mol)	2.016	28.01	16.043	28.03	30.07

Example 9.7 shows how to estimate the HV of syngas from data summarized in Tables 9.6 and 9.7. Example 9.7 also shows how to report the HVs by weight aside from volume basis using the density data. Note that the resulting HV calculations for the syngas are reported in the last column in Table 9.6.

In Table 9.7, the HVs of syngas components in units of MJ/m^3 were taken from Waldheim and Nilsson (2011). The heating density and MWs of the gases were derived from Engineering Toolbox (2012). The HVs in units of MJ/kg were derived from the HVs in units of MJ/m^3 and density (kg/m^3).

When the energy units for biochar and bio-oil are in MJ/kg, it is logical to convert the syngas energy into the same units as well. The HV in MJ/kg is calculated using the density of the individual gases at NTP. Because all energy units are consistent, the overall energy balance may be calculated for microalgae pyrolyzed at 400°C, 500°C, and 600°C, as illustrated by Figures 9.6 through 9.8 for the respective temperatures.

The percentage energy yield (%) was calculated using Equation 9.3.

$$\text{Energy yield (\%)} = \frac{\text{Mass yield}_{\text{prod}} \times \text{HV}_{\text{prod}}}{\text{HV}_{\text{algae}}} \times 100\% \tag{9.3}$$

The HV of microalgae was 24.7 MJ/kg (Table 9.3). If this sample was pyrolyzed at 400°C, one would expect to obtain a yield as summarized in Table 9.4. Thus, at this temperature and for every kilogram of microalgae, one would expect to have 0.518 kg

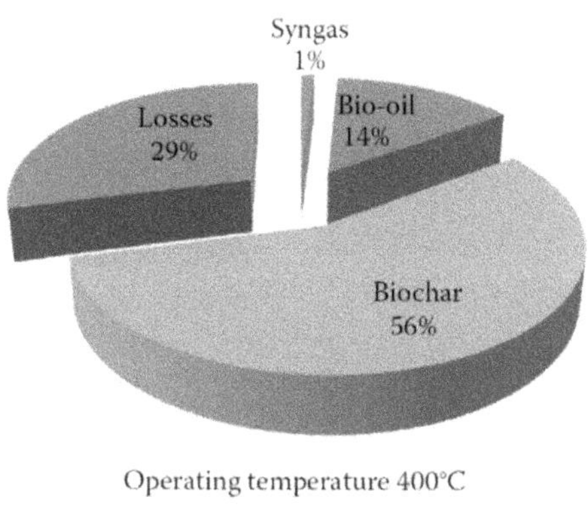

FIGURE 9.6 Energy balance of microalgae pyrolysis at 400°C.

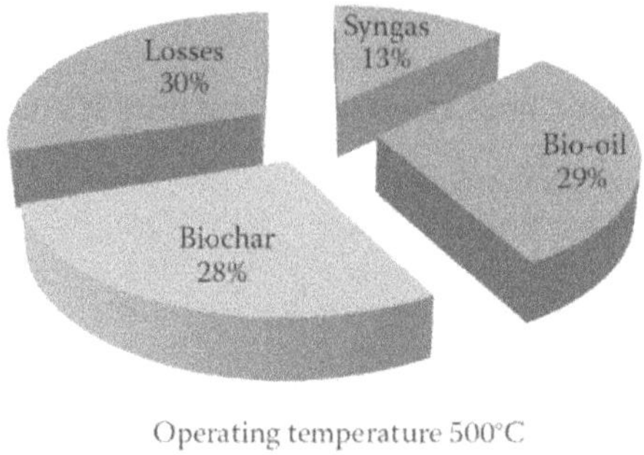

FIGURE 9.7 Energy balance of microalgae pyrolysis at 500°C.

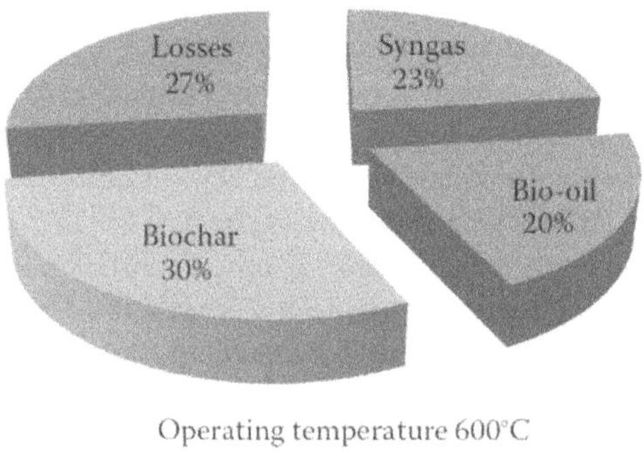

FIGURE 9.8 Energy balance of microalgae pyrolysis at 600°C.

of biochar, 0.19 kg of bio-oil, and 0.04 kg of syngas. This approach leads to the data shown in Figures 9.6 through 9.8.

An example on how to calculate the yields of bio-oil, biochar, and syngas per unit of biomass weight is shown in Example 9.8.

Example 9.7: Calculating the HV of Syngas from Experiments

Estimate the HV of syngas at NTP from the values given in Table 9.5 and the gas yield in Table 9.6 at 500°C. Do the same calculations at temperatures of 400°C and at 600°C. Convert the units on weight basis. Use gas density data from Table 9.7.

Solution:

a. The HV of syngas is simply the weighted average HV of individual gases as shown. Note that for every cubic meter (m^3) of syngas produced, there will be 0.153 m^3 of H_2 (or 15.3/100), 0.036 m^3 of CO, etc.

$$HV = \frac{15.3 \times 12.8 + 3.6 \times 12.6 + 21 \times 38.8 + 3.4 \times 63.0 + 11.6 \times 69.7}{100} = 20.5 \text{ MJ/m}^3$$

b. Likewise, the HV of syngas at other temperatures are calculated accordingly for 400°C:

$$HV = \frac{2.4\times12.8+6\times12.6+5.8\times38.8+1.1\times63.0+3.7\times69.7}{100}$$

$$= 6.6\ MJ/m^3$$

c. For 600°C, the HV of syngas produced is calculated as follows:

$$HV = \frac{14.5\times12.8+4.3\times12.6+44.6\times38.8+1.1\times63.0+11.6\times69.7}{100}$$

$$= 28.5\ MJ/m^3$$

d. The energy units by weight are simply converted using the density of individual gases as shown at 500°C:

$$HV = \frac{12.8\times0.153}{0.0899}+\frac{12.6\times0.036}{1.165}+\frac{38.8\times0.21}{0.668}$$

$$+\frac{63\times0.034}{1.178}+\frac{69.7\times0.116}{1.264} = 42.6\ MJ/kg$$

e. The HV in units of MJ/kg at 400°C is as follows:

$$HV = \frac{12.8\times0.024}{0.0899}+\frac{12.6\times0.06}{1.165}+\frac{38.8\times0.058}{0.668}$$

$$+\frac{63\times0.011}{1.178}+\frac{69.7\times0.037}{1.264} = 10.1\ MJ/kg$$

f. Finally, the HV for 600°C is calculated as follows:

$$HV = \frac{12.8\times0.145}{0.0899}+\frac{12.6\times0.043}{1.165}+\frac{38.8\times0.446}{0.668}$$

$$+\frac{63\times0.011}{1.178}+\frac{69.7\times0.116}{1.264} = 54\ MJ/kg$$

Example 9.8: Calculating Yields per Unit Weight of Biomass

Estimate the biochar, bio-oil, and syngas yields at 400°C, 500°C, and 600°C operating temperatures for microalgae in units of kg/tonne, L/tonne, and L/tonne, respectively. Use the data in Tables 9.4 and assume the average density of bio-oil to be about 840 kg/m^3 and that of the syngas to be 0.661 kg/m^3. Convert the bio-oil yield in gallons/ton at the three reaction temperatures.

Solution:

a. At 400°C, the calculations are as follows:

$$\text{Biochar yield (kg/tonne)} = \frac{0.518\,\text{kg}}{\text{kg biomass}} \times \frac{1{,}000\ \text{kg}}{\text{tonne}} = 518\ \text{kg/tonne}$$

$$\text{Bio-oil yield (L/tonne)} = \frac{0.19\,\text{kg}}{\text{kg biomass}} \times \frac{\text{m}^3}{840\,\text{kg}} \times \frac{1{,}000\ \text{L}}{1\,\text{m}^3} \times \frac{1{,}000\ \text{kg}}{\text{tonne}}$$

$$= 226\ \text{L/tonne}$$

$$\text{Syngas yield (L/tonne)} = \frac{0.04\,\text{kg}}{\text{kg biomass}} \times \frac{\text{m}^3}{0.661\,\text{kg}} \times \frac{1{,}000\ \text{L}}{\text{m}^3} = 60.5\ \text{L/kg}$$

b. At 500°C, the calculations are illustrated as follows:

$$\text{Biochar yield (kg/tonne)} = \frac{0.344\,\text{kg}}{\text{kg biomass}} \times \frac{1{,}000\ \text{kg}}{\text{tonne}} = 344\ \text{kg/tonne}$$

$$\text{Bio-oil yield (L/tonne)} = \frac{0.276\ \text{kg}}{\text{kg biomass}} \times \frac{\text{m}^3}{840\ \text{kg}} \times \frac{1{,}000\ \text{L}}{1\ \text{m}^3} \times \frac{1{,}000\ \text{kg}}{\text{tonne}}$$

$$= 329\ \text{L/tonne}$$

$$\text{Syngas yield (L/tonne)} = \frac{0.109\,\text{kg}}{\text{kg biomass}} \times \frac{\text{m}^3}{0.661\,\text{kg}} \times \frac{1{,}000\,\text{L}}{1\,\text{m}^3} = 165\,\text{L/kg}$$

c. At 600°C, the calculations are shown next:

$$\text{Biochar yield (kg/tonne)} = \frac{0.358\,\text{kg}}{\text{kg biomass}} \times \frac{1{,}000\ \text{kg}}{\text{tonne}} = 358\ \text{kg/tonne}$$

$$\text{Bio-oil yield (L/tonne)} = \frac{0.232\ \text{kg}}{\text{kg biomass}} \times \frac{\text{m}^3}{840\ \text{kg}} \times \frac{1{,}000\ \text{L}}{1\ \text{m}^3} \times \frac{1{,}000\ \text{kg}}{\text{tonne}}$$

$$= 276\ \text{L/tonne}$$

$$\text{Syngas yield (L/tonne)} = \frac{0.148\ \text{kg}}{\text{kg biomass}} \times \frac{\text{m}^3}{0.661\,\text{kg}} \times \frac{1{,}000\,\text{L}}{\text{m}^3} = 224\ \text{L/kg}$$

d. Converting the metric values for syngas into the English system, we have the following:

$$\text{Bio-oil yield (L/tonne)} = \frac{226\ \text{L}}{\text{tonne}} \times \frac{\text{gal}}{3.785\text{L}} \times \frac{0.909\ \text{tonne}}{1\ \text{ton}} = 54\ \text{gal/ton}$$

$$\text{Bio-oil yield (L/tonne)} = \frac{329\text{ L}}{\text{tonne}} \times \frac{1\text{ gal}}{3.785\text{ L}} \times \frac{0.909\text{ tonne}}{1\text{ ton}} = 79\text{ gal/ton}$$

$$\text{Bio-oil yield (L/tonne)} = \frac{276\text{L}}{\text{tonne}} \times \frac{1\text{ gal}}{3.785\text{L}} \times \frac{0.909\text{ tonne}}{\text{ton}} = 66\text{ gal/ton}$$

9.3.3 Overall Product Yields

Shown in the Figure 9.9 are the relationships between product yields from the pyrolysis of microalgae using a batch pyrolysis reactor. As temperature is increased, the biochar energy yield decreases, or it levels off between 500°C and 600°C. The syngas energy yield increases as the operating temperature is increased, while the bio-oil energy yield maximizes at around 500°C.

Some authors prefer to report the yield per unit weight of biomass used, and these are easily calculated once the densities of the individual components are known.

9.4 APPLICATIONS OF PRODUCTS FROM FAST PYROLYSIS

There are various ways of utilizing the solid, liquid, and gaseous products from fast pyrolysis of biomass. The various applications are schematically shown in Figure 9.10. The biochar has many uses, including soil amendment, fertilizer, charcoal, and co-firing fuel. The biochar may be upgraded to activated carbon for use in odor control and wastewater treatment. In the future, it will have a role in carbon sequestration.

The syngas products also have numerous applications. It has high energy value and thus should be easily combusted to generate thermal energy. Syngas may be used to generate steam and as an auxiliary fuel for the pyrolysis process (since pyrolysis requires external heat), and for drying biomass. Usually, for relatively dry agricultural residues, the energy from syngas is enough to bring the moisture down to the

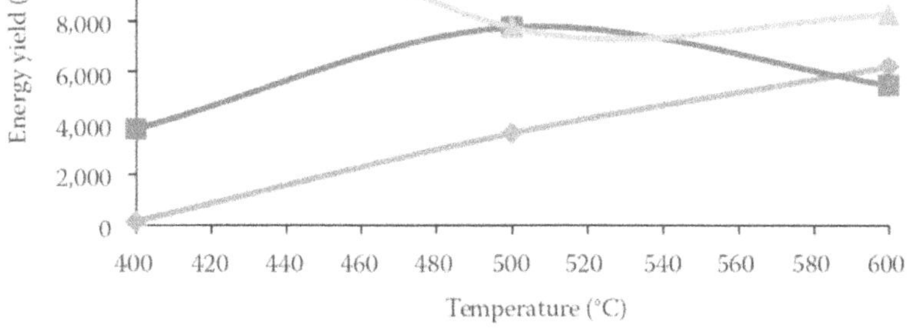

FIGURE 9.9 Energy distribution of solid, liquid, and gaseous products from pyrolysis of microalgae samples.

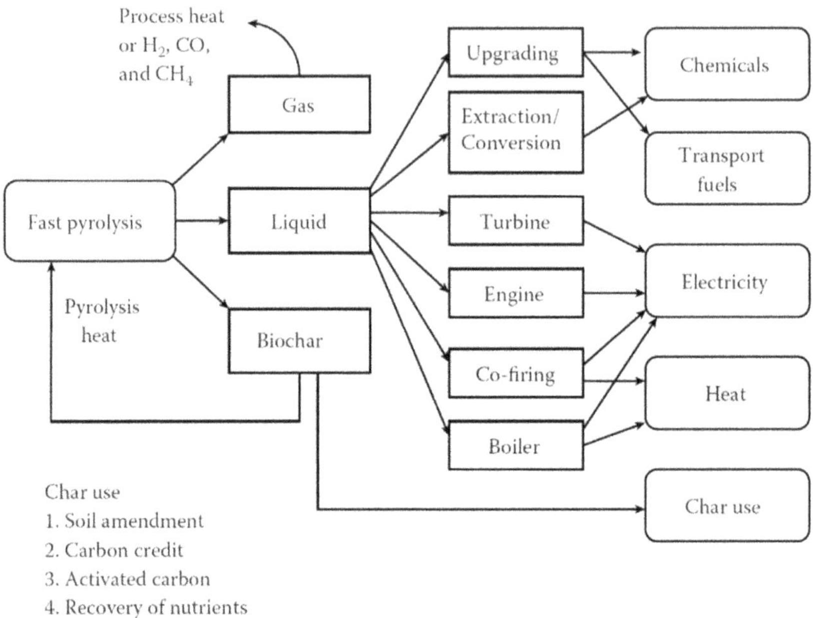

FIGURE 9.10 Various applications of products of pyrolysis process.

ideal 10% for the pyrolysis process. At higher levels of moisture in the biomass, the energy from syngas may not be sufficient for drying. However, additional energy may be obtained from the biochar. The bio-oil when condensed contains non-condensibles and is also useful for combustion and for generating heat for other purposes. Storing this gas may require a large vessel. If compressed, it has enormous amounts of pressure and is not very economical.

The liquid bio-oil component of syngas has the most uses. It also has the most constituents. There are hundreds of compounds present in the bio-oil, ranging from acids to aldehydes, ketones, and alcohols. In fluidized bed systems, when the cyclone particulate removal is not properly designed, some solid biochar may be found in the bio-oil. The organic portion of the bio-oil has very high energy content and can be combusted in a simple burner that is coupled with a boiler. Bio-oil, in small amounts, may be co-fired with diesel. This blend normally requires solvents to make them miscible. Attempts to use bio-oil directly in turbine burners have been made and reported. Other tests used mixed fuel in internal combustion engines.

Perhaps the most exciting use of bio-oil is the upgrade of this dark and viscous liquid product into refinery-grade crude oil. Bio-oil looks exactly like crude oil in many instances. However, the chemical composition of bio-oil is quite different than that of crude oil. Crude oil is composed mainly of hydrocarbons and their derivatives. Bio-oil has other compounds that are not considered hydrocarbon fuels. The three main components that must be removed from the raw bio-oil sample to make it chemically equal to crude oil are moisture, organic acids, and oxygenates (or compounds containing oxygen). The crude oil must be void of moisture, acids, and oxygen so that it can be marketed and blended with commercial crude oil. This blend can then be refined into gasoline, diesel, or JP-8 components. Alkanes and alkenes are present in some bio-oil samples, and all that is required is simple fractionation or distillation in both normal temperatures and in a vacuum. Some chemicals may be extracted from the bio-oil samples, such as phenols or higher alcohols. Finally, the raw bio-oil sample may be catalytically upgraded to generate mostly hydrocarbon fuels.

9.5 BIO-OIL CHARACTERIZATION PROCESSES

It is important to learn the simple process of characterizing bio-oil organic fraction. In some bio-oil samples that have moisture content (MC) greater than 15%, there is usually a phase separation. The organic phase and the water phase are sometimes called the water-soluble and water-insoluble components. They may be separated by simple decanting or with a centrifuge. Once these phases are separated, a carefully selected solvent must be used to further separate the compounds with different affinities to the various solvents, thus providing better separation of the various types of compounds.

The raw bio-oil may have hundreds of compounds. Consequently, it will be necessary to identify these compounds if there are any plans to upgrade the bio-oil. Although a GC with a mass spectrometer (GC-MS) may be used to identify the range of compounds, the bio-oil may have components that can ruin the GC column. Thus, the GC may not be an appropriate protocol for analysis. One should use analytical equipment that would not be destructive, such as Fourier transform infrared (FTIR) spectrometers. An FTIR instrument can scan a sample and produce spectra for components, which, if one has an extensive database of component signatures, can be used to identify groups of compounds found in the bio-oil. When components have been fractionated or distilled with various solvents, the GC-MS can be used to identify the specific compound found in the bio-oil. Corn stover bio-oil was fractionated following the process shown in Figure 9.11, and the various fractions were analyzed and found to have the list of compound groups listed in Table 9.8. Note that the

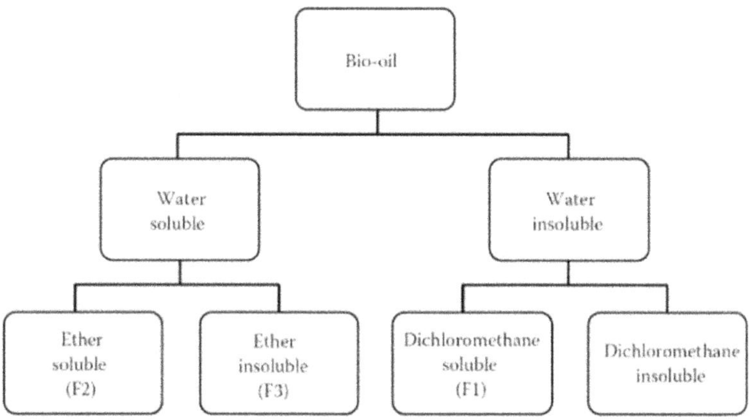

FIGURE 9.11 Bio-oil characterization pathway for product identification and analysis.

TABLE 9.8
Types of Compounds Identified in Corn Stover Bio-Oil Sample

Compounds	% Relative Content			
	Bio-Oil	F1	F2	F3
Alkanes	11.9	7.12	1.29	
Alkenes	8.66	9.13	0.18	2.32
Alcohols	3.06	1.08	9.43	19.19
Ketones	13.41	14.04	21.05	32.84
Phenols	26.7	29.58	38.52	3.65
Benzene/derivatives	20.24	17.78	1.62	
Acids	1.62	6.51	8.81	2.8
Heterocyclics	11.06	9.98	18.58	35.93
Naphthalene	1.68	3.13		
Furans	1.55	0.96		1.26
Esters/aldehyde	0.12	0.69	0.52	0.11
Others				1.9
Total	100	100	100	100

Note: F1: water insoluble—dichloromethane soluble fraction; F2: water soluble—ether soluble fraction; F3: water soluble—ether insoluble fraction.

fractions F1, F2, and F3 are those marked in Figure 9.11 and defined in the footnote at the end of Table 9.8. Compare the list of compounds in column 1 (raw bio-oil) with those of the other fractions. These comparisons show that fractionation with a solvent generated a varied set of compounds, indicating that this process is not a simple separation of compounds present in the original bio-oil, but that some reaction with solvents may have occurred. Thus, one will have to be careful in using solvents that may alter the composition of compounds. Another method of separating the compounds found in raw bio-oil is by simple distillation using atmospheric conditions, or in a vacuum. This procedure will not significantly alter the composition of the bio-oil through chemical reactions but simply separate the compounds by virtue of their different boiling points. Vacuum conditions are used so that the temperature used for distillation can be lower; less energy is required for separation. Note that a large fraction of alkanes, alkenes, phenols, and alcohols are already found in the bio-oil. The catalytic upgrade process may be made for other compounds that are difficult to convert into hydrocarbons, such as organic acids and aldehydes. The dichloromethane-insoluble fractions are not listed in Table 9.8 because of the lower amounts generated.

Distillation studies (100°C–180°C) were also made to determine important hydrocarbon groups present in the bio-oil. Table 9.9 summarizes the results of analysis for corn stover bio-oil. A large portion of fractionates are ketones, benzene and derivatives, alcohols, cyclic compounds, and a few alkanes and alkenes. These compounds comprised more than 60% of the compounds present. These compounds may be easily upgraded into gasoline fuel using catalytic processes.

TABLE 9.9
Compounds Separated through Simple Distillation of Bio-Oil between 100°C and 180°C

Compound	% Relative Content
Alkanes	5.88
Alkenes	4.15
Alcohols	12.01
Ketones	35.8
Phenols	3.5
Benzene and its derivatives	17.2
Cyclic compounds	8.03
Furans	1.88
Esters	10.46
Alkyne	1.09
Total	100

9.6 BIO-OIL UPGRADE PROCESSES

Upgrading bio-oil samples requires careful processing, including the judicious use of catalysts. Bio-oil has the following three main characteristics:

1. High moisture content
2. High composition of acids
3. High oxygen content

Thus, the upgrade process may require removing moisture, bringing the acid number to a certain level, and removing all oxygenated compounds. Although the first two parameters may be easy to handle, the last parameter is difficult to process without the use of catalysts. However, the cost of catalysts can be quite high. The current prices of novel and rare metals that are used during the catalytic processes ranged from a low of $31.85 per troy ounce for silver and as high as $1,653 for gold. One of the most expensive novel metals used in catalysts is platinum, being sold at around $1585 per troy ounce (BASF, 2012). Note that a troy ounce converts to about 31.1 g. The cost of this catalyst may be easily translated into $51/g or $50,965/kg. The use of a particular catalyst must be justified by the percentage recovery of that catalyst, as well as its effectiveness. If catalysts are poisoned and rendered unusable, the economics of the process may not be justified. The amounts of catalysts being used in bio-oil upgrade processes is usually small (<5%).

9.7 STUDIES ON PYROLYSIS OF VARIOUS BIOMASS RESOURCES

Various initial pyrolysis studies have been completed on the following biomass materials:

1. Cotton gin trash (Aquino et al., 2010)
2. Switchgrass (Iman and Capareda, 2012)
3. Corn stover (Capunitan and Capareda, 2012)
4. Sorghum biomass (Santos and Capareda, 2012)
5. Microalgae (Maguyon and Capareda, 2012)

The results will be summarized in this section. Results for microalgae were presented earlier in this chapter.

9.7.1 Cotton Gin Trash Pyrolysis (Aquino et al., 2010)

Cotton gin trash (CGT) is a ubiquitous by-product of cotton ginning (removal of lint from cottonseeds). CGT is generated at a rate between 90 and 270 kg per bale of cotton module (200–600 lb/bale), depending on how the cotton was harvested. A stripper gin would have more trash produced than a picker gin. The trash remains unutilized at the cotton gins and would have to be converted into usable material.

CGT was pyrolyzed at different temperatures (500°C, 600°C, 700°C, and 800°C) using an externally heated batch reactor. The average yields of output products (solid/char, liquid/bio-oil, and gas) were determined. The HV of CGT was measured to be around 15.5 MJ/kg (7,433 Btu/lb). Gas yield increased while char yield decreased as temperature was increased. The maximum char yield of 38 wt% was obtained after 30 min of pyrolysis at 500°C. The char had the largest fraction of energy output (44%–51%), followed by gas (6%–15%) and bio-oil (4%–5%). The maximum gas yield of 35 wt% was observed at 800°C. The average yield of CO, H_2, and total hydrocarbons (THCs) generally increased with the increased temperature, whereas CO_2 production decreased. Methane, ethane, and propane were the dominant constituents of the THCs. The highest bio-oil yield of about 30 wt% was observed at 600°C. The HV of bio-oil was low (2–3 MJ/kg) due to minimal nonhydrocarbon compounds and high MC. The HV of the organic portion of the bio-oil was measured to be 6.1 MJ/kg.

9.7.2 Switchgrass Pyrolysis Studies (Iman and Capareda, 2010)

Bio-oil, syngas, and biochar yields were measured at three different pyrolysis temperatures (400°C, 500°C, and 600°C), and the results are presented in Figure 9.12. Bio-oil and syngas yields increased with an increase in the pyrolysis temperature, whereas biochar content decreased with an increase in the temperature. The bio-oil yield at 400°C was 22% and increased to 37% at 600°C. Similarly, syngas yields increased from 8% to 26% when the temperature was raised from 400°C to 600°C. The char yield decreased from 48% to 25% when the temperature was raised from 400°C to 600°C.

For optimum production of the switchgrass bio-oil and upgrading processes, the higher temperature of 600°C was recommended. On the other hand, an increase in gaseous products at the higher 600°C pyrolysis temperature was also observed. This effect is due primarily to secondary cracking of the pyrolysis vapors and secondary decomposition of the char.

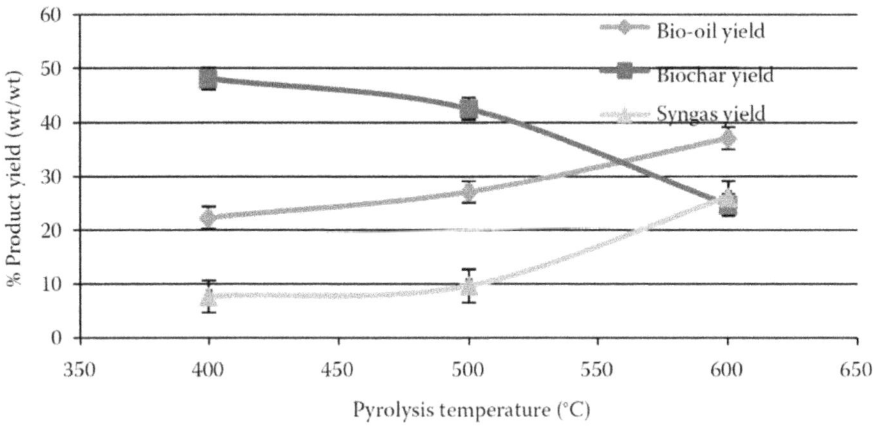

FIGURE 9.12 Results of pyrolysis of switchgrass samples at various temperatures.

Bio-oil (organic fraction) was analyzed using a GC-MS. The results revealed a complex mixture composed of alcohols, esters, ketones, lignin-derived phenols, long-chain alkanes, aldehydes, fatty acid methyl esters, furans, naphthalene, amides, and various aromatics. A large fraction of the components identified were phenols or phenol-related compounds, long-chain hydrocarbons, and most functional groups showing the presence of oxygen. Analysis of both bio-oil and the aqueous phase was divided into various hydrocarbon groups to better understand their chemical properties and for upgrading purposes. The results revealed 36% of branched and long chains of alkanes; 20% of methyl, methoxy, and propenyl attached phenols; 8.7% mixture of acids and methyl ester fatty acids; 14.1% aromatics: indene, methylindene, benzene, toluene, and methyl-naphthalene; 7.4% methyl and methoxy attached alcohols; and the rest included mostly furans and ketones.

The distribution of some compounds detected in the aqueous phase of the bio-oil from switchgrass pyrolysis showed the presence of abundant aldehydes and ketones. Thus, the aqueous phase of the oil showed highly hydrated hydrophilic compounds, which made it difficult to eliminate water from the bio-oil. This aqueous phase was composed mainly of a large amount of acetic acid (28.2%), various branched ketones (42%), small amounts of methyl and methoxy phenols (4.7%), and aromatic hydrocarbons.

The results of a syngas analysis from pyrolysis of a sample of switchgrass are summarized in Table 9.10. Syngas compositions vary with pyrolysis temperature. With an increase in temperature, the CO_2, CO, C_2H_4, and C_2H_6 compositions increased, whereas the H_2 and CH_4 compositions decreased. The overall production of syngas increased from 8% to 10% when the temperature was raised from 400°C to 500°C, and then from 10% to 26% when the temperature was raised from 500°C to 600°C. This increase in composition may be explained by the oxidation of the carbonized char at higher pyrolysis temperatures.

Biochar analysis showed the yield decreased from 48% to 43% from 400°C to 500°C and 43% to 25% from 500°C to 600°C. The biochar from different pyrolysis

TABLE 9.10
Syngas Composition as Function of Pyrolysis Temperature for Switchgrass Sample

Pyrolysis T (°C)	H_2 (%)	CO (%)	CH_4 (%)	CO_2 (%)	C_2H_4 (%)	C_2H_6 (%)
400	28.0	21.7	39.2	9.1	1.2	0.8
500	25.4	21.5	39.4	9.4	2.3	1.9
600	9.7	27.7	17.6	33.2	4.3	7.0

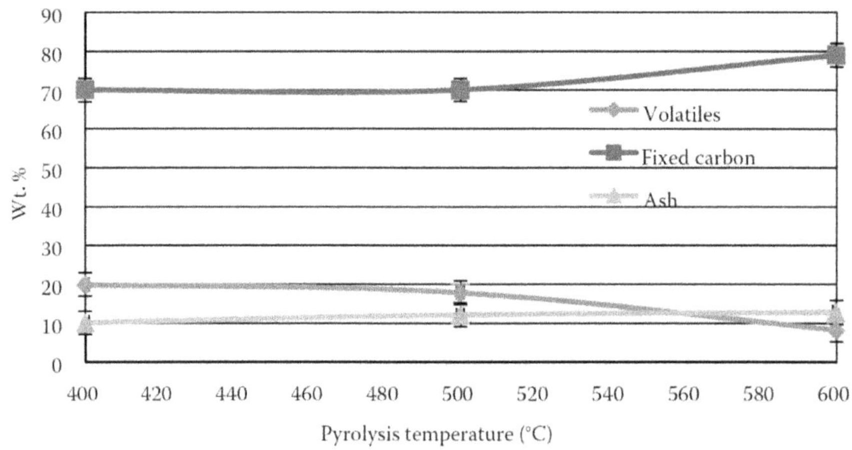

FIGURE 9.13 Proximate analyses of biochar with pyrolysis temperatures.

temperatures were characterized in terms of their proximate analysis and shown in Figure 9.13. The ultimate analysis for the biochar is presented in Table 9.11. The volatiles present in char decrease from 20% to 17.8% from 400°C to 500°C, respectively, and from 17.8% to 8.2% from 500°C to 600°C, respectively. The fixed carbon content remains similar between 400°C and 500°C but increases to 79% at 600°C. The ash content increases slightly with increasing pyrolysis temperature. This can be explained by the higher pyrolysis temperature removing volatiles, and as a result, the carbon content is slightly higher.

The biochar elemental analysis showed that, for an increase in pyrolysis temperature, the carbon content increased while oxygen and hydrogen content decreased (see Table 9.11). Losses in hydrogen and oxygen may be explained by breaking the weaker bonds within the char's structure with the consequence that the char becomes highly carbonaceous at higher pyrolysis temperatures. The HV of char increases slightly with an increase in pyrolysis temperature. This may be due to the slight increase in the carbon content of biochar with increased temperature.

The reactivity and combustion behavior of biochar is strongly affected by its surface area. By increasing the pyrolysis temperature from 400°C to 600°C, the surface area of the char is increased from 0.1 to 1.0 m^2/g (see Table 9.12). Higher pyrolysis

TABLE 9.11
Elemental Composition, HV, Surface Area, and Total Pore Volume of Char

Pyrolysis T (°C)	C (%)	H (%)	N (%)	O (%)	S (%)	HV (MJ/kg)	Surface Area (m^2/g)	Pore Volume (cm^3/g)
400	75.2	4.9	1.9	17.7	0.3	28.9	0.1	0.6
500	78.3	3.6	1.3	16.5	0.3	29.0	0.9	0.7
600	82	2.4	1.2	14.1	0.3	29.4	1.0	0.7

TABLE 9.12
Complete Corn Stover Biomass Characterization before Pyrolysis Studies

Biomass Characteristics	Experimental Value
Proximate Analysis (%w/w, d.b.)	
Ash	6.62 ± 0.27
Volatile matter (VM)	78.7 ± 1.4
Fixed carbon (FC)	14.7 ± 1.7
Moisture content (%w/w)	6.18 ± 0.14
Ultimate Analysis (%w/w, d.b.)	
C	44.3 ± 0.4
H	6.28 ± 0.09
N	0.80 ± 0.06
S	0.23 ± 0.06
O (by difference)	41.8 ± 0.6
Compositional Analysis (%w/w, d.b.)	
Lignin	13.1 ± 0.1
Extractives	18.1 ± 0.4
Glucan	31.9 ± 0.3
Xylan	19.6 ± 0.2
Galactan	1.0 ± 0.1
Arabinan	2.3 ± 0.1
Protein	5.8
HV (MJ/kg)	17.0 ± 0.7

temperatures cause higher devolatilization, resulting in additional pore volume in the samples, and, in turn, greater surface area for reaction or adsorption activities. The total pore volume per gram of sample is also shown to increase with temperature.

Example 9.9 shows how to estimate the biochar yields per unit weight of biomass.

Example 9.9: Calculating Biochar Yields per Unit Weight of Biomass

Estimate the biochar yields at 400°C, 500°C, and 600°C operating temperatures for switchgrass in units of kg/tonne (lb/ton). Use the data where the biochar yields are 49%, 42%, and 25% by weight (w/w), respectively. Convert this into the English system unit (lb/ton) at the three reaction temperatures.

Solution:

a. At 400°C, the calculations are as follows:

$$\text{Biochar yield (kg/tonne)} = \frac{0.49\,\text{kg}}{\text{kg biomass}} \times \frac{1{,}000\,\text{kg}}{\text{tonne}} = 490\ \text{kg/tonne}$$

$$\text{Biochar yield (lb/ton)} = \frac{490\,\text{kg}}{\text{tonne}} \times \frac{2.21\,\text{b}}{1\,\text{kg}} \times \frac{0.909\ \text{tonne}}{\text{ton}} = 980\ \text{lb/ton}$$

b. At 500°C, the calculations are illustrated as follows:

$$\text{Biochar yield (kg/tonne)} = \frac{0.42\,\text{kg}}{\text{kg biomass}} \times \frac{1{,}000\,\text{kg}}{\text{tonne}} = 420\ \text{kg/tonne}$$

$$\text{Biochar yield (lb/ton)} = \frac{420\ \text{kg}}{\text{tonne}} \times \frac{2.21\ \text{b}}{1\,\text{kg}} \times \frac{0.909\ \text{tonne}}{\text{ton}} = 840\ \text{lb/ton}$$

c. At 600°C, the calculations are shown next:

$$\text{Biochar yield (kg/tonne)} = \frac{0.25\ \text{kg}}{\text{kg biomass}} \times \frac{1{,}000\ \text{kg}}{\text{tonne}} = \frac{250\ \text{kg}}{\text{tonne}}$$

$$\text{Biochar yield (lb/ton)} = \frac{250\,\text{kg}}{\text{tonne}} \times \frac{2.21\,\text{b}}{1\,\text{kg}} \times \frac{0.909\ \text{tonne}}{\text{ton}} = 500\ \text{lb/ton}$$

9.7.3 Corn Stover Pyrolysis Studies (Capunitan and Capareda, 2012)

Pyrolysis of corn stover was carried out in a pressurized batch reactor to investigate the biofuel production potential of this biomass waste. Product yields and characteristics and conversion efficiencies were evaluated at temperatures of 400°C, 500°C, and 600°C, and at moderate pressure. Maximum biochar yield of 37.3 wt% and liquid product yield of 31.4% were obtained at 400°C, whereas the gas yield was maximum at 600°C (21.2 wt%). Biochar characterization through its energy content, proximate analysis, and ultimate analysis indicated its potential as an alternative solid fuel, as revealed in the Van Krevelen diagram showing a high atomic H/C ratio and a low atomic O/C ratio. FTIR and GC-MS analyses of the bio-oil identified major components as phenolic compounds, with significant proportions of aromatic and aliphatic compounds. The gas product consisted mainly of methane, hydrogen, and carbon dioxide, with energy content ranging from 10.1 to 21.7 MJ/m^3. Mass and energy conversion efficiencies were evaluated and indicated thatmost of the mass and energy contained in the feedstock was transferred to the biochar. Results revealed that pyrolysis of corn stover produced potentially valuable sources of fuel and chemical feedstock.

The overall product yield is illustrated in Figure 9.14. The complete biomass characterization is summarized in Table 9.12. The bio-oil was analyzed. The results are summarized in Table 9.13, revealing a quite high carbon content. However, the oxygen content and acid numbers were also quite high, indicating the need to upgrade

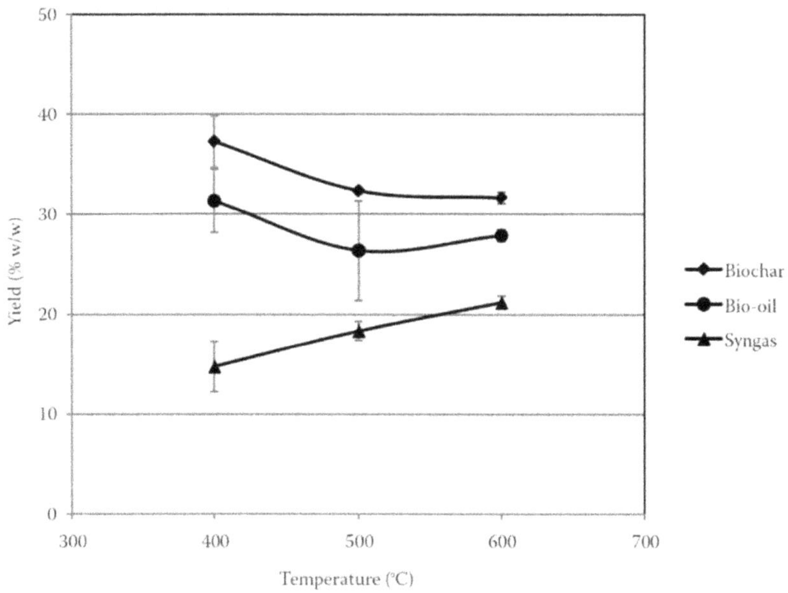

FIGURE 9.14 Yields of biochar, bio-oil, and syngas from batch pyrolysis of corn stover.

TABLE 9.13
Some Corn Stover Bio-Oil Properties

Parameter	Value
Ultimate Analysis (%w/w)	
C	78.0 ± 1.1
H	9.02 ± 0.12
N	1.86 ± 0.03
S	0.38 ± 0.10
Ash	0.18 ± 0.06
O (by diff.)	10.6 ± 1.4
H/C molar ratio	1.38
O/C molar ratio	0.10
Acid number (mg KOH/g)	23.8 ± 3.7
Moisture content (%w/w)	12.8 ± 0.8

the processes. The HVs of the bio-oil ranged from 33.8, 32.6, and 27.9 MJ/kg (14,563, 14,046, and 12,020 Btu/lb) at temperatures of 400°C, 500°C, and 650°C, respectively. Example 9.10 illustrates how to calculate the bio-oil yield from corn stover per unit of biomass weight.

Example 9.10: Calculating Bio-Oil Yield per Unit Weight of Biomass

Convert the bio-oil yield of corn stover at 400°C temperature of 31.4% into L/tonne (and gal/ton) assuming a bio-oil density of 0.9 kg/L. Compare this value with yields from switchgrass at the same temperature.

Solution:

a. At 400°C, the calculations are as follows:

$$\text{Bio-oil yield (kg/tonne)} = \frac{0.314\text{ kg}}{\text{kg biomass}} \times \frac{1{,}000\text{ kg}}{\text{tonne}} = 314\text{ kg/tonne}$$

$$\text{Bio-oil yield (L/tonne)} = \frac{314\text{ kg}}{\text{tonne}} \times \frac{\text{L}}{0.90\text{ kg}} = 249\text{ L/tonne}$$

b. At 400°C, in English system, the calculations are illustrated as follows:

$$\text{Bio-oil yield (lb/ton)} = \frac{314\text{ kg}}{\text{tonne}} \times \frac{2.2\text{ lb}}{1\text{ kg}} \times \frac{0.909\text{ tonne}}{\text{ton}} = 628\text{ lb/ton}$$

$$\text{Bio-oil yield}\left(\text{gal/ton}\right) = \frac{628\text{ lb}}{\text{ton}} \times \frac{\text{L}}{0.90\text{ kg}} \times \frac{1\text{ kg}}{2.2\text{ lb}} \times \frac{\text{gal}}{3.785\text{ L}} = 84\text{ gal/ton}$$

c. At 400°C, the switchgrass bio-oil yield was 22% and the calculations are shown below:

$$\text{Bio-oil yield}\left(\text{kg/tonne}\right) = \frac{0.22\text{ kg}}{\text{kg biomass}} \times \frac{1{,}000\text{ kg}}{\text{tonne}} = 220\text{ kg/tonne}$$

$$\text{Bio-oil yield (L/tonne)} = \frac{220\text{ kg}}{\text{tonne}} \times \frac{\text{L}}{0.90\text{ kg}} = 244\text{ L/tonne}$$

$$\text{Bio-oil yield (lb/ton)} = \frac{220\text{ kg}}{\text{tonne}} \times \frac{2.2\text{ lb}}{1\text{ kg}} \times \frac{0.909\text{ tonne}}{\text{ton}} = 440\text{ lb/ton}$$

$$\text{Bio-oil yield}\left(\text{gal/ton}\right) = \frac{440\text{ lb}}{\text{ton}} \times \frac{\text{L}}{0.90\text{ kg}} \times \frac{1\text{ kg}}{2.2\text{ lb}} \times \frac{\text{gal}}{3.785\text{ L}} = 59\text{ gal/ton}$$

d. At this temperature, the corn stover bio-oil yield is higher.

9.7.4 Sorghum Biomass Pyrolysis Studies (Santos and Capareda, 2012)

Conventional pyrolysis of high tonnage energy sorghum was performed at an elevated pressure and at different temperatures, from 400°C to 600°C, with heating rates of 4°C/min. The changes in temperature did not affect the product distribution but significantly affected product characteristics. The H/C molar ratio of bio-oil was at 1.51, whereas the O/C molar ratio was six times lower than the O/C molar ratio of the sorghum biomass (0.11 vs. 0.67). About 50% of the bio-oil comprised aliphatic hydrocarbons (i.e., alkanes and alkenes) and aromatic compounds (i.e., phenolic and benzene derivatives). The carbon distribution lies between C_8 and C_{44}, whereas the distribution of straight-chain alkanes exhibited a maximum in the range of C_8 to C_{21}, which is found to be in the range of gasoline and diesel fuels. The product biochar from sorghum biomass is a carbon-rich material with C concentrations of 49–52 wt% and classified under the low to medium volatile bituminous coal group according to ASTM standard D388. The produced non-condensible gas consisted mainly of CH_4 and CO_2, with small amounts of CO, H_2, and C_2H_6.

The characteristics of the sorghum biomass used in the study are summarized in Table 9.14. The non-condensible gas, bio-oil, and biochar yields are shown in

TABLE 9.14
Proximate, Ultimate, HV, and Structural Analyses of High-Tonnage Sorghum Biomass

Parameter	As Received	Dry Basis	Dry Ash-Free Basis
Proximate Analysis (wt.%)			
Moisture	12.80		
Ash	8.52	9.77	
Volatile matter	74.95	85.95	95.26
Fixed carbon (by difference)	3.73	4.28	4.74
Ultimate Analysis (wt.%)			
Moisture	12.80		
C	38.10	43.69	48.42
H	5.57	6.39	7.08
N	0.91	1.04	1.16
S	0.18	0.21	0.23
O (by difference)	33.92	38.90	43.11
Ash	8.52	9.77	
Gross HV (MJ/kg)	14.8	17.0	18.9
Composition of Lignocellulosic Material (wt.%)			
Cellulose	29.26		
Hemicellulose	19.97		
Lignin	19.27		
Extractives	18.8		

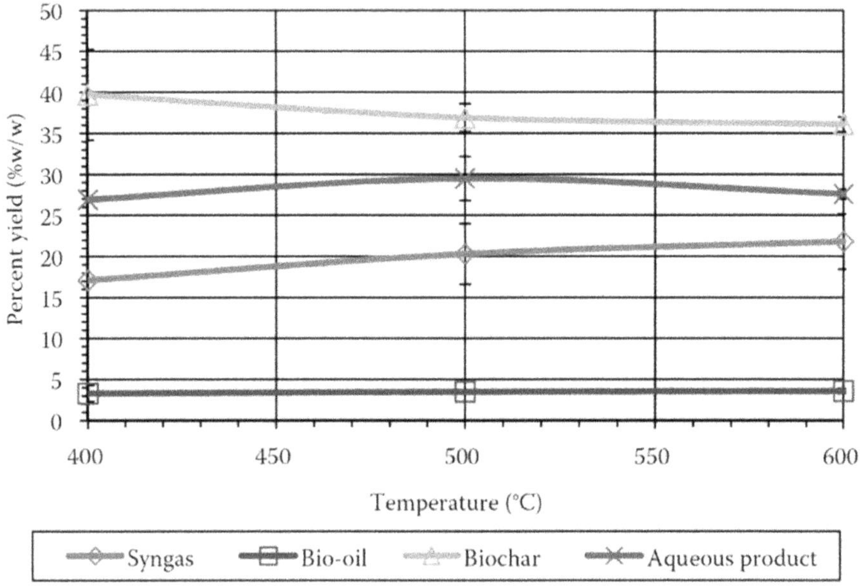

FIGURE 9.15 Non-condensible (NC) gas, bio-oil, biochar, and aqueous products as function of pyrolysis temperature.

Figure 9.15. The graph displayed a trend like other biomass, such as increased syngas yield as temperature is increased and decreased biochar yield as temperature is increased. For bio-oil, the peak yield is at 500°C. The proximate analysis of the biochar at various operating temperatures is summarized in Table 9.15. Similarly, the report discussed in corn stover pyrolysis, more than half of the energy is still contained in the biochar samples. The bio-oil, biochar, and syngas yields are summarized in Table 9.16 by weight basis. Note that in all the processes discussed, the operating unit was a batch reactor; thus, the results are quite conservative when they are compared with those using a fluidized bed pyrolysis system.

TABLE 9.15
Proximate Analysis of Biochar as Function of Pyrolysis Temperature

Property	Temperature (°C)		
(%wt.)	400	500	600
Moisture content	3.33 ± 0.58	1.00 ± 0.00	1.67 ± 0.58
Volatile matter	22.67 ± 1.53	14.67 ± 1.53	9.00 ± 1.73
Ash	36.67 ± 2.52	33.00 ± 4.00	35.00 ± 3.00
Fixed carbon	37.33 ± 4.04	51.00 ± 5.29	54.67 ± 4.04

TABLE 9.16
Yields of Pyrolysis Coproducts as Function of Pyrolysis Temperature

	Yields (wt.%) at Given Temperature (°C)		
Coproduct and Losses	**400**	**500**	**600**
Biochar	39.7	36.9	36.1
Bio-oil	30.2	33.0	31.2
Syngas	17.1	20.3	21.8
Losses	13	9.8	10.9

9.8 CONCLUSION

Thermal pyrolysis of biomass is becoming a very important and popular conversion process because of the wide range of biofuels that can be produced. The quantities of solid, liquid, and gaseous products can be changed by simply varying the heating rate or the reaction temperature. It was shown that a high temperature will likely yield higher gaseous content, whereas low-temperature conditions lead to increased biochar production. The liquid bio-oil product is produced at medium temperature in the range between 400°C and 600°C and must take place at the most rapid heating rate of a few seconds or less.

However, the liquid bio-oil product is composed of hundreds of compounds with varying chemical groups of classes. The produced bio-oil must be upgraded in order to qualify as "refinery-grade" bio-oil. Refinery-grade bio-oil products must be void of moisture, low in acid content, and contain minimal oxygenates, necessitating bio-oil upgrade processes.

There are various approaches to upgrading raw bio-oil. Most upgrade processes involve a series of solvent extraction or distillation steps, whereby some hydrocarbons may be separated. Ultimately, the bio-oil must undergo catalytic processing to bring the remaining compounds into hydrocarbons that meet the requirements of transport fuels, or what one may call "drop-in" fuels. These are quite intensive processes due to the high cost of some novel-metal-based catalysts. It was reported that these catalysts cost around $1,600/troy ounce for palladium. Note that a troy ounce is around 31.1 g, and this amounts to about $48/g or $48,000/kg ($21,800/lb). The total percentage of these metals in catalysts is rather small—around 5% at the most. However, the catalysts must be recovered during the upgrading process. There are now serious efforts to find catalysts that can be easily recovered during the conversion process or catalysts that would not be poisoned (or coked) during the conversion process. There are various homogeneous catalysts that can be easily recovered during the conversion process, by changing the conditions of the reaction, thus making them potential candidates for bio-oil upgrading.

Several companies are pooling their resources to investigate pyrolysis as a process for converting biomass into hydrocarbon fuels, such as MetrohmAG (Florida, USA), Dynamotive (Vancouver, BC, Canada), RLC Technologies, Inc. (Virginia, USA), and Pyrolyx AG (Indiana, USA). More research needs to be undertaken. There is hope for continued funding by government agencies such as the USDOE and Department of Defense; also, US National Laboratories such as the Pacific Northwest National Laboratories and others will provide additional research funding for the more difficult job of removing oxygenates from bio-oil by using the most cost-effective catalytic processes. The situation is analogous to the economy of ethanol production from lignocellulosics, which is linked to the cost of enzymes per liter or gallon of biofuels produced. Similarly, bio-oil upgrade costs to turn this valuable liquid into hydrocarbon depends on the availability of cost-effective catalysts that can be easily recovered during the process.

9.9 PROBLEMS AND DISCUSSION ISSUES

9.9.1 Effect of Temperature on Syngas Production in Pyrolysis

P9.1 Explain why gas production is increased as pyrolysis temperature and residence times are increased.

9.9.2 Pyrolysis Conditions that Favor Biochar and Bio-Oil Production

P9.2 Discuss the process condition that will favor the production of (a) char and (b) bio-oil. What are the expected yields for the other by-products from each process?

9.9.3 Advantages and Disadvantages of Pyrolysis Process

P9.3 List the associated advantages and disadvantages of using pyrolysis as a thermal conversion system for biomass feedstocks.

9.9.4 Problems and Issues of Pyrolysis Process at Extreme Temperatures

P9.4 Enumerate the problems and issues associated with the operation of a biomass pyrolyzer at extremely high temperatures.

9.9.5 Energy Balance in Pyrolysis Process

P9.5 Suppose that 0.454 kg (1 lb) of biomass with an HV of 17.4 MJ/kg (7,500 Btu/lb) was placed in a batch reactor and the furnace temperature was set at 550°C. About 0.15 kg (0.33 lb) of char was produced with an HV of 20.9 MJ/kg (9,000 Btu/lb). Determine the percentage of energy retained in the char from the original input material. Report all the units in English system as well.

9.9.6 Estimation of Residence Time in Fluidized Bed Reactor

P9.6 A fluidized bed pyrolyzer has an internal diameter of 0.3048 m (12 in). Finely ground biomass at a feed rate of 1 kg/min (2.2 lb/min) was introduced with an actual flow rate of 2.834 m^3/min (100 cfm). Determine the residence time of the biomass in a 91.44 cm (36 in) tall vertical reactor. Comment on the value of the residence time.

9.9.7 Estimation of Residence Time in Auger Reactor

P9.7 A 7.62 cm (3 in) auger pyrolyzer has a length of 91.46 cm (3 ft). Finely ground biomass at a feed rate of 0.909 kg/min (2 lb/min) was introduced at 1.2 rpm. Determine the residence time of the biomass if the linear flow velocity of the biomass is 1.418 m/min (4.65 ft/min). Is this considered a fast pyrolyzer?

9.9.8 Calibration Curve for Augers

P9.8 The following data was gathered while calibrating the feed rate of switchgrass in an auger in a pyrolyzer. Develop an equation relating the rpm of the auger with the feed rate. Estimate the regression coefficient. Do not forget to include an intercept. Report the relationship with the highest R^2. What would be the rpm so that a feed rate of 80 kg/h will be fed through the reactor?

Rpm	Mass Flow Rate (kg/h)
11	3.9
21	12.4
42	18.7
56	23.3
69	34.2
82	40.4
99	45.1
113	50.5
142	64.5
168	70

9.9.9 Calculation of HV of Syngas from Experiments

P9.9 Estimate the HV of syngas at NTP (normal temperature and pressure) from gas yield values (volumetric percentages) given in the table below for 400°C. Report units in MJ/m^3. Do the calculations for all trials and convert the units on weight basis (MJ/kg).

Temp 400°C	H_2	CO	CH_4	C_2H_4	C_2H_6	O_2	N_2	CO_2
Trial 1	3.02	6.93	6.78	1.16	4.40	0.68	5.36	59.93
Trial 2	1.87	5.02	4.91	0.95	3.08	0.51	7.49	63.27
Trial 3	1.30	4.87	3.68	1.04	2.31	2.17	13.02	62.46
HV (MJ/m^3)	12.8	12.6	38.8	63.0	69.7			
Density (kg/m^3)	0.0899	1.165	0.668	1.178	1.264			

9.9.10 Calculating Yields per Unit Weight of Biomass

P9.10 Estimate the biochar, bio-oil, and syngas yields at 400°C, 500°C, and 600°C operating temperatures for sorghum biomass in units of kg/tonne, L/tonne, and L/tonne, respectively. Use the data in Table 9.16 and assume the average density of bio-oil to be about 900 kg/m^3 and that of the syngas as 0.65 kg/m^3. Convert the bio-oil yield in gal/ton at the three reaction temperatures.

9.9.11 Ideal Circular Bioeconomy Feedstock

P9.11 The atmospheric carbon sequestration in plants through photosynthesis is part of the circular economy in capturing carbon dioxide in the atmosphere and converted into biomass that will replace fossil fuels. If the criteria is the capturing of highest carbon, which biomass listed below is the ideal feedstock for this carbon capture and sequestration project?

a. rice straw b. woodchips c. switchgrass d. cotton gin trash

9.9.12 Ideal Feedstock for Bio-Oil Production

P9.12 Hydrocarbon fuels are those with no oxygen in their chemical formula. Biomass feedstock is a highly oxygenated material. If one were to select a feedstock with the potential to produce the least oxygenated compounds, which of the following feedstock should one select?

a. poultry manure b. almond hulls c. sorghum biomass
d. alligator grass

9.9.13 Accurate Syngas Measurement

P9.13 Name the device that accurately measures and identifies the syngas components of a pyrolyzer reporting the percentage in volume of individual gases that may also be used to estimate its heating value.

a. gas chromatograph b. Campbell Stokes c. calorimeter
d. brix meter

9.9.14 Pyrolysis of Dairy Manure

P9.14 Develop a project to convert dairy manure into bio-oil, biochar, and syngas. The facility is a 1,000-head dairy farm. Using data from literature and this textbook, estimate the solid, liquid, and gaseous products formed including their potential commercial use.

9.9.15 Pyrolysis of Microalgae Biomass

P9.15 Among numerous other biomass resources that are land-based, microalgae had the highest reported biomass yield per unit of area. Propose and develop a project to grow microalgae in an open pond to generate biodiesel from its oil and pyrolyze the remaining cellulosic biomass to generate bio-oil. List all assumptions made and report biodiesel and bio-oil yield per acre.

9.10 CASE STUDIES

Evaluate a project to pyrolyze sorghum biomass into bio-oil, biochar, and syngas. These products will be sold on a commercial scale. Additional revenue comes from government tax credit. Develop a spreadsheet software and perform simple economic and sensitivity analyses. Data is listed in the table below.

Sorghum Acreage	**2024.30 ha**		
Collection Efficiency (%)	90%	Product Selling Prices	
Yearly Operation	350 days	Bio-oil	\$0.5284/L
Biomass Price	\$66/tonne	Biochar	\$110/tonne
Facility Cost	\$55,000/tonne/day	Syngas	\$176.44/1,000 m^3
Financial Parameters		Other Variable Costs	
Project Life	20 years	Labor Hours	10 h/day
Interest on Investment	8%	# of Personnel	6 full-time
Tax on Capital	2%	# of Manager	1 full-time
Discount Rate	2.36%	Labor Wage Rate	\$15/hr
Tax Credit	30%	Manager Salary	\$80,000/pax/yr
Product Yields (%W)		R&M Cost	10% of capital
Bio-oil	30%	Utilities Cost	15% of capital
Biochar	35%	Bio-oil Upgrade Cost	\$0.0664/L
Syngas	20%	Efficiency of Upgrade	80%
Biomass Yield	4.11 tonnes/ha/yr	Bio-oil Density	0.841 kg/L
Biomass:Waste Ratio	1:3	Waste Initial Moisture	25%
Syngas Density	0.95 kg/m^3	Waste Final Moisture	10%

Required:

a. Determine the product yields by weight and by volume.
b. What is the total plant cost?
c. Report on the yearly feedstock tonnage (tonnes/yr), actual waste yield (tonnes/yr), biomass input (tonnes/day) including wet residue weight (tonnes/ha/yr) and final wet waste weight prior to pyrolysis (tonnes/ha/yr).
d. Develop a simple spreadsheet to estimate economic parameters, such as internal rate of return (IRR), payback period (PBP), benefit cost ratio (BCR), and net present value (NPV), and report on the viability of this project given the data.
e. Perform sensitivity analyses to determine the impact of changing input parameters on the economic indicators. A graph for the relationship would be beneficial visually with predictive equation and with R^2 value n the chart. The following are suggested changes:
   i. acreage on NPV (1012.15–8097.17 ha)
   ii. biochar selling price versus IRR and PBP ($27.50–$330/tonne)
   iii. bio-oil selling price versus IRR and PBP ($0.03963–$0.9247/L)
   iv. syngas selling price versus IRR and PBP ($52.93/1,000–$352.88/1,000 m^3)
   v. tax credit on IRR and PBP (0%–40%)
   vi. capital cost on IRR and PBP ($27,500–$220,000/tonne/day)
   vii. interest on investment on IRR and PBP (5%–30%)
   viii. feedstock cost on IRR and PBP ($16.50–$110/tonne)

REFERENCES

Ammann, J. R., P. F. Britt and A. C. Buchanan. 1999. Pyrolysis of biomass model compounds using flash vacuum pyrolysis: Mechanisms of polycyclic aromatic hydrocarbon formation. USDOE Project Report under contract DE-AC05-960R22464 through Oak Ridge National Laboratory and Managed by Lockheed Martin Energy Research Corporation, Golden, CO.

Aquino, F. L., S. C. Capareda and C. B. Parnell, Jr. 2010. Elucidating the solid, liquid and gaseous products from batch pyrolysis of cotton gin trash. *Transactions of the ASABE* 53 (3): 651–658.

BASF. 2012. *EIB Precious Metal Pricing*, BASF Corporation, North America, Florham Park, NJ. Also available at: https://apps.catalysts.basf.com/apps/eibprices/mp/ (downloaded 4-19-2012).

Bridgewater, A. V. 2012. Review of fast pyrolysis of biomass and product upgrading. *Biomass and Bioenergy* 38: 68–94.

Capunitan, J. and S. C. Capareda. 2012. Assessing the potential for biofuel production of corn stover pyrolysis using a pressurized batch reactor. *Fuels* 95: 563–572.

Del Remedio Hernandez, M., A. N. Garcia, A. Gomez, J. Agullo and A. Marcilla. 2006. Effect of residence time on volatile products obtained in the HDPE pyrolysis in the presence and absence of HZSM-5. *Industrial & Engineering Chemistry Research* 45: 8770–8778.

Engineering Toolbox. 2012. Gases-Densities. Resources, tools and basic information for engineering and design of technical applications. Available at: https://www.engineeringtoolbox.com/gas-density-d_158.html. Last viewed on July 27, 2012.

Garcia-Perez, M., A. Chaala and C. Roy. 2002. Vacuum pyrolysis of sugarcane bagasse. *Journal of Analytical and Applied Pyrolysis* 65 (2): 111–136.
Iman, T. and S. C. Capareda. 2012. Characterization of bio-oil, syngas and biochar from switchgrass pyrolysis at various temperatures. *Journal of Analytical and Applied Pyrolysis*. 93: pp. 170–177.
Maguyon, M. and S. C. Capareda. 2012. Fast pyrolsis as a conversion pathway for algal biomass into biofuels. Paper presented at the 3rd International Conference on Algal Biomass, Biofuels and Bioproducts held on June 10–14, 2012 in San Diego, CA.
Meier, D., S. Schocll and H. Klaubert. 2005. New developments in ablative fast pyrolysis at PYTEC, Germany. *Aston University Birmingham, ThermalNet Newsletter* 1: 4.
Peacocke, G. V. C. and A. V. Bridgewater. 1994. Ablative plate pyrolysis of biomass for liquids. *Biomass and Bioenergy* 7: 147–154.
Santos, B. S. and S. C. Capareda. 2012. Energy sorghum pyrolysis using a pressurized batch reactor. Paper submitted for review in *Fuel*. Under review. Manuscript No. JFUE-D-12-01046.
Waldheim, L. and T. Nilsson. 2011. Heating value of gases from biomass gasification. Report prepared for IEA Bioenergy Agreement, Task 20-Thermal Gasification of Biomass. Paris, France. May 2011.
Zhang, X. H., T. J. Wang, L. L. Ma and J. Chang. 2008. Vacuum pyrolysis of waste tyres with basic additives. *Waste Management* 28 (11): 2301–2310.

ASSESS YOUR KNOWLEDGE

1. What percentage of liquid is typically produced in a fast pyrolysis process with moderate temperature and short hot vapor residence time of about a second?
   a. 5
   b. 30
   c. 75
   d. 50

2. In pyrolysis process, when the conditions are slow heating and very low temperatures, what should be the dominant product?
   a. Bio-oil
   b. Syngas
   c. biochar
   d. ash

11. One lb of rice straw with a heating value of 6,550 Btu/lb was pyrolyzed at a temperature of 932°F. About 0.35 lb of char was produced with an energy content of 8,000 Btu/lb. Determine the percentage of energy retained in the char.
    a. 42.75
    b. 36.4
    c. 73.6
    d. 25.9

12. Determine the bio-oil yield in gallons per ton of biomass if during the slow pyrolysis experiment, the density of the resulting bio-oil was found to be about 8.33 lbs/gal. The bio-oil yield was 33% by weight.
    a. 88.5
    b. 660
    c. 79.2
    d. 7.91

3. This type of pyrolysis process has heating rates of about 1,000°C/min with reaction times of a few seconds.
   a. Slow pyrolysis
   b. Flash pyrolysis
   c. Fast pyrolysis
   d. Ablative pyrolysis

4. This is a type of pyrolysis process whereby biomass material is heated in a pressure below atmospheric to reduce boiling point or reaction temperature and reduce adverse chemical reactions. Name this pyrolysis process.
   a. Slow pyrolysis
   b. Ablative pyrolysis
   c. Flash pyrolysis
   d. Vacuum pyrolysis

13. A fluidized bed pyrolyzer has an internal diameter of 3 inches and finely ground biomass was pyrolyzed and fed at an actual flow rate of 5 CFM. Determine the residence time (in seconds) of this biomass in a 24-inch tall reactor.
   a. 1.31
   b. 2.62
   c. 4.25
   d. 1.18

14. About 1 lb of microalgae sample was pyrolyzed and the following data was obtained. The density of syngas was found to be 0.04 lb/ft^3. The heating value of dry algae samples was 11,500 Btu/lb, the syngas yield was 10% by weight with an energy content of syngas of 500 Btu/ft^3. The biochar yield was 35% by weight with a biochar heating value of 9,000 Btu/lb. The bio-oil yield was 30% by weight and its heating value was 18,000 Btu/lb. Determine the amount of energy (in Btu) that went to the bio-oil.
   a. 3,150
   b. 5,400
   c. 1,250
   d. 3,450

5. This is a type of pyrolysis process that uses dielectric heating. Heating is because of flipping of the orientation of the electric dipoles in the material being heated as the AC field of the microwave changes direction. The process does not require drying and heat transfer occurs directly within the feedstock by convection and not conduction. Name this pyrolysis process.
   a. Catalytic pyrolysis
   b. Flash pyrolysis
   c. Plasma pyrolysis
   d. Microwave pyrolysis

6. This is a type of pyrolysis process that uses methane at very high pressure and temperature. Contact time should be less than 10 s and temperature close to 1,000°C/min and occurs very quickly in a flash mode.
   a. Plasma pyrolysis
   b. Flash hydro pyrolysis
   c. Flash methanolysis
   d. Microwave pyrolysis

15. Convert the bio-oil yield in gallons per ton of biomass if during the slow pyrolysis experiment, the density of the resulting bio-oil was found to be 0.95 g/mL and the bio-oil yield was 35% by weight.
   a. 60
   b. 88.5
   c. 35.5
   d. 120

16. Determine the heating value of the gaseous product from pyrolysis process with the following data by volume: H_2 = 9.765%, with heating value of H_2 = 344 Btu/ft^3; CO = 11.203% with heating value of CO = 338 Btu/ft^3; CH_4 = 2.492% with heating value of CH_4 = 1042 Btu/ft^3; C_2H_4 = 0.791% with heating value of C_2H_4 = 1692 Btu/ft^3; C_2H_6 = 0.200% with heating value of C_2H_6 = 1872 Btu/ft^3. For this question, what is the heating value of the syngas in units of Btu/ft^3?
   a. 114.6
   b. 350.5
   c. 254.8
   d. 560

7. This is a type of pyrolysis process that uses hydrogen at very high pressure and usually at around 20 Mpa. Contact time should be less than 10 s and temperature above 700°C or close to 1,000°C/min and occurs very quickly in a flash mode.
   a. Plasma pyrolysis
   b. Flash hydro pyrolysis
   c. Flash methanolysis
   d. Microwave pyrolysis

8. Name a catalyst that lowers the temperature required for pyrolysis.
   a. $ZnCl_2$
   b. Zeolite
   c. Sulfide nickel
   d. Platinum

17. Switchgrass was pyrolyzed at 400°C and the following syngas compositions were found and with the following data by volume: H_2=28%, with heating value of H_2 = 344 Btu/ft^3; CO = 21.7% with heating value of CO = 338 Btu/ft^3; CH_4 = 39.2% with heating value of CH_4 = 1042 Btu/ft^3; C_2H_4 = 1.2% with heating value of C_2H_4 = 1692 Btu/ft^3; C_2H_6 = 0.8% with heating value of C_2H_6 = 1872 Btu/ft^3. For this question, what is the heating value of the syngas in units of Btu/ft^3?
   a. 613.5
   b. 114.6
   c. 254.8
   d. 560

18. Switchgrass was pyrolyzed at 400°C. The products yield by weight at this temperature are as follows: The bio-oil yield was 22% (HV=15,619 Btu/lb), syngas yield was 8% (HV=613.5 Btu/ft^3 and density of 0.05 lbs/ft^3), and biochar yield was 48% (HV=12,452 Btu/lb). The high heating value of the switchgrass used in this experiment was 12,374 Btu/lb. For this question, determine the energy content of the biochar in units of Btu/lb.
   a. 3436
   b. 981.6
   c. 4221
   d. 5977

9. This is a thermal depolymerization process used to convert wet biomass into crude-like oil, sometimes referred to as bio-oil or biocrude under moderate temperature and high pressure. It was originally developed at the Pittsburgh Energy Research Center (PERC) and later demonstrated (at the 100kg/h scale) at the Albany Biomass Liquefaction Experimental Facility at Albany, Oregon, USA. Name this thermal conversion process.
   a. HTL.
   b. Thermolysis
   c. Methanolysis
   d. Flash pyrolysis

10. Name a type of horizontal reactor used for pyrolysis process.
    a. Vortex
    b. Downflow
    c. Fluidized bed
    d. Lurgi

19. About 1 lb of microalgae sample was pyrolyzed, and the following data was obtained. The density of syngas was found to be 0.04 lb/ft^3. The heating value of dry algae samples was 11,500 Btu/lb, the syngas yield was 10% by weight with an energy content of syngas of 500 Btu/ft^3. The biochar yield was 35% by weight with a biochar heating value of 9,000 Btu/lb. The bio-oil yield was 30% by weight and its heating value was 18,000 Btu/lb. For this question, determine the amount of energy (in Btu) went to the biochar.
    a. 1,250
    b. 5,400
    c. 3,150
    d. 3,450

20. Determine the heating value of the gaseous product from pyrolysis of switchgrass at 500°C with the following data by volume: H_2=25.4%, with heating value of H_2= 344 Btu/ft^3; CO=21.5% with heating value of CO=338 Btu/ft^3; CH_4=39.4% with heating value of CH_4=1042 Btu/ft^3; C_2H_4= 2.3% with heating value of C_2H_4= 1692 Btu/ft^3; C_2H_6=1.9% with heating value of C_2H_6= 1872 Btu/ft^3. What is the heating value of the syngas in units of Btu/ft^3?
    a. 114.6
    b. 645.1
    c. 254.8
    d. 560

Answers: 1c, 2c, 3b, 4d, 5d, 6c, 7b, 8a, 9a, 10d, 11a, 12c, 13d, 14b, 15b, 16a, 17a, 18d, 19c, 20b

ADDITIONAL LEARNING RESOURCES

- Access the Lecture Notes at the Instructor and Student Resources: https://routledgetextbooks.com/textbooks/instructor_downloads/
- The Case Study Videos can be found in the ebook+ version of this book.

10 Gasification

LEARNING OBJECTIVES

Upon completion of this chapter, one should be able to

1. Describe the concept of gasification and the products formed during the biomass conversion process.
2. Identify various types of biomass gasifiers and the various conversion zones in a gasifier.
3. Discuss how synthesis gas is formed through the gasification process and classify the basic syngas composition.
4. Describe the temperature profiles in various types of gasifiers.
5. Differentiate the various applications of gasification technology.

10.1 INTRODUCTION

Gasification is a thermal biomass conversion process. At an elevated temperature, a controlled amount of oxidant reacts with the biomass to generate a low calorific value (LCV) and combustible synthesis gas. The synthesis gas is comprised of primarily carbon monoxide and hydrogen. This gas is also termed "syngas" or "producer gas." For proper operation of a gasifier, it is necessary to know the exact amount of air that will be used during the conversion process. The elemental composition of the biomass (that is to be gasified) must be known to determine the process stoichiometry. This can be accomplished by evaluating the ultimate analysis of the biomass. Data already available for certain feedstocks may be used for determining the amount of air required for the gasification reaction to proceed. For example, from the elemental composition and ash content given for sorghum feedstock (see Table 10.1), the number of moles of each element can be determined by dividing the dry weight

TABLE 10.1
Table of Calculations for Establishing Chemical Formula for Sorghum Stalks

Composition	% Dry Weight	MW	Mol/wt
C	40.0	12	3.33
H	5.2	1	5.2
O	40.7	16	2.54
N	1.4	14	0.1
S	0.2	32	0.006
Ash	12.5	-	
Total	100%		

DOI: 10.1201/9781003294306-10

percentage by its atomic weight. The mole ratio of the elements C, H, O, N, and S in sorghum stalks can thus be obtained to arrive at a "chemical formula" for sorghum stalks, which can be used as a basis to estimate the air-to-fuel (A/F) ratio from the stoichiometric calculations.

Thus, the chemical formula for sorghum stalks is calculated to be $C_{3.3}H_{5.2}O_{2.54}N_{0.1}S_{0.006}$ (see Table 10.1). This formula may also be written as $CH_{1.6}O_{0.77}N_{0.03}S_{0.002}$ to be more consistent with the convention in chemical engineering textbooks. This simplification is made by dividing all the other subscripts by that of the carbon (i.e., 3.3). The balanced chemical equation for complete combustion of sorghum stalks is thus as shown in Equation 10.1. This equation assumes that air is composed of 21% oxygen and 79% nitrogen; hence, the coefficient for N_2 is 3.76 (i.e., 79/21).

$$\begin{aligned} C_{3.33}H_{5.2}O_{2.54}N_{0.1}S_{0.006} + 3.36\left[\left(O_2 + 3.76\,N_2\right)\right] &\rightarrow 3.33\,CO_2 \\ &+ 2.6\,H_2O + 3.36 \times 3.76\,N_2 + 0.006\,S \end{aligned} \tag{10.1}$$

The A/F ratio, defined as the mass of air ($O_2 + N_2$) used, divided by the mass of fuel consumed, may be calculated as shown in Equation 10.2. The number of moles of each element is multiplied by its molecular weight to estimate the mass of the air and fuel consumed.

$$\begin{aligned} \frac{\text{Air}}{\text{Fuel}} &= \frac{3.36 \times (32) + 12.634 \times (28)}{(12) \times 3.33 + (1) \times 5.2 + (16) \times 2.54 + (14) \times 0.1 + (32) \times 0.006} \\ &= \frac{461.27}{87.39} = 5.28 \end{aligned} \tag{10.2}$$

Thus, if 1 kg of sorghum biomass is to be completely combusted, about 5.28 kg of air is required. The total heat generated for this combustion process is equivalent to the heating value of the biomass and measured by an appropriate calorimeter. Gasification occurs when not enough oxygen is present in the system: in this case, well below 5.28 kg of air per kilogram of biomass. Normally, 30%–70% of the stoichiometric air requirement is used to carry out a gasification process. Thus, for sorghum stalks, the amount of air necessary for gasification should be in the range of 1.6–3.7 kg of air per kilogram of sorghum stalks. Gasification is an endothermic continuous process, because a portion of biomass is used to provide heat for the process. For pyrolysis, external heat is required for a continuous process. However, gasification does not require a continuous supply of heat. Example 10.1 shows the procedure in preparing cotton gin trash for the gasification process.

Other parameters used in gasification processes are the fuel-to-air (F/A) ratio and the equivalence ratio. The F/A ratio is the inverse of the A/F ratio defined above, whereas the fuel-to-air equivalence ratio (ϕ) is defined as the ratio of the actual F/A ratio divided by the stoichiometric F/A ratio (Equation 10.3). Thus, if the equivalence ratio is less than 1, the process is said to be on a fuel-lean condition, whereas at an equivalence ratio of greater than 1, the reaction is said to be on a fuel-rich condition.

$$\phi = \frac{(\text{F/A ratio})_{\text{actual}}}{(\text{F/A ratio})_{\text{stoichiometric}}} \tag{10.3}$$

Example 10.1: Determining Chemical Formula for Cotton Gin Trash

The ultimate analysis for cotton gin trash is given as follows: C=42%, H=5.4%, N=1.4%, S=0.5%, O=35%, and the ash content is 14.5%. Determine the chemical formula for cotton gin trash.

Solution:

a. Using the procedure outlined above, assuming 100 g sample, the mole fractions are as follows:

$$C = 42/12 = 3.5$$

$$H = 5.4/1 = 5.4$$

$$N = 1.4/14 = 0.10$$

$$S = 0.5/32 = 0.016$$

$$O = 35/16 = 2.2$$

b. The "chemical formula" for cotton gin trash is shown in the suggested format:

$$C_{3.5}H_{5.4}O_{2.2}N_{0.10}S_{0.016}$$

$$CH_{1.5}O_{0.63}N_{0.029}S_{0.0045}$$

This chapter discusses the basic chemistry of biomass gasification, the various types and designs of biomass gasifiers, the advantages and disadvantages of each system, and their applications. Advanced gasification processes will be discussed in Chapter 11.

10.2 CHEMISTRY OF BIOMASS GASIFICATION

The chemistry of biomass gasification can be easily explained by delineating the thermal conversion zones in an updraft gasifier, as shown in Figure 10.1 with corresponding temperature profile. As air enters through the gasifier, it encounters the

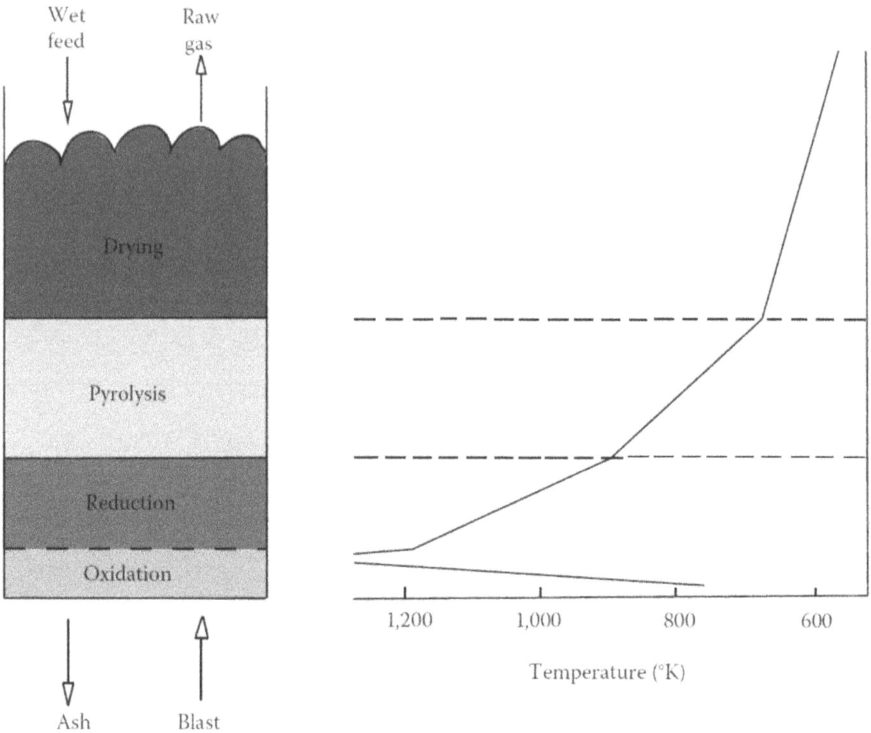

FIGURE 10.1 Various conversion zones in a gasifier and corresponding temperature profile.

heated biomass. Carbon in the biomass reacts with the oxygen to produce CO_2 and thermal energy, in the form of heat. This initial part of the gasifier is called the oxidation region, or sometimes, the hearth zone or combustion zone. The oxidation reaction is illustrated by Equation 10.4.

$$C + O_2 \rightarrow CO_2 + \text{Heat} \tag{10.4}$$

In the (next zone) reduction zone, the carbon dioxide produced reacts with carbon in the biomass to produce a major component of synthesis gas, carbon monoxide (Equation 10.5), followed by the reaction of moisture with carbon (Equation 10.6). This is a very important and desirable reaction in a gasifier.

$$CO_2 + C \rightarrow CO \tag{10.5}$$

$$H_2O + C \rightarrow CO + H_2 \tag{10.6}$$

Equation 10.6 shows that the amount of char formed in a gasifier is relatively small. Although hydrogen (H_2) gas and carbon monoxide (CO) (also called producer gas, synthesis gas, or syngas in short) are the most abundant gases formed, some other intermediate hydrocarbon gases are also formed. These include the following: methane (CH_4), having the most abundant component of all the hydrocarbons, followed by ethylene (C_2H_4), ethane (C_2H_6), propane (C_3H_8), and propylene (C_3H_6). If air is used for the gasification process, the most predominant (but having no energy content) is nitrogen gas (N_2). As much as 50% nitrogen gas may be in the synthesis gas. Some excess oxygen (O_2) may be formed during the conversion process. These intermediate gases are formed if the A/F ratio is properly controlled. Example 10.2 shows how to calculate A/F ratio for a biomass resource. Example 10.3 shows the calculation of the amount of air required for gasification to occur. If the amount of air exceeded that required for complete combustion, the thermal conversion process is no longer considered gasification but full combustion.

Example 10.2: Determining Fuel-to-Air Ratio of Biomass Combustion

Determine the A/F ratio for the complete combustion of cotton gin trash with the ultimate analysis given in Example 10.1.

Solution:

a. The chemical formula for the cotton gin trash has already been calculated as shown here:

$$C_{3.5}H_{5.4}O_{2.2}N_{0.10}S_{0.016}$$

b. A balanced combustion equation is derived as follows:

$$C_{3.5}H_{5.4}O_{2.2}N_{0.1}S_{0.016}+3.75\left\lfloor\left(O_2+3.76N_2\right)\right\rfloor\rightarrow 3.5CO_2+2.7H_2O$$
$$+3.75\times 3.76N_2+0.016S$$

c. Then, the F/A ratio is calculated as follows:

$$\frac{\text{Air}}{\text{Fuel}}=\frac{3.75\times(32)+14.1\times(28)}{(12)\times 3.5+(1)\times 5.4+(16)\times 2.2+(14)\times 0.1+(32)\times 0.016}=\frac{514.8}{89.12}=5.78$$

Example 10.3: Determining Amount of Air for Biomass Gasification

Cotton gin trash is to be gasified. Determine the amount of air required in cubic meters per kilogram of material using the data in Examples 10.1 and 10.2. Assume that only 30% of the stoichiometric amounts of air will be used during the conversion process. Assume the density of air to be 1.2 kg/m^3.

Solution:

a. The stoichiometric amount of air is already known to be 5.78 kg air/kg biomass. Thus, if only 30% air is used, the amount of air required is just 1.73 kg (i.e., $5.78\times 0.30=1.73$).

b. Thus, the amount of air required in cubic meter units is as follows:

$$\frac{\text{Air}}{\text{Fuel}} = \frac{1.73\,\text{kg air}}{\text{kg biomass}} \times \frac{\text{m}^3}{1.2\,\text{kg air}} = 1.44\frac{\text{m}^3\ \text{air}}{\text{kg biomass}}$$

10.3 VARIOUS TYPES OF GASIFIERS

Gasification was developed more than 150 years ago (Bridgewater, 1984). It was used primarily to convert the chemical energy of solid fuel into gas by heating the solid fuel with partial oxygen. An LCV gas is produced, which can be combusted immediately on another location to generate sensible heat. The energy content is usually low and typically about 5 MJ/Nm^3 (134 Btu/ft^3). "Nm^3" stands for "normal" cubic meter, referring to standard or normal conditions. Air is normally used for the process, and thus, the main gaseous product has a high nitrogen content, thereby lowering its overall energy content. If a higher energy-value syngas is required, sometimes pure oxygen is used.

Gasifiers can be classified as either fixed-bed or moving-bed systems. The fixed-bed gasifiers are further classified according to the way the air is fed through the system. Thus, we may have downdraft, updraft, or cross-draft gasifiers. The moving beds can also be further classified into rotary kiln reactors, multiple hearths, entrained flow, fluidized bed, pressurized fluidized bed, or recirculating fluidized bed types. Fluidized bed types may further be classified into atmospheric and pressurized systems. In addition, the fluidizing medium used may be air, oxygen, steam, or a combination of these.

10.3.1 Updraft Gasifier

The simplest air gasifier is the updraft gasifier, schematically shown in Figure 10.2. As the name implies, air enters through the bottom of the fixed bed and exits from the upper portion of the reactor. The sensible heat of gas produced is used to dry and preheat the biomass on top of the hearth zone. The gas leaves uncracked with some tar formation. After the hearth zone is the reduction zone, then the pyrolysis or distillation zone, and the drying of biomass. The temperature profile shown in Figure 10.3 also exemplifies the processes involved as they relate to the amount of oxygen present across each thermal zone.

There are various ways of starting up this system. First, the reactor could be filled with biomass and start-up is done by burning part of the biomass at the bottom of the reactor in the hearth zone. Thus, provisions must be made in this area for this purpose. As soon as a large portion of the biomass is combusted, the blower is turned on to generate the synthesis gas and controlled effectively to generate consistent high-quality synthesis gas. Another way of starting up the gasifier is to burn part of the biomass in the hearth zone, and as soon as a large portion is combusted, the remaining biomass is loaded to fill the whole reactor, then the top lid is closed and the blower at the bottom of the reactor is turned on. There must also be a provision for

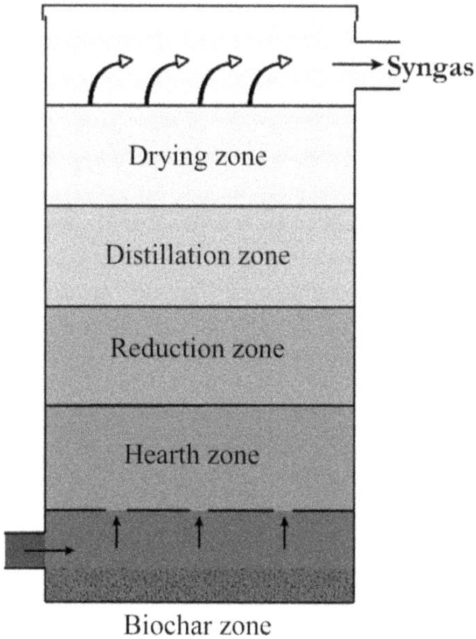

FIGURE 10.2 Schematic of updraft gasification process.

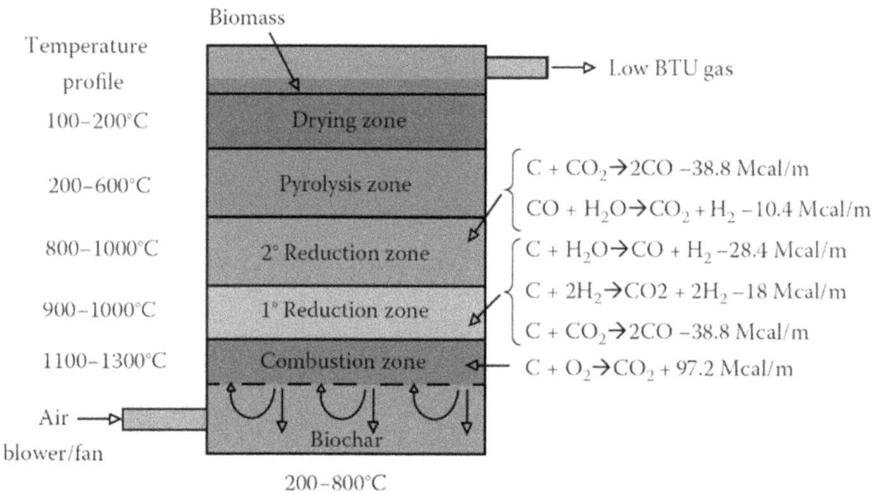

FIGURE 10.3 Temperature profile in an updraft gasifier and reactions occurring during gasification process.

proper control of the air flow to ensure the process is in gasification mode instead of combustion mode. The total amount of air to be used must be below the stoichiometric requirements discussed in Section 10.1. There will be excess oxygen or air at the exit, which must be subtracted from the total air used during the process.

The only disadvantage of the updraft gasifier is the inferior quality of synthesis gas produced. There is very minimal thermal cracking, and other high energy content hydrocarbons, such as methane, ethylene, ethane, and others, are not formed.

10.3.2 Downdraft Gasifier

The basic schematic diagram of a downdraft gasifier is shown in Figure 10.4. In this design, air is introduced into a downward flowing bed of biomass fuel and the synthesis gas outlet is at the bottom of the reactor. The tar oils are cracked and reduced to non-condensible gaseous products, thereby increasing the energy content of the gas (Figure 10.5). Therefore, the gas quality is enhanced, allowing for its use in internal combustion engines.

10.3.3 Cross-Draft Gasifier

In a cross-draft gasifier, the flow of air and gas is across the gasifier, usually along the same level. Cross-draft gasifiers operate at very high temperatures and confine their combustion/reduction zones by using a small diameter air inlet nozzle (Wereko-Brobby and Hagen, 1996). Temperatures of up to 2,000°C are obtained, and water cooling of the metal reactor is usually necessary. The reactors are usually not insulated using the biomass fuel as the heat shield for heat losses. Nozzle velocities used are up to 80 m/s.

Fixed-bed gasifiers are characterized by their very simple design and simple construction. In developing countries, a very common crude oil drum is used to make a fixed-bed gasifier. This crude oil drum is normally about 200 L in volume (52 gal)

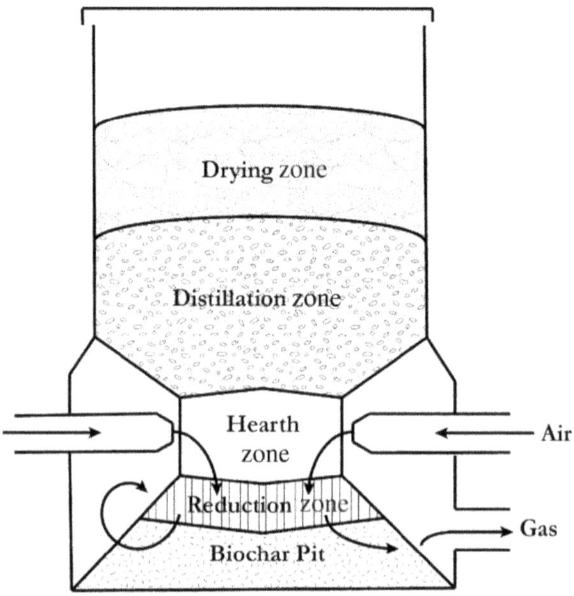

FIGURE 10.4 Schematic of downdraft gasifier.

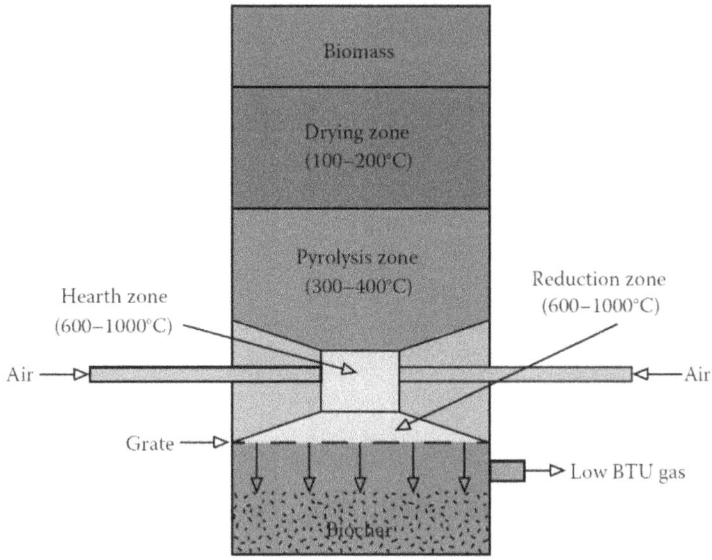

FIGURE 10.5 Temperature profile within a downdraft gasifier.

with a diameter of about 58.5 cm and a height of about 90 cm. The hole on the top cap is normally used for the outlet gas for the updraft gasifier, and the bottom cap is normally cut out so that a grate can be welded. A small blower is fabricated and driven by a 186.5–73 W motor (1/4–1/3 hp motor).

10.3.4 Fluidized Bed Gasifier

One of the most versatile of all moving-bed gasifiers is the fluidized bed gasifier (FBG). Figure 10.6 shows the schematic of the FBG. The basic component parts include the following:

1. Biomass fuel bin
2. Fuel feed system with interlock (to prevent back flow)
3. Fluidized bed reactor with inert fluidizing medium
4. Cyclone particulate removal system
5. Fluidizing air and air distribution system
6. Preheating system for the bed material

The operation starts by preheating the bed material in the reactor. There are various ways to accomplish this operation, such as the use of electrical heaters, natural gas burners, and induction preheaters. The simplest preheating system uses a natural gas burner with an associated flame sensor. Once an appropriate bed temperature is achieved, normally around 800°C, the biomass may be fed. The biomass must be sufficiently dried to around 10%–20% of moisture so that the reaction temperature can be easily maintained. Otherwise, extra amounts of heat energy will be required

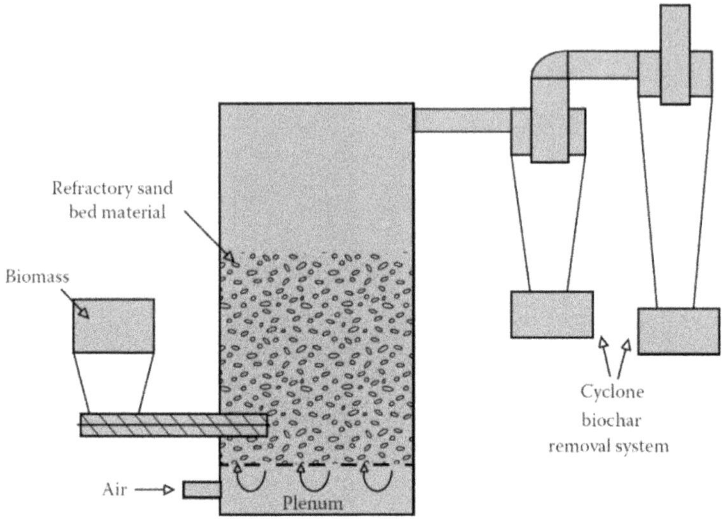

FIGURE 10.6 Schematic diagram of a fluidized bed gasifier.

to dry the biomass material, and it will be difficult to maintain the reaction temperature. The A/F ratio is maintained in the gasification mode. Note that the control of gasification is the opposite to that of a combustion system. In a combustion system, for example, if a higher temperature is required, more fuel will be injected relative to air to bring this reaction temperature higher. In the gasification mode, if a lower temperature is required, more fuel is injected relative to air. One must take care that the flow rate used is such that fluidization is maintained or that elutriation of the bed material does not occur (i.e., at higher flow rates). The procedure for calculating the minimum and maximum fluidizing velocities will be illustrated in Chapter 11.

Figure 10.7 shows the schematic of the patented FBG (US patent no. 4,848,249). A stage combustor is included in this schematic for the efficient combustion of synthesis gas resulting in reduced NO_x emissions. The unit is also attached to a fire-tube boiler for subsequent steam production. The steam may be used to generate heat or electrical power in a steam turbine. More detailed discussions about operation of the FBG will be presented in Chapter 11 on advanced gasification.

The comparison of average gas composition and gasifier operating conditions in a FBG is summarized in Table 10.2. This data was taken from the operation of a 305-mm diameter FBG with a series cyclone particulate removal system. The advantages of the fluidized bed systems are as follows:

1. Accurate control of reaction temperature.
2. High throughput (measured in MJ/m^2 or $MBtu/h/ft^2$).
3. Can handle a wide range of fuels, including the use of various biomass fuels in one operation.
4. Large sizes may be designed without increasing overall footprint.
5. The system may be automatically controlled.

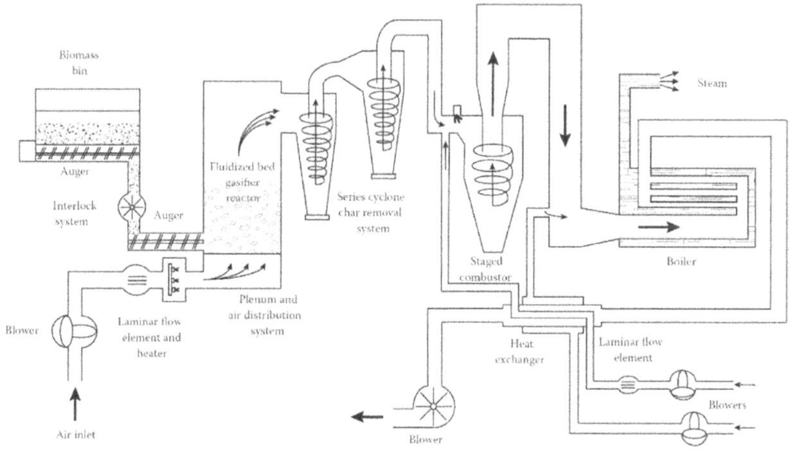

FIGURE 10.7 Schematic of a patented fluidized bed gasifier.

TABLE 10.2
Data for Gasification of Various Fuels Using Fluidized Bed Gasifier

Gasifier	Gin Trash	Manure	Shavings
Bed temperature (°C)	762	762	756
Air flow (m^3/min)	0.40	0.42	0.39
Fuel feed rate (g/min)	306	339	300
F/A ratio (kg/kg)	0.625	0.651	0.625
Carbon efficiency (%)	65.74	51.35	56.10
Cold gas efficiency (%)	45.61	27.84	46.29
Carbon saturation ratio	0.376	0.401	0.357
H/C ratio	1.217	1.011	1.056
Carbon in ash (%)	20.3	6.1	12.20
LCV gas			
Hydrogen (H_2)	9.765	3.999	4.377
Methane (CH_4)	2.492	1.643	3.549
Ethylene (C_2H_4)	0.791	0.239	1.116
Ethane (C_2H_6)	0.200	0.031	0.246
Propane (C_3H_8)	0.180	0.292	0.279
Propylene (C_3H_6)	0.016	0.007	0.012
Carbon monoxide (CO)	11.203	5.744	12.098
Carbon dioxide (CO_2)	12.548	8.494	9.797
Nitrogen (N_2)	56.629	71.786	60.348
Oxygen (O_2)	6.536	7.771	8.188
Production rate (m^3/min)	0.56	0.47	0.51
Heating value (MJ/m^3)	4.25	2.22	4.41

Source: LePori W. A. and E. J. Soltes, In: *Biomass Energy: A Monograph* (E. A. Hiler and B. A. Stout, eds.). Texas Engineering Experiment Station, The Texas A&M University Press, College Station, TX, 1985.

Calculating the amount of air to be used during the gasification process is also quite easy. This is shown clearly in Example 10.5. One must specify the sub-stoichiometric amount of air to use for the gasification process. The procedure is true for whatever gasifier type is used. In many developing countries, a simple 200L drum is used for updraft or downdraft gasifiers. The blower requirement for these systems is very minimal, as shown in the example. A simple blower powered by a single 1/4 or 1/2 hp electric motor is normally sufficient. Only about 400L (14 cfm standard) is required for small-scale systems. For biomass cookstoves, only 2 W (0.0027 hp) is required, which is quite a small power requirement for household use.

The use of a fluidized bed system may require additional understanding of fluidization air requirements and the calculations are a little different. These will be shown in Chapter 11. Data for fluidized bed gasification systems are summarized in Table 10.2. Note that the air requirements are not very different for various fuels; between 390 and 420 L/min of air for gasification will be used, as listed in the table. Of course, this is for a 0.305 m (1 ft) diameter unit, compared with larger diameter updraft or downdraft gasifiers having a reactor diameter of about twice as much, or about 0.585 m (1.92 ft).

In designing gasifiers, additional parameters are used, for example, the proper residence times such that sufficient time is available to generate good-quality synthesis gas. The residence time may be estimated by calculating the superficial velocity through the reactor as shown in Equation 10.7.

$$\text{Superficial velocity}\left(V_s,\ \text{m/s}\right) = \frac{\text{Gas production rate}}{\text{Cross-sectional area}} = \frac{\text{m}^3}{\text{min} - \text{m}^2} = \frac{\text{m}}{\text{min}}\left(\text{or } \frac{\text{m}}{\text{s}}\right) \tag{10.7}$$

Gas residence times of 0.1–0.5 s have been used for designing the combustion-reduction zone lengths in a gasifier (Mande, 2009). Example 10.4 shows how to calculate the superficial velocity in a FBG using the data presented in Table 10.2.

Example 10.4: Determining Superficial Velocity in Gasifier

Determine the superficial velocity in a 0.0305-m-diameter fluidized bed gasifier with a gas production rate of 0.56 m^3/min.

Solution:

a. Calculate the cross-sectional area of the gasifier as follows:

$$A\left(\text{m}^2\right) = \frac{\pi D^2}{4} = \frac{\pi\left(0.0305\ \text{m}\right)^2}{4} = 0.073\ \text{m}^2$$

b. Thus, the superficial velocity in units of m/s is as follows:

$$V_s\left(\frac{\mathrm{m}}{\mathrm{s}}\right)=\frac{0.56\,\mathrm{m}^3}{\mathrm{min}}\times\frac{1}{0.073\,\mathrm{m}^2}\times\frac{1\,\mathrm{min}}{60\,\mathrm{s}}=0.13\frac{\mathrm{m}}{\mathrm{s}}$$

Example 10.5: Determining Amounts of Air to Use in Gasifier

A 200 L updraft gasifier is used to gasify cotton gin trash. Determine the flow rate to use during gasification (in L/min) assuming only 30% of stoichiometric air is required. Assume an air density of 1.2 kg/m³. About 25 kg of cotton gin trash is used and the gasification is to last for approximately 2 h. For simplification, 10% excess air is measured on the syngas; thus, use 40% for air requirement. Determine the flow rate in units of standard cfm of air.

Solution:

a. From previous examples, about 5.78 kg of air is required for stoichiometric reactions, and if 40% of this air is targeted for gasification runs, then about 2.3 kg of air will be required per kilogram of biomass (i.e., $5.78\times0.40=2.3$).
b. The remaining calculation will just be simple conversion of units as follows:

$$\text{Liters of air}=\frac{2.3\,\text{kg air}}{\text{kg biomass}}\times\frac{25\,\text{kg biomass}}{2\,\text{h}}\times\frac{\mathrm{m}^3}{1.2\,\text{kg air}}$$

$$\times\frac{1{,}000\,\mathrm{L}}{\mathrm{m}^3}\times\frac{1\,\mathrm{h}}{60\,\mathrm{min}}=399\frac{\mathrm{L}}{\mathrm{min}}$$

c. In the English system of units, this is equivalent to around 14 cfm of standard air as shown:

$$\text{cfm of air}=\frac{399\ \mathrm{L}}{\mathrm{min}}\times\frac{\mathrm{m}^3}{1{,}000\,\mathrm{L}}\times\left(\frac{3.28\,\mathrm{ft}}{1\,\mathrm{m}}\right)^3=14\,\mathrm{cfm}$$

10.4 APPLICATIONS OF BIOMASS GASIFIERS

There are various applications of biomass gasifiers. The first known gasifier units were built in France during the 1850s (Mande, 2009). The widespread development of biomass gasifiers, however, was out of necessity during the Second World War, when Germany ran out of hydrocarbon fuels for the internal combustion engines that ran their trucks. German engineers developed several versions of the downdraft gasifier with improved synthesis gas quality: syngas was used to run internal combustion engines. By the end of the war in 1945, more than a million vehicles worldwide were in operation because of gasifiers (Mande, 2009). Today, gasifiers are used to generate heat, electricity, and mechanical power, as well as to produce liquid biofuel from synthesis gas. Synthesis gas can be used to generate and separate individual gases, such as hydrogen and methane. Synthesis gas may also be used as input for fermentation to produce bioethanol.

10.4.1 Production of Thermal Sensible Heat

The simplest use of a biomass gasifier is to produce heat used for household cooking and heating purposes. Figure 10.8 shows a gasifier stove developed by Tom Reed in the 1990s. A 2-W DC-powered micro-blower (purchased from Radioshack) was used to supply adequate air for the gasification reaction. The DC blower can be powered by solar energy. The performance of the unit is such that a typical teapot of water boils in 3–4 min. Less than 500 g (1 lb) of wood or other biomass is used per hour, and the simmer mode may last up to 2 h. The system efficiency was estimated to be about 50% (Reed, 1999). Many cookstoves using biomass would normally operate in the combustion mode. This unit is perhaps the only reported small-scale biomass gasifier cookstove that was popularized in the late 1990s.

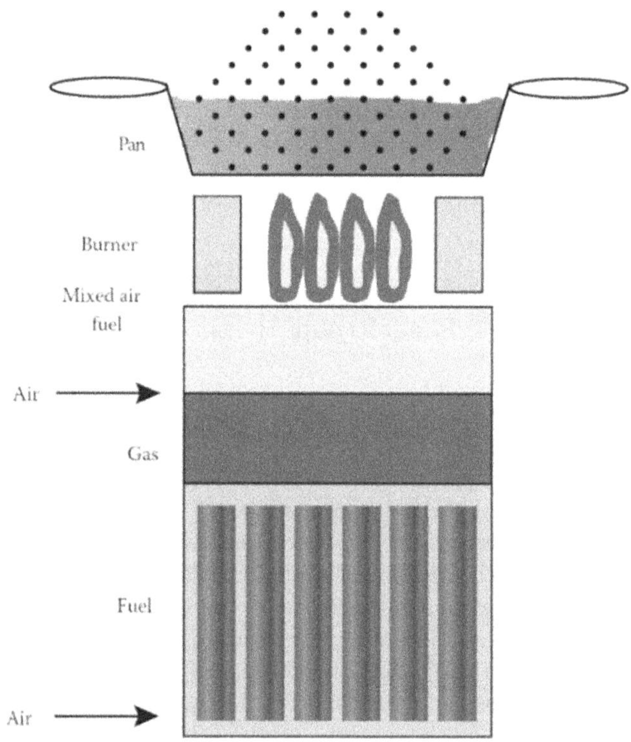

FIGURE 10.8 Schematic diagram of turbo gasifier stove developed by Tom Reed.

10.4.2 Electrical Power Generation

Biomass gasifiers are also used to produce electrical power. The most popular commercial scale gasifier sold in the United States, the BioMax Power System, is manufactured by the Community Power Corporation (CPC) from Littleton, Colorado. The first unit had an electrical output of around 50 kWe and a turn-key operation (i.e., fully automatic start-up, operation, and shutdown). The unit included an automatic dryer/feeder module, a gas production module, and a power generation module.

The system utilizes a microprocessor-based PLC (programmable logic controller) system to ensure continuous production of electrical power. The unit can use a variety of woody biomass fuels, such as woodchips, pellets, scraps, and nut shells. The footprint is around 5 m × 5 m and weighs about 1,500 kg. The synthesis gas has a reported low heating value of 5 MJ/m^3, with less than 5 ppm of tars and particulates. The fuel conversion is about 1.5 kg of wood per kilowatt-hour equivalent. The turndown ratio (TDR, to be discussed later) is greater than 10:1. Thus, the electrical power may range from 5 to 50 kWe. If thermal power is the only energy type required, the unit will have a thermal output ranging from 14.6 to 146 kW (50,000–500,000 Btu/h). The reported gas composition has about 20% hydrogen, 20% carbon monoxide, 7% carbon dioxide, and 2% methane, with the balance being nitrogen gas. The start-up fuel is liquefied petroleum gas, propane, or butane gas.

The overall biomass power-generating industry in the United States has approximately 11,000 MW of summer-operating capacity and produces about 1.4% of the US electrical supply (EIA, 2010). Currently, the largest biomass power plant in the United States is the New Hope Power Partnership, located in South Bay, Florida. The 140 MW facility uses sugarcane bagasse and recycled urban wood as fuel to generate this much power for its milling and refining operations, as well as to supply electricity to nearly 60,000 nearby homes (*Scientific American* Magazine, 2009). Example 10.6 shows the calculations for power generation.

Example 10.6: Determining Amount of Biomass Required to Produce Power

Determine the amount of biomass required in kilograms per hour to generate 1 MW of electrical power. Assume the following: woodchips are used with a heating value of 20.4 MJ/kg; the overall conversion efficiency is 15%.

Solution:

a. This is a very simple and straightforward calculation using the efficiency equation introduced in earlier chapters as follows:

$$\text{Input energy or power} = \frac{\text{Output energy or power}}{\text{Efficiency}\left(\text{in decimal}\right)}$$

b. The input energy or power is that coming from the biomass and calculated as follows:

$$\text{Input power} = \frac{1\,\text{MW}}{0.15} = 6.67\,\text{MW}$$

c. Thus, the input biomass power, using the heating value of biomass, will be used to determine the amount of biomass energy to use per hour as follows:

$$\text{Biomass input (kg/h)} = \frac{6.67\,\text{MW}}{20.4\,\text{MJ}} \times \frac{\text{kg}}{1\,\text{MW}-\text{s}} \times \frac{3{,}600\,\text{s}}{\text{h}}$$

$$= 1{,}177\ \text{kg/h}$$

d. More than a tonne per hour of biomass is required to produce 1 MW of electrical power using a gasifier with an overall conversion efficiency of only 15%.

10.4.3 Mechanical Power Generation

A gasification system may also be designed to generate mechanical power to run an internal combustion engine. A system is shown in Figure 10.9. The unit is a continuous flow downdraft gasifier with a mechanical output of 60 kWe (80 hp). A downdraft gasifier is operated in a continuous mode using a bottom grate to continuously remove the char. The gas cleanup system consists of a simple low-volume water spray, an activated carbon filter, and an additional filter-dryer before injection into an internal combustion engine. A Brigs & Stratton gasoline engine was used. If electrical power generation is required, a generator may be coupled to this engine.

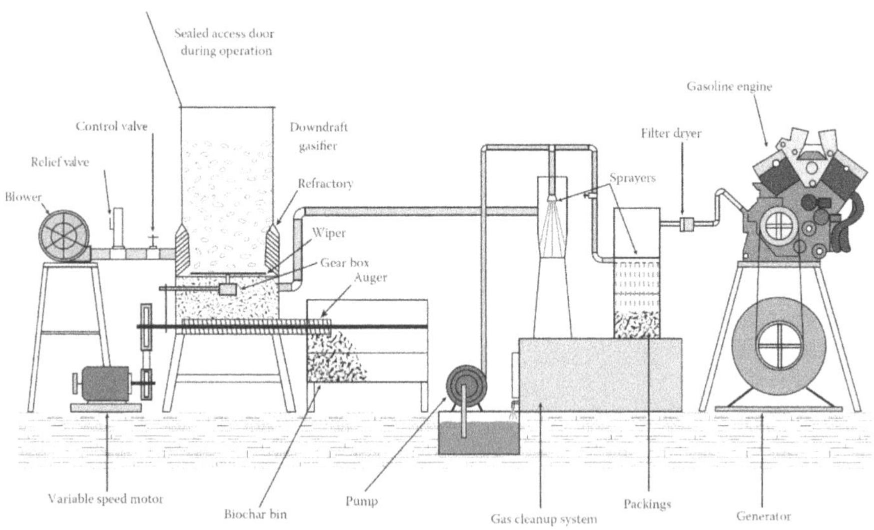

FIGURE 10.9 Schematic diagram of a downdraft gasifier with gas cleanup system to generate power from internal combustion engines.

Example 10.7: Determining Area Required for Biomass Electrical Power

How many acres would be required to build a 50 MW (energy output) biomass power plant using fluidized bed gasification (operated 350 days a year and 24 h per day) if the heating value of the biomass (dry) is 7,500 Btu/lb? The biomass yield in tons (dry) per acre per year is about 15. Assume a conversion efficiency of 15%.

Solution:

a. The energy input must be equal to 50 MW/0.15 = 333.3 MW.
b. This is a simple conversion process using dimensional analysis to the correct units of acres as follows:

$$\frac{333.3\,\text{MW}}{\text{MW}} \times \frac{1\times 10^6\ \text{W}}{\text{MW}} \times \frac{\text{J}}{\text{W}-\text{s}} \times \frac{\text{Btu}}{1{,}055\,\text{J}} \times \frac{\text{lb}}{7{,}500\,\text{Btu}} \times \frac{3{,}600\,\text{s}}{\text{h}}$$

$$\times \frac{24\,\text{h}}{\text{day}} \times \frac{350\ \text{days}}{\text{yr}} \times \frac{\text{ton}}{2{,}000\,\text{lb}} \times \frac{\text{acre}}{15\ \text{tons}} = 42{,}465\ \text{acres}$$

c. The answer is about 42,500 acres.

10.4.4 Liquid Fuel Production from Synthesis Gas

Synthesis gas produced by gasification may be processed into liquid biofuel using the Fischer–Tropsch (FT) technology (Steynberg and Dry, 2004). The FT technology was discovered in the early 1900s. It became popular during the Second World War as a process for generating gas-to-liquid biofuels. Synthesis gas from any gasification process may be passed through novel metal catalysts to generate liquid fuel. The known catalysts of commercial significance are of three broad classes, as follows:

1. Fused iron catalysts
2. Precipitated iron catalysts
3. Supported cobalt catalysts

Note that the FT technology is always associated with iron and cobalt catalysts. Franz Fischer and Hans Tropsch were two chemists working at the Kaiser Wilhelm Institute for Coal Research in Germany. Their process was created in 1913, although significant results were not obtained until the 1920s. The goal was to use coal-derived synthesis gas to produce hydrocarbon molecules from which fuels and chemicals could be produced. Thus, the origin of the work was based on coal. Biomass-based materials were also used later. Today, numerous catalysts are being used to convert synthesis gas from the gasification process into liquid fuels. More detailed discussions will be presented in the biomass liquefaction chapter (Chapter 12).

10.5 GASIFIER TDR AND THROUGHPUT

The TDR, also known as the turndown capacity, is an inherent property of most conversion processes. When applied to the gasification processes, the TDR may be defined as the ratio of the maximum gasification rate to the minimum gasification rate (see Equation 10.8). It was suggested in earlier publications (Reed, 1981) that TDR should be recognized as an important parameter for gasifiers and should be included in their measurements and specifications.

$$\text{TDR} = \frac{\text{Maximum gasification rate}}{\text{Minimum gasification rate}} \tag{10.8}$$

In flow measurements, TDR indicates the range of volumetric flow that a certain flow meter can deliver with acceptable accuracy. Thus, if a certain flow meter can be regulated from 2.8 to 28 m^3/min (100–1,000 ft^3/min), its TDR is 10. Devices that convert electrical energy into light energy and are controlled by an on-and-off switch have a TDR of 1, unless a dimmer is used. With a dimmer, the TDR could vary.

Fixed-bed gasifiers have a high TDR of at least 5 (Reed, 1981). This property is very useful when the gasifier is used for various applications, such as heating, mechanical, or electrical power production. The TDR is adjusted by using different amounts of air during gasification while maintaining the A/F ratio at the gasification mode. The A/F ratio may change as long as it is still within the gasification mode (i.e., less than stoichiometric amounts of air). To make the A/F ratio constant, simply increase the fuel feed rate as the air flow rate is increased. This phenomenon could be easily explained when using a FBG whose TDR could be controlled as well. In FBGs, however, increasing the air flow rate requires the use of larger bed material. Using a higher flow rate causes the bed material to be entrained into the cyclones and may deplete this important FBG component. This situation will be discussed further in Chapter 11.

Gasifier throughput is defined as the rate of energy input into the gasifier per unit of area and normally described in units of GJ/h-m^2 (or Btu/h/ft^2). A FBG throughput may be as high as 9 GJ/h/m^2 (or 0.8 MMBtu/h/ft^2). Example 10.8 shows how to specify the cross-sectional area of a FBG assuming a certain throughput. With improved understanding of fluidized bed gasification systems, the unit may now be operated over a wide range of throughputs, thus increasing its TDR as well. The FBG can then be operated on low throughput and a higher throughput without changing the dimensions of the gasifier simply by changing the bed material sizes. Detailed calculations will be presented in Chapter 11.

Example 10.8: Turndown Ratio Calculations for Fluidized Bed Gasifier

A fluidized bed gasifier can be operated from 9 GJ/h/m^2 to as high as 20 GJ/h/m^2. Calculate the TDR for this case. The minimum gasifier feed rate is about 15 kg/h using a 0.3048 m diameter gasifier on a biomass with a heating value of 20 MJ/kg. Recalculate the throughput in units of GJ/h/m^2 and the new TDR.

Solution:

a. Using the relationships above (Equation 10.8), we have the following calculations:

$$\text{TDR} = \frac{\text{Maximum gasification rate}}{\text{Minimum gasification rate}} = \frac{20\ \text{GJ/h/m}^2}{9\ \text{GJ/h/m}^2} = 2.2$$

b. Using the new assumptions on the second case, we have the following calculations for gasification rates:

$$\text{Minimum gasification rate} = \frac{15\ \text{kg}}{\text{h}} \times \frac{20\ \text{MJ}}{\text{kg}} \times \frac{4}{\pi\left(0.3048\ \text{m}\right)^2}$$

$$\times \frac{\text{GJ}}{1{,}000\ \text{MJ}} = 4.1\ \text{GJ/h/m}^2$$

c. The new TDR is calculated as follows:

$$\text{TDR} = \frac{\text{Maximum gasification rate}}{\text{Minimum gasification rate}} = \frac{20\ \text{GJ/h/m}^2}{4.1\ \text{GJ/h/m}^2} = 4.88$$

10.6 GASIFICATION STUDIES

Gasification studies began during the onset of the energy crisis in the late 1970s. During this period, the first reported attempts at thermal conversion of biomass concentrated on the use of animal manure. Subsequently, successful combustion systems were demonstrated that used wood wastes or forest products as fuel, with quite minimal problems.

A fluidized bed combustor is an ideal system for thermal conversion, and aside from animal manure, cotton gin trash was also tested. However, the initial process showed that it was quite difficult to gasify animal manure. At the end of the combustion run, the material found in the reactor is shown in Figure 10.10. The micron-sized bed material is placed in the reactor at the beginning of each run, and the reactor ends up with numerous stones or rocks as shown.

FIGURE 10.10 Photo of material found in fluidized bed combustor using animal manure.

It was concluded that high-ash biomass feedstocks, such as animal manure, have the tendency to slag and to foul the reactor, thereby making the combustion process infeasible. When wood was combusted, bed agglomeration was not observed. This was primarily because most wood wastes have very low ash content. Consequently, wood wastes do not have the propensity to slag. There was renewed focus on using the same fluidized bed combustion system, focusing on operating it in gasification mode (i.e., with incomplete amounts of oxygen).

Fluidized bed technology offers several important characteristics for biomass gasification compared with fixed-bed systems. The turbulent, fluidized state of inert particles in the bed creates a near-isothermal zone and enables accurate control of reaction temperatures (LePori and Soltes, 1985). These inert bed material particles store energy. The energy is then transferred to the biomass during its rapid insertion into the reactor, while a stable temperature is maintained. The turbulent movement of the bed material creates good interaction with the biomass, allowing immediate efficient conversion at nearly isothermal temperatures. Thus, biomass of different sizes may be used in fluidized bed systems. A broad range of operating conditions are possible in a fluidized bed by changing the A/F ratios, flow velocity, operating pressure, and both the type and size of bed material. One of the greatest advantages of fluidized bed systems is their ability to accept a wide variety of biomass with very little preparation or pretreatment.

Numerous studies made use of fuels with varying ash levels (cotton gin trash, rice hulls, rice straw, corn cobs, animal manure, woodchips, and sorghum stalks). It was found that the output of synthesis gas varies in composition and heating values for different fuels when processed at the same operating temperature. Consequently, operating temperatures were investigated within the range of 700°C–900°C (1,292°F–1,652°F). F/A ratios ranged from 0.7 to 1.8 kg/kg (LePori and Soltes, 1985). The significant results of these studies were reported in Chapter 2 of the book *Biomass Energy: A Monograph* (Hiler and Stout, 1985). The results of these studies, including additional recent experiments, are summarized in the following list.

Summary of Fluidized Bed Gasification of Various Biomass

1. The heating value of the synthesis gas was independent of temperature and primarily a function of F/A ratios. As the F/A ratio was decreased, the heating value of syngas increased. This illustrates the dilution effect of nitrogen in the gas.
2. Using cotton gin trash, the overall conversion efficiency for steam production ranged from 42% to 56%.
3. Using cotton gin trash, the tar concentration of the LCV gas ranged from 5 to 42.4 g of tar/m^3 for gas at a fuel feed rate between 0.99 and 1.3 kg/min.
4. Using cotton gin trash, the tar concentration of the LCV gas ranged from 1.9 to 3.4 g or tar/m^3 of gas at fuel feed rates between 0.45 and 0.95 kg/min.
5. For cotton gin trash and cottonseed hulls, the heating value of LCV gases ranged from 166.86 to 212.73 MJ/m^3, respectively.
6. The efficiency of particulate removal using series cyclones ranged from a low of 65.6% for cotton gin trash to a high of 80.7% for cattle feedlot manure in the secondary cyclones; the efficiency of particulate removal ranged from

a low of 84.1% using cotton gin trash to a high of 92.4% using feedlot manure in the primary cyclones. Using newer cyclone designs, the char collection efficiencies of the primary and secondary cyclones exceed 97%. These cyclone designs will be presented in Chapter 11 (Advanced Gasification).

7. In the secondary cyclones, the emitted concentration of particulates ranged from 1.7 to 3.3 g/m^3 for cattle feedlot manure and cotton gin trash, respectively. In the primary cyclones, the emissions' concentrations ranged from 5.3 to 7.1 g/m^3 for cotton gin trash and cattle feedlot manure, respectively.

The actual composition of LCV gas with the 305 mm FBG using cottonseed hulls and cotton gin trash is summarized in Table 10.3.

In recent years, from 2005 to 2020, studies using an improved FBG have been completed. A new full US patent was issued in 2021 (US 11,186,779 B2, Date of Patent: November 30, 2021). The improvements were associated with the following components:

1. A new air distribution system reduces pressure drop to less than half.
2. A fully dedicated mobile gasification system operating from natural gas lines was built. The system operates independently and does not require any power or heat source.
3. Using National Instrument's LabVIEW software, a new computer process control program was developed for various feedstocks (Maglinao, 2009).
4. New computer programs were developed and prepared for the following feedstocks: cotton gin trash, switchgrass, dairy manure, feedlot manure, rice straw, sorghum biomass, peat, poultry litter, woodchips, and municipal solid waste.

TABLE 10.3
Composition of LCV Gas Using a 305-mm Gasifier

	Cottonseed Hulls		Cotton Gin Trash	
Components	**Average Mol%**	**Standard Dev.**	**Average Mol%**	**Standard Dev.**
Hydrogen	9.71	1.13	14.78	1.83
Carbon monoxide	15.51	0.85	19.43	2.59
Carbon dioxide	20.10	0.57	15.84	1.20
Methane	4.14	0.35	2.33	0.43
Ethylene	1.07	0.16	0.76	0.15
Ethane	0.39	0.04	0.23	0.04
Other hydrocarbons	2.70	0.11	1.20	0.18
Oxygen	1.17	0.33	0.50	0.22
Nitrogen	45.09	1.94	44.54	2.08
Heating value (MJ/m^3)	212.73		166.86	

Source: LePori W. A. and E. J. Soltes, In: *Biomass Energy: A Monograph* (E. A. Hiler and B. A. Stout, eds.). Texas Engineering Experiment Station, The Texas A&M University Press, College Station, TX, 1985.

Most of the results presented in this section were made from the gasification runs using a new FBG. Figure 10.11 is a typical temperature profile for a gasification run starting from the preheating stage up to the gasification stage. The preheating stage would normally happen in 30 min. However, preheating may take longer, depending on the ambient temperature, as well as system control by the operator. Figure 10.12 shows the pressure profile during a typical gasification runs.

The gasification temperature is normally maintained at an average of 775°C (1,425°F), as shown in Figure 10.11. The temperature is maintained by proper adjustment of the fuel feed rate and the air flow rate using the process control system in either the manual or closed-loop system mode. In Example 10.9, the air flow rate was maintained at an average of 1.99 kg/min, while the fuel flow rate was maintained at an average of 0.95 kg/min. These flow rates will result to an A/F ratio of 2.10 kg/kg (see Example 10.9). The typical pressure profile from gasification start-up to operation is shown in Figure 10.17.

Example 10.9: A/F Ratio Calculations in Fluidized Bed Gasifier

Determine the A/F ratio during a gasification system if the air flow rate is 1.99 kg/min and the fuel flow rate is 0.95 kg/min. At the beginning of the gasification runs, the air flow rate was 2.16 kg/min and the mass flow rate was 0.94 kg/min. Compare the air-to-fuel ratios used at each stage.

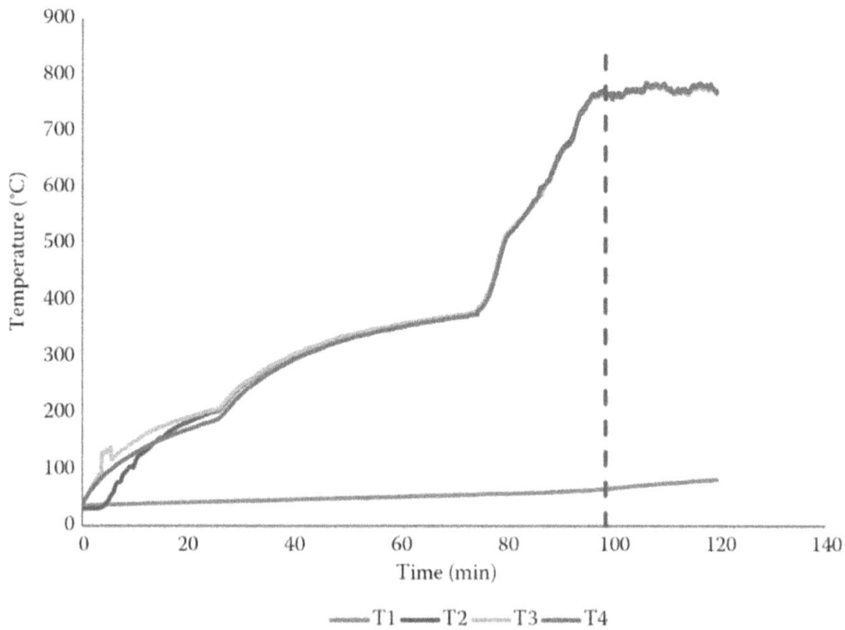

FIGURE 10.11 Typical temperature profile during gasification, start-up, and operation.

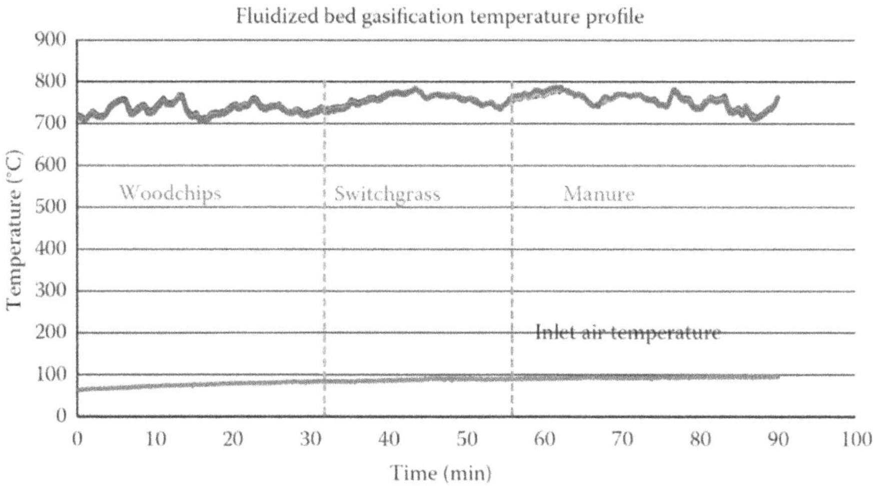

FIGURE 10.12 Temperature profile when successive biomass feedstocks (woodchips, switchgrass, and animal manure) were gasified using a fluidized bed gasifier.

Solution

a. This is a simple ratio and proportion to arrive at the proper units for A/F ratio as shown below:

$$\text{A/F ratio} = \frac{1.99\,\text{kg}}{\text{min}} \times \frac{\text{min}}{0.95\,\text{kg}} = 2.1\,\text{kg/kg}$$

b. The A/F ratio at the beginning of the process is calculated as follows:

$$\text{A/F ratio} = \frac{2.16\,\text{kg}}{\text{min}} \times \frac{\text{min}}{0.94\,\text{kg}} = 2.30\,\text{kg/kg}$$

The operational versatility of the FBG is depicted by the temperature profile shown in Figure 10.12. The figure shows a stable temperature throughout the gasification process for woodchips, switchgrass, and animal manure. This graph was accomplished to test the viability of the control systems specifically developed to handle various biomass feedstocks during one continuous run (without system shutdown). Note that some biomass resources have different A/F ratios if combustion equations are compared. The differences are only slight, as shown in Example 10.10. In fact, switchgrass and dairy manure have the same A/F ratios (5.0) with similar controls.

In summary, gasification of biomass resources is very similar to an engine's operation, where an exact F/A ratio must be specified for continuous sustainable operation.

Example 10.10: Determining A/F Ratio of Biomass Combustion

Determine the A/F ratios for the complete combustion of woodchips, switchgrass, and dairy manure. Use the data in Table 3.5.

Solution:

a. The ultimate analysis table is summarized here. To calculate the subscripts, divide the value by the MW of each element.

Biomass	C (MW=12)	H (MW=1)	O (MW=16)	N (MW=14)	S (MW=32)
Woodchips	46.1	6.6	46.6	0.2	0.1
Switchgrass	42.0	6.1	47.4	0.4	0.1
Dairy manure	30.5	4.4	33.4	1.5	0.4

b. The chemical formulas for the feedstock named are written here.

Woodchips: $C_{3.8}H_{6.6}O_{2.9}N_{0.01}S_{0.003}$

Switchgrass: $C_{3.5}H_{6.1}O_{3.0}N_{0.03}S_{0.003}$

Dairy manure: $C_{2.5}H_{4.4}O_{2.1}N_{0.11}S_{0.013}$

c. A balanced combustion equation is derived as follows:

$$C_{3.8}H_{6.6}O_{2.9}N_{0.01}S_{0.0.3}+4\lfloor(O_2+3.76N_2)\rfloor\rightarrow 3.8CO_2+3.3H_2O$$
$$+4\times 3.76N_2+0.003S$$

$$C_{3.5}H_{6.1}O_3N_{0.03}S_{0.003}+3.525\lfloor(O_2+3.76N_2)\rfloor\rightarrow 3.5CO_2+3.05H_2O$$
$$+3.525\times 3.76N_2+0.003S$$

$$C_{2.5}H_{4.4}O_{2.1}N_{0.11}S_{0.013}+2.55\lfloor(O_2+3.76N_2)\rfloor\rightarrow 2.5CO_2$$
$$+2.2H_2O+2.55\times 3.76N_2+0.013S$$

d. Then, the A/F ratio is simply calculated as follows:

$$\frac{\text{Air}}{\text{Fuel}}=\frac{4\times\lfloor(32)+3.76\times(28)\rfloor}{(12)\times 3.8+(1)\times 6.6+(16)\times 2.9+(14)\times 0.01+(32)\times 0.003}$$
$$=\frac{549.12}{98.836}=5.6$$

$$\frac{\text{Air}}{\text{Fuel}}=\frac{3.525\times[(32)+3.76\times(28)]}{(12)\times 3.5+(1)\times 6.1+(16)\times 3+(14)\times 0.03+(32)\times 0.003}$$
$$=\frac{483.912}{96.616}=5.0$$

$$\frac{\text{Air}}{\text{Fuel}} = \frac{2.55 \times [(32) + 3.76 \times (28)]}{(12) \times 2.5 + (1) \times 4.4 + (16) \times 2.1 + (14) \times 0.11 + (32) \times 0.013}$$

$$= \frac{350.064}{69.956} = 5.0$$

e. Note the differences are not very great. The control system must be able to adjust the A/F ratios appropriately for proper gasifier operation with a stable operating temperature.

10.7 CONCLUSION

The development of the gasification technology has a very long history and dates to several centuries ago. The technology has not yet been perfected and more research is required to have widespread commercialization and use. There are not many companies commercializing this technology except for a few fixed-bed type gasifiers (such as the downdraft systems). The development of the gasification systems for liquid fuel production came about as a necessity during the Second World War because of scarcity of gasoline and diesel fuel. The application has again diminished in use with the availability of commercial transport fuel. The revival of this technology in the 1970s came about due to the oil embargo and the rising cost of transport fuels. The recent interest in gasification is a combination of the rising cost of fossil fuels and the increased necessity for renewable resources such as agricultural biomass and wastes.

The future of gasification technology may not be limited to providing heat. Gasification systems will enable a wide range of applications, including power generation and liquid fuel production. The continuous production of highly valuable synthesis gas may facilitate the economical production of pure hydrogen gas. Synthesis gas from gasifiers has now been used to generate ethanol in combination with biochemical conversion processes. In addition, syngas can be condensed to generate liquid bio-oil from biomass. Because of the low energy density of biomass, the development of a decentralized gasification system may be beneficial.

Gasification by-products that satisfy real market requirements include activated carbon, briquettes, biochar (for soil amendment), and many other uses.

One of the most versatile moving-bed gasifiers is the fluidized bed system. Moving-bed systems are no different than fluidized bed combustion systems, except that they should be operated in the gasification mode (i.e., deficient amounts of oxygen). The use of deficient air gives rise to intermediate gaseous compounds, such as synthesis gas, which may then be used for various applications. The advantages of moving beds, such as the FBG, include high throughput, the ability to gasify a wide range of biomass feedstocks, and the ability to continue operating while using a succession of different feedstocks. It is not necessary to shut down the system when using different feedstocks if the A/F ratios are correct and if the operating temperature is stable. In the future, gasifiers will provide electricity for many biorefineries, as well as heat and liquid fuels once the syngas has condensed. It will not be surprising if gasifiers become ubiquitous in many biorefineries worldwide, because of their varied uses and applications and moderate costs.

10.8 PROBLEMS AND DISCUSSION ISSUES

10.8.1 Empirical Chemical Formula of Biomass

P10.1 Determine the approximate chemical formula for rice hulls. The ultimate analysis is as follows: C = 38.5, H = 5.7, O = 39.8, N = 0.5, S = 0.0, and ash = 15.5. These values are in % by weight.

10.8.2 Air-to-Fuel Ratio of Biomass for Thermal Conversion

P10.2 Determine the A/F ratio (kg/kg) for the complete combustion of rice hulls. The ultimate analysis is as follows: C = 38.5, H = 5.7, O = 39.8, N = 0.5, S = 0.0, and ash = 15.5. These values are in % by weight.

10.8.3 Air Requirement for Gasification

P10.3 Determine the amount of air required to gasify rice hull in units of m^3/kg, assuming 70% of stoichiometric air is used at standard conditions. Note: assume density of air at standard condition is 1.2 kg/m^3. The ultimate analysis is as follows: C = 38.5, H = 5.7, O = 39.8, N = 0.5, S = 0.0, and ash = 15.5. These values are in % by weight.

10.8.4 Volumetric Air Requirement in Fixed-Bed Gasifiers

P10.4 Determine the volumetric air flow rate (m^3/min) to be used in gasifying rice hull in a fixed-bed downdraft gasifier, assuming 10 kg of biomass is used and only 70% of stoichiometric air is utilized and the normal completion time for this biomass to be gasified is about an hour. Assume all the oxygen is used during the reaction. Use an air density of 1.2 kg/m^3.

10.8.5 Acreage Required to Build Power Plants

P10.5 How many hectares (acres) would be required to build a 50 MW (output) biomass power plant if the heating value of the biomass (as received) is 19.4 MJ/kg (8358.5 Btu/lb) with a moisture content of 10%. The power plant will be operated for 5,000 h in a year. The biomass yield in (raw) tonnes per hectare per year is 45 (20 dry tons/acre per year) with around 50% moisture content after harvest. Assume an overall electrical power conversion efficiency of 50% and the biomass will be dried to 10% moisture prior to conversion.

10.8.6 Equivalence Ratios Calculations in a Gasifier

P10.6 Calculate the equivalence ratio for the gasification of rice hull. The ultimate analysis is as follows: C = 38.5, H = 5.7, O = 39.8, N = 0.5, S = 0.0, and ash = 15.5. These values are in % by weight. The A/F ratio requirement for gasification is 70% of the stoichiometric value.

10.8.7 Power from Manure Gasification

P10.7 Estimate the power (MW) that could be derived continuously from the thermal gasification of cattle manure from a 40,000-head facility. The heating value of cattle manure used for conversion was estimated at 17.4 MJ/kg (7,500 Btu/lb). The recoverable manure amounts to approximately 2.3 kg/hd/day (5 lb/hd/day) of relatively dried manure. If manure is to be used in a thermal conversion system with overall conversion efficiency of 15%, what would be the generated power in MW? Calculate the amount of char produced (carbon and ash) in tonnes/day (tons per day) and on a yearly basis, assuming 8,760 h of operation and a 20% yield left as char.

10.8.8 Superficial Velocity in Updraft Gasifiers

P10.8 Determine the superficial velocity in an updraft gasifier with a diameter of 0.585 m and a gas production rate of 0.400 m^3/min.

10.8.9 Syngas Requirement in Internal Combustion Engines

P10.9 A downdraft gasifier is to be used to run a four-cylinder four-stroke internal combustion engine with the following data:

Piston diameter = 60 mm
Piston stroke = 60 mm
Engine rpm = 6,500
Volumetric efficiency = 80%

Determine the minimum amount of gas that is required by this engine if 80% of the gas would come from synthesis gas from the gasifier and the remaining 20% will be combustion air.

10.8.10 Design of Downdraft Gasifier

P10.10 Design the dimensions (in m) of a downdraft gasifier with a required syngas output of 2.5 m^3/min (standard conditions) using a cylindrical cross section. The A/F ratio should be 70% of the stoichiometric requirements and sawdust will be used as fuel. The ultimate analysis is as follows: C = 49.7%, H = 6.2%, O = 42.0%, N = 0.7%, S = 0.7%, and ash = 0.7%. The superficial velocity should be 0.03 m/s and the air density at standard conditions is equal to 1.2 kg/m^3. About 25 kg of biomass is used every hour.

10.8.11 Exothermic Versus Endothermic Reactions

P10.11 An exothermic process releases heat, causing the temperature of the intermediate surrounding to rise. An endothermic process absorbs heat and cools the surroundings. How do you consider gasification in relation to these terms?

a. endothermic   b. exothermic   c. It depends   d. No idea

10.8.12 Gasification Reactor Types

P10.12 Name a moving-bed gasification system.
a. Updraft b. downdraft c. cross-draft d. fluidized bed

10.8.13 Capacities of Gasifiers

P10.13 This parameter refers to the ratio of the maximum capacity of a system to its minimum capacity.
a. equivalence ratio b. turn down ratio c. capacity factor d. availability

10.8.14 Gasifier in a Cotton Gin

P10.14 Design the dimensions (in m) of a fluidized bed gasifier suitable for a 50 bale per hour cotton gin-processing cotton bale with average weight of 455 kg [1,000 lbs] and having a trash content of 227 kg/bale [500 lbs/bale]. The energy content of cotton gin trash is 16.25 MJ/kg [7,000 Btu/lb]. State all the assumptions made.

10.8.15 Gasifier in a Poultry Farm

P10.15 A poultry farm manager has a population of 15,000 heads. He wanted to convert this manure into electrical power. Design a fluidized bed gasification system suitable for this facility and estimate the maximum power that could be produced from this biomass resource. State all the assumptions made.

10.9 CASE STUDIES

Design a fluidized bed gasification system for a 500-head dairy farm whose revenue comes from sale of electric power, biochar, and government tax credit. Perform simple economic and sensitivity analyses. Data is shown in the table.

Dairy Population	500 heads	Biochar Selling Cost	$110/tonne
Manure Moisture (%W)	87%	Electricity Selling Cost	$0.12/kWh
Interest on Investment	8% of capital	Generator Set Cost	$325/kW
TSI	2% of capital	Repair & Maintenance	2% of capital
Yearly Operation	350 days/yr	Feedstock Cost	$1/tonne
Labor & Management	20% of capital	Miscellaneous Cost	5% of capital
Biochar Yield	20% of dry feed	Discount Rate	2.36%
Tax Credit	30% of capital	% Recoverable Manure	90%
Capital Cost	$1,000/kW	Manure Heating Value	9.1 MJ/kg
Gas Cleanup Cost	10% of gasifier cost	Syngas Heating Value	5.58 MJ/m^3
Gasifier Maintenance	10% of capital	Overall Efficiency	15%
Gasifier Throughput	22,700 MJ/h/m^2	Dryer Cost	$100/kg/h
Manure Production	68 kg/head/day	% Phosphorus	

TSI = taxes, shelter, and insurance

Required:

a. Determine the power output of the gasification facility (kW).
b. Determine the size of the gasifier assuming circular diameter (m^2).
c. Enumerate all system component costs (gasifier, genset, gas cleanup, and dryer and sand separation system costs) and the total capital cost of the gasification system ($).
d. Report on daily and yearly fresh and dry manure generation, manure recovered, the gasifier federate (kg/h), biochar generation rate.
e. Develop a simple spreadsheet to estimate economic parameters, such as internal rate of return (IRR), payback period (PBP), benefit cost ratio (BCR), and net present value (NPV), and report on the viability of this project given the data.
f. Perform sensitivity analyses to determine the impact of changing input parameters on the economic indicators. A graph for relationship would be beneficial visually with predictive equation and with R^2 value n the chart. The following are suggested changes:
    i. dairy population on NPV (350–10,000 heads)
    ii. biochar selling price versus IRR and PBP ($27.50/tonne to $330/tonne)
    iii. net metering cost on IRR and PBP ($0.05/kWh to $0.15/kWh)
    iv. tax incentive percentage on IRR and PBP (1%–30%)
    v. heating values of manure on IRR and PBP (5.5 MJ/kg to 17.8 MJ/kg)
    vi. initial capital cost on IRR and PBP ($500/kW to $2,000/kW)
    vii. interest on investment on IRR and PBP (2% to 30%)
    viii. manure feedstock cost on IRR and PBP ($1.10/tonne to $33/tonne)

REFERENCES

Bridgewater, A. V. 1984. *Thermochemical Processing of Biomass*. Butterworths and Company Publishers, Ltd., London.

EIA. 2010. *US Energy Information Administration Annual Energy Review 2009*. Office of Energy Markets and End Use, US Department of Energy, Washington, DC.

Hiler, E. A. and B. A. Stout (eds.). 1985. *Biomass Energy: A Monograph*. Texas Engineering Experiment Station, The Texas A&M University Press, College Station, TX.

LePori, W. A. and E. J. Soltes. 1985. Thermochemical conversion for energy and fuel. In: *Biomass Energy: A Monograph* (E. A. Hiler and B. A. Stout, eds.). Texas Engineering Experiment Station, The Texas A&M University Press, College Station, TX. pp. 9–75.

Maglinao, A. L., Jr. 2009. Instrumentation and Evaluation of a Pilot Scale Fluidized Bed Biomass Gasification System. MS Thesis. Department of Biological and Agricultural Engineering, Texas A&M University, College Station, Texas.

Mande, S. 2009. Thermochemical conversion of biomass (Chapter 13). In: *Renewable Energy Engineering and Technology: Principles and Practice* (V. V. N Kishore, ed.). EarthScan Publishing, London. pp. 673–762.

Reed, T. B. 1981. *Biomass Gasification: Principles and Technology*. Noyes Data Corporation, Park Ridge, NJ.

Reed, T. B. 1999. Turbo wood stove gasifier (3 kW thermal) handout. Personal communication during his visit to the Biomass Energy Laboratory of the University of the Philippines at Los Baños, College, Laguna, Philippines.

Scientific American Magazine. 2009. Biomass: Can renewable power grow on trees? March 26, 2009 Issue, *Scientific American*, A Division of Nature America, New York.
Steynberg, A. and M. Dry. 2004. *Fischer-Tropsch Technology*. Elsevier Science Ltd., London, UK. 700 p.
Wereko-Brobby, C. Y. and E. B. Hagen. 1996. *Biomass Conversion and Technology*. John Wiley & Sons, New York.

ASSESS YOUR KNOWLEDGE

1. What are the two most important primary gases produced during gasification of biomass?
   a. CO and H_2.
   b. CO_2 and H_2O
   c. CH_4 and H_2
   d. CO_2 and H_2

2. Give an example of a moving-bed gasifier.
   a. Updraft gasifier
   b. Downdraft gasifier
   c. Recirculating fluidized bed gasifier
   d. Cross-draft gasifier

3. Give an example of a fixed-bed gasifier.
   a. fluidized bed gasifier
   b. Updraft gasifier.
   c. entrained flow gasifier
   d. Recirculating bed gasifier

11. The ultimate analysis for corn cob is given as follows: C = 46.2%; H = 7.6%; O = 42.3%; N = 1.2%; and S = 0.3% and the ash content is 2.4%. Determine the chemical formula for this biomass material.
    a. $C_{3.85}H_{7.6}O_{2.64}N_{0.086}S_{0.009}$
    b. $C_{46.2}H_{7.6}O_{1.2}N_{1.2}S_{0.3}$
    c. $C_1H_{7.6}O_{2.64}N_{0.086}S_{0.009}$
    d. $C_{3.7}H_{5.7}O_{4.1}N_1S_{0.2}$

12. Determine the amount of biomass required in lbs per hour to generate one megawatt (1 MW) of electrical power through gasification. Assume the following: municipal solid waste is used with a heating value of 10,000 Btu/lb; the overall conversion efficiency is 13%.
    a. 2625
    b. 2.62
    c. 4,500
    d. 1.31

13. Design the dimensions of a fluidized bed gasifier (assuming circular cross section) with a throughput of 2 MMBtu/h/ft^2. The heating value of the biomass to be used was 9,500 Btu/lb. The feed rate for biomass was 3,000 lbs/h. Report your answer in ft.
    a. 2.15
    b. 8.6
    c. 6.6
    d. 4.3

4. Syngas is also used to produce methanol and hydrogen and into synthetic fuel such as gasoline or aviation fuel using what known process developed by German Engineers?
   a. TL process
   b. Fischer–Tropsch
   c. Pyrolysis
   d. Plasma

5. Name the simplest fixed-bed air gasifier.
   a. Updraft gasifier
   b. fluidized bed
   c. entrained bed
   d. downdraft gasifier

14. Determine the amount of air required (in units of cubic feet of air per lb of biomass) when municipal solid waste (MSW) is gasified. First determine the stoichiometric air-to-fuel ratio for the complete combustion of municipal solid waste (MSW) (in units of lb/lb) with a chemical formula of $C_4H_6O_2N_{0.08}S_{0.009}$ and then determine the air-to-fuel ratio (lb/lb) during gasification if only 70% of stoichiometric air is to be used instead of 100% air. Convert this A/F value into units of cubic feet of air per lb of fuel assuming air density at STP of 0.075 lbs/ft^3.
    a. 33
    b. 9
    c. 4.5
    d. 66

15. How many acres would be required to build a 1 MW (energy output) biomass power plant through fluidized bed gasification (operated 350 days a year and 24 h/day) if the heating value of the biomass (dry) is 9,500 Btu/lb? The biomass yield in tons (dry) per acre per year in dry tons is about 15. Assume an overall conversion efficiency of 13%.
    a. 77,341
    b. 7,734
    c. 773
    d. 77.3

6. This term is the ratio of the practical syngas generation rate to the lowest generation rate. Identify this gasification term.
   a. Turn down ratio
   b. Swirling constant
   c. Fluidization constant
   d. Boiler plate ratio

7. This is a process of combusting the synthesis gas produced from a gasifier to bring the NO_x emissions down. Here is usually a delay by one second to introduce secondary air.
   a. Knock enhancement
   b. Swirling process
   c. Staged combustion.
   d. Slagging

8. Give an example of a fixed-bed gasifier.
   a. fluidized bed gasifier
   b. entrained flow gasifier
   c. downdraft gasifier.
   d. Recirculating bed gasifier

16. Determine the amount of air required (in units of lbs air per lb of biomass) when corn stover is gasified. First, determine the stoichiometric air-to-fuel ratio for the complete combustion of this material (in units of lb/lb) with a chemical formula of $C_{3.3}H_{5.9}O_{2.9}N_{0.09}S_{0.006}$ and then determine the air-to-fuel ratio (lb/lb) during gasification if only 70% of stoichiometric air is to be used instead of 100% air.
   a. 5.17
   b. 7.07
   c. 3.3
   d. 2.12

17. Determine the stoichiometric air-to-fuel ratio for the complete combustion of corn stover units of lb/lb). Assume the chemical formula for corn stover is $C_{3.3}H_{5.9}O_{2.9}N_{0.09}S_{0.006}$.
   a. 5.17
   b. 4.73
   c. 4.5
   d. 4.94

18. The ultimate analysis for corn stover is given as follows: C = 40%; H = 5.9%; O = 46.2%; N = 1.3%; and S = 0.2% and the ash content is 6.3%. Determine the chemical formula for this biomass material.
   a. $CH_{1.77}O_{0.87}N_{0.028}S_{0.0018}$
   b. $C_{40}H_{5.9}O4_{6.2}N_{1.3}S_{0.2}$
   c. $C_1H_{7.6}O_{2.64}N_{0.086}S_{0.009}$
   d. $C_{3.7}H_{5.7}O_{4.1}N_1S_{0.2}$

9. Give an example of a moving-bed gasifier.
   a. Updraft gasifier
   b. Downdraft gasifier
   c. Crossflow gasifier
   d. Entrained flow gasifier.

10. This process is the melting of the ash in biomass with very high ash content. Hence, during gasification, operating temperature is usually lower than the melting point of the ash of the biomass to prevent this unwanted process from forming. Name this ash-melting process.
    a. Slagging
    b. Sublimation
    c. Condensation
    d. ashing

19. Determine the amount of biomass required in lbs per hour to generate one megawatt (1 MW) of electrical power through gasification. Assume the following: corn stover is used with a heating value of 6,721 Btu/lb; the overall conversion efficiency is 15%.
    a. 3,386
    b. 2,621
    c. 1.7
    d. 1.31

20. Determine the flow rate of air exiting the gasifier if the fluidizing temperature was 704°C for a fluidized bed gasification system in units of cubic feet per minute (CFM), assuming 30% of stoichiometric air is used for the gasification of a biomass. The 100% stoichiometric air-to-fuel ratio was found to be 5.7 lbs air/lb fuel and the feed rate was found to be 3,500 lbs per hour of biomass. Assume the air density at STP of air as 0.075 lbs/ft^3 for the entering air and 0.0225 lbs/ft^3 at the gasification temperature for the exit air.
    a. 1,330
    b. 3,562
    c. 2,450
    d. 888

Answers: 1a, 2c, 3b, 4b, 5a, 6a, 7c, 8c, 9d, 10a, 11a, 12a, 13d, 14d, 15c, 16c, 17b, 18a, 19a, 20b

ADDITIONAL LEARNING RESOURCES

- Access the Lecture Notes at the Instructor and Student Resources: https://routledgetextbooks.com/textbooks/instructor_downloads/
- The Case Study Videos can be found in the ebook+ version of this book.

11 Advanced Gasification

LEARNING OBJECTIVES

Upon completion of this chapter, one should be able to

1. Describe state-of-the-art and advanced gasification processes.
2. Design a fluidized bed gasifier from the selection of biomass material for the gasification process to the applications of synthesis gas.
3. Specify the types of bed materials to use in various gasification systems.
4. Differentiate gasifier throughputs and turndown ratios for various gasifier types.
5. Outline the various state-of-the-art applications of gasification technology.

11.1 INTRODUCTION

This chapter deals with the design and operation of a fluidized bed gasification system. Design steps include selecting a biomass fuel, determining the gasifier's dimensions, selecting the bed material, specifying the series cyclone system, and making provisions for char removal. The proper way to burn low calorific value gas will also be discussed to ensure that the NO_x and SO_x concentrations are below their permitted operating limits. Finally, various ways to utilize synthesis gas and other coproducts will be discussed.

The advantages of using a fluidized bed are as follows:

1. The throughput (volume of feedstock such as manure) is relatively high, providing a smaller footprint requirement for the system.
2. The unit can handle a wide range of high-moisture biomass feedstocks.
3. There is a high heat transfer rate in the reactor caused by the high thermal inertia of the inert bed material particles in the reactor.
4. There is a high turndown ratio (TDR).

Fluidization is the key concept to high throughput and high TDR. Fluidization is the operation by which the solids are transformed into a fluid-like state through contact with a gas or solid (Kunii and Levenspiel, 1977). The reactor in a fluidized bed gasifier is usually filled with solid bed material that will be heated before the biomass fuel is fed (into the reactor). This bed material, brought into a fluidized state, enables the continuous flow of biomass particles through the reactor and enhances the reaction. Rapid mixing leads to nearly isothermal conditions throughout the reactor and the operation can be controlled properly. Fluidization also enables the transport of vast quantities of biomass over a short period that would be infeasible in a fixed-bed reactor. Fluidized bed systems are suited for large-scale operations. The heat transfer

DOI: 10.1201/9781003294306-11

rates in the reactor are quite high, although the diameter of a typical fluidized bed reactor is not large. The events occurring in a fluidized bed reactor as the velocity of fluidizing medium is increased are as follows (see Figure 11.1 for the behavior of fluidizing media in a fluidized bed reactor):

1. At the initial lowest velocity, the bed material is a solid bed and no amount of biomass can be moved appreciably (leftmost setup).
2. As the velocity is increased, the bed material's virtual bed height is increased, allowing void spaces for the biomass to pass through.
3. During smooth fluidization, there is relative ease in the passage of biomass in the reactor. The heated bed material will then transfer the energy onto the biomass and begin initiating the thermal conversion of the biomass into gaseous synthesis products. Too much velocity must be prevented, lest the bed material will be entrained out of the reactor. The operator must be familiar with the minimum and maximum fluidizing velocity allowed.
4. As gas velocity is further increased, aggregate or bubbling fluidization will occur.
5. Further increase in gas velocity will cause the reactor to have a slugging effect (large pockets of air or bubbles are formed like a slug).
6. Finally, as velocity is increased, the condition is called "lean phase" fluidization, and the chance of the bed material being thrown out of the reactor is increased.

In real applications, the quality of fluidization is affected by the choice of bed material and type of gas distribution system. Figure 11.2 shows the quality of fluidization as influenced by the type of gas distribution plates. If the gas distribution plate is only a single orifice plate, fluidization will be of poor quality. In this design, there will be too much fluctuation in density with some channeling and intermittent movement of particles called "slugging." As numerous holes are drilled in the distributor plate, fluidization is improved. The best quality is obtained when a sintered plate is used, with numerous holes in the distribution plate drilled at random. Unfortunately, this type of air distribution system will cause very high pressure drop through the plate, thereby requiring higher power requirements for the blower system.

There are several examples of gas distribution systems for fluidized beds. These are shown in Figure 11.3. The most common distribution plate was the use of bubble caps (second row, last column in the figure). This design is very easy to fabricate. All that is required is a pipe nipple with an end cap attached to one end of the thread. The other threaded end is attached to the flat distributor plate. Holes are drilled on the side of the nipple, just below the cap, thereby minimizing the entry of sand when air is not introduced into the reactor. There are many other air distribution types as follows and shown in the figure:

1. Single perforated plate
2. Staggered perforated plate
3. Sandwich-packed bed
4. Grate bars and slits

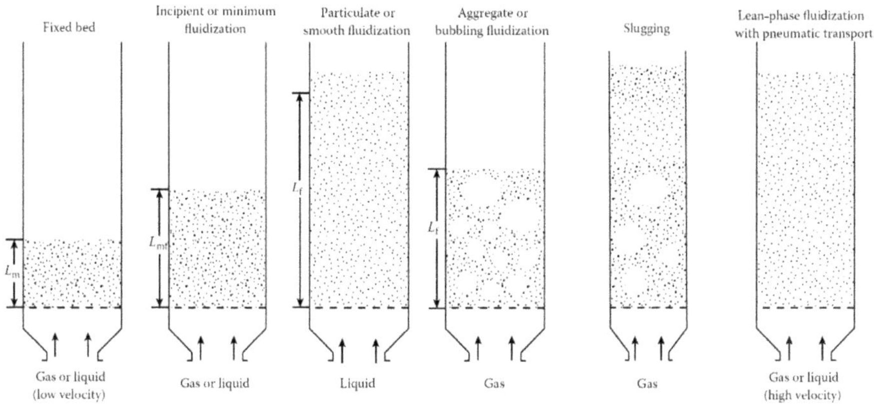

FIGURE 11.1 Behavior of fluidizing media in fluidized bed reactor. (Adapted from Kunii, D. and O. Levenspiel, *Fluidization Engineering*, Robert E. Krieger Publishing Company, Huntington, NY, 1977.)

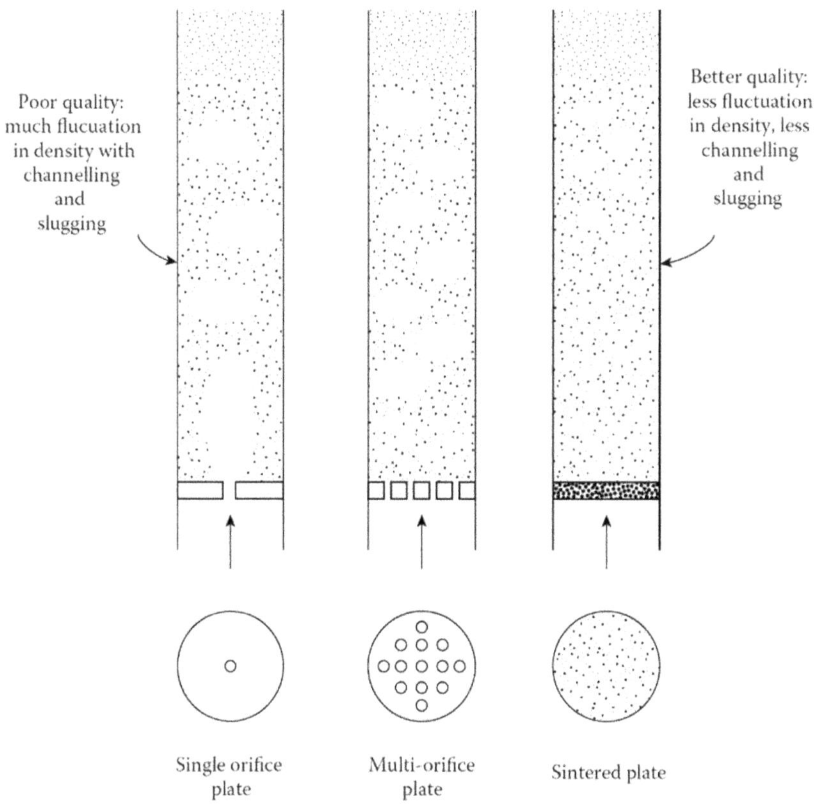

FIGURE 11.2 Quality of fluidization as influenced by type of gas distribution plates. (Adapted from Kunii, D. and O. Levenspiel, *Fluidization Engineering*, Robert E. Krieger Publishing Company, Huntington, NY, 1977.)

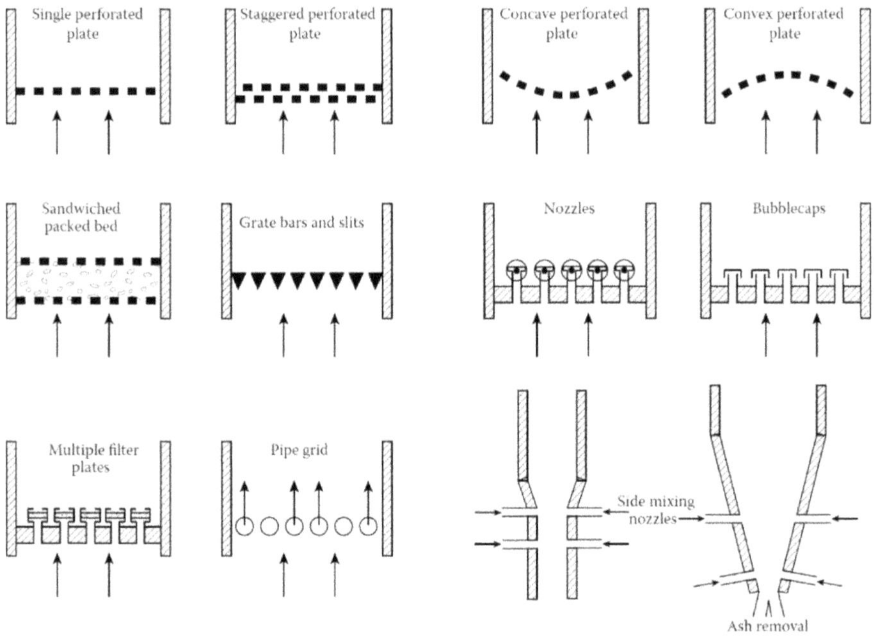

FIGURE 11.3 Several examples of gas distribution systems for fluidized beds. (Adapted from Kunii, D. and O. Levenspiel, *Fluidization Engineering*, Robert E. Krieger Publishing Company, Huntington, NY, 1977.)

5. Multiple filter plates
6. Pipe grid
7. Concave or convex perforated plate
8. Nozzles

Fluidized bed systems have been commonly used in combustion of coal, in fluid catalytic cracking reactors, and in reforming of hydrocarbons using solid catalysts. The catalysts act as the bed material in this case. Some fluidized bed systems have also been used to activate coal and produce activated carbon.

11.2 DETERMINING AVERAGE PARTICLE SIZE OF BED MATERIAL

The first step in designing a fluidized bed system is the specification of a bed material to use. There are numerous sources of bed material and the most common is the use of quarry sand. The bed material used for combustors may also be used for gasification reactors. Once the bed material is selected, the average particle size, d_p, is determined. However, it is difficult to make this estimate because most bed materials are not spherical. Thus, we define the particle size, d_p, to be the diameter of a sphere that would have the volume of the particle. This leads one to defining the sphericity of the bed material, as shown in Equation 11.1, where sphericity is defined as the ratio of the surface of a sphere to the surface of the particle, both having the same volume. Data on sphericity for fluidizing materials has been reported, especially for sand. For example, if quarry sand is used, the most common values of sphericity range from 0.6 to 0.86 (Kunii and Levenspiel, 1977). Some other materials and catalysts would have different sphericity values. These values should be reported by the manufacturer of those materials.

$$\Phi_s = \left(\frac{\text{surface of sphere}}{\text{surface of particle}}\right)_{\text{both of same volume}} \tag{11.1}$$

To determine the particle size of the bed material, a certain amount of representative sample is weighed and sieved, using carefully selected sieve sizes. The different sizes of sieves to use vary according to the particle sizes in question. Table 3.1 lists the mesh number for Tyler Standards Screens, as well as the corresponding apertures in microns.

In the pilot-scale fluidized bed gasification systems used for research purposes, one would normally use the following set of sieves: 8, 10, 14, 20, and 35. The typical weight distribution for every 500 g sample is summarized in Table 11.1, as well as the percentages of material captured at each sieve range. The average particle size is calculated using Equation 11.2. Example 11.1 shows the step-by-step calculations necessary to establish the average particle size, d_p.

$$\bar{d}_p = \frac{1}{\sum^{\text{all } i} \left(x/d_p\right)_i} \tag{11.2}$$

where

$\bar{d}_p$ = the mean particle size
x = weight fraction within a certain sieve size range
d_p = average particle size within a certain sieve size range

TABLE 11.1
Common Particle Size Data of Bed Material for Pilot-Scale Gasification Research

Mesh Size	Size Range (microns)	Material Percentage (%)	Weight (g)
20	840	0	0
28	595–840	29	145
35	420–595	45	225
48	297–420	23	115
Pan	<297	3	15
		100%	500 g

Example 11.1: Calculation of Average Particle Size for Bed Material

Given the data in Table 11.1, estimate the average particle size for this bed material size distribution.

Solution:

a. The table is converted into the calculations of parameters in Equation 11.2, as follows:

Diameter Range (μ)	d_{pi} (μ)	Weight Fraction in Interval $(pd_p) = x_i$	$(x/d_p)i$
840	–	0	0
595–840	717.5	(145/500) = 0.29	(0.29/717.5) = 0.000404
420–595	507.5	(225/500) = 0.45	(0.45/507.5) = 0.000887
297–420	358.5	(115/500) = 0.23	(0.23/358.5) = 0.000642
<297	148.5	(15/500) = 0.03	(0.03/148.5) = 0.000202
		1	$\Sigma(x/d_p)_i = 0.002135$

b. Thus, the average is calculated as follows:

$$\bar{d}_p = \frac{1}{\sum^{\text{all } i}(x/d_p)_i} = \frac{1}{0.002135} = 468 \text{ microns}$$

11.3 MINIMUM FLUIDIZING VELOCITY, TERMINAL VELOCITY, AND PRESSURE DROP IN FLUIDIZED BED REACTORS

The empirical relationship to estimate the minimum fluidizing velocities for small bed material particles in a fluidized bed is given by Equations 11.3 and 11.4 for larger particles (Kunii and Levelspiel, 1977). The Reynolds number (Re) will have to be checked to verify the ideal equation to use, and this is given by Equation 11.5.

$$v_{mf} = \frac{(\Phi_s d_p)^2}{150} * \frac{\rho_s - \rho_g}{\mu} g\left(\frac{\varepsilon_{mf}^3}{1-\varepsilon_{mf}}\right) \text{ for } \mathrm{Re}_p < 20 \tag{11.3}$$

$$v_{mf}^2 = \frac{\Phi_s d_p}{1.75} \frac{\rho_s - \rho_g}{\rho_g} * g * \varepsilon_{mf}^3 \text{ for } \mathrm{Re} > 1{,}000 \tag{11.4}$$

$$\mathrm{Re} = \frac{d_p v \rho_{air}}{\mu_{air}} \tag{11.5}$$

where

Φ = sphericity

ρ = density of solids (s) and gas (g)

TABLE 11.2
Voidage at Minimum Fluidizing Conditions

	Size (mm)						
Particles	**0.02**	**0.05**	**0.07**	**0.10**	**0.20**	**0.30**	**0.40**
Sharp sand, $\Phi_s = 0.67$	–	0.60	0.59	0.58	0.54	0.50	0.49
Round sand, $\Phi_s = 0.86$	–	0.56	0.52	0.48	0.44	0.42	–
Mixed round sand	–	–	0.42	0.42	0.41	–	–
Coal and glass powder	0.72	0.67	0.64	0.62	0.57	0.56	–
Anthracite coal, $\Phi_s = 0.63$	–	0.62	0.61	0.60	0.56	0.53	0.51
Absorption carbon	0.74	0.72	0.71	0.69	–	–	–
Fischer–Tropsch catalyst, $\Phi_s = 0.67$	–	–	–	0.58	0.56	0.55	–
Carborundum	–	0.61	0.59	0.56	0.48	–	–

Source: Kunii, D. and O. Levenspiel, *Fluidization Engineering*, Robert E. Krieger Publishing Company, Huntington, NY, 1977.

μ = viscosity of gas
ε_{mf} = voidage
d_p = mean particle size of bed material (also equal to $\bar{d}_p$)

The voidage, ε_{mf}, is the space between the particles and may be found experimentally. For sharp sand, the reported value is 0.67 and for round sand, the value to use is around 0.86 (Kunii and Levenspiel, 1977). Example 11.2 shows the procedure for calculating the minimum fluidizing velocity at cold state. Table 11.2 presents some reported values of voidage.

The equation to estimate the terminal velocity during fluidization is given in Equation 11.6. This equation is a function of the particle densities and the viscosity of the fluidizing medium. Example 11.3 shows how to calculate the terminal velocity of particles in a fluidized bed gasification system.

$$v_t = \left[\frac{4}{225} * \frac{(\rho_s - \rho_g)^2 g^2}{\rho_g \mu}\right]^{1/3} d_p \text{ for } \quad 0.4 < \mathrm{Re}_p < 500 \tag{11.6}$$

Example 11.2: Determining Minimum Fluidizing Velocity

Calculate the minimum fluidizing velocity for particles with the following data at 21°C.

- $d_p = 0.059$ cm; $\mu = 1.78 \times 10^{-4}$ g/cm-s (air at 21°C)
- $\rho_g = 0.0012$ g/cm^3 (air at 21°C); $\rho_s = 2.4$ g/cm^3
- $g = 980$ cm/s^2; $\varepsilon_{mf} = 0.50$; $\Phi_s = 0.60$

Solution:

a. Equation 11.3 will be used as the assumed equation as follows:

$$v_{mf} = \frac{\left[(0.60)(0.059\,\text{cm})\right]^2}{150} \times \frac{(2.4-0.0012)\,\text{g/cm}^3}{1.78\times10^{-4}\ \text{g/cm}-\text{s}} \times 980\,\text{g/cm}^2 \left(\frac{(0.5)^3}{1-0.5}\right) = 27.6\,\text{cm/s}$$

b. Now that the velocity is known, Re may be calculated to see if the use of this equation is valid:

$$\text{Re} = \frac{d_p v \rho_{air}}{\mu_{air}} = \frac{(0.059)(27.6)(0.0012)}{1.78\times10^{-4}} = 10.98 < 20\,\text{ok}$$

Example 11.3: Terminal Velocity Calculations

Determine the terminal velocity for the particles given in the previous example (i.e., Example 11.2).

Solution:

a. Only the relevant parameters are given, as listed here:
   - $d_p = 0.059\,\text{cm}$
   - $\mu = 1.78\times10^{-4}$ g/cm-s (air at 21°C)
   - $\rho_g = 0.0012$ g/cm^3 (air at 21°C); $\rho_s = 2.4$ g/cm^3
   - $g = 980$ cm/s^2

b. Thus, using Equation 11.6, the calculation is done as follows:

$$v_t = \left[\frac{4}{225} \times \frac{(2.4-0.0012)^2(980)^2}{(0.0012)(1.78\times10^{-4})}\right]^{1/3} (0.059) = 455\ \text{cm/s}$$

Gasifier velocities are not reported nor are they displayed in data-logging systems or display monitors during the operation of a fluidized bed gasifier. The parameters that are commonly displayed or monitored are the volumetric flow rates and the temperature or pressures within the gasifier. As such, the minimum and maximum fluidizing velocities must be converted into these more practical units, if the dimensions of the fluidized bed gasifier are known, as well as temperatures. Example 11.4 illustrates how these minimum and maximum fluidizing velocities are converted into volumetric flow rates at the actual operating temperature of the gasifier. At elevated temperatures, data for air density should be taken from an appropriate air properties table.

Fluidized bed gasifiers usually have minimum throughput of approximately 9 GJ/h/m^2 (0.8 MMBtu/h/ft^2). These throughput values may be used in the design of a fluidized bed gasifier in the succeeding sections.

Example 11.4: Volumetric Flow Rate Calculations

Calculate the minimum required volumetric flow rates in m^3/min (or cfm) for a 0.3048 m fluidized bed gasifier at the operating temperature of 816°C. Compare this value at cold conditions. The other parameters are provided as follows:

- $d_p = 0.059$ cm
- $\mu = 4.42 \times 10^{-4}$ g/cm-s (air at 816°C) (from air properties table)
- $\rho_g = 3.25 \times 10^{-4}$ g/cm^3 (air at 816°C) (from air properties table)
- $\rho_s = 2.4$ g/cm^3
- $g = 980$ cm/s^2
- $\varepsilon_{mf} = 0.50$
- $\Phi_s = 0.60$

Solution:

a. The minimum fluidizing velocity is calculated as in Example 11.2 using new values for density and viscosity at elevated temperatures as follows:

$$v_{mf} = \frac{\left[(0.60)(0.059\,\text{cm})\right]^2}{150} \times \frac{\left(2.4 - 3.25 \times 10^{-4}\right)\text{g/cm}^3}{4.42 \times 10^{-4}\ \text{g/cm} - \text{s}} \times 980\,\text{g/cm}^2 \left(\frac{(0.5)^3}{1 - 0.5}\right) = 11.1\,\text{cm/s}$$

b. This velocity is then used to calculate the required volumetric flow rate as follows (in units of m^3/min units or cfm):

$$Q\left(\frac{\text{m}^3}{\text{h}}\right) = \frac{11.1\,\text{cm}}{\text{s}} \times \frac{\text{m}}{100\,\text{cm}} \times \frac{3{,}600\,\text{s}}{\text{h}} \times \frac{\pi^*(0.3048\,\text{m})^2}{4} = 29\,\text{m}^3/\text{h}$$

$$Q\left(\frac{\text{m}^3}{\text{min}}\right) = 29\frac{\text{m}^3}{\text{h}} \times \frac{1\ \text{h}}{60\ \text{min}} = 0.48\frac{\text{m}^3}{\text{min}}$$

$$Q(\text{cfm}) = \frac{11.1\,\text{cm}}{\text{s}} \times \frac{60\,\text{s}}{\text{min}} \times \frac{1\,\text{in}}{2.54\,\text{cm}} \times \frac{\text{ft}}{12\,\text{in}} \times \frac{0.7854\,\text{ft}^2}{} = 17.2\,\text{cfm}$$

c. Compared with cold conditions the values are as follows:

$$Q\left(\frac{\text{m}^3}{\text{h}}\right) = \frac{27.6\,\text{cm}}{\text{s}} \times \frac{\text{m}}{100\,\text{cm}} \times \frac{3{,}600\,\text{s}}{\text{h}} \times \frac{\pi^*(0.3048\ \text{m})^2}{4} = 72\,\text{m}^3/\text{h}$$

$$Q\left(\frac{\text{m}^3}{\text{min}}\right) = 72\frac{\text{m}^3}{\text{h}} \times \frac{1\ \text{h}}{60\ \text{min}} = 1.2\frac{\text{m}^3}{\text{min}}$$

$$Q(\text{cfm}) = \frac{27.6\,\text{cm}}{\text{s}} \times \frac{60\,\text{s}}{\text{min}} \times \frac{1\ \text{in}}{2.54\ \text{cm}} \times \frac{\text{ft}}{12\,\text{in}} \times \frac{0.7854\,\text{ft}^2}{} = 42.7\,\text{cfm}$$

d. There is about 2.5 times as less flow or velocity at operating conditions than at cold conditions.

The pressure drop through the gasifier bed reactor is calculated from Equation 11.7.

$$p = L_{mf}(1-\varepsilon_{mf})(\rho_s - \rho_g)\frac{g}{g_c} \tag{11.7}$$

where

L_{mf} = bed height at minimum fluidization
g = acceleration due to gravity
g_c = conversion factor [980 g-cm/(g-wt)s^2]
ρ = density of solids (s) and gas (g)
ε_{mf} = voidage

Example 11.5 shows the estimated pressure drop through the gasifier bed for a given bed height.

Example 11.5: Pressure Drop through Fluidized Bed Gasifier

Estimate the pressure drop through the fluidized bed gasifier using the following assumptions:

- L_{mf} = bed height at minimum fluidization = 50 cm
- g, g_c, ρ_s, and ρ_s, μ, and ε_{mf} are the same as previous examples

Solution:

a. The pressure drop is calculated as follows:

$$\Delta p = 50\,\text{cm}(1-0.5)(2.4-0.0012)\text{g/cm}^3 \times \frac{980\ \text{cm/s}^2}{980\,\text{g}-\text{cm}/(\text{g}-\text{wt})\text{s}^2} = 60\ \text{g}-\text{wt/cm}^2$$

b. This pressure drop is also equivalent to 60 cm water (23.6 in water).

11.4 OPERATION OF 0.3048 FLUIDIZED BED GASIFIER

This section presents the results of synthesis gas production using a 0.3048-m (1 ft) fluidized bed gasifier operating on three biomass feedstocks. The first step in the operation of a fluidized bed gasifier is to understand the characteristics of the biomass to be tested. The important physical properties are bulk density and a good calibration curve for the different feedstock. Once this is understood, the actual calibration is made to determine how much biomass feedstock is used during the test. The most common feeding system is the screw conveyor. The mass flow rate of the solids in the feeding system is then calculated. A regular helicoid screw conveyor may be used following the equation proposed by Woodstock and Mason (1987), as shown in Equation 11.8 and Figure 11.4.

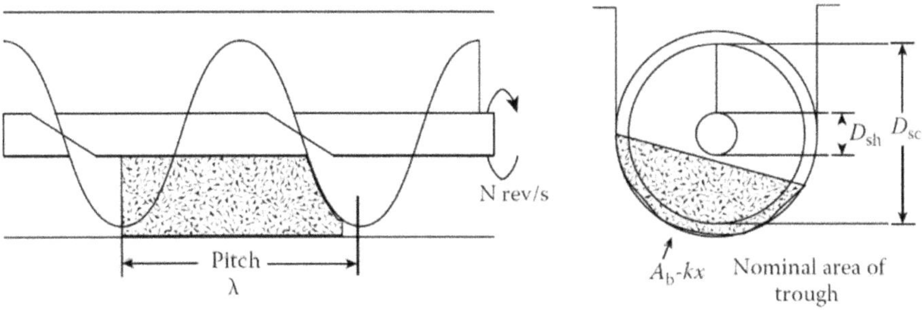

FIGURE 11.4 Capacity of screw conveyor. (Modified from Woodstock, C. R. and J. S. Mason, *Bulk Solids Handling: An Introduction to the Practice and Technology*, Chapman & Hall Inc., New York, 1987.)

$$\dot{m}_s = \rho_b \frac{1}{4}\pi\left(D_{sc}^2 - D_{sh}^2\right)k\lambda N \tag{11.8}$$

where

ρ_b = bulk density, kg/m^3
D_{sc} = trough of flight diameter, m
D_{sh} = shaft diameter, m
k = loading factor (dimensionless)
λ = pitch, m
N = rpm (revolutions per minute) of the shafting/auger

Values for the biomass bulk density and loading factor are summarized in Table 11.3 for the biomass that will be described in this section.

Example 11.6 shows the calculation of the mass flow rate for a biomass with the given parameters. The mass flow rate may then be established as a function of the auger's rpm. To control gasification, it is important to establish the instantaneous mass rate of flow. In addition, the actual air flow rate must be established such that an actual air-to-fuel ratio (AFR), or equivalence ratio, is maintained. The AFR sets the given fluidizing operating temperature. Note that the control of a gasifier is opposite to that of a combustion system. In a combustion system, if a higher temperature is required, more fuel is used, assuming the air flow rate is the same. On the other hand,

TABLE 11.3
Bulk Density and Loading Factors for Various Biomasses Used for Gasification in a Fluidized Bed Gasifier

Biomass	Bulk Density (kg/m^3)	Loading Factor
Sorghum	163	0.0942
Manure	264	0.4876
CGT	189	0.1621

in gasification mode, if a lower temperature is required, more fuel is used relative to the air flow.

Example 11.6: Capacity of Screw Conveyor

Calculate the mass flow rate of cotton gin trash (CGT) with the screw conveyor at 60 rpm. The bulk density of CGT was 190 kg/m³ and the approximate loading factor, *k*, was 0.16. The following are the given dimensions of the screw conveyor:

- $D_{sc} = 10\,\text{cm}$
- $D_{sh} = 7.75\,\text{cm}$
- $\lambda = 7.4\,\text{cm}$

Solution:

a. Using the formula for a regular helicoid screw conveyor, the calculation is as follows:

$$\dot{m}_s\left(\frac{\text{kg}}{\text{min}}\right) = \frac{190\,\text{kg}}{\text{m}^3}\times\frac{1}{4}\pi\left(\left(10\text{ cm}\times\frac{\text{m}}{100\text{ cm}}\right)^2 - \left(7.75\text{ cm}\times\frac{\text{m}}{100\text{ cm}}\right)^2\right)(0.16)$$

$$\left(7.4\text{ cm}\times\frac{\text{m}}{100\text{ cm}}\right)\left(60\frac{\text{rev}}{\text{min}}\times\frac{2\times 3.1416}{\text{rev}}\right)$$

$$= 29\frac{\text{kg}}{\text{min}}$$

The next sets of parameters required are the proximate and ultimate analyses of the feedstock. These parameters should be measured to quantify the properties that are relevant to gasification. Table 11.4 summarizes the properties of the three biomasses that will be described in this section.

The proximate analysis showed there are high amounts of volatiles in all of the biomass fuels used. However, the high ash content, especially that of the manure sample, indicates that slagging and fouling may occur during the gasification process. The ultimate analysis did not show much variability among the three feedstocks used. The calculated stoichiometric AFRs for the different biomasses used are shown here and did not differ. Note that these are on a dry ash-free basis (as opposed to "as-received" basis, as illustrated in earlier chapters).

Sorghum:

$$C_{3.74}H_{6.37}O_{2.89}N_{0.03}S_{0.003} + 3.89(O_2 + 3.76N_2) \rightarrow 3.74CO_2 + 3.19H_2O$$
$$+14.63N_2 + 0.003S$$

Cotton gin trash (CGT):

$$C_{3.72}H_{6.15}O_{2.95}N_{0.12}S_{0.012} + 3.785(O_2 + 3.76N_2) \rightarrow 3.72CO_2$$
$$+3.08H_2O + 14.23N_2 + 0.012S$$

TABLE 11.4
Properties of Three Biomasses Used in Gasification Experiments

Biomass	Sorghum	CGT	Manure
Moisture content, %	22.11 ± 1.44	9.01 ± 0.05	13.08 ± 0.54
Heating value (dry basis), MJ/kg	19.58 ± 0.13	16.67 ± 0.35	15.93 ± 0.26
Proximate analysis (db), %			
Volatile combustible matter	71.40 ± 0.12	71.20 ± 2.10	59.09 ± 0.39
Fixed carbon	14.16 ± 0.90	13.02 ± 0.46	29.80 ± 2.80
Ash	14.45 ± 0.80	15.78 ± 1.65	11.15 ± 2.92
Ultimate analysis (daf basis), %			
C	44.92	44.58	43.50
H	6.37	6.15	6.19
O	46.17	47.25	47.63
N	0.47	1.63	2.17
S	0.08	0.38	0.49
Stoichiometric AFR (daf)			
mol/mol	19.31	17.96	17.49
kg/kg	5.52	5.13	5.00

Source: Maglinao, A. L., Jr., Instrumentation and evaluation of a pilot-scale fluidized bed biomass gasification system, MS Thesis, Department of Biological and Agricultural Engineering, Texas A&M University, College Station, TX, 2009.

Manure:

$$C_{3.63}H_{6.19}O_{2.98}N_{0.16}S_{0.015} + 3.69\left(O_2 + 3.76N_2\right) \rightarrow 3.63CO_2 + 3.1H_2O$$

$$+13.87N_2 + 0.015S$$

The quality and production rates for synthesis gas during gasification of sorghum, CGT, and manure as feedstocks are summarized in Table 11.5. Comparable yields of methane, nitrogen, and carbon dioxide were observed in the gasification of the three biomasses. Sorghum exhibited a lower production of hydrogen than either CGT or manure, each of which showed higher concentrations. The gas yields and their heating values did not differ much but were considered relatively low. The gasification system produced gas calorific values ranging from 4.09 to 4.19 MJ/m^3 and yields ranging from 1.8 to 2.5 N m^3/kg. Gil et al. (1999) showed similar values of gas yield of between 1.4 and 2.4 Nm^3/kg of biomass for similar gasification experiments using air.

11.4.1 Fluidized Bed Gasifier Operational Protocol

The discussion that follows describes a fluidized bed gasifier operational protocol implemented when running a fluidized bed gasifier system. The focus is on

TABLE 11.5
Data for Gasification of Various Fuels Using a Fluidized Bed Gasifier

Synthesis Gas Production	Sorghum	CGT	Manure
Hydrogen (H_2)	5.24	7.99	7.72
Methane (CH_4)	4.11	4.70	4.38
Carbon monoxide (CO)	13.56	11.02	10.92
Ethane (C_2H_6)	0.42	0.33	0.43
Nitrogen (N_2)	58.61	56.31	56.67
Oxygen (O_2)	2.93	3.25	3.40
Carbon dioxide (CO_2)	14.06	14.26	14.18
Heating value, MJ/m^3	4.09	4.28	4.19
Synthesis gas yield, m^3/kg	2.04	1.81	2.11
Synthesis gas production, kg/min	3.24	1.30	5.35
Carbon conversion efficiency, %	82.28	90.46	82.29
Cold gas efficiency, %	49.99	44.68	51.05
Char proximate analysis			
VCM	12.19	14.25	16.20
Ash	61.31	59.11	81.18
Fixed carbon (FC)	26.50	26.64	2.62

Source: Maglinao, A. L., Jr., Instrumentation and evaluation of a pilot-scale fluidized bed biomass gasification system, MS Thesis, Department of Biological and Agricultural Engineering, Texas A&M University, College Station, TX, 2009.

preparations that need to be done before the gasification test, the start-up procedure, monitoring for sustained operation, and system shutdown.

11.4.1.1 Gasifier Preparation

1. The fuel bin must be full of biomass before the start of each test.
2. The reactor must be loaded with proper bed material (refractory sand) and with the proper size distribution.
3. The start-up natural gas or propane tanks must be readily available.
4. The operator must be aware of the minimum and maximum fluidizing velocity for the type of bed material being used.

11.4.1.2 *Start-Up*

1. The blower is the first to be turned on to start fluidizing the sand and to ensure the flow rate is set for the fluidizing state. The operator will notice that the pressure increases as the flow rate is increased, then, there will be a hump in pressure just before fluidization is initiated. After this hump in pressure, the pressure drop will level off, indicating that the bed material has been fluidized.

2. The burner is then lit to heat up the sand. The readout must display all temperatures in all of the zones in the gasifier, including the inlet air to the exhaust portion.
3. The temperature must reach around 500°C–600°C (930°F–1,112°F) before the biomass is to be fed into the gasifier through the auger. A split auger design with interlocking is the best design for biomass feeding. This must be done very slowly until the reactor temperature starts to increase.
4. Once there is a gradual and steady rise in the reactor temperature, the preheating may be turned off. However, doing so decreases the amount of fluidizing air into the reactor.
5. Continue feeding the biomass until the proper fuel-to-air ratio is within the gasification range (i.e., more fuel than air) and well below combustion conditions, while observing the rise in temperature to around 704°C–760°C (1,300°F–1,400°F). At this temperature range, the AFR must be fine-tuned to keep the temperature within the appropriate range. Note the control of gasification is exactly the opposite of combustion; that is, if one would like to decrease the operating temperature, more biomass is fed relative to air and vice versa. If the temperature keeps rising, increase the biomass fuel while maintaining the air flow rate. (In a combustion system, if one wanted to increase the temperature, more fuel would be added relative to air.)
6. Once the temperature has stabilized, the system is in monitoring mode. Then, the control system takes over to manage and sustain the operation.

11.4.1.3 Sustained Operation

1. Temperature control is fundamental to the gasification operation. An accurate temperature must be maintained.
2. In a fluidized bed system, the control system not only monitors temperature but other system conditions, some of which require action.
3. The pressure drop through the bed must be maintained to ensure that the proper amount of bed material is still available in the reactor. Take note that some of the bed material may be elutriated out of the reactor and into the cyclone particulate removal system. If this happens, some amount of bed material must be added to the system to keep the change in pressure (ΔP) through the reactor bed constant.
4. The auger's calibration assumes a full fuel bin. Consequently, a partially full fuel bin would result in an incorrect feed rate. For proper operation, the fuel bin sensor must detect and effectively maintain the proper level of biomass in the fuel bin.
5. The char bin, when full, must also be unloaded and the temperature must be monitored here as well.
6. The computer monitor must also check and provide velocity control to ensure the process stays within the minimum-to-maximum fluidizing velocity range.
7. If possible, gas quality may be monitored to optimize syngas production. Normally, the syngas heating value is increased as the solid fuel-to-air

ratio is increased. This action will correspondingly increase the operating temperature.

8. It gets complicated with some feedstock, such as sorghum, where the heating value is increased at lower fuel-to-aid ratios (e.g., 0.6 kg/kg) and the bed temperature will rise incrementally. However, at higher fuel-to-air ratios (e.g., 1.25 kg/kg), the heating value will decrease as the temperature is increased.
9. The operation of a fluidized bed gasifier is highly dependent on a given feedstock. The control system must be adjusted accordingly, based on data and experience.

11.4.1.4 Shutdown Procedure

1. The shutdown procedure is accomplished by sequentially eliminating the four elements that might cause an explosion: fuel, air, high temperature, and containment pressure. The fuel feed rate must be cut to zero.
2. This action will slightly increase the temperature, but it will not rise further as biomass is being depleted. If this is just a temporary shutdown, the blower may be shut off; then, all pressurized systems must be de-pressurized by means of an inert gas. This procedure conserves energy in the reactor so that at start-up, it will be easy to continue with the process.
3. However, if this is a full shutdown, then the air flow will be continued until the reactor temperature is below the auto-ignition temperature of the biomass or of the gaseous combustibles that may be produced. Once the two elements have been removed, then the blower may be shut off and the valves may be opened.
4. Note the char will have a high temperature at the end of the gasification run. Care must be taken so that char will not be in contact with the air, or it will turn into ash.

A surge tank is normally placed after the cyclone particulate removal system. The surge tank is used to even out the syngas production and to eliminate cycling of gas output. This is important if syngas is to be used in an engine that requires continuous positive flow. It was observed that without the surge tank, there will be cycling of syngas production at some point, because gasification reactions fluctuate with time, even if the biomass feed rate is constant.

11.4.2 Proper Use of Bed Material

Proper selection of the bed material to be used in a fluidized bed system is very important. The size used will dictate the range of flow rate that may be used. A large bed material will provide higher flow rates through the system. If the gasifier does not have sufficient freeboard, the bed material may be elutriated out of the system. Table 11.6 summarizes an example of a type of material that may be used in a fluidized bed. A corresponding range of sizes is available from the manufacturer. This bed material, called Vanguard Refractory Grade Aggregate, is produced from olivine, a high-purity, naturally occurring forsterite (Mg_2SiO_4) and fayalite (Fe_2SiO_4). The bed

TABLE 11.6
Particle Size Analysis and Properties of Olivine Bed Material Used for Fluidized Bed Gasifiers

Mesh Size		Vanguard Refractory Grade Aggregates (Aurora, IN)			
ASTM	**Microns**	**3/8″×6**	**4×18**	**6×20**	**10×18**
3/8″	9.5 mm	1.7	-	-	
1/4″	6.3 mm	34.5	-	-	-
4	4.75 mm	54.7	5.5	6.5	-
6	3.35 mm	6.7	36.1	27.3	-
8	2.36 mm	0.9	37.1	24.9	-
12	1.07 mm	0.2	15.0	18.5	28.3
16	1.18 mm	-	3.4	13.5	43.9
20	850	-	0.9	4.8	23.2
30	600	-	0.3	1.4	1.3
50	300	-	0.5	1.6	1.2
−50	−300	1.3	1.4	1.9	1.8

material is distributed by Unimin Corporation (New Canaan, CT), the American Subsidiary of Sibelco (Aaheim, Norway) and the leading US producer of silica sand. Unimin Corporation (Unimin Corporation, 2010) specializes in industrial minerals and supplies olivine for various 342fluidized bed applications.

These bed material grades are mined, processed, and screened with rigid adherence to the QIP^{SM} quality assurance programs. As a result, the materials are consistent in grain fineness. Appropriate chemistries guarantee the MgO units meet performance requirements for high-temperature environments. The fusion point temperatures range between 2,800°F and 3,200°F (1,538°C and 1,760°C) following ASTM C-24 standards. These temperatures are well above the temperatures encountered in gasification systems and therefore will not deteriorate, slag, or melt. The free silica content is less than 0.1% by using x-ray diffraction methods, while the thermal conductivity is 0.0025 cal/cm^3-s°C at 1,000°C. The bulk density is around 90–109 lb/ft^3 (1,444–1,748 kg/m^3), and the specific gravity is between 3.2 and 3.4 g/cm^3 (199 and 212 lb/ft^3).

The first two columns in Table 11.6 list the typical mean percentage retained on individual mesh size sieves. If one were to use any of these grades, the information shown in Example 1.1 may be used to estimate the mean particle size, d_p. Once the d_p is known, the minimum and maximum fluidizing velocity will be estimated. These values will need to be corrected by using the actual operating temperatures, which will be affected by the air density values. Note that a better protocol, after calculating the fluidizing velocities, is to test the bed material in actual cold fluidization tests.

11.4.3 Cold Fluidization Tests

It is recommended that a cold fluidization setup be available to validate the calculations for the minimum and maximum fluidizing velocity to be used with a selected bed material in a gasification system. The system needs a blower whose flow rate can be incrementally controlled. A laminar flow element is used between the blower and blower control. This configuration allows one to estimate the actual flow rate flowing through the test bed gasifier. The reactor is made from clear plexiglass with a design similar to the fluidized bed. In summary, the test setup air distribution system must be the same as that of the actual gasifier. Note the presence of several pressure taps through the reactor column at various heights. These taps allow one to ascertain the effect of different bed heights.

11.5 DESIGNING DIMENSIONS OF FLUIDIZED BED GASIFIER

In designing the dimensions of a fluidized bed gasifier, the process could start with the design output or the throughput. As discussed in previous sections, the minimum throughput of a fluidized bed gasifier is approximately 9 GJ/h/m^2 (0.8 MMBtu/h/ft^2) to as high as 22.7 GJ/h/m^2 (2 MMBtu/h/ft^2). Thus, if the output in units of GJ/h (Btu/h) is desired, the cross-sectional area is easily calculated by dividing the gasifier energy output with the assumed gasifier throughput as shown in Equation 11.9.

$$\text{Gasifier area}\left(\text{m}^2\right) = \frac{\text{Gasifier output}\left(\text{GJ/h}\right)}{\text{Gasifier throughput}\left(\text{GJ/h/m}^2\right)} \tag{11.9}$$

If 20 GJ/h (18.96 MMBtu/h) is required, the minimum area of a gasifier should be 2.22 m^2 (23.9 ft^2). If the gasifier's cross section is circular, this is equivalent to about 1.68 m (5.5 ft). See Example 11.7 to check how these numbers were derived. For a given size of a fluidized bed gasifier, the throughput may be increased by increasing the biomass feed rate. The limitation will be on the capability of the system for the range of feed rate desired as well as limitations in the amount of fluidizing air available. In this case, there will be instances where the bed material will have to be changed to accommodate a higher air flow rate. These changes are what enabled the fluidized bed gasifier to improve its TDR.

Example 11.7: Determining Dimensions of Fluidized Bed Gasifier

The desired throughput of the gasifier is 20 GJ/h and the throughput to use is 9 GJ/h/m^2 (0.8 MMBtu/h/ft^2). Estimate the size of this fluidized bed gasifier. If the cross section is a circle, calculate the appropriate diameter for this problem.

Solution:

a. Using Equation 11.9, we have the following:

$$\text{Gasifier area}\left(\text{m}^2\right) = 20\frac{\text{GJ}}{\text{h}} \times \frac{\text{h}-\text{m}^2}{9\,\text{GJ}} = 2.22\,\text{m}^2$$

b. If the cross section is a circle, then the diameter is calculated using the following equation:

$$\text{Area}\left(\text{m}^2\right) = \frac{\pi d^2}{4} \quad \text{or} \quad \text{d(m)} = \sqrt{\frac{4\,\text{A}}{\pi}}$$

c. The diameter is determined as follows:

$$\text{d(m)} = \sqrt{\frac{4\,\text{A}}{\pi}} = \sqrt{\frac{4 \times 2.22\,\text{m}^2}{\pi}} = 1.68\text{ m or }(5.5\,\text{ft})$$

Once the dimensions of the gasifier have been designed, the next step would be to design the cyclone char removal system. Note that for a given throughput and diameter, the required air flow rate for the gasification mode to occur may easily be calculated, if the ultimate analysis of the biomass in question is known, along with the biomass feed rate. This has already been introduced in Chapter 10. For example, if the required air for the gasification mode is $1.5\,m^3/kg$ of biomass, and the biomass feed rate to be used is 1 kg/min, the air flow rate to use during the gasification (at cold state) will be $1.5\,m^3/min$ (52.9 cfm). Use this air flow rate at standard conditions to design the cyclone char removal system.

11.6 DESIGNING DIMENSIONS OF SERIES CYCLONE CHAR REMOVAL SYSTEM

A cyclone separator is a cost-effective device that removes char particulates from a syngas stream in fluidized bed gasifiers. In designing the cyclones for char particulate removal, it is customary to use two cyclones in series. The first cyclone is usually a less efficient cyclone, the purpose of which is to remove the larger particulate char. The less efficient cyclone is followed by a more efficient cyclone that removes the fine particles. A cyclone design procedure was outlined in Cooper and Alley (1994). This procedure was referred to as the classic cyclone design (CCD) process that was originally developed by Lapple in the 1950s. The CCD process was considered the standard method for cyclone design. However, research resulted in improved design procedures, which have subsequently been adopted. The most recent summary of publications for cyclone design is the dissertation by Wang (2004), which will be used in this chapter.

11.6.1 Primary (Low-Efficiency) Cyclone

The cyclones used in the agricultural industry are appropriate for use in a fluidized bed gasifier, where usually higher air flow rates are used. The most common cyclones used are the 2D2D (Shepherd and Lapple, 1939) and the 1D3D (Parnell and Davis, 1979). The letter "D" in the designation refers to the cyclone barrel diameter. The first number preceding the letter D refers to the height of the barrel and the second number refers to the cone length. So, the 2D2D has a barrel length equal to twice the diameter and the cone length also equal to twice the diameter. The 1D1D and the 2D2D are considered low-efficiency cyclones and are commonly used in the first stage. However, this "rule" is not inviolable. For example, the author conducted research to produce activated charcoal with a fluidized bed gasifier that used a 1D3D cyclone for the first stage and a more efficient 1D5D for the second stage (Capareda, 1990). The unconventional series cyclones removed as much char from the synthesis gas streams as the more classic types. However, care must be taken when employing nontraditional designs because they usually result in a higher pressure drop. The schematic of the 1D2D cyclone, including all the associated dimensions, is shown in Figure 11.5. Example 11.8 illustrates the simplicity of this design.

The primary cyclone will be designed based on the appropriate flow rate. Wang (2004) recommended a flow velocity of 12.2 ± 2 m/s (2,400 ± 400 ft/min) for the 1D2D cyclones. Thus, if the flow rate used for the gasification was 1.5 m³/min (standard conditions), the diameter of the cyclone would be 12.8 cm (5 in), as illustrated in Example 11.8. The simple formula to use is shown in Equation 11.10, for the metric system of units.

$$D_c\,(\text{cm}) = 100 \times \sqrt{\frac{8 \times Q\left(\text{m}^3/\text{s}\right)}{V\left(\text{m/s}\right)}} \tag{11.10}$$

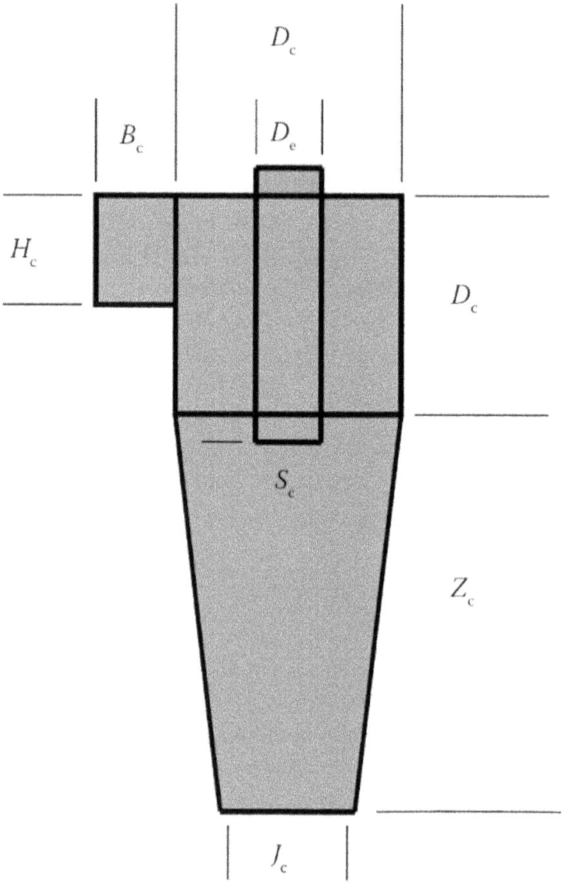

FIGURE 11.5 Dimensions nomenclature for cyclone design.

where
Q = design flow rate, m³/s
V = recommended flow velocity, 1.22 m/s

The simple formula to use is shown in Equation 11.11 for the English system of units.

$$D_c(\text{in}) = 12 \times \sqrt{\frac{8 \times Q\left(\text{ft}^3/\text{min}\right)}{V\,(\text{ft/min})}} \tag{11.11}$$

where
Q = design flow rate, ft/min
V = recommended flow velocity, 2,400 ft/min

Once the barrel diameter has been estimated, the other dimensions may be calculated using the relationships for a 1D2D cyclone that are listed in Table 11.7.

Example 11.8: Determining Dimensions of 1D2D Cyclone

Design a 1D2D cyclone using a design flow rate of 1.5 m³/min (52.9 cfm). Calculate the dimensions for the other nomenclatures.

Solution:

a. Using Equation 11.11, we have the following after converting 1.5 m³/min into 0.025 m³/s:

$$D_c(\text{cm}) = 100 \times \sqrt{\frac{8 \times 0.025\left(\text{m}^3/\text{s}\right)}{12.2(\text{m/s})}} = 12.8\ \text{cm}\left(5\,\text{in}\right)$$

$$D_c(\text{in}) = 12 \times \sqrt{\frac{8 \times 52.9\left(\text{ft}^3/\text{min}\right)}{2{,}400(\text{ft/min})}} = 5\ \text{in}\left(12.8\ \text{cm}\right)$$

b. The other nomenclatures should have the following dimensions:

$$B_c = D_c/4 = 3.2\ \text{cm} \qquad S_c = \text{Dc}/8 = 1.6\ \text{cm}$$
$$H_c = D_c/2 = 6.4\ \text{cm} \qquad Z_c = 2\text{Dc} = 25.6\,\text{cm}$$
$$D_e = D_c/1.6 = 8\ \text{cm} \qquad J_c = \text{Dc}/4 = 3.2\,\text{cm}$$

TABLE 11.7
Nomenclatures for Design of 1D2D Cyclone

Parameter	Actual Dimensions	Parameter	Actual Dimensions
B_c	$D_c/4$	S_c	$D_c/8$
H_c	$D_c/2$	Z_c	$2D_c$
D_e	$D_c/1.6$	J_c	$D_c/4$

TABLE 11.8
Nomenclatures for Design of 1D3D Cyclone

Parameter	Actual Dimensions	Parameter	Actual Dimensions
B_c	$D_c/4$	S_c	$D_c/8$
H_c	$D_c/2$	Z_c	$3D_c$
D_e	$D_c/1.6$	J_c	$D_c/4$

11.6.2 Secondary (High-Efficiency) Cyclone

In a two-stage (series) cyclone system, the secondary cyclone is usually more efficient relative to the first. The secondary cyclone should be designed based on the recommended flow velocity of 16.3 ± 2 m/s (3,200+400 ft/min). Generally, a 1D3D secondary cyclone is used (Wang, 2004). The same procedure will be used as in the first-stage cyclone, except for a different flow velocity. Standard conditions must also be used. Thus, if a volumetric flow rate of 1.5 m^3/min (52.9 ft^3/min) is used, the appropriate diameter of the secondary cyclone will be 17.7 cm (4.5 in), as shown in Example 11.9. For ease of fabrication, the dimensions are usually rounded off to the nearest common fraction of an inch. As in the previous Section 11.6.1, once the barrel diameter is estimated, the other dimensions may be calculated using the relationships for a 1D3D cyclone that are listed in Table 11.8.

Example 11.9: Determining Dimensions of 1D3D Cyclone

Design a 1D3D cyclone using a design flow rate of 1.5 m^3/min (52.9 cfm). Calculate the dimensions for the other nomenclatures listed in Table 11.8.

Solution:

a. Using Equation 11.11, we have the following after converting 1.5 m^3/min into 0.025 m^3/s and a flow velocity of 16.3 m^3/s:

$$D_c(\text{cm}) = 100 \times \sqrt{\frac{8 \times 0.025(\text{m}^3/\text{s})}{16.3(\text{m/s})}} = 11.1\ \text{cm}(4.4\ \text{in})$$

$$D_c(\text{in}) = 12 \times \sqrt{\frac{8 \times 52.9(\text{ft}^3/\text{min})}{3,200(\text{ft/min})}} = 4.4\ \text{in}\ (11.1\ \text{cm})$$

b. A nominal size of 11.43 cm (4 ½ in) will be used for ease of fabrication. Thus, the other nomenclatures should have the following dimensions:

$$B_c = D_c/4 = 2.8575\,\text{cm} \qquad S_c = D_c/8 = 1.4288\,\text{cm}$$

$$H_c = D_c/2 = 5.715\,\text{cm} \qquad Z_c = 3D_c = 34.29\,\text{cm}(13.3'')$$

$$D_e = D_c/1.6 = 7.1438\ \text{cm} \qquad J_c = D_c/4 = 2.8575\,\text{cm}$$

11.6.3 Pressure Drop in Cyclones

The cyclone design process, called TCD by Wang (2004), also provided an empirical model for cyclone pressure drop calculations. This model is shown in Equation 11.12:

$$\Delta P = K \times (VP_i + VP_o) \tag{11.12}$$

where

ΔP = pressure drop through the cyclone, mm H_2O
K = a dimensionless empirical constant, 5.1 for 1D3D and 3.4 for 1D2D
VP_i = inlet velocity pressure, mm H_2O
VP_o = outlet velocity pressure, mm H_2O

This pressure drop calculation is much simpler than the CCD design outlined in Cooper and Alley (1994). Note that this pressure drop is independent of the cyclone size. Example 11.10 shows how this equation is used to estimate pressure drop through the cyclones. For a small 15.24 cm (6 in) cyclone, the pressure drop is only around 388 Pa (1.56 in H_2O) and for a large 91.44 cm (36 in) 1D2D cyclone, the pressure drop is about 1,071 Pa (4.3 in H_2O). Please refer to the dissertation of Wang (2004) for a more detailed discussion of pressure drop in cyclones. Note that the pressure drop is not dependent on the dimensions of the cyclones, it is very much influenced by the density of air used during the test. Consequently, this density-related pressure drop must be accounted for in the calculations.

Example 11.10: Predicting Pressure Drop in Cyclones

Predict the pressure drop in small 1D2D and 1D3D cyclones with the following data: The inlet velocity pressure for a 15.24 (6 in) 1D3D cyclone was 0.64 in H_2O and the outlet velocity pressure was 0.20 in H_2O. The inlet velocity pressure for a 91.44 cm (36 in) 1D2D cyclone was 0.36 in H_2O and the outlet velocity pressure was 0.10 in H_2O. Convert the units into Pa.

Solution:

a. Using Equation 11.12, we have the following calculations for the 1D2D cyclone:

$$\Delta P = K \times (VP_i + VP_o) = 3.4 \times (0.36 + 0.10) = 1.56 \text{ in } H_2O$$

b. For the 1D3D cyclone, the calculations are as follows:

$$\Delta P = K \times (VP_i + VP_o) = 5.1 \times (0.64 + 0.2) = 4.3 \text{ in } H_2O$$

c. Converting the units into Pa is shown here. One inch of water is equivalent to 249.1 Pa, as shown:

$$\Delta P = 1.56 \text{ in } H_2O \times \frac{249.1\,\text{Pa}}{1 \text{ in } H_2O} = 388 \text{ Pa}$$

$$\Delta P = 4.3 \text{ in } H_2O \times \frac{249.1 \text{ Pa}}{1 \text{ in } H_2O} = 1{,}071\,\text{Pa}$$

11.7 DIRECT USE OF SYNTHESIS GAS FOR HEAT AND STEAM PRODUCTION

The synthesis gas produced from a fluidized bed gasifier may be used immediately for heat or steam power production. Data in Table 11.5 may be used to estimate the amount of energy that may be utilized during the operation. For example, if sorghum biomass is used during the gasification process, the amount of synthesis gas produced on a 305-mm-diameter fluidized bed gasifier will be around 2.04 m^3/kg (32.72 ft^3/lb) of biomass used. If the biomass feed rate is around 1.1364 kg/min (2.5 lb/min), some 2.32 m^3/min (81.9 ft^3/min) of synthesis gas will be produced. If the heating value of the synthesis gas is 4.09 MJ/m^3 (110 Btu/ft^3), the energy output will be equivalent to around 158 kW (539,147 Btu/h). This is the energy contained in the synthesis gas that is produced. If this heat is used in a steam boiler, the steam output can be easily calculated, depending on the combustion efficiency of the fuel as well as the thermal efficiency of the steam boiler system. The typical net boiler efficiencies vary between a low of 65% to a high of 87% for several passes. The overall steam conversion efficiency for biomass systems may be around 70% (Council of Industrial Boiler Owners, 2003). Thus, for the gasifier described above, the steam energy output could be around 110 kW (375,355 Btu/h). This is still quite a lot of energy from a mere 305-mm-diameter fluidized bed gasifier.

11.8 ELECTRICAL POWER PRODUCTION FROM FLUIDIZED BED GASIFICATION

Synthesis gas may also be used to generate electrical power. Although the steam cycle is the most common system used for coal, it may not be suitable for a gasification system, primarily due to the high cost associated with the steam power cycle and the need for enormous amounts of water for the operation. Perhaps the most practical use of syngas for power generation, especially for smaller and modular system of less than 6 MW, is the use of internal combustion engines that are suited for the synthesis gas from gasifiers. There are now engine manufacturers that design engines to handle low Btu gas or lower energy synthesis gas from gasification systems. These systems were originally designed to use methane from landfills. It has now been expanded to include synthesis gas from biomass gasifiers (Packham, 2007). There are now large reciprocating generator sets than can run reliably on alternative fuels, such as low Btu methane gas from landfills and synthesis gas from biomass gasifiers. Combustion turbines are also being used. However, because of the high cost, reciprocating engines are now the most popular choice. There are also engines that can handle the bifuel systems, that is, the use of one conventional fuel, such as diesel or natural gas, and a low Btu fuel from syngas or landfill gas. In this case, a bifuel control system is required. There are various manufacturers of bifuel systems in the United States, including the GTI Bifuel system from Altronics (Girard, OH), the ComAP Bifuel system (Roscoe, IL), and the Generac Bifuel system from Generac Industrial Power (Waukeshia, WI).

A bifuel system is defined as one that operates using the simultaneous combustion of two fuels. The main fuel could be diesel and the secondary fuel could be methane from landfill or synthesis gas from gasifiers. If two conventional fuels are used (e.g., natural gas and diesel), these bifuel systems after the conversion can operate on either 100% diesel fuel or alternatively on a mixture of diesel fuel and natural gas (or other methane-based fuels such as landfill gas). No modifications are made to the engine. The bifuel systems are mainly retrofit subsystems that are mounted externally on the engine and, therefore, do not require alteration of any critical engine part. Some companies claim a minimal loss in power while operating on the bifuel mode. The system is supposed to maintain the compression ratio values for the engine. In fact, there will usually be some deration (reduction in power) due to composition and quality of the gas supply. If diesel and pure natural gas are used, the replaced diesel fuels will be compensated for by the increased volume of natural gas displaced. For example, for each 3.785 L (1 gal) of diesel displaced during the bifuel operation, there will be a corresponding consumption of approximately 3.967 m^3 (140 ft^3) of natural gas introduced into the engine combustion chamber. The energy equivalents will be the same, assuming natural gas has an energy content of 34.6 MJ/m^3 (930 Btu/scf); and the No. 2 diesel fuel has an energy content of 36 MJ/L (129,000 Btu/gal); then, 3.785 L will have a total energy of 136 MJ. This will result in similar total energy input of $34.6\ \text{MJ/m}^3 \times 3.967\ \text{m}^3 = 137\ \text{MJ}$ (i.e. $930\ \text{Btu/ft}^3 \times 140\ \text{ft}^3 = 130{,}200\ \text{Btu}$).

The major components of a bifuel system are as follows:

1. Air–fuel mixer
2. Gas power valve
3. Gas train
4. Control panel

Air will have to be mixed with the secondary fuel before it is introduced into the combustion chamber because the secondary fuel is normally injected into the air intake valve of the engine. The gas power valve regulates the pressure to ensure that an appropriate amount of secondary fuel is injected equal to the energy content of displaced primary fuel. The gas train conditions the secondary fuel before admission and includes the secondary fuel filter and the shut-off valve, in case of over pressure or for shutdown purposes. The control panel monitors and controls the overall operational process, including a safety protocol during operation. This panel monitors several temperature and pressure control points within the system to ensure that all systems are operating under normal conditions. Alarms are also included to warn of impending errors in the system operations.

Example 11.11: Conversion of Biomass to Electrical Power Using Fluidized Bed

How many acres would be required to build a 1 MW (energy output) biomass power plant (operated 350 days a year and 24 h per day) if the heating value of the biomass (dry) is 8,500 Btu/lb? The biomass yield in tons (dry) per acre per year is about 5. Assume a conversion efficiency of 15% using a fluidized bed gasifier.

Solution:

a. The energy input must be equal to (1 MW/0.15 = 6.67 MW).
b. This is a simple conversion process using dimensional analysis to the correct units of acres as follows:

$$6.67\,\text{MW} \times \frac{1\times10^6\,\text{W}}{\text{MW}} \times \frac{\text{J}}{\text{W}-\text{s}} \times \frac{\text{Btu}}{1{,}055\,\text{J}} \times \frac{\text{lb}}{8{,}500\,\text{Btu/lb}} \times \frac{3{,}600\,\text{s}}{\text{h}}$$

$$\times \frac{24\,\text{h}}{\text{day}} \times \frac{350\,\text{days}}{\text{yr}} \times \frac{\text{ton}}{2{,}000\,\text{lb}} \times \frac{\text{acre}}{10\,\text{tons}} = 1{,}125\,\text{acres}$$

c. Answer = 1,125 acres.

11.9 CONCLUSION

In this chapter, the complete steps in designing a fluidized bed gasification system from feedstock selection to utilization of the synthesis gas were explained. The versatility of the fluidized bed gasification system has been illustrated for various applications of producing sensible heat, mechanical power, and electrical power. A broad range of operating conditions can be obtained from the fluidized bed gasification unit in terms of throughput and quality of synthesis gas produced. Various operating parameters could be varied that include the volumetric flow of air, operating pressure, and temperature. The throughput can be varied by replacing the bed material in the reactor and adjusting the appropriate fuel-to-air ratio such that it is within the gasification mode.

Various gasification conditions may be used as follows:

1. Instead of air, pure oxygen may be used to improve the synthesis gas quality.
2. Synthesis gas quality can be further enhanced by using steam.
3. Pressurized fluidized bed gasification systems have also been developed.
4. Operating the system at high fuel-to-air ratios could result in maximizing the char content.

The TDR for a fluidized bed gasifier is quite high and may be varied even during continued operation. In addition, for the same dimensions of a gasifier, the throughput may also be varied and doubled depending on the ability to feed more biomass per unit of time. In the operation of a fluidized bed gasifier, the temperature is maintained by a set fuel-to-air ratio. If this F/A ratio is maintained, the temperature should remain constant. Thus, if the fuel feed rate is increased, the air flow rate must be increased accordingly to keep the F/A ratios the same. This way, the temperature will be maintained. The only limitation would be the ability to feed more biomass and the ability to force more air flow through the system. One has to be aware of the minimum and maximum terminal velocity such that the bed material will not be elutriated out of the reactor. This is where the importance of additional freeboard in a fluidized bed gasifier becomes important.

If the synthesis gas is to be used in internal combustion engines, there are a few other things to note. These include the proper adjustment of temperature of the synthesis gas before injection into the combustion chamber of the engine. If the engine would require that the synthesis gas must be cooled, then the issue of tar generation becomes important and may have to be addressed. If the syngas is not cooled before injection into the engine, there should not be any tar accumulation. Some engines may also tolerate some particulates if they are on the micron level. Improper design of the cyclone char removal system will cause an additional solid fuel load on the engine, which some engines may not be able to tolerate, and the filtering system may be affected, resulting in additional regular maintenance procedures.

11.10 PROBLEMS AND DISCUSSION ISSUES

11.10.1 Calculation of Bed Material Average Particle Size

P11.1 Given the data in the accompanying table, estimate the average particle size for this bed material size distribution.

Mesh Size	Size Range (microns)	Material Percentage (%)	Weight (g)
8	2,380	0	0
10	1,680–1,190	4	20
14	1,190–1,680	47	235
20	840–1,190	35	175
28	595–840	12	60
Pan	<595	2	10
		100%	500 g

11.10.2 Determining Minimum Fluidizing Velocity

P11.2 Calculate the minimum fluidizing velocity (cm/s) for particles with the following data at 21°C.

- $d_p = 0.072\,\text{cm}$; $\mu = 1.78 \times 10^{-4}$ g/cm-s (air at 21°C)
- $\rho_g = 0.0012$ g/cm^3 (air at 21°C); $\rho_s = 2.4$ g/cm^3
- $g = 980$ cm/s^2; $\varepsilon_{mf} = 0.50$; $\Phi_s = 0.60$

11.10.3 Terminal Velocity Calculations

P11.3 Determine the terminal velocity (cm/s) for the particles with the following data at 21°C.

- $d_p = 0.072\,\text{cm}$; $\mu = 1.78 \times 10^{-4}$ g/cm-s (air at 21°C)
- $\rho_g = 0.0012$ g/cm^3 (air at 21°C); $\rho_s = 2.4$ g/cm^3
- $g = 980$ cm/s^2

11.10.4 Volumetric Flow Rate Calculations

P11.4 Calculate the minimum required volumetric flow rates in m^3/min (or cfm) for a 0.3048 m fluidized bed gasifier at the operating temperature of 816°C. Compare this value at cold conditions. The other parameters are provided as follows:

- $d_p = 0.072$ cm
- $\mu = 4.42 \times 10^{-4}$ g/cm-s (air at 816°C) (from air properties table)
- $\rho_g = 3.25 \times 10^{-4}$ g/cm³ (air at 816°C) (from air properties table)
- $\rho_s = 2.4$ g/cm³
- $g = 980$ cm/sec²
- $\varepsilon_f = 0.50$
- $\Phi_s = 0.60$

11.10.5 Pressure Drop through Fluidized Bed Gasifier

P11.5 Estimate the pressure drop through the fluidized bed gasifier using the following assumptions:

- L_{mf} = bed height at minimum fluidization = 100 cm
- The following are the data at 21°C:
- $d_p = 0.072$ cm; $\mu = 1.78 \times 10^{-4}$ g/cm-s (air at 21°C)
- $\rho_g = 0.0012$ g/cm³ (air at 21°C); $\rho_s = 2.4$ g/cm³
- $g = 980$ cm/s²; $\varepsilon_{mf} = 0.50$; $\Phi_s = 0.60$

11.10.6 Capacity of Screw Conveyor

P11.6 Calculate the mass flow rate of CGT with the screw conveyor at 100 rpm. The bulk density of CGT was 190 kg/m³ and the approximate loading factor, k, was 0.16. The following are the given dimensions of the screw conveyor:

- $D_{sc} = 10$ cm
- $D_{sh} = 7.75$ cm
- $\lambda = 7.4$ cm

11.10.7 Determining Dimensions of Fluidized Bed Gasifier

P11.7 The desired throughput of the gasifier is 30 GJ/h and the throughput to use is 22.7 GJ/h/m² (2 MMBtu/h/ft²). Estimate the size of this fluidized bed gasifier. If the cross section is a circle, calculate the appropriate diameter for this problem.

11.10.8 Determining Dimensions of 1D2D Cyclone

P11.8 Design a 1D2D cyclone using a design flow rate of 47.61 m³/min (1,680 cfm). Calculate the dimensions for the other nomenclatures shown in Figure 11.5.

11.10.9 Determining Dimensions of 1D3D Cyclone

P11.9 Design a 1D3D cyclone using a design flow rate of 47.61 m^3/min (1,680 cfm). Calculate the dimensions for the other nomenclatures shown in Figure 11.5.

11.10.10 Predicting Pressure Drop in Cyclones

P11.10 Predict the pressure drop in small 1D2D and 1D3D cyclones with the following data: The design inlet velocity for the 91.44 cm (36 in) 1D3D cyclone was (2,500 ft/min). The measured inlet velocity pressure for this 1D3D cyclone was 0.39 in H_2O and the outlet velocity pressure was 0.28 in H_2O. The design inlet velocity for the 1D2D cyclone was 16.3 m/s (3,200 ft/min). The measured inlet velocity pressure for a 15.24 cm (6 in) 1D2D cyclone was 0.64 in H_2O and the outlet velocity pressure was 0.19 in H_2O. Convert the units into Pa.

11.10.11 Conversion of Biomass into Electrical Power Using Fluidized Bed

P11.11 How many acres would be required to build a 3 MW (energy output) biomass power plant (operated 333 days a year and 24 h per day) if the heating value of the biomass (as received at 10% moisture) is 18 MJ/kg [7,755 Btu/lb]? The raw biomass yield is 16 tonnes/ha/yr. The moisture content of raw biomass right after harvest is 50%. Assume a conversion efficiency of 15% using a fluidized bed gasifier.

11.10.12 Fluidization Issues

P11.12 A problematic aspect that can arise from bubble growth in deep fluidized bed system is when bubbles grow so big that they almost cover the entire cross section of the reactor. Name this technical term in fluidized bed systems.

a. fouling b. slugging c. bubbling d. turbulence

11.10.13 Factors Affecting Fluidization

P11.13 True or False. Temperature is a strong function to estimate the minimum and maximum fluidization velocity in fluidized bed systems.

a. True b. False c. No effect d. No idea

11.10.14 Pressure Drop in Cyclone Particulate Biochar Removal

P11.14 True or False. The pressure drop in a cyclone is dependent on the cyclone size.

a. True b. False c. It depends d. No idea

11.10.15 Cyclone Design

P11.15 You are to design the first- and second-stage cyclone for a fluidized bed system having a size of 1 MW. Cotton gin trash is used as biomass. Specify the diameter of the 1D2D low-efficiency cyclone and a 1D3D high-efficiency cyclone for this application. State all assumptions made including references used.

11.10.16 Sizing of Fluidized Bed Gasifier

P11.16 Specify the dimensions of a 1 MW fluidized bed gasification system using cotton gin trash. Assume the reactor is circular and having a freeboard twice the diameter and a height twice the diameter. List all assumptions made including references.

REFERENCES

Capareda, S. C. 1990. Studies on activated carbon produced from thermal gasification of biomass wastes. PhD Dissertation, Department of Agricultural Engineering, Texas A&M University, College Station, TX.

Cooper, C. C. and G. C. Alley. 1994. *Air Pollution Control: A Design Approach.* Waveland Press Inc., Prospect Heights, IL.

Council of Industrial Boiler Owners (CIBO). 2003. Energy efficiency and industrial boiler efficiency: An industry perspective white paper. March 2003. CIBO, Warrenton, VA.

Gil, J., J. Â. Corella, M. P. Aznar and M. A. Caballero. 1999. Biomass gasification in atmospheric and bubbling fluidized bed: Effect of the type of gasifying agent on the product distribution. *Biomass and Bioenergy* 17: 389–403.

Kunii, D. and O. Levenspiel. 1977. *Fluidization Engineering.* Robert E. Krieger Publishing Company, Huntington, NY.

Maglinao, A. L., Jr. 2009. Instrumentation and evaluation of a pilot scale fluidized bed biomass gasification system. MS Thesis. Department of Biological and Agricultural Engineering, Texas A&M University, College Station, TX.

Packham, K. 2007. The case for waste to energy: Utilizing low-Btu reciprocating gas engine generators. Technical Power Topic #6015, Cummins Power Generation, Ramsgate, England, UK.

Parnell, C. B., Jr. and D. D. Davis. 1979. Predicted effects of the use of new cyclone designs on agricultural processing particle emissions. ASAE Paper No. SWR-79-040, presented at the 1979 Southwest Region Meeting of the ASAE, Hot Springs, AK.

Shepherd, C. B. and C. E. Lapple. 1939. Flow pattern and pressure drop in cyclone dust collectors. *Industrial and Engineering Chemistry* 31 (8): 872–984.

Unimin Corporation. 2010. *Product Brochure.* Vanguard Refractory Grade Aggregates, New Canaan, CT.

Wang, L. 2004. Theoretical study of cyclone design. PhD Dissertation, Biological and Agricultural Engineering Department, Texas A&M University, College Station, TX.

Woodstock, C. R. and J. S. Mason. 1987. *Bulk Solids Handling: An Introduction to the Practice and Technology.* Chapman & Hall Inc., New York.

ASSESS YOUR KNOWLEDGE

1. Design the dimensions of a fluidized bed gasifier (assuming circular cross section, give diameter in feet) with a throughput of 2 MMBtu/h/ft^2. The heating value of the biomass to be used was 5,435 Btu/lb. The feed rate for biomass was 3,000 lbs/h.
   a. 3.23
   b. 4.1
   c. 6.1
   d. 8.2

2. Design the diameter (in feet) of the 1D2D primary cyclone for a gasification system using an inlet design velocity of 2,400 ft/min and using a design flow rate of 1,500 CFM of standard air. Use the formula given in lecture class for this problem as shown.

$$D_c(\text{in}) = 12 \times \sqrt{\frac{8 \times Q\left(\frac{\text{ft}^3}{\text{min}}\right)}{V\left(\frac{\text{ft}}{\text{min}}\right)}}$$

   a. 4.5
   b. 26.83
   c. 2.24
   d. 1.5

11. Calculate the minimum fluidizing velocity (cm/s) for particles with the following data: $d_p = 500$ microns (0.05 cm); $\mu = 1.78 \times 10^{-4}$ g/cm-s (air at 21°C); $\rho_g = 0.0012$ g/cm^3; $\rho_s = 2.5$ g/cm^3; g = 980 cm/s^2; $\varepsilon_{mf} = 0.50$; $\Phi_s = 0.80$.
    a. 42.6
    b. 36.7
    c. 980
    d. 12.4

12. Design the diameter (in feet) of the 1D3D secondary cyclone for a gasification system using an inlet design velocity of 3,200 ft/min and using a design flow rate of 1,500 CFM of standard air. Use the formula given in lecture class for this problem.
    a. 2.24
    b. 23.24
    c. 4.82
    d. 1.94

3. Design the diameter (in inches) of the 1D3D secondary cyclone for a gasification system using an inlet design velocity of 3,200 ft/min and using a design flow rate of 1,100 CFM of standard air. Use the formula given in lecture class for this problem.
   a. 1.66
   b. 23.24
   c. 4.82
   d. 20

4. If the minimum fluidizing velocity for a given bed material is 40 cm/s, what would be the volumetric flow rate in CFM if the reactor has a diameter of 2 ft.
   a. 42.7
   b. 126
   c. 309
   d. 247

13. Design the diameter (in inches) of the 1D2D primary cyclone for a gasification system using an inlet design velocity of 2,400 ft/min and using a design flow rate of 1,100 CFM of standard air. Use the formula given in lecture class for this problem as shown.

$$D_c(\text{in}) = 12 \times \sqrt{\frac{8 \times Q\left(\frac{\text{ft}^3}{\text{min}}\right)}{V\left(\frac{\text{ft}}{\text{min}}\right)}}$$

   a. 4.5
   b. 26.83
   c. 24
   d. 1.9

14. Cotton gin trash was used as biomass fuel in a gasification facility. The ultimate analysis by weight is as follows: C = 41.6%; H = 5.8%; O = 43.5%; N = 1%; and S = 0.3%; Ash = 7.8% (i.e., for every 100 g of material, 41.6 g is carbon and so on). Determine the air-to-fuel ratio in units of lb/lb if only 50% of the stoichiometric air is used in the gasification process.
   a. 1.6
   b. 2.7
   c. 5.4
   d. 7.2

5. Estimate the diameter (in inches) of a 1D2D cyclone using a design inlet velocity of 2,400 ft/min and a design flow rate of 250 CFM.
   a. 15
   b. 11
   c. 21
   d. 60

6. How much biomass is to be inputted (in tons per h) if one would design a 1 MW power plant fueled by biomass. The overall conversion efficiency is 15%, the high heating value of biomass is 7,500 Btu/lb [17.4 MJ/kg], and the power plant is running continuously for 365 days in a year.
   a. 1.5
   b. 0.91
   c. 5.0
   d. 8.0

7. Gasifiers are designed based on throughput (Btu/h/ft^2) and the feed rate of fuel. Calculate the cross-sectional area of a gasifier (in square feet) required to convert 1 ton/h of biomass if the throughput is 2 million Btu/h/ft^2. The high heating value of the biomass is 7,500 Btu/lb [17.4 MJ/kg].
   a. 7.5
   b. 24.1
   c. 32.6
   d. 60

15. Determine the stoichiometric air-to-fuel ratio for the complete combustion of municipal solid wastes (MSW) (in units of lb/lb). Assume the chemical formula for MSW is $C_4H_6O_2N_{0.08}S_{0.009}$.
   a. 5.17
   b. 7.07
   c. 4.5
   d. 4.94

16. Determine the amount of air required (in units of lbs air per lb of biomass) when municipal solid wastes (MSW) is gasified. First, determine the stoichiometric air-to-fuel ratio for the complete combustion of municipal solid wastes (MSW) (in units of lb/lb) with a chemical formula of $C_4H_6O_2N_{0.08}S_{0.009}$ and then determine the air-to-fuel ratio (lb/lb) during gasification if only 70% of stoichiometric air is to be used instead of 100% air.
   a. 5.17
   b. 7.07
   c. 4.5
   d. 4.95

17. Determine the amount of biomass required in lbs per hour to generate one megawatt (1 MW) of electrical power through gasification. Assume the following: municipal solid waste is used with a heating value of 10,000 Btu/lb; the overall conversion efficiency is 13%.
   a. 1.31
   b. 2.62
   c. 4,500
   d. 2,625

8. The ultimate analysis for corn cob is given as follows: C = 46.2%; H = 7.6%; O = 42.3%; N = 1.2%; and S = 0.3% and the ash content is 2.4%. Determine the chemical formula for this biomass material.
   a. $C_{3.85}H_{7.6}O_{2.64}N_{0.086}S_{0.009}$
   b. $C_{46.2}H_{7.6}O_{1.2}N_{1.2}S_{0.3}$
   c. $C_1H_{7.6}O_{2.64}N_{0.086}S_{0.009}$
   d. $C_{3.7}H_{5.7}O_{4.1}N_1S_{0.2}$

9. How many acres would be required to build a 1 MW (energy output) biomass power plant using fluidized bed gasification (operated 350 days a year and 24 h/day) if the heating value of the biomass (dry) is 9,500 Btu/lb? The biomass yield in tons (dry) per acre per year in dry tons is about 15. Assume an overall conversion efficiency of 13%.
   a. 77,341
   b. 7,734
   c. 773
   d. 77.3

18. Determine the amount of air required (in units of cubic feet of air per lb of biomass) when municipal solid waste (MSW) is gasified. First, determine the stoichiometric air-to-fuel ratio for the complete combustion of municipal solid wastes (MSW) (in units of lb/lb) with a chemical formula of $C_4H_6O_2N_{0.08}S_{0.009}$ and then determine the air-to-fuel ratio (lb/lb) during gasification if only 70% of stoichiometric air is to be used instead of 100% air. Convert this A/F value in units of cubic feet of air per lb of fuel assuming air density at STP of 0.075 lbs/ft^3.
   a. 33
   b. 66
   c. 4.5
   d. 4.94

19. Determine the amount of biomass required in tons per hour to generate one megawatt (1 MW) of electrical power through gasification. Assume the following: municipal solid waste is used with a heating value of 10,000 Btu/lb; the overall conversion efficiency is 13%.
   a. 1.31
   b. 2.62
   c. 4,500
   d. 2,625

10. Design the dimensions of a fluidized bed gasifier (assuming circular cross section) with a throughput of 2 MMBtu/h/ft^2. The heating value of the biomass to be used was 9,500 Btu/lb. The feed rate for biomass was 3,000 lbs/h.
    a. 2.15
    b. 4.3
    c. 6.6
    d. 8.2

20. Estimate the average particle size (in microns) for the bed material whose weight fractions are provided in the table.

Diameter range (u)	Weight of Material Captured
50–75	50
75–100	120
100–125	200
125–150	70
150–175	60
Total	500 g

    a. 103.6
    b. 85.9
    c. 207
    d. 965

Answers: 1a, 2c, 3d, 4d, 5b, 6b, 7a, 8a, 9c, 10b, 11b, 12d, 13c, 14b, 15b, 16d, 17d, 18b, 19a, 20a

ADDITIONAL LEARNING RESOURCES

- Access the Lecture Notes at the Instructor and Student Resources: https://routledgetextbooks.com/textbooks/instructor_downloads/

12 Biomass Liquefaction

LEARNING OBJECTIVES

Upon completion of this chapter, one should be able to

1. Discuss ways to convert biomass into liquid hydrocarbon fuels by advanced thermochemical conversion processes.
2. Distinguish between direct and indirect liquefaction processes.
3. Enumerate the advantages and disadvantages of various biomass liquefaction processes.
4. Describe the different catalysts used in both direct and indirect liquefaction processes.
5. Compare the various technical advantages of biomass liquefaction processes over the other conversion processes discussed in previous chapters.

12.1 INTRODUCTION

A great number of biomass conversion processes exist for converting solid biomass directly into liquid fuels. Although biological conversion processes are always noted to produce liquid fuels, thermal conversion processes are always associated with the production of gaseous fuels such as synthesis gas (syngas). For most thermal conversion processes, additional steps are required to convert gaseous fuel into liquid fuel. Biomass liquefaction is normally associated with the production of liquid fuels directly or indirectly from thermal conversion of solid biomass. Sometimes, this process is also referred to as the "gas-to-liquid" or GTL process. There are basically two routes for biomass liquefaction processes: direct liquefaction, where biomass slurry is directly converted into hydrocarbon fuels, and indirect liquefaction, where biomass is first gasified before conversion into liquid fuels. Syngas becomes the primary feed input in the indirect conversion process, and the efficiency of the final product will depend on the heating value of the syngas. In previous chapters, it was shown that the syngas from pyrolysis has a higher heating value than those from gasification. Although the heating value of syngas from pyrolysis is higher, the amount is lessened if the bio-oil is to be recovered. Most biomass liquefaction processes are based on gasification systems to generate an abundant amount of syngas. In the operation of gasifiers, liquid product production is not favored. The energy content of syngas may be estimated following Example 12.1. Notice that (as discussed previously) the heating value of syngas from gasification systems is about half this value.

 DOI: 10.1201/9781003294306-12

Example 12.1: Heating Value of Synthesis Gas Produced from Pyrolysis

Determine the energy content of the syngas (in MJ/m³ based on major components) produced from the pyrolysis process with the following amounts: Total syngas collected was 120L with the following composition: H_2 = 20%, CH_4 = 10%, and CO = 15% by volume. The heating values of the gases at NTP are as follows: H_2 = 12.8 MJ/m³, CH_4 = 36.4 MJ/m³, and CO = 11.6 MJ/m³. Convert the HV in units of Btu/ft³.

Solution:

a. The energy content may be easily calculated by determining the actual volume of gas produced and multiplied by its heating value as follows:

$$H_2(\text{MJ}) = 120\,\text{L} \times 0.20 \times \frac{12.8\,\text{MJ}}{\text{m}^3} \times \frac{1\,\text{m}^3}{1{,}000\,\text{L}} = 0.307\,\text{MJ}$$

$$CH_4(\text{MJ}) = 120\,\text{L} \times 0.10 \times \frac{36.4\,\text{MJ}}{\text{m}^3} \times \frac{1\,\text{m}^3}{1{,}000\,\text{L}} = 0.437\,\text{MJ}$$

$$CO(\text{MJ}) = 120\,\text{L} \times 0.15 \times \frac{11.6\,\text{MJ}}{\text{m}^3} \times \frac{1\,\text{m}^3}{1{,}000\,\text{L}} = 0.209\,\text{MJ}$$

b. Collectively, the energy content of the syngas is about 0.953 MJ/120L or about 7.94 MJ/m³ (213 Btu/ft³) as shown below:

$$\text{Syngas}\left(\frac{\text{MJ}}{\text{m}^3}\right) = \frac{(0.307 + 0.437 + 0.209)\,\text{MJ}}{120\,\text{L}} \times \frac{1{,}000\,\text{L}}{\text{m}^3} = 7.94\frac{\text{MJ}}{\text{m}^3}$$

$$\text{Syngas}\left(\frac{\text{Btu}}{\text{ft}^3}\right) = \frac{7.94\,\text{MJ}}{\text{m}^3} \times \frac{\text{Btu}}{1055\,\text{J}} \times \left(\frac{1\,\text{m}}{3.28\,\text{ft}}\right)^3 \times \frac{1 \times 10^6\,\text{J}}{\text{MJ}} = 213\frac{\text{Btu}}{\text{ft}^3}$$

12.2 INDIRECT LIQUEFACTION PROCESSES

Indirect liquefaction of biomass normally starts with the use of syngas as a feedstock. A majority of indirect liquefaction processes was originally based on the conversion of natural gas or syngas derived from coal. Although syngas is composed mostly of hydrogen and carbon monoxide and not very high in methane concentration, the conversion of syngas into liquid fuels should work the same way. New energy companies are now developing biomass liquefaction plants using indirect liquefaction processes (e.g., Sasol in South Africa). The following indirect liquefaction processes have already been established:

1. Methanol production from syngas
2. Conversion of methanol into synthetic petroleum using the Mobil process
3. Fischer–Tropsch (FT) process from syngas
4. China Lake process

To produce large quantities of synthetic petroleum hydrocarbons, the concentration of syngas, particularly the ratio of carbon monoxide and hydrogen, must be adjusted to take advantage of the "water gas shift reaction." In this reaction, carbon monoxide reacts with steam to produce carbon dioxide and hydrogen. In most cases, excess carbon monoxide is produced when pure oxygen is used instead of air. The important reactions that lead to the highest enthalpy values of products include CO_2 reforming, steam reforming, and the Boudouard equation. The Boudouard reaction is the reaction of solid carbon with carbon dioxide to produce carbon monoxide. The basic steam-reforming process is the reaction of methane with water to produce carbon monoxide and hydrogen gas, while carbon dioxide reforming is the reaction of methane with carbon dioxide to produce carbon monoxide and hydrogen gas.

The various indirect liquefaction processes will be discussed individually, beginning with the most popular biomass liquefaction process of converting syngas ($CO+H_2$) into methanol.

12.2.1 Methanol Production from Synthesis Gas

Methanol has commonly been produced from syngas derived from the gasification of coal. This process is used extensively by the petrochemical industry. There are a variety of possible methods, but the most common approach is to combine hydrogen with carbon monoxide at high pressure (normally around 100 atm, 1,470 psi) and at around 300°C, using specially made catalysts such as copper oxide and zinc oxide (CuO-ZnO) combinations (Ren et al., 2003). Other catalysts include Cu-Zn-Al_2O_3 and CuZnO-Cr_2O_3. The basic reaction is shown in Equation 12.1. Note that this syngas may come from gasification or pyrolysis of biomass or other carbonaceous materials such as coal or lignite.

$$\begin{array}{ccc} CO+2H_2 & \rightarrow & CH_3OH \\ \text{Synthesis gas} & \rightarrow & \text{Methanol} \end{array} \tag{12.1}$$

Methanol yields of as high as 360 kg/dry tonne of wood have been reported (Bhattacharya, 1983). This corresponds to a theoretical conversion efficiency of between 30% and 40% as shown in Example 12.2. At present, however, syngas is produced from the methane component of natural gas rather than coal. These processes are commercially practiced at pressures of 40 atm and temperatures of around 850°C using a nickel catalyst, according to Equation 12.2. This reaction is commonly called steam-methane reforming or SMR, an endothermic process. An endothermic process is a process whereby energy is required for the reaction to proceed.

$$CH_4+H_2O \rightarrow CO+3H_2 \tag{12.2}$$

Methane can also undergo partial oxidation (with oxygen) to produce syngas, as shown in Equation 12.3. This reaction is exothermic (meaning heat is released) and may be used for the endothermic steam-reforming equation above (Equation 12.2).

$$2CH_4+O_2 \rightarrow 2CO+4H_2 \tag{12.3}$$

If Equations 12.2 and 12.3 are combined, the final resulting equation is called "autothermal reforming," because of the balanced energy generated for the endothermic and exothermic reactions. The CO and H_2 produced from the above reactions then react with another catalyst to produce methanol as in Equation 12.1. These three equations were used by companies such as the Imperial Chemical Industries (ICI, a British chemical company now called Syngenta [Wilmington, DE]) to produce syngas using pressures of between 50 and 100 atm and a temperature of 250°C (Trifiro, 2009). Note that the production of methanol uses 2 mol of hydrogen per mole of carbon monoxide, whereas the production of syngas from methane produces 3 mol of hydrogen. Therefore, excess hydrogen gas is generated. This excess hydrogen may be further converted into methanol, following Equation 12.4. In this equation, carbon dioxide reacts with hydrogen to generate methanol.

$$CO_2 + 3H_2 \rightarrow CH_3OH + H_2O \qquad (12.4)$$

Although natural gas is still the most economical feedstock for methanol production, many other feedstocks are now being used to produce synthesis gas using steam reforming. Coal is increasingly utilized, followed by biomass (primarily woodchips), the latter being close to carbon neutral because most CO_2 produced during the conversion processes is being used to generate biomass through photosynthesis. Some newer processes use homogeneous catalyst systems such as those carried out in the $Ni(CI)_4$-KOMe catalytic system (Li and Jiang, 1999). The influence of various factors such as the reaction temperature, the concentration of potassium methoxide and of nickel tetracarbonyl, and the reaction pressure on the catalytic activity has been investigated. Results revealed that the most preferred conditions for production of methanol using the above catalysts are at a temperature of 378 K and pressure of 1.8 MPa. The schematic for methanol production from syngas is shown in Figure 12.1.

Example 12.2: Methanol from Wood Conversion Efficiency

Determine the efficiency of converting 1 tonne of wood waste with a high heating value of 20 MJ/kg into pure methanol having a yield of 360 kg for every tonne. The energy content of pure methanol is 19.8 MJ/kg. Convert all units in the English system.

Solution:

a. This is the energy of methanol divided by total energy input from wood as follows:

$$\text{Input energy (MJ)} = 1 \text{ tonne wood} \times \frac{1{,}000 \text{ kg}}{1 \text{ tonne}} \times \frac{20 \text{ MJ}}{\text{kg}} = 20{,}000 \text{ MJ}$$

$$\text{Ouput energy (MJ)} = \frac{360 \text{ kgMeOH}}{} \times \frac{19.8 \text{ MJ}}{\text{kg}} = 7{,}128 \text{ MJ}$$

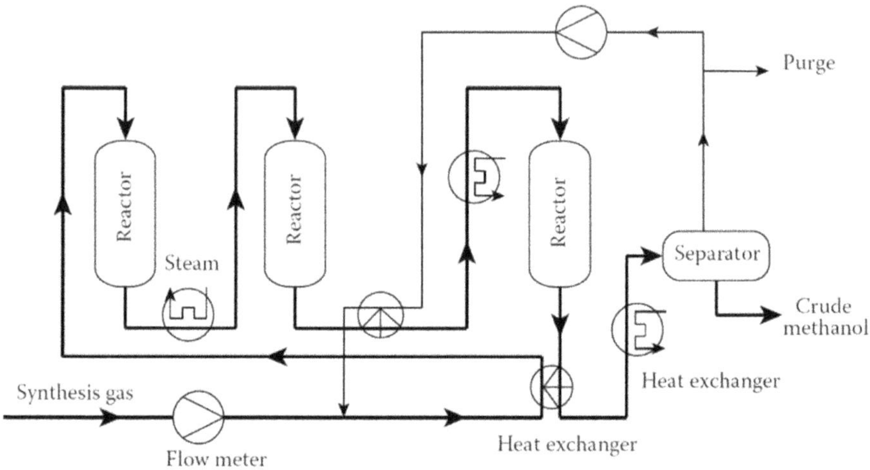

FIGURE 12.1 Schematic diagram of producing methanol from syngas. (Modified from Trifiro, F., From syngas to methanol and dimethylether. Lecture presented in Summer School 2009, University of Bologna, Spain. Sponsored by the International Center for Science and High Technology, UNIDO, Rome, Italy, 2009.)

$$\text{Efficiency} = \frac{\text{output}}{\text{input}} \times 100\% = \frac{7{,}128\ \text{MJ}}{20{,}000\ \text{MJ}} \times 100\% = 35.64\%$$

b. The English system of units is shown below:

$$\text{Input energy (MMBtu)} = 20{,}000\ \text{MJ} \times \frac{\text{Btu}}{1{,}055\ \text{J}} \times \frac{1 \times 10^6\ \text{J}}{1\ \text{MJ}} = 18.96\ \text{MMBtu}$$

$$\text{Output energy (MMBtu)} = 7128\ \text{MJ} \times \frac{\text{Btu}}{1{,}055\ \text{J}} \times \frac{1 \times 10^6\ \text{J}}{1\ \text{MJ}} = 6.76\ \text{MMBtu}$$

12.2.2 Mobil Process

Methanol can be further converted into high-octane gasoline using the process developed by Mobil Corporation (Chang and Silvestri, 1977). This process is based on the use of a shape-selective zeolite catalyst and patented by Mobil. Research was conducted at the Central Research Division of Mobil Research and Development Corporation in Princeton, NJ. The conversion efficiencies are reportedly in the order of 90% (see Example 12.3 for calculations), and a plant in New Zealand produces around 12,500 tonnes of synthetic petroleum per day using this technique (Bhattacharya, 1983). The methanol used is derived from the conversion from natural gas as discussed in the previous Section 12.2.1, or from other similar processes. The underlying equation is governed by that shown in Equation 12.5, where the methanol is first dehydrated to produce dimethyl ether. This dimethyl ether is then dehydrated further using a zeolite catalyst, ZSM-5, to give gasoline with 80% (by weight based on the organics in the product stream) C5+ hydrocarbon products.

$$\begin{array}{lll} 2CH_3OH \rightarrow CH_3OCH_3 & & + H_2O \\ \text{Methanol} \rightarrow \text{dimethyl ether} + \text{water} & & \\ 64\,\text{kg} \qquad 46\,\text{kg} & & + 18\,\text{kg} \end{array} \tag{12.5}$$

ZSM-5 is deactivated by the buildup of carbon on its active surface (called "coking") after several uses and may need to be regenerated. Regeneration is normally accomplished by burning off the coke in a stream of hot air at around 500°C (930°F). The number of regeneration cycles, however, is limited. Another device that may be used for the same purpose is called a petroleum hydrotreater (Wise and Silvestri, 1976). In this process, coal is initially used to produce syngas, then the syngas is used to produce methanol, which is subsequently converted into gasoline using the same catalyst. The quality of gasoline is quite high with octanes exceeding the then current requirements for both unleaded regular and leaded premium gasoline. The typical process conditions are described as above. This process is sometimes called the methanol-to-gasoline or MTG process, which was discovered accidentally (Kam et al., 1984). Zeolite ZSM-5 was being used to convert methanol into a fuel additive. Instead, the process produced dimethyl ether, which, with increasing space and contact time, produced olefins (alkanes), then paraffins (alkenes) and aromatics, with a mixture that is the same as the gasoline profile (Chang et al., 1981). Note that gasoline hydrocarbons are in the carbon range from C4 to C10 boiling range. Details of the MTG process may be found in the cited references, but the process steps are solely owned by Mobil Corporation.

Example 12.3: Mobil Process Conversion Efficiency

Determine the efficiency of converting methanol into gasoline using the governing equation (Equation 12.5), using the unit weights shown. The energy content of pure methanol is 19.8 MJ/kg and that of gasoline is 46.8 MJ/kg. Assume that only 30 kg of liquid organics are produced instead of 46 kg for every 64 kg of methanol, with 80% in the gasoline range. Convert all units to the English system.

Solution:

a. This is a simple energy output divided by energy input calculations as follows:

$$\text{Input energy (MJ)} = 64\,\text{kg methanol} \times \frac{19.8\,\text{MJ}}{\text{kg}} = 1267.2\,\text{MJ}$$

$$\text{Ouput energy (MJ)} = \frac{30\,\text{kg organics}}{\text{kg organics}} \times \frac{0.8\,\text{kg gasoline}}{} \times \frac{46.8\,\text{MJ}}{\text{kg gasoline}} = 1123.2\,\text{MJ}$$

$$\text{Efficiency} = \frac{\text{output}}{\text{input}} \times 100\% = \frac{1123.2\,\text{MJ}}{1267.2\,\text{MJ}} \times 100\% = 88.64\%$$

b. The English system of units is shown below.

$$\text{Input energy}\,(\text{MMBtu}) = 1267.2\,\text{MJ} \times \frac{\text{Btu}}{1{,}055\,\text{J}} \times \frac{1\times 10^{6}\,\text{J}}{1\,\text{MJ}} = 1.2\,\text{MMBtu}$$

$$\text{Output energy}\,(\text{MMBtu}) = 1123.2\,\text{MJ} \times \frac{\text{Btu}}{1{,}055\,\text{J}} \times \frac{1\times 10^{6}\,\text{J}}{1\,\text{MJ}} = 1.06\,\text{MMBtu}$$

12.2.3 Fischer–Tropsch Process

Syngas can also be used to produce hydrocarbon fuels using the FT process. The FT process was developed in the early 1920s and used extensively during the Second World War to produce synthetic gasoline and diesel fuels. Franz Fischer and Hans Tropsch were two chemists working at the Kaiser Wilhelm Institute for Coal Research in Mulheim, Ruhr, Germany, and were responsible for the development of this process (Steynberg, 2004). The FT process is always associated with the use of metal catalysts (fused or precipitated iron or cobalt catalysts), which are quite expensive. Another issue associated with the FT process is that a great variety of products are formed that are not exactly hydrocarbons. Products include olefins, alcohols, and waxes (Steynberg and Dry, 2004). The current research in FT technology is the search for catalysts that would result in a narrow product range and a cleaner product. Many FT technologies in the past used coal as the feedstocks and this has caused several problems, particularly coking or poisoning of catalysts with sulfur. This is where agricultural residues have an advantage. Agricultural biomass is normally low in sulfur and this would reduce poisoning problems during the catalytic step. The other major disadvantages are the high cost of equipment, the complexity of the process, and the required economies-of-scale (Steynberg, 2004).

The three basic steps in the production of liquid hydrocarbon fuels using the FT process is as follows:

1. Syngas preparation
2. FT synthesis
3. Product upgrading

The schematic diagram for the basic FT process is shown in Figure 12.2.

Syngas may be produced through pyrolysis or gasification of carbonaceous material such as agricultural biomass. Coal is the commonly used feedstock; however, recently, various biomasses are utilized, with CO and H_2 as the primary products of interest for the next step. In FT synthesis, there are two routes: (a) the low-temperature FT process (called LTFT) and (b) the high-temperature FT process (called HTFT). The LTFT is normally conducted at around 220°C–240°C and the HTFT at around 340°C. Iron catalysts are normally used with the high-temperature processes. Both iron and cobalt are commonly used with the low-temperature processes. For

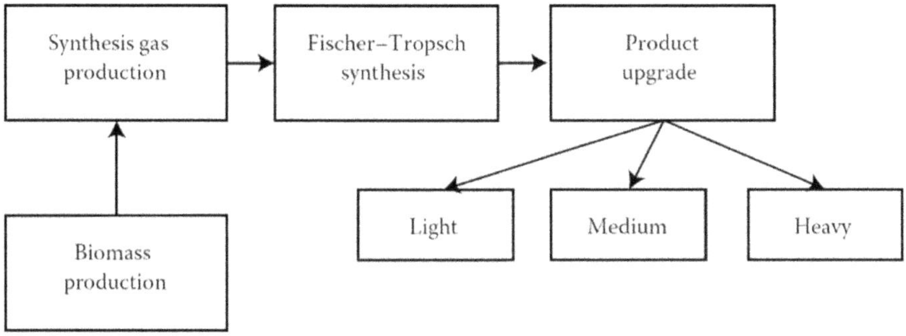

FIGURE 12.2 Schematic diagram of FT process.

example, using the HTFT and cobalt catalyst at 220°C, the percentages of alkanes to alkenes (in the C5–C12 carbon boiling point range) are 60:39 with 1% oxygenates. Under the same conditions, using the HTFT process with iron catalysts yields a ratio of 13:70 alkanes to alkenes, with 5% aromatics and 12% oxygenates (Dry, 2004). Unfortunately, these numbers are relative percentages of output within those carbon ranges. The other coproducts of the process include C13–C18 cut and some C24–C35 cut or waxes.

Note that the range of products is affected by the quality of syngas produced, as well as the ratio of CO to H_2. These gas qualities are defined (Aasberg-Petersen et al., 2004) in three classes as follows:

1. Low calorific value gas or LCV gas with heating values between 3.8 and 7.6 MJ/m^3
2. Medium calorific value gas or MCV gas with heating values between 10.5 and 16 MJ/m^3
3. High calorific value gas or HCV gas with heating values greater than 21 MJ/m^3

Example 2.4 shows the comparison of three types of syngas and their calorific value categories. There is an ideal H_2/CO ratio to be used for the FT process, which should be greater than 2.1. Most gasifier syngas has the H_2/CO ratio close to 1, as shown in Example 12.4. Thus, there is a need to upgrade the syngas, if necessary, and this would require additional H_2 gas from other sources. The syngas heating value must also be quite high to have a better product. The final product may be separated or fractionated into light, medium, or heavy hydrocarbon cuts, including the presence of heavy materials, such as wax.

There are several reported commercial FT plants in operation, and some are listed below (Dry and Steynberg, 2004):

1. Sasol (Sasolburg and Secunda, South Africa)
2. PetroSA/Mossgas (Mossel Bay, South Africa)
3. Shell SMDS (Bintulu, Malaysia)

There are additional similar facilities being built in China, Nigeria, the United States, Iran, Australia, and Indonesia, using combinations of HTFT and LTFT. The most popular is the Sasol plant in South Africa. Until 2004, the raw syngas for all three Sasol plants was derived from coal. However, at present, Sasol uses methane piped from Mozambique (Dry and Steynberg, 2004). The Sasol plant uses LTFT reactors that are operated to maximize the production of linear alkanes/alkenes and waxes. Sasol is the world's largest producer of linear paraffinic waxes. The Mossgas/PetroSA plant mainly produces gasoline and diesel fuel. Syngas is derived from an offshore natural gas plant. Note if methane is used in the FT process, it will have to be broken down into CO and H_2, as required by the FT process. This facility uses a circulating fluidized bed to produce the syngas required by the FT reactors. Alcohols, ketones, and aldehydes are produced; they are further hydrogenated to form alcohols. The succeeding processes involve isomerization, alkylation, and oligomerization to produce high-octane gasoline and diesel cuts. In this latter process, shape-selective acid zeolite catalysts are used to produce either the diesel range or gasoline range hydrocarbons. The total fuel production for this facility is approximately one million tonnes per year (Dry and Steynberg, 2004).

The Shell SMDS (Shell Middle Distillate Synthesis) plant mainly produces waxes. The initial gas feedstock is methane coming from offshore drills. The syngas is produced by a noncatalytic partial oxidation process at high pressure and about 1,400°C temperature. The reported carbon efficiency at the plant is greater than 95%. The H_2/CO ratio of the syngas is about 1.7, below the usual usage ratio of 2.1, which is required for the cobalt catalysts. Consequently, additional hydrogen gas is required. It is noted that the Shell facility is a low-efficiency unit with high operational costs.

The operating pressure and temperature of this Shell facility is 3 MPa and between 200°C and 300°C, respectively. Conversion of about 80% and C5+ selectivities of about 85% are reported (Dry and Steynberg, 2004). New FT plants are envisioned that will use a range of new catalysts, such as the now popular ZSM-5, and impregnated with novel metals such as Co/Mo, $CoSiO_2$/Mo-Pd-Pt-ZSM-5, Ni, and others (Teiseh, 2012).

Example 12.4: Fischer–Tropsch Syngas Calculations

The composition of syngas produced from two types of gasifiers is given below on a volume basis. Determine whether these gases are LCV, MCV, and HCV syngas. Also, estimate the H_2/CO ratios. Use the heating values of gases as follows: $H_2 = 12.8\ MJ/m^3$, $CH_4 = 36.4\ MJ/m^3$, and $CO = 11.6\ MJ/m^3$. Comment on whether which syngas is appropriate for the FT process.

	H_2	CO	CO_2	CH_4
Gasifier A	26%	31%	6%	0.3%
Gasifier B	31%	44%	18%	6%

Solution:

a. As in Example 12.1, the heating values are calculated using the volumetric fractions of syngas as follows, for Gasifier A (assuming 1 m^3 of syngas):

$$H_2(MJ) = 0.26\,m^3 \times \frac{12.8\,MJ}{m^3} = 3.328\,MJ$$

$$CH_4(MJ) = 0.003\,m^3 \times \frac{36.4\,MJ}{m^3} = 0.1092\,MJ$$

$$CO(MJ) = 0.31\,m^3 \times \frac{11.6\,MJ}{m^3} = 3.596\,MJ$$

b. Thus, the syngas heating value is the sum of heating values of the individual components (3.328 + 0.1092 + 3.596) = 7.0 MJ/m^3 and considered LCV gas.
c. Gasifier B syngas heating value is calculated the same way to give the answer below:

$$H_2(MJ) = 0.31\,m^3 \times \frac{12.8\,MJ}{m^3} = 3.968\,MJ$$

$$CH_4(MJ) = 0.06\,m^3 \times \frac{36.4\,MJ}{m^3} = 2.184\,MJ$$

$$CO(MJ) = 0.44\,m^3 \times \frac{11.6\,MJ}{m^3} = 5.104\,MJ$$

d. Thus, the syngas heating value for Gasifier B is (3.968 + 2.184 + 5.104) = 11.256 MJ/m^3 and considered MCV.
e. The H_2/CO ratios are the ratios of volumetric H_2 and CO values as follows:

$$\text{Gasifier A} = 26/31 = 0.84$$

$$\text{Gasifier B} = 31/44 = 0.70$$

f. Gasifier B syngas is appropriate for the FT process.

12.2.4 China Lake Process

The China Lake Process was developed several years ago by the US military in its establishment at the China Lake Military Facility in California. The process makes use of syngas directly and converts this into liquid hydrocarbon fuels employing an advanced "fast pyrolysis" step. The difference between this process and the fast pyrolysis process (discussed in Chapter 9) is that the syngas products contain larger amounts of olefins than the standard fast pyrolysis methods. These improved gases can be subsequently polymerized under high pressure to form a mixture of higher molecular weight hydrocarbons, which can then be refined to yield what are called "drop-in" fuels (either gasoline, aviation fuel, or diesel fuel and the like). In the early stages of its development, the conversion efficiencies were in the order of about 22% (Bhattacharya, 1983). More recent hydrothermal conversion processes claim to have raised the conversion efficiencies significantly higher.

The China Lake process employs an adaptation of the petrochemical pyrolysis process used to crack hydrocarbon oils to make primary ethylene with some propylene, butylene, and aromatic by-products. Instead of using crude oil, highly lignocellulosic material is used in the fast pyrolysis mode to generate a similar product distribution of hydrogen, ethylene, methane, propylene, butylene, and aromatic compounds with carbon oxides present. This gas is then purified to separate the olefinic products (ethylene, propylene, butylene, etc.). These products are then used in the same manner as if they were derived from crude oil. This is a process that does not require any catalysts and yet produces synthetic gasoline compounds. Of course, the yield of these olefinic compounds is quite low compared with crude oil. Most of the initial work at China Lake was directed at demonstrating the process on a bench scale to show that gasoline could be made noncatalytically. The initial biomass materials used were those highly cellulosic components of municipal solid wastes. The overall schematic of the China Lake process is shown in Figure 12.3. The process starts with selective fast pyrolysis of finely ground cellulosic biomass to form gases containing the desirable olefins, followed by compression and purification of those gases to concentrate these olefins. The next step is the noncatalytic polymerization of the olefins to make the synthetic crude oil. The crude contains about 90% high-octane gasoline with some fuel oils and lubricating oils as the by-products (Diebold, 1980). All char and by-product gaseous fuels formed, such as hydrogen, carbon monoxide, and methane, are consumed during the process. Pyrolysis gases are compressed to 30 atm (450 psi) during the initial compression and the pressure booster further increases this to 51 atm (750 psi) before the polymerization step. The hydrocarbon absorber generates rich oil that is fed to the hydrocarbon stripper.

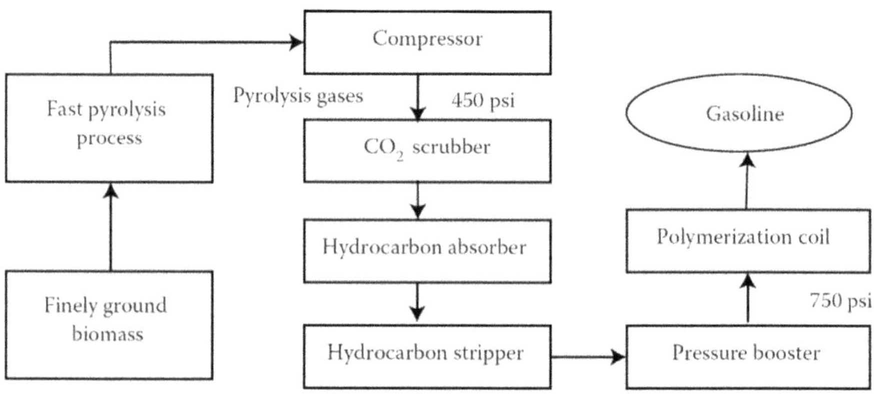

FIGURE 12.3 Schematic diagram of the China Lake process.

Example 12.5: HV of Syngas Produced for China Lake Process

Determine the heating values and H_2/CO ratios of syngas produced from the China Lake process (Diebold, 1980), whose volumetric (%V) composition is shown below. Use the heating values of gases as follows: H_2 = 12.8 MJ/m³, CO = 11.6 MJ/m³, C_2H_4 (ethylene) = 59.5 MJ/m³, $CH_{4=}$ 36.4 MJ/m³, C_3H_6 (propene) = 81.2 MJ/m³, C_2H_2 (acetylene) = 53.9 MJ/m³, C_4H_8 (butene) = 107.1 MJ/m³, and C_2H_6 (ethane) = 60.7 MJ/m³. Comment on the properties of this syngas compared with those used in gasification, pyrolysis, or for the FT process.

	H_2	CO	C_2H_4	CH_4	CO_2	C_3H_6	C_2H_2	C_4H_8	C_2H_6
Syngas (%V)	16.0	45.6	12.8	11.0	7.3	2.1	1.6	1.3	0.9

Solution:

a. The heating values are calculated using the volumetric fractions of syngas as follows (assuming 1 m³ of syngas):

$$H_2(MJ) = 0.16\,m^3 \times \frac{12.8\,MJ}{m^3} = 2.05\,MJ$$

$$CO(MJ) = 0.456\,m^3 \times \frac{11.6\,MJ}{m^3} = 5.29\,MJ$$

$$C_2H_4(MJ) = 0.128\,m^3 \times \frac{59.5\,MJ}{m^3} = 7.62\,MJ$$

$$CH_4(MJ) = 0.11\,m^3 \times \frac{36.4\,MJ}{m^3} = 4.0\,MJ$$

$$C_3H_6(MJ) = 0.021\,m^3 \times \frac{81.2\,MJ}{m^3} = 1.71\,MJ$$

$$C_2H_2(MJ) = 0.016\,m^3 \times \frac{53.9\,MJ}{m^3} = 0.86\,MJ$$

$$C_4H_8(MJ) = 0.013\,m^3 \times \frac{107.1\,MJ}{m^3} = 1.39\,MJ$$

$$C_2H_6(MJ) = 0.009\,m^3 \times \frac{60.7\,MJ}{m^3} = 0.55\,MJ$$

g. Collectively, the heating value for this syngas is 23.47 MJ/m³ (= 2.05 + 5.2 9 + 7.62 + 4.0 + 1.71 + 0.86 + 1.39 + 0.55), considered as an HCV gas and far superior to gasifier gas and typical of syngas from some fast pyrolysis processes described in previous chapters.

h. The H_2/CO ratios are the ratios of volumetric H_2 and CO values as follows: (16.0/45.6) = 0.35, which is quite low. The heating value is quite high and higher than most syngas discussed in previous chapters.

12.3 DIRECT LIQUEFACTION PROCESSES

In direct liquefaction processes, the biomass is converted directly into liquid hydrocarbon fuels without going through the syngas route. The biomass is heated in slurry form at high temperature and high pressure, and the product is subsequently catalyzed during the thermal conversion process. Examples of two direct liquefaction processes are as follows:

1. Bureau of Mines (BOM) or the Pittsburgh Energy Research Center (PERC) process
2. Laurence Berkeley Laboratory (LBL) process

Both the above processes directly hydrogenate biomass at high temperatures and pressures by using carbon monoxide and/or hydrogen as the reducing agents. In the LBL process, the reaction is performed on aqueous cellulose slurry derived from acid hydrolysis of biomass, particularly wood chips. The BOM process blends dried wood powder with recycled oil products in the presence of sodium carbonate or ferric chloride catalysts.

12.3.1 BOM or PERC Process

The US Department of Energy's Biomass Liquefaction Experimental Facility at Albany OR was constructed in 1977 for the purpose of developing biomass-to-oil conversion processes (Elliott and Giacoletto, 1977). The facility and equipment were sized to process 0.909–2.73 tonnes (1–3 tons) per day of wood chips. The original process was called the Bureau of Mines (BOM) process because it was developed by researchers at the former Bureau of Mines facility near Pittsburgh. Some literature reports (Friedman et al., 1977; Yokoyama, 1989) also referred to this process as the Pittsburgh Energy Research Center, or PERC, process because this research center is operated by the Bureau of Mines. The process involves drying and grinding the biomass (woodchips) to form a slurry in a heavy oil medium before it is thermally converted into liquid hydrocarbon fuels. The BOM process is shown schematically in Figure 12.4. In this process, wood chips are dried and ground before being placed in a blender, where it is mixed with recycled bio-oil before being directed to the reactor preheater. The percentage of solids is between 20% and 30% to provide a slurry that can be pumped to the reactor. This slurry, along with carbon monoxide and aqueous sodium carbonate, is pumped at a high pressure of between 137 and 273 atm (2,015 and 4,015 psi) through a scraped-surface preheater and into a stirred tank reactor. The average residence time in the reactor can be varied from 20 to 90 min at temperatures ranging from 300°C to 370°C. After leaving the reactor, the product is moved into a flash tank where it is cooled, the pressure is reduced, and the fixed gases, along with most of the water, are removed.

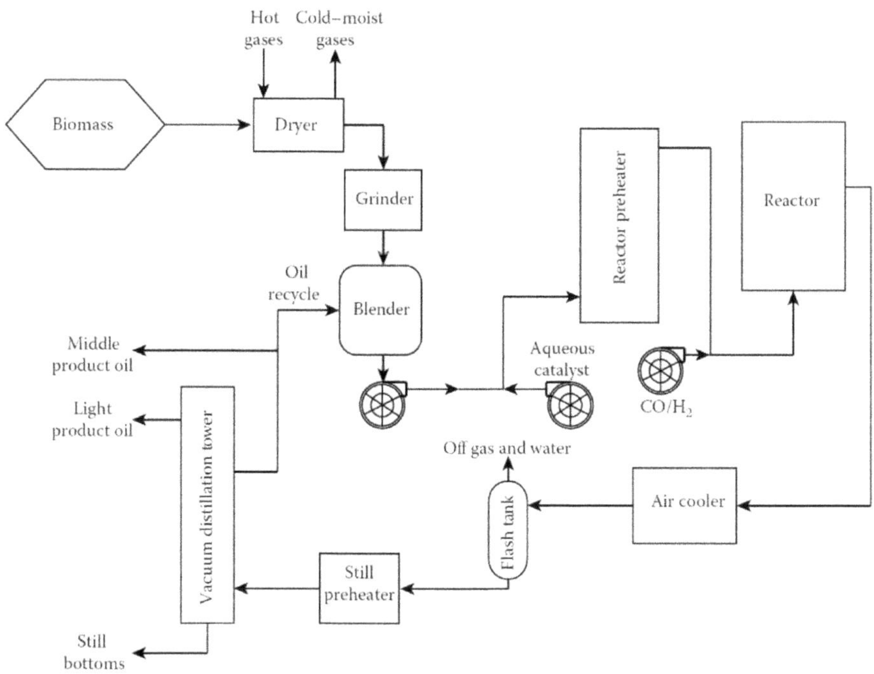

FIGURE 12.4 Schematic of BOM or PERC liquefaction process. (Adapted from Elliott, D. C, In: *Thermochemical Processing of Biomass: Conversion into Fuels, Chemicals and Power* (R. C. Brown, ed.). John Wiley and Sons, Ltd., Chichester, UK. pp. 200–231, 2011.)

A major change in the original process is to use a centrifuge in the product cleanup stage and a vacuum still (Elliott, 1980). After the pressure is reduced, the product is reheated and flashed in the still, then, light product oil is drawn off, the middle fraction is recovered, and a portion is recycled for slurry makeup. The heavy products are then removed from the still bottom. The still bottom is pumped to a gasifier to produce syngas that is required for the liquefaction process. Sodium could be leached from the gasifier char and could be recycled to the process. The heating value of the product oil is around 33.65 MJ/kg (14,500 Btu/lb), with a high heating value of 36.67 MJ/kg (15,800 Btu/lb). Note that this heating value is higher than the regular bio-oil from the pyrolysis process, which is around 24.6 MJ/kg (10,600 Btu/lb) but lower than No. 6 fuel oil, with a heating value of around 42.24 MJ/kg (18,200 Btu/lb). The density of the raw bio-oil is reported at around 1.19 g/mL.

Example 12.6: BOM Process Efficiency Calculations

Determine the bio-oil production from the BOM process in liters per tonne (and gal/ton) if the oil yield by weight is reported at 55%. Use the density of bio-oil of 1.19 g/mL and a heating value of 33MJ/kg. Assume 1 tonne of woodchips was used for the process. Determine the energy conversion efficiency if the heating value of the woodchips was 21 MJ/kg.

Solution:

a. This is a simple conversion of units as follows:

$$\text{Yield}\left(\frac{\text{L}}{\text{tonne}}\right) = \frac{550\,\text{kg bio-oil}}{\text{tonne}} \times \frac{\text{mL}}{1.19\ \text{g}} \times \frac{\text{L}}{1{,}000\,\text{mL}} \times \frac{1{,}000\,\text{g}}{\text{kg}} = 462\frac{\text{L}}{\text{tonne}}$$

$$\text{Yield}\left(\frac{\text{gal}}{\text{ton}}\right) = 462\frac{\text{L}}{\text{tonne}} \times \frac{\text{gal}}{3.785\,\text{L}} \times \frac{1\,\text{tonne}}{1.1\,\text{ton}} = 111\frac{\text{gal}}{\text{ton}}$$

b. The conversion efficiency is based on the energy content of the bio-oil product divided by the energy content of wood chips and calculated per tonne as follows:

$$\text{Output (MJ)} = 550\,\text{kg} \times \frac{33\,\text{MJ}}{\text{kg}} = 18{,}150\,\text{MJ}$$

$$\text{Input (MJ)} = 1\,\text{tonne} \times \frac{21\,\text{MJ}}{\text{kg}} \times \frac{1{,}000\,\text{kg}}{\text{tonne}} = 21{,}000\ \text{MJ}$$

$$\text{Efficiency (\%)} = \frac{\text{output}}{\text{input}} \times 100\% = \frac{18{,}150\,\text{MJ}}{21{,}000\,\text{MJ}} \times 100\% = 86\%$$

12.3.2 LBL Process

The LBL process is much like the BOM process except that drying and grinding is not necessary, eliminating an energy-intensive pretreatment step. Instead, the biomass materials undergo acid hydrolysis at normal low pressures before being pumped into the tubular reactor. By this hydrolysis step, the concentration of solids can be maintained at around 20% so that the slurry can be pumped easily. The schematic of the process is shown in Figure 12.5. The biomass materials are reduced directly into an aqueous slurry that will be pumped into the pressure reactor. The hydrolysis step allows the biomass-to-water ratio to be maintained at a level equivalent to the biomass-to-oil ratio used in the BOM or PERC process. The acidic aqueous slurry must then be made basic using sodium carbonate and then passed through a process like the BOM/PERC scheme. The pressure is then lowered, and the oil and water may be separated by gravity, as shown in Figure 12.5. Off-gas recovery is made on the low-pressure vessels, and these gases may be used later for the front-end of the process after the high-pressure pump. The collected product may then be distilled to recover the hydrocarbons. No recycled oil is used in the LBL process, and this system is essentially a continuous once-through process for the biomass. The aqueous stream will have to be recycled to recover catalyst residues and other soluble organics (Elliott, 1980).

The temperature used for the process ranges from 170°C to 250°C with a residence time of only about 1 h. The aqueous acids have a pH range between 2.5 and

3.0. The catalysts used in the tubular reactor include aqueous sodium carbonate (also acting as base to create neutral slurry) and ferric chloride. The temperature is raised to around 360°C prior to cooling and pressure reduction (Elliott et al., 2010a). Over several years of research, many solvents have been used: methanol, ethanol, propanol, ethylene glycol, glycerol, and phenol. A more detailed process flow diagram is shown in Figure 12.5. The comparison of this process with other fast pyrolysis processes described in earlier chapters is illustrated in Example 12.7. Reducing gases are sometimes required to improve the bio-oil product. It has been theorized that a biomass with a hydrogen (H_2) to carbon (C) atom ratio of between 1.4 and 1.5 would not require any reducing gas agents for the liquefaction process (Yokoyama, 1989). In Example 12.7, one may easily calculate this to show that the H/C ratios are 1.4 and 1.6 for kelp and spent grain, respectively. Thus, for these feedstocks, a reducing gas agent may not be necessary. In practice, however, because of the availability of CO and H_2 gas during the thermal conversion of biomass, reducing gases are still added to further improve the bio-oil product. Note that these H/C atom ratios of 1.4 and 1.6 are very common to wood biomass but not to other major agricultural biomass resources. One must be aware of these differences when conducting biomass liquefaction processes. The ultimate analysis tables presented in Chapter 2 are useful resources for making this estimate.

The bio-oil product of the LBL process has low oxygen content and a heating value close to the heating value of No. 6 fuel oil. However, the density is much higher

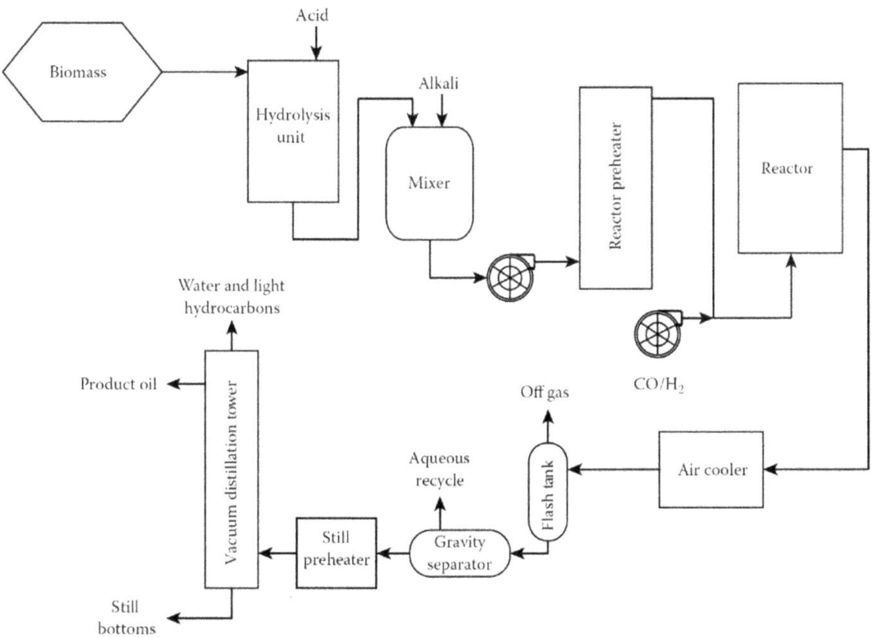

FIGURE 12.5 Detailed schematic of LBL process. (Adapted from Elliott, D. C, In: *Thermochemical Processing of Biomass: Conversion into Fuels, Chemicals and Power* (R. C. Brown, ed.). John Wiley & Sons, Ltd., Chichester, UK. pp. 200–231, 2011.)

(1.2) than that of No. 6 fuel oil (0.89) (Yokoyama, 1989). In succeeding years, the LBL oil has been upgraded to a better liquid product. Further research has diversified the use of catalysts for the process. For example, upgrading studies by Elliott et al. (2010b) made use of CoMo, CuCr, Pd, NiW, and many other novel metals as catalysts. The overall process has changed, with methane used as the most recent gas feedstock. The bio-oil upgrade process is now called hydrothermal liquefaction (HTL) conversion to produce liquid fuels (or the HTL process). The HTL process is being developed by Doug Elliott at the Pacific Northwest National Laboratory (PNNL).

Example 12.7: Determining Bio-Oil Yields by Hydrothermal Processes

Liquefaction experiments with high-moisture biomass produce oil product yields of between 19% and 35% by weight (Elliott et al., 1988). Convert these yields into L/tonne (and gal/ton) and compare the yields with other biomass liquefaction processes. Sodium carbonate was used as a catalyst at a temperature of around 350°C for 30 min and pressurized with carbon monoxide at 500 atm (7,350 psi). Determine the conversion efficiencies. Assume oil density of about 1.2 kg/L and heating value of approximately 35 MJ/kg (15,000 Btu/lb).

	Heating Values			Bio-Oil Ultimate Analysis			
Biomass	Btu/lb	MJ/kg	Oil Yield (%W)	C	H	N	O
Kelp	7,150	16.6	19.2	76.7	8.9	3.5	1.38
Spent grain	9,160	21.3	34.7	75.2	10.2	3.8	1.61

Solution:

a. The volumetric yields are calculated as follows:

$$\text{Kelp bio-oil yield}\left(\frac{\text{L}}{\text{tonne}}\right) = \frac{192\text{ kg}}{\text{tonne}} \times \frac{\text{L}}{1.2\text{ kg}} = 160\frac{\text{L}}{\text{tonne}}\left[38.4\frac{\text{gal}}{\text{ton}}\right]$$

$$\text{Kelp bio-oil yield}\left(\frac{\text{gal}}{\text{ton}}\right) = \frac{160\text{ L}}{\text{tonne}} \times \frac{1\text{ tonne}}{1.1\text{ ton}} \times \frac{\text{gal}}{3.785\text{ L}} = 38.4\frac{\text{gal}}{\text{ton}}$$

$$\text{Spent grain bio-oil yield}\left(\frac{\text{L}}{\text{tonne}}\right) = \frac{347\text{ kg}}{\text{tonne}} \times \frac{\text{L}}{1.2\text{ kg}} = 289\frac{\text{L}}{\text{tonne}}\left[69.4\frac{\text{gal}}{\text{ton}}\right]$$

$$\text{Spent grain bio-oil yield}\left(\frac{\text{gal}}{\text{ton}}\right) = \frac{289\text{ L}}{\text{tonne}} \times \frac{1\text{ tonne}}{1.1\text{ ton}} \times \frac{\text{gal}}{3.785\text{ L}} = 69.4\frac{\text{gal}}{\text{ton}}$$

b. The bio-oil energy yields are calculated assuming 1,000 kg (1 tonne) as follows:

$$\text{Kelp bio-oil energy yield (MJ)} = 192\ \text{kg} \times \frac{35\ \text{MJ}}{\text{kg}} = 6{,}720\ \text{MJ}$$

$$\text{Spent grain bio-oil energy yield (MJ)} = 347\ \text{kg} \times \frac{35\ \text{MJ}}{\text{kg}} = 12{,}145\ \text{MJ}$$

c. Biomass energy content in a tonne:

$$\text{Kelp energy}\left(\frac{\text{MJ}}{\text{tonne}}\right) = \frac{16.6\ \text{MJ}}{\text{kg}} \times 1{,}000\ \text{kg} = 16{,}600\ \text{MJ}$$

$$\text{Spent grain energy}\left(\frac{\text{MJ}}{\text{tonne}}\right) = \frac{21.3\ \text{MJ}}{\text{kg}} \times 1{,}000\ \text{kg} = 21{,}300\ \text{MJ}$$

d. Thus, the conversion efficiencies are as follows:

$$\text{Kelp conversion efficiency (\%)} = \frac{6{,}720\ \text{MJ}}{16{,}600\ \text{MJ}} \times 100\% = 40.5\%$$

$$\text{Spent grain conversion efficiency (\%)} = \frac{12{,}145\ \text{MJ}}{21{,}300\ \text{MJ}} \times 100\% = 57\%$$

12.4 OTHER BIOMASS LIQUEFACTION PROCESSES

There are other biomass liquefaction processes that have been developed over the years and there are numerous variations, including a wide range of catalysts used. The procedural issue is whether syngas is required or whether the biomass slurry can be used directly. In recent years, innovation has mainly focused on the modifications of the China Lake process. The China Lake process has been referred to as the fast pyrolysis process, discussed in Chapter 9. It is characterized by short residence times and relatively moderate temperatures (400°C–600°C), the purpose of which is to obtain high liquid product yields. Another emerging process made use of high-pressure processing systems characterized by lower temperature and longer residence times, with an operating pressure around 200 atm (2,940 psi) (Elliott et al., 1988). Other biomass liquefaction processes combine the conversion processes discussed earlier. These include the combination of thermal, biological, and chemical conversion processes with the single goal of generating liquid hydrocarbon fuels from biomass. Some of these processes and technologies have already been discussed. This chapter will focus on the hydrocarbon fuels as the main product. The delineation has complicated because the industry began to include some oxygenated compounds, such as ethanol or methanol, in the gasoline fuel. Newer research projects are geared toward the production of alcohol-related compounds, such as ethanol, methanol, or butanol, as possible mixes for gasoline. On the other hand, the use of bio-oil as a possible mixer with diesel fuel is also gaining popularity. Although bio-oil is not especially miscible with diesel fuel, solvent additives eliminate the miscibility problem, thereby expanding the use of bio-oil produced by fast pyrolysis or hydrothermal biomass processing.

12.4.1 Hydrothermal Liquefaction (HTL) Process

Both the PERC and the LBL processes have matured in recent years because of varied production methods as well as the use of catalysts. The HTL process is now being developed at the PNNL by Doug Elliot and co-researchers. The biomasses have been diversified to include microalgae samples. In the HTL process, the conversion is still done on wet biomass at elevated temperatures (300°C–360°C or 570°F–680°F) and the pressure is maintained above the vapor pressure of water (15–20 MPa or 2,200–3,000 psi) to facilitate a condensed phase reaction medium. Using HTL, bio-oil yields of about 50% on a carbon basis were achieved from both woody feedstock and corn stover (Holladay et al., 2011) in continuous testing with a 1-L bench-scale, continuous-stirred tank reactor. The resulting bio-oil was readily upgraded to a hydrocarbon product consistent with gasoline and diesel. A schematic of the HTL process is shown in Figure 12.6. The diagram shows how hydrogen is used to upgrade the bio-oil product. This process can be accomplished with proper control of temperature, time, pressure, and the use of a suitable catalyst plus reducing gases. Currently, the process is technically feasible, and some economic analysis has been undertaken. Other issues of concern include the

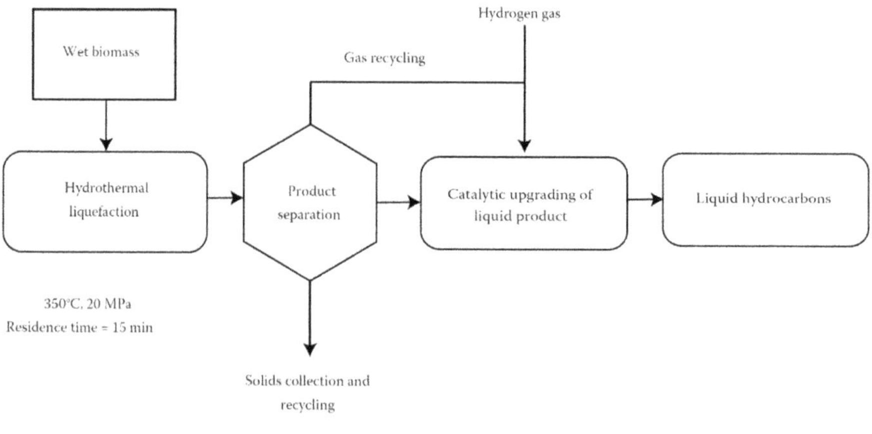

FIGURE 12.6 Schematic of HTL biomass conversion process. (Adapted from Holladay, J. E., A. J. Schmidt and D. C. Elliott, *Hydrothermal Liquefaction—A Route to Improved Biooils*. Pacific Northwest National Laboratory, Richland, WA. #PNNL-SA-78800, 2011. With credit to the National Advanced Biofuels Consortium funded by the USDOE through its Office of the Biomass Program.)

multiphase environment, the details of mass and heat transfer, and the emphasis on heat and catalyst recovery. The long-term operation in a commercial-scale system remains to be accomplished; such a unit will be quite complicated and possibly expensive due to material limitations.

Some other related research on HTL technology is ongoing in Japan, but with oil yields of only 9–34%, while those in Canada have utilized different catalysts such as barium hydroxides and iron sulfate (Elliott, 2011). HTL studies are also underway in China (Song et al., 2004). More recently, HTL processes have used additional catalysts. These include Raney nickel catalysts, palladium (Pd) catalysts, and the less expensive nickel catalysts. Other commercially available catalysts with different active sites, such as Fe_2O_3, NiO, and MoO_3, have been used to convert biomass to bio-oil in the aqueous phase (Zhang, 2010). The popularity of the HTL process is perhaps because it mimics the natural process of fossil fuel formation from biomass feedstocks. Thus, HTL holds great promise as a future biomass conversion process to generate liquid hydrocarbon fuels if the process can overcome the high cost of equipment and the use of diverse and expensive catalysts. Perhaps HTL's greatest advantage is the use of highly lignocellulosic biomass; it does not compete with the world's food requirements.

12.4.2 Hydrolytic and Hydrothermal Upgrading (HTU) Processes

Numerous biomass liquefaction processes are focused on simply upgrading the bio-oil that was produced during thermal conversion. Thus, the HTU process has been developed and has gained popularity, especially with the initiation of a series of experiments and funding provided by Shell Research Laboratory in Amsterdam since 1982 (Behrendt et al., 2008). From 1997 to 2000, a project on the process development unit was initiated, followed by startup and subsequent demonstration in the years 2008–2009, with an annual throughput of 25,000 tonnes/year. In the HTU process, various biomasses with high moisture content are liquefied under high pressure—at temperatures between 300°C and 350°C, pressures between 120 and 180 bars (or atm), and residence times of between 5 and 20 min. Bio-oil is produced and the oxygen is removed by water and CO_2. The products consist of 45% bio-oil, 25% gaseous material (mostly CO_2 of around 90%), 20% moisture, and 10% organic materials (acetic acids, ethanol, etc.). The bio-oil is a heavy organic liquid that becomes solid at around 80°C. The heating value is between 30 and 35 MJ/kg, the H/C ratio is 1:1, and the oxygen content is between 10% and 18%. The theoretical efficiency was estimated to be around 78.6%, with actual results reaching 74.9% (see Example 12.8 for some estimates). The products, separated into light and heavy fractions, are used for various applications. Another variant to the process is the hydrodeoxygenation step (called HDO), which is similar to fluid catalytic cracking (FCC) processes. Still another variation is the addition of hydrogen to the product (called hydrogenation) that may use low or elevated pressure. Engineers can use hundreds of combinations of pressure and temperature, catalysts for deoxygenation, and the use of hydrogen for the hydrogenation step. One would imagine a wide range of products as each parameter is changed.

Example 12.8: Conversion Efficiency for HTU Process

Determine the efficiency of converting 1 tonne of wood waste with a heating value of 21 MJ/kg. The wood waste was converted into bio-oil using the HTU process. The bio-oil yield was 45% by weight, and the heating value of the bio-oil was 35 MJ/kg.

Solution:

a. This example is very similar to previous examples on efficiency calculation as shown below, assuming again 1 tonne of wood waste:

$$\text{Input energy (MJ)} = 1\,\text{tonne wood} \times \frac{1{,}000\,\text{kg}}{1\,\text{tonne}} \times \frac{21\,\text{MJ}}{\text{kg}} = 21{,}000\,\text{MJ}$$

$$\text{Output energy (MJ)} = 450\,\text{kg bio-oil} \times \frac{35\,\text{MJ}}{\text{kg}} = 15{,}750\,\text{MJ}$$

$$\text{Efficiency} = \frac{\text{output}}{\text{input}} \times 100\% = \frac{15{,}750\,\text{MJ}}{21{,}000\,\text{MJ}} \times 100\% = 75\%$$

b. The conversion efficiency of 75% is very close to the reported data from experiments discussed in Section 12.4.2.

12.5 ADVANTAGES AND DISADVANTAGES OF BIOMASS LIQUEFACTION PROCESSES

Direct and indirect liquefaction processes have several advantages over other liquid production processes. The most obvious advantage of the direct liquefaction process is the absence of the drying and grinding steps of the biomass before the conversion step. The main advantage of the indirect liquefaction process is that various feedstocks may be used to produce the syngas using either a gasifier or a pyrolyzer. The option to vary the composition of the syngas to favor a higher H_2/CO ratio is also an advantage of the indirect liquefaction process. Specific advantages and disadvantages will be enumerated in Section 12.5.1–12.5.4.

12.5.1 Advantages of Direct Liquefaction Process

1. Drying and grinding may not be necessary.
2. Catalysts may be mixed with the slurry.
3. Catalysts and water may be recycled.
4. Recycled oil may be used to prepare the slurry.
5. Hydrolysis acid may be used to form the slurry.
6. Base catalyst may be used to make the pH of the slurry neutral.

12.5.2 Disadvantages of Direct Liquefaction Process

1. There is a pumping issue with very high solid slurry.
2. Complex coproducts are produced.
3. The product has high water content.
4. The equipment costs are high.
5. The engineering materials for conveying high-solids slurry are specialized and costly.

12.5.3 Advantages of Indirect Liquefaction Process

1. Syngas may be produced from a wide range of biomass resources.
2. Other gases may be used, such as methane from biogas, or natural gas (i.e., pure methane).
3. The H_2/CO ratios may be easily adjusted during operation.
4. The rate of syngas production is high, thereby improving production throughput.
5. There are various fixed- or moving-bed thermal conversion systems possible in both the gasification and pyrolysis modes.
6. Gas purification is easy and will have less problem with catalyst coking.

12.5.4 Disadvantages of Indirect Liquefaction Process

1. The basic syngas compounds may need to be oligomerized catalytically.
2. The bio-oil produced is quite complex and may need to be upgraded.
3. Condensing the oil or liquid product may be difficult in some designs.
4. Catalysts are necessary for most indirect liquefaction processes.
5. The biomass must be dried and ground.
6. Syngas is not transportable; it must be used on-site or conveyed properly.

12.6 CONCLUSION

Biomass liquefaction processes have evolved from simple direct or indirect conversion of biomass into liquid fuels to much more complicated conversion systems. The earlier versions of the liquefaction processes focused on the use of syngas from gasification. This approach has been expanded with the popularity of the fast pyrolysis process. However, the main process routes are the same. Biomass is either converted into basic gaseous materials before being converted into liquid fuels or it is directly converted into liquid fuels, which must be broken down into simple syngases. The direct liquefaction process, however, has not gained as much popularity as its direct liquefaction counterpart. Although the drying and grinding steps may have been eliminated, the low bio-oil yield, as well as the presence of significant amounts of water, did not offset the reduction in energy usage for grinding and drying.

We have seen numerous FT-type conversion facilities throughout the world that are being used to convert biomass into light, medium, or heavy hydrocarbons, including waxes. Variations of the FT process have also become diverse based on the use of newer catalysts other than cobalt and iron. The recent use of zeolite catalysts (ZSM-5) and its subsequent popularity has also provided significant strides in the way hydrocarbons are produced from biomass. There are many variations, such as the use of other novel metals (e.g., platinum, molybdenum, ruthenium, and even nickel catalysts), in conjunction with ZSM-5.

Recently, the most popular biomass liquefaction process is a variation of the China Lake fast pyrolysis process, developed by the US military. Fast pyrolysis processing is becoming popular because of the high bio-oil yield produced at moderate temperatures. Numerous variations include the use of elevated pressure conditions coupled with hydrothermal processing, that is, the presence of aqueous solvents under both high temperature and high pressure. This process gives rise to valuable chemical products, in addition to the primary energy-related products.

The direction of future research is unclear because of the combination of the biological and chemical pretreatments of the biomass before thermal processing. The syngas may be used by certain microbes to produce diverse liquid fuels as well, with alcohols being the simplest energy-related products. In the future, we will see an influx of newer techniques to generate hydrocarbon fuels through a combination of newer and earlier techniques, with or without the use of novel catalysts. In biological conversion processing, the cost of enzymes is the driving force for the economical production of liquid fuels. For biomass thermal liquefaction processes, the cost of catalysts will also become the limiting step. The race for the elimination or limitation of these costly inputs will continue, until a cost-effective hydrocarbon fuel produced from biomass is found.

There are several advantages in the use of biomass for liquid fuel production over the use of coal. The low sulfur content of biomass allows the use of catalysts that will survive longer. Unfortunately, although sulfur concentration is low, the nitrogen concentration in biomass is relatively high, which may affect the environmental air quality when the liquid fuel coming from biomass is combusted.

12.7 PROBLEMS AND DISCUSSION ISSUES

12.7.1 Heating Value of Synthesis Gas Produced from Pyrolysis

P12.1 Determine the energy content of the syngas (in MJ/m^3 based on major components) produced from the pyrolysis process with the following amounts—The total syngas collected has the following composition by volume: H_2 = 25.4%, CH_4 = 39.4%, CO = 21.5%, CO_2 = 9.4%, C_2H_4 = 2.3%, and C_2H_6 = 1.9%. The heating values of the gases at NTP are as follows: H_2 = 12.8 MJ/m^3, CH_4 = 38.8 MJ/m^3, C_2H_4 = 63 MJ/m^3, C_2H_6 = 69.7 MJ/m^3, and CO = 12.6 MJ/m^3. Convert the HV to units of Btu/ft^3.

12.7.2 Methanol from Biomass Conversion Efficiency

P12.2 Determine the efficiency of converting 1 tonne of rice straw with a high heating value of 17 MJ/kg into pure methanol, having a yield of

320 kg for every tonne. The energy content of pure methanol is 19.8 MJ/kg. Convert all units to the English system.

12.7.3 Mobil Process Conversion Efficiency

P12.3 Determine the efficiency of converting methanol into gasoline using the data provided as follows: 1 tonne of methanol produces 340 kg of gasoline. The energy content of pure methanol is 19.8 MJ/kg and that of gasoline is 46.8 MJ/kg. Convert all units to the English system.

12.7.4 Fischer–Tropsch Syngas Calculations

P12.4 The Sasol-Lurgi fixed-bed coal gasifier has the following data for the synthesis gas used for the subsequent FT process (Dry and Steynberg, 2004). Comment on the following: (a) heating value of the gases and (b) suitability for the FT process based on the H_2/CO ratio. Use the heating values of gases as follows. H_2 = 12.8 MJ/m³, CH_4 = 36.4 MJ/m³, and CO = 11.6 MJ/m³. Comment on which syngas type is appropriate for the FT process.

	H_2	CO	CO_2	CH_4
Sasol A	59.5	24.8	0.9	14.0
Sasol B	55.6	28.5	0.9	14.2
Sasol C	53.6	31.5	0.8	13.3

12.7.5 Heating Value of Syngas Produced for China Lake Process

P12.5 Determine the heating value and H_2/CO ratio of syngas by weight produced from the China Lake Process (Diebold, 1980), whose weight composition (%W) is shown below. Use the heating values of gases as follows: H_2 = 119.8 MJ/kg, CO = 10.14 MJ/kg, C_2H_4 (ethylene) = 47.64 MJ/kg, CH_4 = 49.75 MJ/kg, C_3H_6 (propene) = 45.56 MJ/kg, C_2H_2 (acetylene) = 48.36 MJ/kg, C_4H_8 (butene) = 45.07 MJ/kg, and C_2H_6 (ethane) = 47.10 MJ/kg. Assume 1 kg material. Estimate the density of this syngas.

	H_2	CO	C_2H_4	CH_4	CO_2	C_3H_6	C_2H_2	C_4H_8	C_2H_6
Syngas (%W)	1.3	50.8	14.2	6.9	12.7	3.4	1.6	2.9	1.0

12.7.6 BOM Process Efficiency Calculations

P12.6 Determine conversion efficiency for the conversion of bio-oil from woodchips using the BOM process if the bio-oil yield is 51% by weight. The heating value of bio-oil was found to be 35 MJ/kg and the density is 1.2 kg/L. The woodchips used have a heating value of 20 MJ/kg. Convert the yield to liters per tonne (and to gal/ton).

12.7.7 Determining Bio-Oil Yields by Hydrothermal Processes

P12.7 Hydrothermal liquefaction experiments with high-moisture biomass have oil product yields listed in the table below (Elliott et al., 1988). Convert these yields into L/tonne (and gal/ton) and estimate the conversion efficiencies of bio-oil. Assume an oil density of about 1.2 kg/L and a heating value of approximately 35 MJ/kg (15,000 Btu/lb).

	Heating Values			Oil Ultimate Analysis			
Biomass	**Btu/lb**	**MJ/kg**	**Oil Yield (%W)**	**C**	**H**	**N**	**O**
Water Hyacinth	7,730	17.9	26.0	76.3	9.9	3.3	1.54
Napier Grass	7,870	18.3	34.4	74.5	8.5	0.4	1.36
Sorghum	8,046	18.7	26.6	75.9	8.7	1.7	1.36

12.7.8 Differences among BOM, PERC, and LBL Liquefaction Processes

P12.8 Discuss the differences among the BOM, PERC, and the LBL biomass liquefaction processes.

12.7.9 Hydrogen-to-Carbon Ratios for Biomass Liquefaction Processes

P12.9 Calculate the hydrogen-to-carbon atom ratios for the biomass materials discussed in Example 12.7 as well as Problem 12.7. Discuss the implication of these ratios on the requirements for reducing gases used for biomass liquefaction. Note that H/C ratios between 1.4 and 1.6 are values where suitable reducing agents for the biomass liquefaction processes are not required.

12.7.10 Conversion Efficiency for HTU Process

P12.10 Determine the efficiency of converting 1 tonne of wood waste with a heating value of 20 MJ/kg into pure bio-oil using the HTU process having a yield of 45% by weight and a heating value of 34 MJ/kg.

12.7.11 Catalysts to Produce Methanol from Syngas

P12.11 Identify the catalysts used to produce methanol from synthesis gas.
a. iron-cobalt b. CuO-ZnO c. ZSM-5 d. sulfided Ni

12.7.12 Catalysts for Fischer–Tropsch Process

P12.12 Identify the most common catalysts used for the Fischer–Tropsch process.
a. iron-cobalt b. CuO-ZnO c. ZSM-5 d. sulfided Ni

12.7.13 Catalysts for the Mobil Process

P12.13 Identify the catalysts used for the Mobil Process.
a. iron-cobalt b. CuO-Zn-O c. ZSM-5 d. sulfided Ni

12.7.14 Methanol Production from Fast-Growing Trees

P12.14 You are provided 405 ha [1,000 acres] of land planted with hybrid poplar, a fast-growing bioenergy crop. Estimate the amount of methanol produced, the potential revenue, and compare with bioethanol or gasoline production. State all the assumptions made including references.

12.7.15 Bio-Oil Production from Hydrothermal Liquefaction of Animal Manure

P12.15 A dairy farmer operates a 1,000-head dairy facility in the Northwest of the United States. He plans to convert all manure into bio-oil, upgrade the bio-oil into refinery-grade crude oil, and sell the product to a nearby oil refinery. Estimate the potential bio-oil produced, the revenue as well as other marketed coproducts of the process. State all assumptions made including references.

REFERENCES

Aasberg-Petersen, K., T. S. Christensen, I. Dybkjaer, J. Sehested, M. Ostberg, R. M. Coertzen, M. J. Keyser and A. P. Steynberg. 2004. Synthesis gas production for F-T synthesis. In: *Fischer-Tropsch Technology* (A. Steynberg and M. Dry, eds.). Studies in Surface Science and Catalysis, Vol. 152. Elsevier Science Ltd., London, UK. pp. 258–405.

Behrendt, F., Y. Neubauer, M. Oevermann, B. Wilmes and N. Zobel. 2008. Direct liquefaction of biomass: A review. *Chemical Engineering & Technology* 31 (5): 667–677.

Bhattacharya, S. C. 1983. ET 71: Renewable energy resources lecture notes. Division of Energy Technology, Asian Institute of Technology, Bangkok, Thailand.

Chang, C. D. and A. J. Silvestri. 1977. The conversion of methanol and other O-compounds to hydrocarbons over zeolite catalysts. *Journal of Catalysis* 47: 249–259.

Chang, C. D., W. H. Lang and W. K. Bell. 1981. Molecular shape-selective catalysis in zeolites. In: *Catalysis of Organic Reactions* (W. R. Moser, ed.). Dekker Press, New York. pp. 73–94.

Diebold, J. P. 1980. Gasoline from solid wastes by a noncatalytic thermal process. In: *Thermal Conversion of Solid Wastes and Biomass* (J. L. Jones and S. B. Radding, eds.). American Chemical Society (ACS) Symposium Series No. 130. Based on a Symposium Sponsored by the Division of Environmental Chemistry at the 178th Meeting of the American Chemical Society, Washington, DC, September 10–14, 1979. ACS, Washington, DC.

Dry, M. E. 2004. Chemical concepts used for engineering purposes. In: *Fischer-Tropsch Technology* (A. Steynberg and M. Dry, eds.). Studies in Surface Science and Catalysis, Vol. 152. Elsevier Science Ltd., London, UK. pp. 196–257.

Dry, M. E. and A. P. Steynberg. 2004. Commercial F-T process application. In: *Fischer-Tropsch Technology* (A. Steynberg and M. Dry, eds.). Studies in Surface Science and Catalysis, Vol. 152. Elsevier Science Ltd., London, UK. pp. 406–481.

Elliott, D. C. 1980. Bench-scale research in biomass liquefaction in support of the Albany, Oregon, experimental facility. Quarterly Technical Status Report, December 31, 1979–March 30, 1980. Pacific Northwest National Laboratory, Richland, WA.

Elliott, D. C. 2011. Hydrothermal processing. Chapter 7. In: *Thermochemical Processing of Biomass: Conversion into Fuels, Chemicals and Power* (R. C. Brown, ed.). John Wiley and Sons, Ltd., Chichester, UK. pp. 200–231.

Elliott, D. C. and G. M. Giacoletto. 1977. Bench scale research in biomass liquefaction in support of Albany, Oregon, experimental facility. *Proceedings of the 3rd Annual Biomass Energy Systems Conference held on June 5–7, 1977, at the Solar Energy Research Institute (SERI), and Sponsored by the US Department of Energy*, Golden, CO. pp. 123–130. SERI/TP-33-285.

Elliott, D. C., L. J. Sealock Jr., and R. S. Butner. 1988. Product analysis from direct liquefaction of several high moisture biomass feedstocks. Chapter 17. In: *Pyrolysis Oils from Biomass*. ACS Symposium Series, Vol. 376. The American Chemical Society, Washington, DC. pp. 179–188.

Elliott, D. C., G. Neuenschwander and T. Hart. 2010a. Development in hydroprocessing biooil at PNNL. *Paper presented at the 2010 Symposium on Thermal and Catalytic Sciences for Biofuels and Bio-Based Products and Funded by the DOE-EERE-Office of the Biomass Program*, USDOE. September 22, 2010.

Elliott, D. C., T. R. Hart, G. G. Neuenschwander, L. J. Rotness and A. H. Zacher. 2010b. Catalytic hydroprocessing of biomass fast pyrolysis biooil to produce hydrocarbon products. *Environmental Progress & Sustainable Energy* 28 (3): 441–449.

Friedman, S., A. S. Metha and P. L. Thigpen. 1977. Wood-to-oil process. In: *Energy Technology Handbook* (D. M. Considine, ed.). McGraw Hill Book, Co., New York. pp. 3-168–3-173.

Holladay, J. E., A. J. Schmidt and D. C. Elliott. 2011. *Hydrothermal Liquefaction-A Route to Improved Biooils*. Pacific Northwest National Laboratory, Richland, WA. #PNNL-SA-78800.

Kam, A. Y., M. Schreiner and S. Yurchak. 1984. Mobil methanol-to-gasoline (MTG) process. In: *Handbook of Synfuels Technology* (R. A. Myers, ed.). McGraw Hill Book Co., New York. pp. 2-75–2-111.

Li, K. and D. Jiang. 1999. Methanol synthesis from syngas in the homogeneous system. Journal of Molecular Catalysis A: *Chemical* 147 (1–2): 125–130.

Ren, F., H. Li, D. Wang, J. Wang, W. Pan and J. Li. 2003. Methanol synthesis from syngas in a slurry reactor. *Preprint Papers-American Chemical Society, Division of Fuel Chemistry* 48 (2): 921–922.

Song, C., H. Hu and S. S. Hu. 2004. Nonisothermal catalytic liquefaction of corn stalk in subcritical and supercritical water. *Energy and Fuels* 18: 90–96.

Steynberg, A. P. 2004. Introduction to Fischer-Tropsch technology. In: *Fischer-Tropsch Technology* (A. Steynberg and M. Dry, eds.). Studies in Surface Science and Catalysis, Vol. 152. Elsevier Science Ltd., London, UK. Chapter 1, pp. 1–63.

Steynberg, A. and M. Dry (eds.). 2004. *Fischer-Tropsch Technology*. Studies in Surface Science and Catalysis, Vol. 152. Elsevier Science Ltd., London, UK.

Teiseh, E. A. 2012. Conversion of MixAlco(tm) process sludge to liquid transportation fuels. PhD Dissertation, Department of Biological and Agricultural Engineering, Texas A&M University, College Station, TX.

Trifiro, F. 2009. From syngas to methanol and dimethylether. Lecture presented in Summer School 2009, University of Bologna, Spain. Sponsored by the International Center for Science and High Technology, UNIDO, Rome, Italy.

Wise, J. J. and A. J. Silvestri. 1976. Mobil process for the conversion of methanol to gasoline. *Paper presented at the Annual International Conference on Coal Gasification and Liquefaction held in Pittsburgh, PA*, on August 3–5, 1976, and Co-Sponsored by the University of Pittsburgh, PA. 15 pp.

Yokoyama, S. 1989. Liquefaction. In: *Biomass Handbook* (O. Kitani and C. W. Hall, eds.). Gordon and Breach Science Publishers, New York. pp. 427–433.

Zhang, Y. 2010. Hydrothermal liquefaction to convert biomass into crude oil. Chapter 10. In: *Biofuels from Agricultural Wastes and Byproducts* (H. P. Blaschek, T. C. Ezeji and J. Scheffran, eds.). Blackwell Publishing, New York. pp. 201–232.

ASSESS YOUR KNOWLEDGE

1. What do you call the process of converting biomass directly into liquid fuels?
   a. Torrefaction
   b. Combustion
   c. Biomass liquefaction
   d. Gasification

2. Name an example of an indirect liquefaction process.
   a. LBL Process
   b. PERC Process
   c. Gasification
   d. Mobil Process

11. Determine the heating value (HV) of the syngas (Btu/ft^3) produced from the China Lake process with composition shown by volume, including their heating values in Btu/ft3: H_2, 15%, 344; CO, 45%, 312; C_2H_4, 12%, 1,598; CH_4, 10%, 978; C_3H_6, 2%, 2181; C_2H_2, 2%, 1,448; C_4H_8, 0.5%, 2,877; and C_2H_6, 1%, 1,630.
    a. 292.5
    b. 585
    c. 877
    d. 1,170

12. Determine the efficiency of converting 1 tonne (1,000 kg) of woodwaste with a high heating value of 18 MJ/kg into pure methanol having a yield of 400 kg methanol for every tonne. The energy content of pure methanol is 19.8 MJ/kg. Convert all units in English system.
    a. 99%
    b. 35.64%
    c. 1%
    d. 31.68%

3. Name the catalyst used to convert synthesis gas into methanol.
   a. K_2O
   b. ZnO
   c. ZSM5
   d. $ZnCl_2$

4. Name an example of a direct liquefaction process.
   a. LBL Process
   b. Mobil Process
   c. China Lake Process
   d. Gasification

5. The Mobil process is the conversion of methanol to high-octane gasoline. What was the catalyst used for this process?
   a. ZnO
   b. K_2O
   c. ZSM-5
   d. $ZnCl_2$

13. Gasification process of biomass was used to produce the input synthesis gas for a biomass liquefaction facility. The heating value of synthesis gas produced was 150 Btu/ft^3. What do you consider the class of this synthesis gas?
    a. Low calorific value gas
    b. Medium calorific value gas
    c. High calorific value gas
    d. Ultra-high calorific value gas

14. Determine the efficiency of converting methanol into dimethyl ether and finally to gasoline using the governing equation shown in lecture notes. The energy content of pure methanol is 20 MJ/kg and that of gasoline is 45 MJ/kg. Assume that about 25 kg of dimethyl ether was produced for every 65 kg of methanol used with 85% in the gasoline range.
    a. 80%
    b. 86%
    c. 73.6%
    d. 90%

15. Determine the efficiency of converting methanol into gasoline using the data provided as follows: One tonne of methanol produces 340 kg of gasoline. The energy content of pure methanol is 19.8 MJ/kg and that of gasoline is 46.8 MJ/kg.
    a. 19.8
    b. 42.31
    c. 80.4
    d. 37.3

6. The composition of syngas produced from a gasifier is as follows (by volume): H_2 = 15%; CO = 25%; CH_4 = 5%. Determine whether this syngas is considered LCV, MCV, and HCV gas or appropriate for the FT process. The energy content of H_2 was 344 Btu/ft^3; CO = 312 Btu/ft^3 and CH_4 = 978 Btu/ft^3.
   a. HCV
   b. MCV
   c. LCV
   d. None of the above

7. Determine the efficiency of converting 2 mol of methanol (CH_3OH) into dimethyl ether (CH_3OCH_3), with water (H_2O) as the only product. The energy content of pure methanol is 9,838 Btu/lb and that of dimethyl ether is 13,620 Btu/lb.
   a. 100%
   b. 99.5%
   c. 80.5%
   d. 90.2%

8. On ton of wood chips was used for the BOM process (HV = 8,500 Btu/lb). Determine the bio-oil production from the BOM process (gal/ton) with the following data: liquid product yield is 60% by weight; density of bio-oil = 10 lbs/gal; HV of bio-oil = 12,000 Btu/lb.
   a. 60
   b. 30
   c. 240
   d. 120

16. Determine the efficiency of converting 1 tonne of rice straw with a high heating value of 17 MJ/kg into pure methanol, having a yield of 320 kg for every tonne. The energy content of pure methanol is 19.8 MJ/kg. Convert all units in the English system.
   a. 73.3
   b. 37.3
   c. 85.9
   d. 95

17. Determine the energy content of the synthesis gas (Btu/ft^3) on major components produced from pyrolysis process with the following amounts: Total syngas collected was 75 gal [561 ft^3], with the following composition: hydrogen = 12%, methane = 8%, and carbon monoxide = 10% by volume. The heating values of the gases at standard temperature and pressure (STP) are as follows: H_2 = 344 Btu/ft^3, CH_4 = 970 Btu/ft^3, and CO = 310 Btu/ft^3.
   a. 186
   b. 150
   c. 250
   d. 350

18. Determine the bio-oil production from BOM process in units of gal/ton of biomass. The bio-oil yield by weight was 45%. The density of the bio-oil was 9 lbs/gal, with a heating value of 14,000 Btu/lb. The heating value of the biomass was 10,000 Btu/lb.
   a. 50
   b. 75
   c. 150
   d. 100

9. Determine the conversion efficiency (%) for turning wood chips into methanol. The energy of methanol is 8,530 Btu//lb and that of wood chips is 8,000 Btu/lb. About 35% by weight of methanol is produced for every ton of wood chips.
   a. 37.3%
   b. 74.6%
   c. 0.373%
   d. 24.5%

10. On ton of wood chips was used for the BOM process (HV = 8,500 Btu/lb). Determine the conversion efficiency of the process with the following data: liquid product yield is 60% by weight; density of bio-oil = 10 lbs/gallon; HV of bio-oil = 12,000 Btu/lb. Report also the conversion efficiency.
    a. 84.7%
    b. 42.35%
    c. 60%
    d. 95.5%

19. Determine the conversion efficiency for the conversion of bio-oil from woodchips via the BOM process if the bio-oil yield is 51% by weight. The heating value of bio-oil was found to be 35 MJ/kg and the density is 1.2 kg/L. The woodchips used have a heating value of 20 MJ/kg.
    a. 89.3%
    b. 57.1%
    c. 37.5%
    d. 93.8%

20. Determine the efficiency of converting 1 tonne of woodwaste with a heating value of 20 MJ/kg to begin with into pure bio-oil using the HTU process having a yield of 45% by weight and a heating value of 34 MJ/kg.
    a. 37.5%
    b. 76.5%
    c. 58.8%
    d. 56.7%

Answers: 1c, 2d, 3b, 4a, 5c, 6c, 7b, 8d, 9a, 10a, 11b, 12d, 13a, 14c, 15c, 16b, 17b, 18d, 19a, 20b

ADDITIONAL LEARNING RESOURCES

- Access the Lecture Notes at the Instructor and Student Resources: https://routledgetextbooks.com/textbooks/instructor_downloads/

13 Biomass Combustion

LEARNING OBJECTIVES

Upon completion of this chapter, one should be able to

1. Define biomass combustion and distinguish this conversion process with the other thermal conversion systems described in previous chapters.
2. Enumerate the equipment used for the combustion of carbonaceous biomass materials.
3. Describe slagging and fouling problems in coal combustion and discuss other appropriate slagging and fouling indices for agricultural biomass.
4. Enumerate the various applications of combustion systems.

13.1 INTRODUCTION

Direct combustion of biomass has been the traditional means in many economies to generate useful heat and energy. Combustion is the most efficient thermal conversion process for useful energy, such as for cooking, drying food products, or producing electricity. This pattern of activity is prevalent in many developing countries but is slowly changing, because of concerns for the environment. Deforestation is a common issue with most developing countries. Many countries have started banning the combustion or incineration of biomass for this reason. Combustion is a process of utilizing the carbon and hydrogen in biomass to generate carbon dioxide, water, and heat. The heat may be used to provide energy to a furnace or to produce steam to run a turbine, which generates electricity. For electricity production, the most common process is shown schematically in Figure 13.1. In this schematic diagram, the biomass is first dried before being combusted in a furnace. The heat energy generated is used to boil water and produce steam, which is required to run the steam turbine. The rotating shaft of the turbine is attached to an electrical generator, thereby producing electricity. This is not the only process that is used to produce electricity from biomass. Various other processes are possible, such as utilizing hot gases to run the turbines, instead of steam. Usually, the efficiency of conversion is the ratio of the electrical energy produced divided by the rate at which biomass is used per unit of time. The overall thermal conversion efficiency of an ideal steam cycle (or Rankine cycle) is around 25%–30% (Wark, 1983).

Combustion equations for many biomass resources have been introduced in previous chapters. They will not be repeated here, except to present a general equation that can be used to estimate the combustion stoichiometry for any biomass with known ultimate analysis (see Equation 13.1). The generic equation for any biomass resource is given by the empirical equation $C_xH_yO_zN_aS_b$, where x, y, z, a, and b represent the mole fraction of each element in the biomass.

DOI: 10.1201/9781003294306-13

$$C_xH_yO_zN_aS_b + \left(x + \frac{y}{4} - \frac{z}{2}\right) \times \left[O_2 + 3.76\,N_2\right] \rightarrow xCO_2 + \left(\frac{y}{2}\right)H_2O$$
$$+ \left(x + \frac{y}{4} - \frac{z}{2}\right) * 3.76\,N_2 + b\ S + \text{Heat} \tag{13.1}$$

The constants x, y, z, a, and b result from computing the ratio of the weight percentage of each component with its molecular weight. The air-to-fuel ratio by weight during combustion may be calculated (see Equation 13.2) by using the molecular weight of each major element in units of kg/kg (or lb/lb). The air-to-fuel ratio by volume is simply the coefficient of the air to that of the biomass fuel, as shown in Equation 13.3. In many combustion reactions using biomass, excess air is normally used to ensure that all carbon in the material is utilized. Equation 13.1 is used to estimate the carbon dioxide production during combustion of biomass. For every mole of biomass with a chemical formula $C_xH_yO_zN_aS_b$, there are x mol of carbon dioxide produced and $y/2$ mol of water produced. In terms of weight, a similar procedure for calculating air-to-fuel ratio may be used. Example 13.1 illustrates how much CO_2 is produced for every metric ton (or tonne) of a given biomass with its known ultimate analysis.

$$\frac{\text{Air}}{\text{Fuel}}\left(\frac{\text{kg}}{\text{kg}}\right) = \frac{\left(x + \frac{y}{4} - \frac{z}{2}\right) \times \left[O_2 + 3.76\,N_2\right]}{C_xH_yO_zN_aS_b} = \frac{137.28 * \left(x + \frac{y}{4} - \frac{z}{2}\right)}{12x + y + 16z + 14a + 32b} \tag{13.2}$$

$$\frac{\text{Air}}{\text{Fuel}}\left(\frac{\text{L}}{\text{L}}\right) = \frac{\left(x + \frac{y}{4} - \frac{z}{2}\right)}{1} \tag{13.3}$$

Example 13.1: Amount of CO_2 Produced for Every Metric Ton of Biomass Combusted

Suppose you have certain biomass with the following chemical formula: $C_{4.03}H_{5.6}O_{2.77}N_{0.02}$. Determine the amount of CO_2 produced (in kg) per metric ton of this biomass when combusted completely in air.

Solution:

a. The stoichiometric equation is found by calculating the coefficients for air to give the completed equation as follows:

$$\left(x + \frac{y}{4} - \frac{z}{2}\right) = \left(4.03 + \frac{5.6}{4} - \frac{2.77}{2}\right) = 4.05$$

$$C_{4.03}H_{5.6}O_{2.77}N_{0.02} + 4.05\left(O_2 + 3.76\,N_2\right) \rightarrow 4.03CO_2$$
$$+ 2.8H_2O + 15.23\,N_2$$

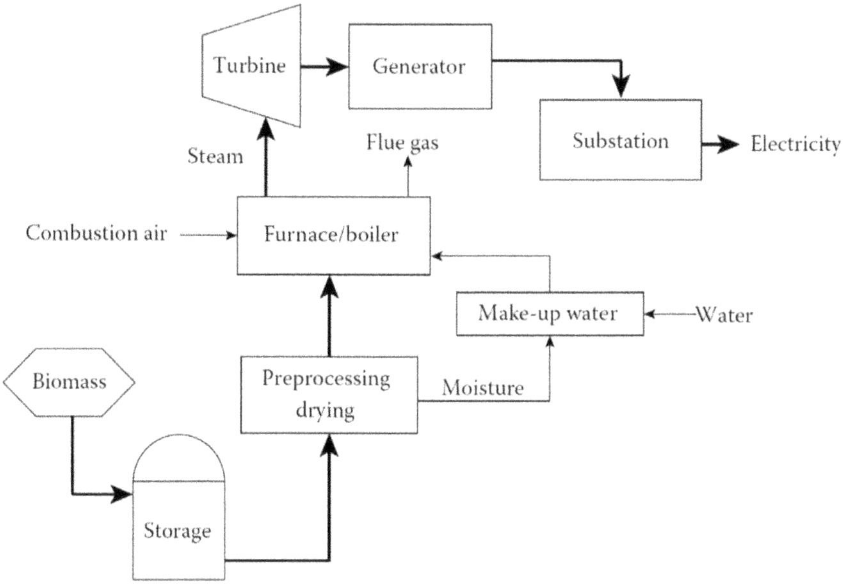

FIGURE 13.1 Schematic of biomass combustion process for electricity production.

b. Thus, the ratio of CO_2 per biomass by weight is given here:

$$\frac{CO_2}{\text{Biomass}}\left(\frac{\text{kg}}{\text{kg}}\right)=\frac{(4.03)^*(12+32)}{(12*4.03)+(1^*5.6)+(16^*2.77)+(14^*0.02)}$$

$$=\frac{177.32}{98.56}=1.8$$

c. As 1 MT is 1,000 kg, this is equivalent to 1,800 kg/MT.

13.2 TYPES OF BIOMASS COMBUSTION SYSTEMS

There are various types of biomass combustion systems. The most basic system is represented by the biomass cookstove used by rural households. These combustors will only generate heat energy for heating, cooking, and drying purposes. A biomass combustion system may be a simple cookstove or a more sophisticated fluidized bed combustion (FBC) system that is used for coal-fired power plants. Perhaps the most common of the larger biomass combustion systems are those used to incinerate municipal solid wastes (MSWs) (when they are still permitted to do so). Spreader stoker systems are perhaps the most popular, followed by multiple-hearth systems. Various biomass combustion systems will be described here. However, it should be noted that some systems have become obsolete, and other combustion systems have been in use for several decades.

13.2.1 Spreader Stoker System

The spreader stoker system is perhaps one of the most popular combustion systems ever developed and one commonly used. The schematic of this system is shown in Figure 13.2. The unit is composed of the following component parts, as numbered in the figure.

1. Refuse charging hopper
2. Refuse charging throat
3. Charging ram
4. Grates
5. Roller bearings
6. Hydraulic power cylinders and control valves
7. Vertical drop off
8. Overfire air jets
9. Combustion air
10. Automatic sifting removal system

This technology is very important for combusting refuse-derived biomass. Most MSW incineration systems in the United States have already been dismantled, giving way to more sophisticated systems. This unit can handle high moisture biomass and normally operates using staged combustion. At the beginning of the process,

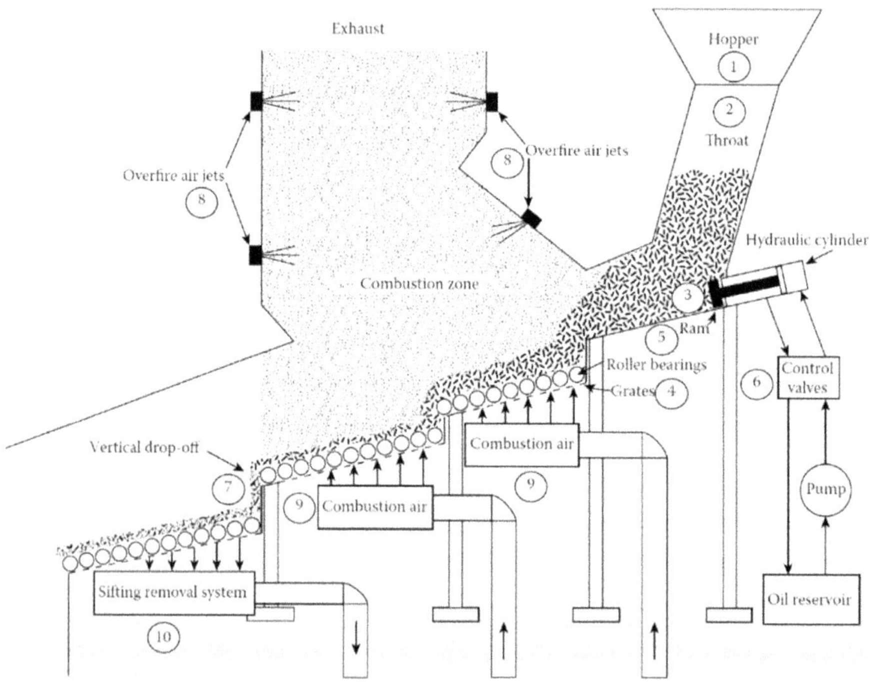

FIGURE 13.2 Schematic of spreader stoker biomass combustion system.

sub-stoichiometric amounts of air are used (often called "starved air"), then, excess air is used when combustible gas is produced, to generate heat. Basically, this system operates in the gasification mode at the beginning of the process. Then, it burns the gas at the top of the reactor, virtually using excess air to make it a combustion system, from the point of view of the total amount of air used. The calculation for air requirements in the combustion of biomass is illustrated in Example 13.2.

Example 13.2: Amount of Air Required for Combustion of MSW

The ultimate analysis of segregated MSW (in percent weight) is presented in the table below. Determine the exact amount of air (m^3/kg and ft^3/lb) required for combustion of this material in air. Assume standard air conditions with an air density of 1.2 kg/m^3 (0.075 lb/ft^3).

Biomass	C	H	O	N	S	Ash	MC	HV (MJ/kg)
MSW	60.8	9.8	25.0	0.12	0.08	2.6	1.6	30.4

Solution:

a. The ultimate analysis is first converted into dry basis so that the hydrogen and oxygen content will not cause a slight error in calculations. Sulfur is ignored because the value is too small. Then, the representative chemical formula of the MSW biomass is first estimated by dividing the weight percentage of each element by its molecular weight assuming 100 g of sample. This value will be the subscript of each element used for the combustion equation.

Biomass	C	H	O	N	S	Ash	MC	HV (MJ/kg)
MSW (wb)	60.8	9.8	25.0	0.12	0.08	2.6	1.6	30.4
MSW (db)	61.8	10	25.4	0.12	0.008	2.6		
MW	12	1	16	14	32			
Subscript	5.2	10	1.6	0.01	0.0025			

$$\text{Chemical formula} = C_{5.2}H_{10}O_{1.6}N_{0.01}$$

b. Once the chemical formula is established, the stoichiometric combustion equation is made and balanced as follows:

$$C_{5.2}H_{10}O_{1.6}N_{0.01} + 6.9\left(O_2 + 3.76N_2\right) \rightarrow 5.2CO_2 + 5H_2O + 6.9 \times 3.76N_2$$

c. Then, the air-to-fuel ratio is calculated as follows:

$$\frac{\text{Air}}{\text{MSW}}\left(\frac{\text{kg}}{\text{kg}}\right) = \frac{(6.9)*(32+105.28)}{(12*5.2)+(1*10)+(16*1.6)+(14*0.01)}$$

$$= \frac{974.2}{98.14} = 9.9\frac{\text{kg}}{\text{kg}}$$

d. Finally, the air requirement per kilogram of the MSW material is calculated using the air density provided:

$$\frac{\text{Air}}{\text{MSW}}\left(\frac{\text{m}^3}{\text{kg}}\right) = \frac{9.9\,\text{kg}}{\text{kg MSW}} \times \frac{\text{m}^3}{1.2\,\text{kg}} = 8.25\frac{\text{m}^3}{\text{kg}}\left[132\frac{\text{ft}^3}{\text{lb}}\right]$$

13.2.2 Fluidized Bed Combustion System

The physical design of the FBC system is very similar to the fluidized bed gasification systems introduced in Chapters 10 and 11, except that excess air is used to produce only heat, instead of combustible syngas. The reactor is exactly the same as the fluidized bed gasification system shown schematically in Figure 13.3. The fluidized bed combustor may be operated under normal atmospheric pressures (FBC), pressurized (PFBC), or on auto-thermal combustion system mode (not requiring any flame or spark to ignite). The FBC system is the most common combustion technology used for converting coal into electrical power. Combustion efficiency calculations are shown in Example 13.3.

Example 13.3: Biomass Combustion Efficiency

Determine the overall efficiency of a biomass combustion system utilizing biomass with the following data: The heating value of the biomass was 18 MJ/kg and fed at a rate of 2 tonnes/h. The output of the process was 3 MW.

Solution:

a. The efficiency is calculated simply as input divided by the output. As the output is known, the easiest procedure will be to calculate the input energy in order to arrive at the same units. Thus, the energy input from biomass in units of MW is given as follows:

$$\text{Biomass (MW)} = \frac{2\text{ tonnes}}{\text{h}} \times \frac{1\text{ h}}{3600\text{ s}}\frac{1{,}000\text{ kg}}{1\text{ tonne}} \times \frac{18\text{ MJ}}{\text{kg}} \times \frac{\text{MW}-\text{s}}{\text{MJ}}$$

$$= 10\text{ MW}$$

b. Thus, the efficiency is easily calculated as follows:

$$\text{Efficiency (\%)} = \frac{3\text{ MW}}{10\text{ MW}} \times 100\% = 30\%$$

FBC used for biomass combustion differs from FBC used for coal combustion when it comes to the use of bed material. In coal combustion systems, limestones or

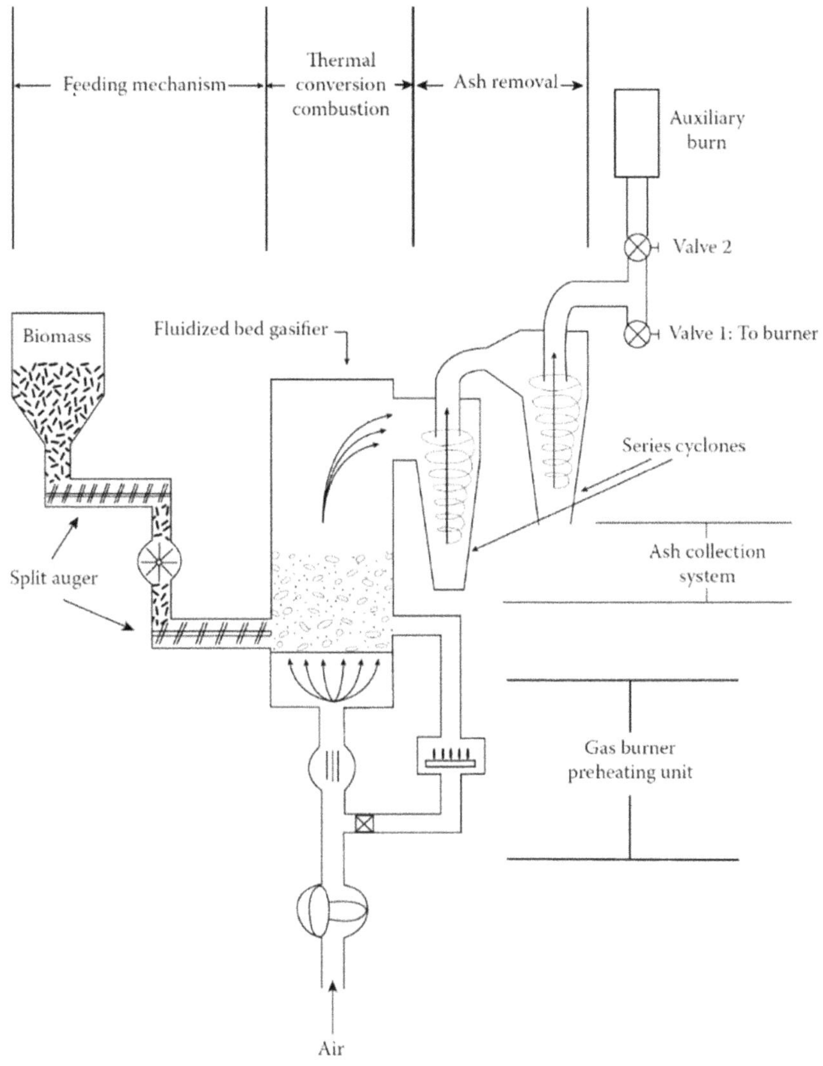

FIGURE 13.3 Schematic of FBC system for biomass.

dolomites are used to capture sulfur because of the high sulfur contents of most coal material. This is not necessary with combustion of agricultural or woody biomass where the sulfur content is quite low. The principles of operation (start-up) are similar to those of the gasification systems, where bed material is heated until a temperature is reached for continued biomass combustion. The control systems for combustors are, however, opposite to those used for gasification. When a higher temperature is required, more fuel is injected into the reactor relative to air and vice versa. The other difference is that the higher operating temperature, which is a necessary condition for combustion, is incompatible with most biomass. As will be demonstrated in later sections of this chapter, most agricultural biomass has very high ash content.

When exposed to the higher operating temperature in combustion systems, the ash tends to melt, creating slag that leads to fouling the system. Consequently, there are few biomass resources suitable for combustion systems. High-ash biomass, such as animal manure, can never be operated in the combustion mode since slag will generate in the reactor, creating serious problems in the conveying systems.

The main advantage of FBC systems (as in gasification systems) is fuel flexibility, or the wide range of biomass fuels that can be used with very limited preprocessing steps (Johnsson, 2007). Note that in combustion systems, the final particulate product is ash or fly ash around 50–100 μm in size. The ash material will also contain fine particles of elutriated bed material that is used for fluidization. The temperature in the reactor is at least 900°C (1,652°F) (Van Loo and Koppejan, 2008), which is always higher than gasification temperatures. At this temperature, biomass ash starts to melt or fuse, creating larger solid particles in the reactor. It is difficult to remove the ash using fluidizing air alone. Consequently, future runs will create clogging.

When a combustor is being designed to burn a particular biomass or the feedstock has changed, the designer or operator should be concerned about predicting whether fouling or slagging will occur in the reactor, which would cause problems during the operation. Numerous parameters are used for judging the quality of biomass material with respect to fouling and slagging potential. Many of these parameters are poorly defined or are difficult to find in the literature. The American Society of Mechanical Engineers (ASME) has published coal fouling and slagging parameters, initially mentioned in Chapter 2, with examples presented in Chapter 3. This chapter will present different slagging and fouling factors that are believed to be applicable to most agricultural biomass and residues. The parameters presented in Chapter 3 were for specific coal-fired biomasses and not for agricultural-based biomasses. Even lignite coal is not suitable, based on slagging and fouling parameters published by the ASME.

The other variations of FBC are differentiated by whether the units are in the circulating mode or the bubbling mode. The principle behind the bubbling FBC system is to drop the fuel down a chute located slightly above the combustion chamber, where a fluidizing medium is placed. This location is just above the fluidizing air distribution plate. In the circulating FBC systems, the materials removed by the cyclones, including the elutriated fluidizing medium (usually refractory sand), are recycled back into the reactor. In Figure 13.4, recycling is achieved when the materials in the ash collection bin are channeled back into the reactor, as shown. The figure also shows the various zones in an FBC.

13.2.3 Biomass Cookstoves

In developing countries, the most popular biomass combustors are the biomass cookstoves. Two improved designs are shown in Figures 13.5 and 13.6 (Capareda, 1995). The design in Figure 13.5 was developed by engineers at the International Rice Research Institute (IRRI) based in Los Baños, Laguna, Philippines. As shown in Figure 13.5, this cookstove has two versions: an elevated model on the left and a tabletop model on the right. Both are operated the same way. The usual biomass feedstock is rice hull. Both models work the same way. The unit is supplied with dry

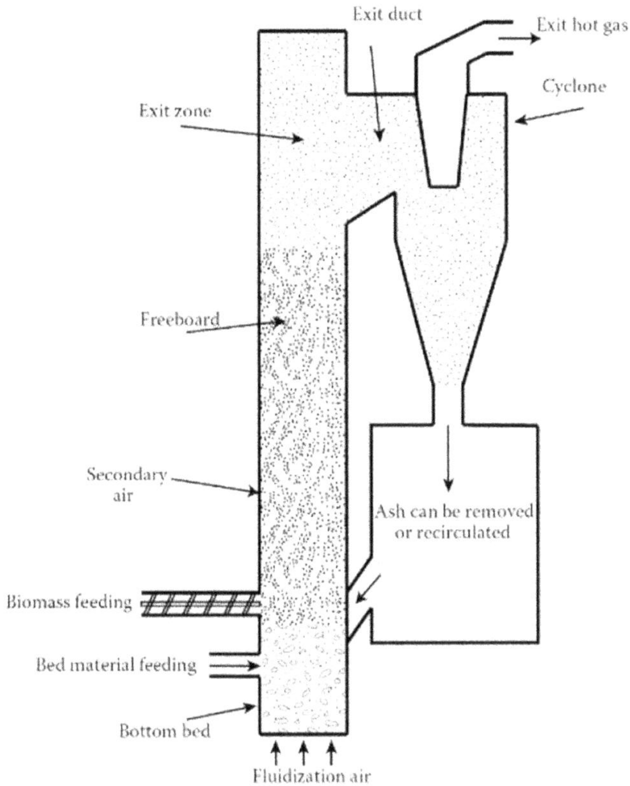

FIGURE 13.4 Schematic of various zones in FBC system, including the option to recycle bed material in the reactor.

rice hull material on the side conical hopper until it is filled to the top rim. Note that the center frustum will have void space in the middle, after the hopper becomes filled with biomass. The void space is where ignition starts. Ignition usually happens by lighting a small portion of a newspaper and dropping it into the middle center frustum. Immediately upon ignition (with good ventilation), a flame will be produced, then, to start the cooking, the pot may be placed on top along the braze. One full load of rice hull material will be able to boil more than a liter of water. If the pot contains raw rice, that rice will be fully cooked using one load of rice hulls. When carefully calibrated, the unit may be able to automatically cook the rice without continuous attention. The flame will be vigorous at first and will heat the pot and the rice–water mixture quickly. Later, after the water has boiled, there will be a smaller flame. When the rice hull fuel is almost depleted, the heat energy is reduced, which is an excellent mode for cooking rice. After the water has boiled and is simmering, there is less possibility of overcooking the rice. In the Philippines, this model is called IRRI-IPA-QALAN (IPA means "rice hull" and QALAN means "cookstove").

An improved version developed in Vietnam is shown in Figure 13.6. This unit is called the "Lo Trau Rice Hull Stove." The unit works the same way as the one used

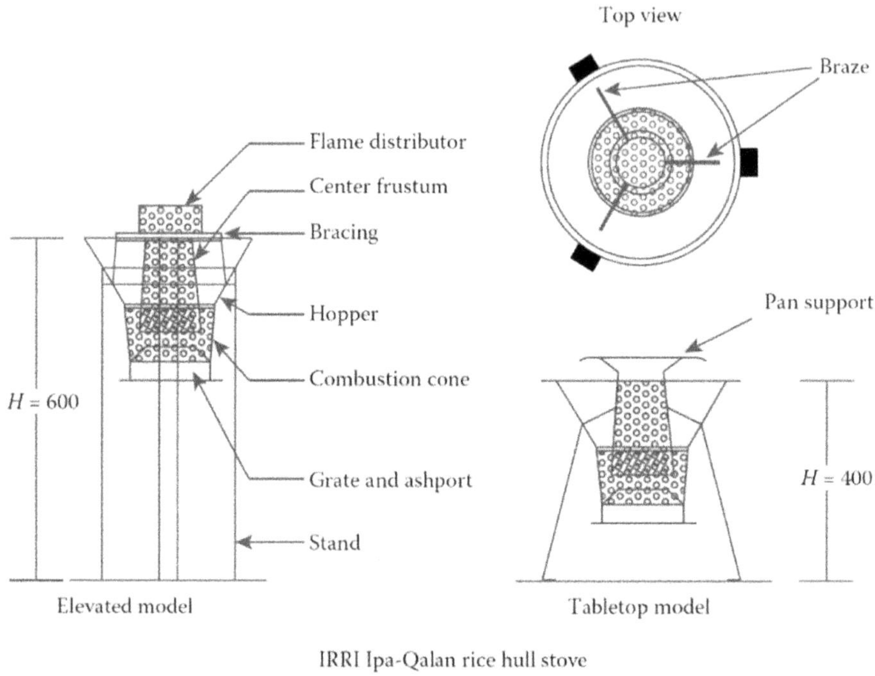

FIGURE 13.5 Designs of cookstoves used for cooking in developing countries (Philippines).

in the Philippines and may well be the basis for the Philippine design, or vice versa. For continuous and longer cooking times, one removes the ash on the lower grate and the unit can be filled with additional biomass as the top level recedes when ash is removed from the bottom grate. Designs in other developing countries are more permanent and use bricks. The brick units have a sophisticated chimney for smoke removal and to encourage convective flow of combustion air. The units presented here are portable systems for use in single households.

Efficiency calculations for this type of cookstove are varied. A given unit would be rated according to the amount of biomass used to boil a given amount of water (called the water-boiling test) and how long boiling is maintained for a given amount of biomass. Two additional tests are the controlled cooking test and the kitchen performance test. Other tests will likely be required to cover all phases of stove design for the third world (Johnson et al., 2010). These tests are quite complicated for anyone who is not familiar with cookstove operations. Note that in these new cookstove standards, thermal efficiency calculations are not included because a number of "stove experts" as far back as 1985 questioned the wisdom of relying solely on thermal efficiency calculations. The experts recommended that the measurements be replaced by a different set of standards (Bailis et al., 2007). To date, there are various standards around the world on evaluating cookstoves, and these are tailored to the cultural norms in a given region.

The water-boiling test is divided into three phases. Phase 1 is called the high-power (cold start) phase, the purpose of which is to determine how long it takes to boil a standard amount of room temperature water and the fuel required, starting with a cold cookstove. Phase 2 is called the high-power (hot start) phase, the purpose of which is to determine how long it takes to boil room temperature water, when starting with a hot cookstove. Phase 3 is called the low-power (simmering) phase (and done immediately after Phase 2), the purpose of which is to determine how much fuel is required to maintain water just below the boiling point, for 45 min.

Note that the boiling point of water may not reach 100°C in some elevated areas. Equation 13.4 is used to determine the boiling temperature of water for a given locality if the elevation (in meters) is known. Example 13.4 shows the variation in boiling point for every 1,000 m of elevation.

$$T_b(°C) = \left(100 - \frac{h}{300}\right) \tag{13.4}$$

where

T_b = local boiling temperature

h = altitude, m

Example 13.4: Effect of Elevation on Boiling Point of Water

Determine the boiling point (in °C) of water if the elevation of a locality is 1,000, 2,000, or 3,000 m.

Solution:

a. This problem is calculated using Equation 13.4 as follows:

$$T_b(°C) = \left(100 - \frac{1{,}000}{300}\right) = 96.7°C$$

b. At 2,000 m, the boiling point is as follows:

$$T_b(°C) = \left(100 - \frac{2{,}000}{300}\right) = 93.3°C$$

c. At 3,000 m, the boiling point is as follows:

$$T_b(°C) = \left(100 - \frac{3{,}000}{300}\right) = 90°C$$

d. Thus, there is about a 3°C decrease in the boiling point of water for every 1,000 m of increase in elevation.

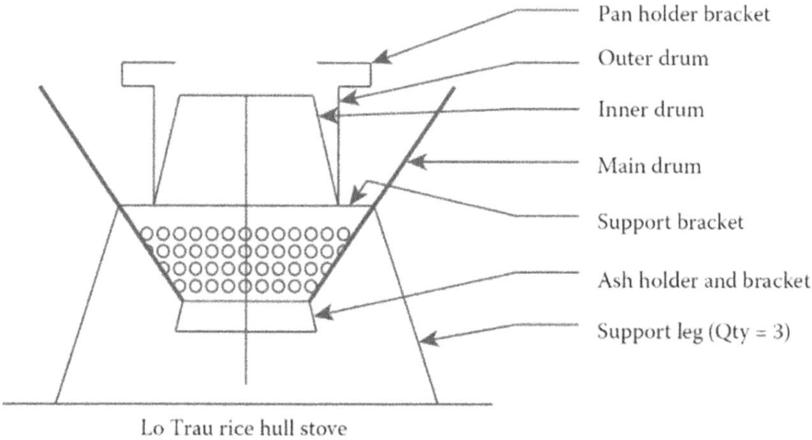

FIGURE 13.6 Design of cookstove used for cooking in developing countries (Vietnam).

13.2.4 Pellet Stoves

Pellet stoves are pellet fuel appliances burning wood pellets or other biomass such as nut shells, corn kernels, and other densified agricultural biomass. They are considered clean residential heating systems. They are certified by the US Environmental Protection Agency (US EPA) and have conversion efficiencies between 70% and 83% (Energy Saver, 2022). A good rule of thumb is that a stove rated at 63.3 MJ (60,000 British Thermal Units (Btu)) can heat a 185.9-m^2 (2,000-ft^2) home while the one rated at 44.31 MJ (42,000 Btu) can heat a 120.8-m^2 (1,300-ft^2) space. Most pellet stoves cost between $1,700 and $3,000 depending upon the size. A large pellet stove can provide heat up to 48 h, whereas most smaller versions can be used for 24 h. The loading rates vary between 16 and 60 kg (35 and 130 lbs).

These wood stove pellets must be maintained on a regular basis. The chimney sweep must be inspected every year. For example, the Chimney Safety Institute of America inspects and certifies wood pellet stove systems. The appliance, hearth, connecting pipe, air inlets, chimney, and all other components should function efficiently and safely. There are after burners (catalytic) sold to ensure clean exhaust and costs between $75 and $160 and may last between 5 and 8 years.

Pellet fuel is normally sold in 18-kg (40-lb) bags at about $3–$4 each or about $198–$275/metric tonnes (MT) (180–250/English ton). Most homeowners who use a pellet appliance as the main source of heat use about 2–3 tons of fuel pellet each year if they live in higher latitudes. Pellet fuel appliances are almost always cheaper to operate compared with electric resistance heating and oil- and propane-fueled heating appliances. Most pellet fuels have a moisture content of between 5% and 10%. There are fuel pellet standards. The Pellet Fuels Institute (PFI) launched the PFI Standards Program for residential and commercial-grade pellet fuels. This standard assures the consumer of the highest quality pellet when certified pellets are purchased.

13.3 CO-COMBUSTION OF BIOMASS AND CO-FIRING WITH COAL

Agricultural biomass can be mixed with conventional fuel, such as coal, to improve its effect on greenhouse gas emissions and to take advantage of carbon credits. When biomass is used with a base fuel such as coal, the technical term is co-firing. Co-firing biomass with coal represents one of the nearest-term and lowest-cost options for reducing CO_2 in the electrical power sector. This process takes advantage of the low cost of the biomass fuel in some areas. The process also takes advantage of the fact that most biomass residues from agricultural processes are very low in sulfur and are important in limiting the formation of smog.

In general, the combustion efficiencies of biomass are lower than coal. However, when biomass is co-fired with coal, the efficiency could be as high as 40% based on percent low heating values of biomass (or termed "as-received" basis) (IEA, 2007). The efficiency of other power generation routes from biomass using various processes varies. The steam cycle efficiencies, for example, are usually within 30%–35% and additional improvement is achieved when biomass is used. The selling cost of electricity will also decrease by about 50% or more than most other conversion routes. The capital investment cost per kilowatt-hour is also lower for most co-firing facilities, especially for large-scale systems between 10 and 50 MWe in size, as shown. Many biomass-fueled power plants in Europe are being converted into combined heat and power (CHP) plants because of these characteristics. The main issue is to find locations where both electricity and heat are required, to take advantage of these benefits.

Receiving carbon credits is an additional benefit to the low sulfur nature of using biomass in power plants. Unfortunately, the carbon credits trade is not yet established. However, if one anticipates more stringent exhaust emissions for coal-fired power plants in the future, biomass co-firing may be the only logical choice. The issue is magnified if this is coupled with increased concern for greenhouse gases, particularly CO_2 emissions from coal.

Example 13.5: Biomass Co-Firing Efficiency

Determine the efficiency of a co-firing system utilizing biomass and coal with the following data: About 16% of biomass was used by weight at a rate of 0.45 tonnes/h and coal at a rate of 2.8 tonnes/h. The heating value of the biomass was 20 MJ/kg and that of coal was 29.5 MJ/kg. The output of the process was 10 MW.

Solution:

a. The efficiency is calculated as input divided by the output. As the output is known, the easiest procedure will be to calculate the energy input to arrive at the same units. Thus, the energy input from biomass in units of MW is given as follows:

$$\text{Biomass}(\text{MW}) = \frac{0.450\ \text{tonnes}}{\text{h}} \times \frac{1{,}000\ \text{kg}}{1\ \text{tonne}} \times \frac{20\ \text{MJ}}{\text{kg}} \times \frac{1\ \text{h}}{3600\ \text{s}} = 2.5\ \text{MW}$$

b. Then the energy input from coal is given as follows:

$$\text{Coal (MW)} = \frac{2.8 \text{ tonnes}}{\text{h}} \times \frac{1{,}000 \text{ kg}}{\text{tonne}} \times \frac{29.5 \text{ MJ}}{\text{kg}} \times \frac{1 \text{ h}}{3600 \text{ s}} = 22.9 \text{ MW}$$

c. The efficiency will be calculated as follows:

$$\text{Efficiency (\%)} = \frac{10 \text{ MW}}{(2.5 + 22.9) \text{ MW}} \times 100\% = 39.4\%$$

13.4 POWER GENERATION CASE STUDIES IN THE UNITED STATES

There are various wood-for-power generation projects being envisioned by many states in the United States. However, most are based on fuelwood resources. Table 13.1 summarizes some estimates of wood biomass production in several southern states where the feasibility study was conducted. The southern United States (see list of 13 states in Table 13.1) produces nearly 60% of the US wood supply. Projections show that production will continue to increase. In this project feasibility study, only wildland–urban interface (WUI) areas were considered for sources of wood (i.e., near urban areas that are close to forest resources). In these areas, human development is interspersed with undeveloped wildland and represents a large-forested land base containing a tremendous volume of available biomass. The woody biomass is available because of thinning and conversion of forestland into urban areas. The other advantage of using WUI areas is the fact that utilities and the source of biomass are geographically closer. The existing road infrastructure for biomass transport is also in place. The cost of the biomass is relatively lower compared with the cost of primary fuelwood used by commercial mills.

For these states, 20- and 40-MW-sized biomass power plants were envisioned and evaluated for economic feasibility. The analyses showed that all power plant sizes will provide positive economic impacts to all counties but varied widely among selected counties depending on the local economy. As in many other power plant projects, the fuelwood cost and its utilization represent one of the largest expenditures for the project. Likewise, because of economies of scale, larger facilities had better positive impacts, but the results were not linear. The biggest barrier for this biomass-to-power program is not the technical issue but rather the economic feasibility (Hoffmann and Weih, 2005). High costs are associated with harvesting, collecting, and transporting biomass fuel, even if the biomass resources are near the urban areas where power plants are to be established. In addition, the initial capital investment is also quite high if there is no government support or subsidy. Example 13.6 shows the acreage required to build a 40-MW power plant from low yield resulting from thinning and wood wastes only. Note that all locations in the states listed will be able to satisfy the acreage requirement for a 40-MW biomass-fueled power plant. Some states may have multiple units of 40-MW size or larger.

TABLE 13.1
Woody Biomass Power Development Case Studies

County, State	Area (mi²)	Population (2003)	Forest Area (acres)	Biomass Available (t/a)	Per Capita (t)	Power Generation (MWh)
Lee, AL	600	119,561	393,991	18	NA	25
Shelby, AL	798	159,445	418,341	20	NA	74
Saline, AR	737	87,554	571,054	26	NA	0
Union, AR	1,061	44,829	675,441	29	NA	58
Alachua, FL	976	223,578	604,986	11	0.05	8
Clay, FL	652	157,502	410,846	16	0.02	0
Leon, FL	704	242,577	448,549	16	0.13	4
Nassau, FL	612	61,625	463,899	18	0.03	9
Santa Rosa, FL	1,045	133,092	738,121	16	0.13	1
Coweta, GA	454	101,395	285,445	19	NA	57
Douglas, GA	200	102,015	126,809	27	NA	0
Murray, GA	355	39,446	214,373	26	NA	45
Union, GA	335	19,119	186,122	26	NA	1
Laurel, KY	447	55,488	282,223	26	0.02	2
Trimble, KY	152	8,759	99,998	25	0.02	498
Livingston, LA	681	102,046	449,851	25	NA	0
DeSoto, MS	499	124,378	317,949	15	0.17	5
Warren, MS	639	48,993	396,049	29	0.05	19
Buncombe, NC	653	212,672	416,687	30	0.06	10
Orange, NC	400	118,183	255,316	26	0.11	1
Le Flore, OK	1,639	48,896	1,007,238	15	NA	42
Oconee, SC	683	68,523	413,423	18	0.08	265
Anderson, TN	342	71,904	220,692	32	0.03	75
Blount, TN	569	111,510	267,666	24	0.04	5
Sevier, TN	605	75,503	255,145	27	0.06	7
Montgomery, TX	1,106	344,700	685,399	21	NA	5
Chesterfield, VA	437	276,840	278,264	25	NA	30
Fluvanna, VA	288	23,078	185,753	23	NA	60
Totals				629		1,306

Source: Hodges, A., In: *Wood to Energy: Using Southern Interface Fuels for Bioenergy* (C. Staudhammer, A. L. Hermansen-Baez, D. R. Carter and E. A. Macie, eds.). USDA Forest Service General Technical Report GTR-132, 2011.

Example 13.6: Conversion of Biomass into Electrical Power

How many acres would be required to build a 40 MW (energy output) biomass power plant (operated 350 days a year and 24 h per day) if the heating value of the biomass (dry) is 20 MJ/kg (8,617 Btu/lb)? The biomass productivity or yield is only 9 dry tonnes/ha/yr (4 tons (dry) per acre per year). Assume an overall conversion efficiency of 30%. Convert the units into the English system.

Solution:

a. The energy input is first calculated using the efficiency equation as follows:

$$\text{Efficiency}(\%) = \frac{\text{output}}{\text{input}} \times 100\%$$

$$\text{Input} = \frac{\text{output}}{\text{efficiency}(\%)} \times 100\% = \frac{40\,\text{MW}}{30} \times 100 = 133.3\,\text{MW}$$

b. This input energy must come from biomass at the following rates as follows:

$$\text{Biomass input rate}\left(\frac{\text{tonnes}}{\text{h}}\right) = 133.3\,\text{MW} \times \frac{\text{MJ}}{1\,\text{MW}-\text{s}} \times \frac{\text{kg}}{20\,\text{MJ}}$$

$$\times \frac{\text{tonne}}{1{,}000\,\text{kg}} \times \frac{3{,}600\,\text{s}}{\text{h}} = 24\frac{\text{tonnes}}{\text{h}}$$

$$\text{Biomass input rate}\left(\frac{\text{tons}}{\text{h}}\right) = \frac{24\ \text{tonnes}}{\text{h}} \times \frac{1.1\ \text{tons}}{1\ \text{tonne}} = 26.4\frac{\text{tons}}{\text{h}}$$

c. This biomass input rate is converted into acreage required using the biomass productivity value provided:

$$\text{Acreage (ha)} = \frac{24\,\text{tonnes}}{\text{h}} \times \frac{24\,\text{h}}{\text{day}} \times \frac{350\,\text{days}}{\text{y}} \times \frac{\text{ha-yr}}{9\,\text{tonnes}} = 22{,}400\,\text{ha}$$

$$\text{Acreage (acres)} = 22{,}400\,\text{ha} \times \frac{2.47\,\text{acres}}{\text{ha}} = 55{,}328\,\text{acres}$$

d. Thus, at least 22,400 ha (55,328 acres) of land is required if the overall conversion efficiency is around 30%.

13.5 SLAGGING AND FOULING ISSUES WITH AGRICULTURAL BIOMASS

Agricultural biomass resources are excellent feedstock for waste-to-energy facilities, such as the production of heat and electrical power (Maglinao et al., 2008). However, there is difficulty in the thermal conversion of biomass residues due to high ash content, especially in the complete combustion mode (Nutalapati et al., 2007). The inorganic components of biomass feedstock cause problems such as slagging, fouling, bed agglomeration, and corrosion-related problems (Bryers, 1996).

Inorganic materials in the biomass melt when exposed to high reaction temperatures in combustion and form slag. Slag sticks to heat transfer-conveying surfaces upon cooling, thereby decreasing the cross-sectional areas; eventually, fouling (or clogging) occurs. Very limited studies have been conducted to evaluate the slagging and fouling phenomena in most common agricultural wastes, such as animal manure and crop residues.

Formation of deposits on heat transfer surfaces, referred to as slagging and fouling, is one of the biggest problems for all solid fuel-fired boilers, especially in biomass combustion (Tortosia-Masia et al., 2005). Slagging occurs in the boiler sections that are directly exposed to flame irradiation, and its formation involves stickiness, ash melting, and sintering. Fouling deposits, on the other hand, form in the convective parts of the boiler and are due mainly to the condensation of volatile species that have been vaporized in previous boiler sections and are loosely bonded. The slagging and sintering behavior of different types of fuel ash varies widely. Formation of deposits depends mainly on fuel quality, boiler design, and boiler operation. Although all biomass fuels exhibit fouling behavior, their rates differ depending on the content and composition of the ash. For instance, wood tends to foul at a lower rate than straw because, for the same fuel-firing rate, wood has less ash than straw.

Predicting the ash behavior before a fuel is used is desirable to avoid problems in the combustion operation (Zevenhoven-Onderwater et al., 2001). Reducing slagging and fouling reduces investment and operational costs and increases performance efficiency of combustors. Although numerous indices have been developed for coal and other fuels relating composition to fouling and slagging behavior, these have proved for most part to be of limited value as predictors for biomass (Jenkins et al., 1998). Despite their shortcomings, these empirical indices have been widely used to predict ash behavior and deposition tendencies for biomass-type ashes.

Vamvuka and Zografos (2004) used the alkali index (AI), the base-to-acid ratio, and the bed agglomeration index (BAI) to predict deposition tendencies in four types of agro residues: olive kernel, olive tree wood, citrus tree wood, and vine shoots. They found that the removal of troublesome elements by leaching these biomass materials with water reduced slagging and fouling in furnaces. Leaching with water resulted in significant ash reduction (up to 40%). In addition, there was improved fusibility behavior due to removal of problematic elements K (up to 93%), Na (up to 96%), P (up to 85%), Cl (up to 97%), and, finally, the heavy metals Co, U, Mo, Pb, and As. Either a lower combustion temperature or water leaching substantially reduced the fouling potential due to alkali. Ash pellet compressive strength measurements have likewise been used in earlier studies to develop predictions of ash agglomeration during pulverized coal combustion (Skrifvars et al., 1994). More recently, modeling techniques have been applied to visualize slagging and fouling tendencies using mass and energy balances (Tortosia-Masia et al., 2005). They developed an online monitoring system that models heat transfer between flue gases, water/steam cycle, and deposit formation on heat exchanger surfaces. The various indices for evaluating slagging and fouling issues in combustion systems will be discussed individually and they will be estimated for some biomass ash.

13.5.1 Coal Slagging and Fouling Indices

The slagging (R_s) and fouling factors (R_f) for coal were introduced and discussed in Chapter 3. Originally developed for judging the quality of coal in burners, these factors were only applicable for certain types of coal. Consequently, when the same factors are used for agricultural biomass, many of the calculations provide unreliable or meaningless results. For example, lignitic-type ash is associated with those materials whose CaO and MgO portions are greater than Fe_2O_3 (Winegartner, 1974). However, the reverse is true for bituminous-type coal. Most Western-type coals have lignitic-type ash, whereas Eastern-type coals have bituminous-type ash. This rule is not sacrosanct. The rule applies only to general behavior as a function of source. The real test of melting ash is to determine fusion temperatures of ash. Fusion temperatures are subdivided as follows: the initial temperature of ash deformation, the softening temperature, the hemispherical temperature, and the fluid temperature. There is also a standard test method to evaluate these conditions as contained in ASTM D1857 (Standard Test Method for Fusibility of Coal and Coke Ash). When performed in a laboratory, the initial deformation temperature is that temperature at which a prepared standard cone of ash just begins to fuse or shows evidence of deformation at the top of the cone as it is being heated in a standard test furnace. The softening temperature, also commonly called the fusion temperature, corresponds to an observed condition between the initial deformation and fluid temperature. This temperature usually occurs when the test cone height is equal to its width. Note that the cone mold is usually 19 mm (3/4 in) high and 6.4 mm (1/4 in) wide. The hemispherical temperature is that temperature at which the cone has fused down to a hemispherical lump, where the height is one-half the width of the base. The fluid temperature is that temperature at which the fused mass has spread out in a nearly flat layer with a maximum height of 1.6 mm (1/16 in). Note that the tests may be performed in either an oxidizing or a reducing atmosphere. The type of test depends on the type of thermal conversion system used. For example, combustion systems normally use an oxidizing atmosphere.

The ASME publication on coal slagging and fouling protocol also cautions the operator about the use of this procedure and the calculations with lignitic-type ash. The R_f and R_s should not be applied to lignite because it is very similar to agricultural biomass. However, in the absence of a suitable standard, R_f and R_s are used to make an initial guess about the propensity of the biomass ash to form slag.

13.5.2 Other Slagging and Fouling Indices

The other slagging and fouling indices normally used for biomass resource ash are the AI, the base-to-acid ratio ($R_{b/a}$), and the BAI discussed by Vamvuka and Zografos (2004).

13.5.2.1 Alkali Index

The AI expresses the quantity of alkali oxides in the fuel per unit of fuel energy (kg alkali/GJ or lb alkali/MMBtu). The alkali oxides are those from potassium oxide (K_2O) and sodium oxide (Na_2O). The AI is computed using Equation 13.5 for the metric system and Equation 13.6 for the English system of units:

$$\mathrm{AI} = \frac{\mathrm{kg}\left(\mathrm{K_2O} + \mathrm{Na_2O}\right)}{\mathrm{GJ}} \tag{13.5}$$

In the English system of units, the equation is slightly different as follows:

$$\frac{1\times10^6}{\text{Btu/lb}}\times\%\text{ash}\times\%\text{alkali of the ash}=\frac{\text{lb alkali}}{\text{MMBtu}} \quad (13.6)$$

When the alkali value is below 0.17 kg/GJ (0.4 lb MMBtu), the biomass is said to have a low slagging risk. For values above 0.34 kg/GJ (0.8 lb/MMBtu), slagging is high and very likely to occur. However, within the range of 0.17–0.34 kg/GJ (0.4–0.8 lb/MMBtu), fouling or slagging may or may not occur. Note that complete inorganic analysis is necessary, as well as determining the heating value of the biomass. Cotton gin trash (CGT) and animal manure are completely combusted in a controlled furnace and the ash collected for complete inorganic analysis. Data for CGT and animal manure ash is presented in Table 13.2. The values summarized in the table are on a per unit weight basis, as well as the heating value of the original biomass. Example 13.7 shows how the AI is calculated.

Example 13.7: Alkali Index for CGT and Animal Manure

Calculate the alkali index for CGT samples using the data in Table 13.2. Comment on the values generated.

TABLE 13.2
Raw Data for CGT and Manure Ash for Calculating Various Slagging and Fouling Indices

Ash Analysis (%)	CGT Ash	Manure Ash
Al_2O_3	3.46	3.12
CaO	23.30	27.41
Fe_2O_3	1.11	1.84
MgO	5.69	10.90
MnO	0.06	0.14
P_2O_5	2.25	4.98
K_2O	24.62	5.28
SiO_2	21.70	32.46
Na_2O	0.76	1.82
SO_3	7.40	6.12
TiO_2	0.25	0.22
Total	90.60	94.29
Heating value (MJ/kg)	16.67	15.93
Heating value (Btu/lb)	7,182	6,863
% Ash in biomass	11.86	18.62
% Alkali in ash	25.38	7.10

Solution:

a. The percent ash in this biomass is 11.86%. Thus, for every 1 kg of this biomass, the amount of ash is 0.1186 kg.
b. The percentage of K_2O and Na_2O is 24.62% and 0.76%, respectively, by percentage of weight. The total alkali is 25.38% (24.62 + 0.76). Thus, for every kilogram of CGT biomass, the amount of alkali is 0.0301 kg (i.e., 0.1186 × 0.2538). The AI for CGT is calculated as follows:

$$\mathrm{AI}\left(\mathrm{CGT}\frac{\mathrm{kg\,alkali}}{\mathrm{GJ}}\right)=\frac{\mathrm{kg}(\mathrm{K_2O}+\mathrm{Na_2O})}{\mathrm{GJ}}=\frac{0.0301\,\mathrm{kg}}{1\,\mathrm{kg\,CGT}}$$

$$\times\frac{\mathrm{kg\,CGT}}{16.67\,\mathrm{MJ}}\times\frac{1{,}000\,\mathrm{MJ}}{\mathrm{GJ}}=1.8\frac{\mathrm{kg}}{\mathrm{GJ}}$$

c. In the English system, the index is calculated is calculated as follows:

$$\mathrm{AI}\left(\mathrm{CGT}\frac{\mathrm{lb\,alkali}}{\mathrm{MMBtu}}\right)=\frac{1\times10^6}{7182\,\mathrm{Btu/lb}}\times\frac{11.86\%}{100}\times\frac{25.38\%}{100}=4.19\frac{\mathrm{lb\,alkali}}{\mathrm{MMBtu}}$$

d. Thus, slagging and fouling is certain to occur if this biomass is combusted.

13.5.2.2 *Base-to-Acid Ratio*

The base-to-acid ratio is much simpler. This is calculated by determining all the acid and base components in the ash and taking their ratios as shown in Equation 13.7. The acid components of the biomass are the sum of the following compounds that are considered acidic: Fe_2O_3, CaO, MgO, K_2O, and Na_2O. The basic compounds in the ash are the SiO_2, TiO, and Al_2O_3 components. If the base-to-acid ratio is more than unity, slagging and fouling is certain to occur during combustion of this biomass.

$$R_{\mathrm{b/a}}=\frac{\%\left(\mathrm{Fe_2O_3}+\mathrm{CaO}+\mathrm{MgO}+\mathrm{K_2O}+\mathrm{Na_2O}\right)}{\%\left(\mathrm{SiO_2}+\mathrm{TiO_2}+\mathrm{Al_2O_3}\right)} \tag{13.7}$$

Example 13.8 shows how to calculate the base-to-acid ratio using data in Table 13.2.

Example 13.8: Base-to-Acid Ratio for CGT Ash Sample

Calculate the base-to-acid ratio for CGT samples using the data in Table 13.2. Comment on the values generated.

Solution:

a. The total percentage of the basic ash component in CGT is determined as follows:

$$\%\left(\mathrm{Fe_2O_3}+\mathrm{CaO}+\mathrm{MgO}+\mathrm{K_2O}+\mathrm{Na_2O}\right)=\%(1.11+23.30+5.69+24.62+0.76)=55.48$$

b. The total percentage of acid ash component in CGT is determined as follows:

$$\%\left(SiO_2 + TiO_2 + Al_2O_3\right) = \%\left(21.70 + 0.25 + 3.46\right) = 25.41$$

c. Thus, the $R_{b/a}$ for CGT is calculated as follows:

$$R_{b/a} = \frac{55.48}{25.41} = 2.18$$

d. Thus, slagging and fouling is also certain to occur for this biomass if combusted.

13.5.2.3 *Bed Agglomeration Index*

The BAI relates ash composition to agglomerations in fluidized bed reactors. It is calculated using Equation 13.8. Bed agglomeration is said to occur when the BAI values are lower than 0.15.

$$BAI = \frac{\%Fe_2O_3}{\%\left(K_2O + Na_2O\right)} \tag{13.8}$$

Example 13.9 shows how the agglomeration index is calculated.

Example 13.9: BAI of CGT Ash Sample

Calculate the BAI for CGT samples using the data in Table 13.2. Comment on the values generated.

Solution:

a. The total percentage of iron oxide in the CGT sample is 1.11%; thus, the calculation is straightforward as shown below. From Table 13.3, the K_2O for CGT is 24.62% and Na_2O is 0.76%.

$$BAI = \frac{\%Fe_2O_3}{\%\left(K_2O + Na_2O\right)} = \frac{1.11}{24.62 + 0.76} = 0.04$$

b. Thus, slagging and fouling is almost certain to occur for this biomass if combusted, because the BAI is less than 0.15.

13.6 DETERMINING EUTECTIC (OR MELTING) POINT OF BIOMASS ASH PELLETS

No single index has thus far been developed that reliably describes the behavior of ash under all combustion conditions, especially the operating temperature. One method that was used for ash behavior prediction is the measurement of the compressive strength of ash pellets that were previously heated at different temperatures. The method does not adequately predict the ash-sintering behavior of woody fuels and other biomasses with a lower content of alkaline elements. However, the compressive strength measurement method seems to offer consistent results under the conditions tested, when compared with the disintegrating and fusibility methods for biomass with high alkaline oxide contents (Fernandez Llorente and Carrasco Garcia, 2005).

A procedure was developed (Maglinao and Capareda, 2010) to measure the compressive strength of biomass ash pellets that had been exposed to various combustion temperatures. The process starts by placing the biomass sample in a furnace to generate appreciable amounts of ash. The ash is then molded into pellets. The dimensions of the pellets vary. However, for this experiment, the pellets had a diameter of 2.54 cm and a height of 1.65 cm. These pellets were exposed to an oxidizing atmosphere at various furnace temperatures. After each exposure, the pellets were cooled, and a compressive test measurement was made. The compressive strength of the pellets was determined using an MTS Model 810 Material Testing System (Eden Prairie, MN).

The compressive strengths of dairy manure and CGT ash pellets are summarized in Table 13.3 and shown graphically in Figure 13.7. The compressive strength of CGT ash was lowest at the lowest temperature of 550°C; however, it continued to increase sharply until about 800°C, after which its strength decreased rapidly. On the other hand, manure ash pellets exhibited their highest compressive strength at 600°C.

Ash pellets reduce their breakdown pressure as the heating temperature increases due to the volatilization of CO_2 from the ash carbonates. In turn, the carbonates increase the ash pellet porosity and fragility at higher heating temperatures (Maglinao, 2009). The temperature at which the compressive strength is highest, before any decrease with increasing temperature, is suggested to be the maximum combustion temperature at which slagging and fouling of ash could be avoided. Based on the results, it can be suggested that manure can be combusted at an operating temperature of 600°C without slagging and CGT can be burned with no ash deposition at a higher temperature of 800°C. These temperatures indicate the melting point of the ash, at which the pellets become brittle. The results indicate that manure is not a suitable fuel for combustion, because it does not provide any flexibility in

TABLE 13.3
Measured Compressive Strengths of Treated Ash Pellets of Dairy Manure and CGT Measured Compressive Strengths of Treated Ash Pellets of Dairy Manure and CGT

	Compressive Strength (kPa)	
Treatment Temperature (°C)	CGT	Manure
550	808.6 ± 43.6	1,428.6 ± 177.6
600	1,365.2 ± 115.7	1,778.1 ± 83.2
700	2,785.7 ± 290.5	1,159 ± 85.4
800	4,129.6 ± 178.4	1,037.4 ± 100
900	1,968 ± 351.7	(Not tested)

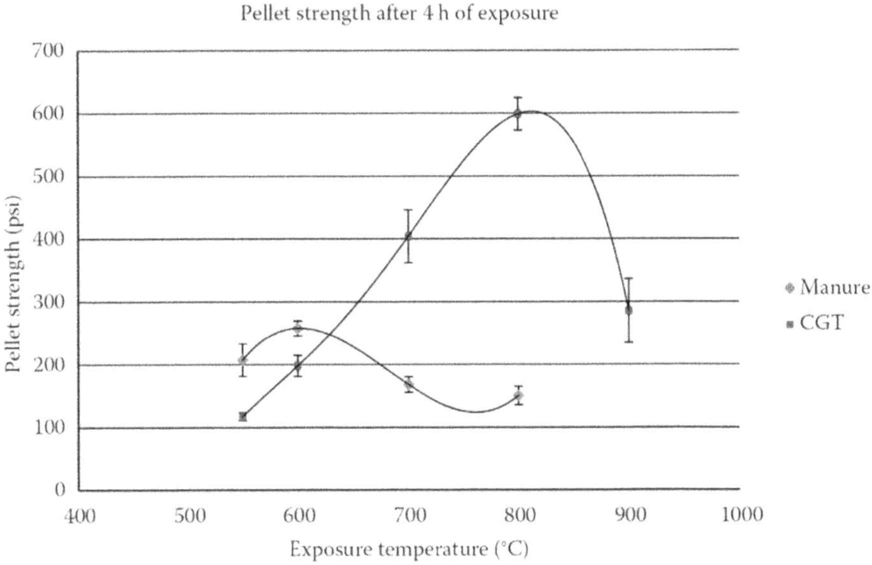

FIGURE 13.7 Graph showing eutectic points of manure and CGT biomass ash.

terms of operating temperature. Because of the higher compressive strength values and higher melting temperature for the CGT, this feedstock is expected to present fewer problems during thermal conversion (than animal manure). The normal operating temperature for the gasification process is approximately 760°C (Maglinao et al., 2008). Thus, CGT can be gasified conveniently at this operating temperature, with minimal slagging. Animal manure will be difficult to gasify at this operating temperature because of the production of slag. Conventional atmospheric FBC temperatures are normally within the range from 800°C to 900°C (Levy et al., 1981). CGT and animal manure feedstock are not suitable for atmospheric combustion because of the requirement for a higher operating temperature. Dairy manure was combusted in a 0.3048 m (1 ft) diameter FBC system. After operating for 1 day, the container of slag shown in Figure 13.8 was collected. Based on the volume of slag, it is clear that combustion is not the appropriate thermal conversion system for dairy manure. The calculation of compressive strengths for pellets is illustrated in Example 13.10.

Example 13.10: Compressive Strength of CGT Ash Pellet Sample

Determine the compressive strength of CGT ash pellets (in units of kPa) if the load measured just before crumbling is 2,093 N. The dimensions of the pellets are as follows: 2.54 cm in diameter by 1.65 cm in height. Note that 1 Pa = 1 N/m^2.

Solution:

a. The compressive stress or strength of a material is equal to the load force divided by area. The cross-sectional area (in m^2) of the pellet is first calculated as follows:

FIGURE 13.8 Photo of about 18 kg (40 lb) of slag collected after full-day gasification of animal manure using a fluidized bed gasifier.

$$\text{Area}\left(\text{m}^2\right) = \frac{\pi\left(\frac{2.54\,\text{cm}}{100\,\text{cm}}\right)^2}{4} = 0.000507\,\text{m}^2$$

b. Then, the compressive strength is calculated as follows:

$$\text{Compressive strength}\left(\frac{\text{N}}{\text{m}^2}\right) = \frac{2093\,\text{N}}{0.000507\,\text{m}^2} = 4{,}128{,}205\,\text{Pa}$$

c. This value is also equal to 4,128 kPa.

13.7 APPLICATIONS OF BIOMASS COMBUSTION SYSTEMS

The most common application of biomass combustion is heat. Recently, we have seen the proliferation of biomass-fueled power plants of various configurations using combustion. However, these applications are for biomass that do not slag or foul. For biomass with high ash content, and for those with inorganic ash components that melt at the high temperatures of combustion, gasification systems may be the only possible option. Enumerated below are the various applications of biomass combustion systems:

1. Cooking and heating
2. Mechanical power
3. Electrical power
4. CHP applications

13.7.1 Cooking, Heating, Drying, and Steam Production

The cooking and heating applications of biomass are clearly the most common. The use of fuelwood for cooking is still the predominant use of biomass, followed by household heating in cold climates. Biomass is also predominantly used for drying crops and other commodities, such as tobacco leaves. The combustion of animal manure in small-scale systems is also prevalent in some countries. However, large-scale utilization of biomass with high inorganic ash components is a challenge because of slagging and fouling issues. Wood has been used for this purpose primarily because of its low ash content. There is continuing research on the use of improved cookstoves around the world. The new challenge is to have the best cookstoves that will not provide harmful exhaust emissions in terms of NO_x, smog, and particulate matter. The US Department of Energy (USDOE) continually supports the improvement of the design of biomass cookstoves around the world. Biomass use for steam production is still prevalent in wood-processing plants where it is used for heat or wood pretreatment. Some cotton textile mills are still using biomass wastes to provide this steam of heat. Many animal-processing plants, in need of cheaper fuel for meat processing, also rely on wood or biomass waste for this purpose. Sugar mills use bagasse for this purpose. Bagasse from sugarcane as well as cottonseed hulls are biomass resources that may be combusted without slagging and fouling problems. In the palm and coconut industry, most wastes are residues that are still being used for processing and drying.

13.7.2 Mechanical Power Generation

In the past, when mechanical power was required for sawmills and other milling and grinding applications, steam generation plants were used. These steam plants are becoming rare because of other available conversion routes. Perhaps the most interesting research is in the use of wood biomass for mechanical power generation. Wood powder is combusted to generate heat for use with the Stirling engine (Nishiyama et al., 2007). Stirling engines operate using external heat. Biomass powder is normally combusted using a separate burner, where the heat generated operates a Stirling engine, which generates mechanical power. In 2004, a demonstration plant of this nature with an output of 55 kWe was started in Japan. This plant promoted the application of biofuels as an element of the Japanese Renewable Portfolio Standards initiative. Initial test results revealed that heat from wood powder, with high conversion efficiencies above 90%, is excellent for Stirling engines. In Japan, work is underway for a CHP application. Several years ago, there was renewed interest in the use of external combustion engines, such as the Stirling engine types, for mechanical power generation; however, the cost to manufacture the engines is still an important barrier for commercialization. Countries without a sophisticated manufacturing industry may not be able to follow this renewed route.

13.7.3 Electrical Power Generation

Biomass electrical power generation systems have been in operation for many decades in most countries where forestry resources and water are still abundant.

Currently, this approach is being pursued by many developed countries such as the United States, as shown by numerous techno-economic feasibility projects that are being initiated by states with abundant forestry reserves. In a report commissioned by the USDOE (Wiltsee, 2000), several lessons learned were discussed from the analysis of 18 existing biomass-fueled power plants in the United States and other units in Finland and Canada. The majority of the 18 units in the United States use spreader stoker systems. However, some are FBC systems, and a few are circulating fluidized bed systems. The issues and problems discussed focused on the cost of fuel and logistics of fuel sourcing, including problems with on-site storage as well as fuel feed systems' issues, reliability, and dependability. The final issue discussed is the need to have stable financial support from the government (e.g., subsidies) or support for appropriate pricing on the sale of electrical power from the private sector and from local utilities.

The issues and problems presented, however, are minor compared with the environmental and economic benefits that were discussed for most of the projects. These benefits include pollution improvement in rural settings, elimination of open burning, providing for the use of additional heat energy from the process, effective management of forest residues, and the prevention of many wildfires when forest land covers are removed on a routine basis. Note that most of the projects discussed focused on the use of wood fuel with very limited use of agricultural residues.

13.7.4 Combined Heat and Power Applications

The most important benefit of utilizing biomass for energy purposes is the opportunity to use excess energy for multiple purposes. Thus, electrical power plants fueled by biomass must look for applications of their excess heat. In this way, the facility can become a fully dedicated CHP plant. Likewise, if heat is the main product, the excess energy may also be utilized to generate small amounts of electrical power, thereby increasing the overall plant conversion efficiency. This is especially true if the biomass-fueled power plant would like to take advantage of the second law of thermodynamics. Cogeneration plants must have close relationships with either the local utilities or local industries that require heat or steam. This is the only way they will improve the overall economics of the system. A lot of CHP plants are being planned using biomass resources from large animal farms, the so-called confined animal farm operations or CAFOs. Unfortunately, many of these plants have not addressed the slagging and fouling issues for these types of biomasses and will likely fail, especially if the thermal conversion system selected is that of combustion. CHP plants should become a popular option for biomass resources in the future. CHP produces electrical and thermal energy. At best, the operating efficiency of a conventional power plant is around 40%. A CHP plant can achieve an overall efficiency above 80% (Obernberger and Thek, 2004). The overall annual efficiencies of biomass power plants (CHP) investigated range from 83% to 92%.

13.8 CONCLUSION

The combustion of biomass, particularly fuelwood, is the simplest and oldest thermal conversion process worldwide. However, if the ash content of the biomass is high, there may be slagging and fouling problems. Unlike coal or some wood waste, the ash in most agricultural biomass tends to melt at the high operating temperatures of combustion systems. This is especially true for large-scale operations. One will not experience this event in the operation of small biomass cookstoves, primarily because the ash temperature at the bottom of the stove is lower compared with the higher combustion temperatures above it. The ash residues are quickly cooled, and the gaseous products produced are burned in the combustion zone above it. This is where the high temperature is experienced. However, in many high-throughput and continuous combustion systems, particularly in fluidized bed systems, the ash and the hot gases are both in the combustion zone where the temperature is relatively high. The ash generated during the combustion process could melt and form slag in conveyance systems, leading to fouling.

Various slagging and fouling indices or parameters have been developed, particularly for coal combustion systems (such as the slagging, R_s, and fouling, R_f, factors). It was shown that these factors are not applicable to most agricultural biomass residues, or even to lignite. However, additional factors that have been evaluated, such as the BAI, base-to-acid ratio ($R_{b/a}$), and AI, may be applicable to most agricultural biomasses. If these additional parameters are still inappropriate, then the best method is to expose the biomass ash at high temperatures and observe its melting properties. This test may be conducted using inert gas or an air system. A procedure was presented in this chapter to guide the operator or combustion systems designers regarding the suitability of the use of biomass resources in combustion systems. When the biomass ash is exposed to the range of temperatures encountered in combustion systems and melting is observed, it is likely that this biomass resource is not appropriate for use in combustion systems. In this case, the appropriate thermal conversion processes are gasification or pyrolysis systems that operate at lower temperatures. The eutectic (or melting) point analysis of biomass ash becomes an important step before any planned combustion system operation.

Perhaps the simplest and the most appropriate way of using agricultural biomass in large-scale combustion systems is to use a small portion of the biomass and mix it with conventional fuel, such as with coal in co-firing or co-combustion systems. High-ash biomass may also be mixed with low-ash biomass to have the same effect.

If biomass resources are used for power generation systems through combustion, large tracks of land may be required to satisfy the requirements of these facilities. The amount of biomass required is dependent primarily on the conversion efficiencies and the biomass productivity per unit of area. The amount of acreage required may range between 500 and more than 1,000 ha/MW of electrical power output. One will realize that a 50 MW facility may require about 15,000 ha (37,000 acres) of land, depending on the combination of the overall conversion efficiency and land productivity. The daily biomass requirement for this facility is nearly 1,000 tonnes/day. Twenty-four trucks will be making a delivery to this facility each day if the truck capacity is 40 tonnes or about one truck per hour. This will be quite a logistic problem if the area of coverage is not contiguous. There must be a sophisticated biomass logistics and scheduling program in place to minimize downtime for the power plant. Contingency plans must be in place in case there are untoward climate events, such as typhoons, storms, or drought, in the biomass collection area.

13.9 PROBLEMS AND DISCUSSION ISSUES

13.9.1 Amount of CO_2 Produced for Every Metric Ton of Biomass Combusted

P13.1 Suppose you have certain biomass with the following chemical formula: $C_{3.33}H_{5.2}O_{2.56}N_{0.10}$. Determine the amount of CO_2 produced (in kg) per metric ton of this biomass when combusted completely in air.

13.9.2 Amount of Air Required for Combustion of Sorghum Biomass

P13.2 The ultimate analysis of sorghum biomass (in percent weight) is taken from Table 3.5 and copied below (HV was taken from Table 3.2). Determine the exact amount of air (m^3/kg and ft^3/lb) required for combustion of this material in air. Assume standard air conditions with an air density of 1.2 kg/m^3 (0.075 lb/ft^3).

Biomass	C	H	O	N	S	Ash	HV (MJ/kg)
Sorghum	38.1	5.6	47.8	0.9	0.2	8.5	17.0

13.9.3 Biomass Combustion Efficiency and Acreage Needed

P13.3 How much biomass is required in tonnes/h (and tons/h) for a 50 MW electrical facility if the overall conversion efficiency is 25%? The facility is to be operated 333 days in a year for 24 h per day. How much biomass is required per year? The heating value of the biomass was 20 MJ/kg. If the biomass productivity is 11.23 dry tonnes/ha/yr (5 dry tons/acre/year), what is the area required to satisfy this requirement?

13.9.4 Boiling Water at Mile High Stadium

P13.4 Determine the boiling point (in °C) of water if one is at the Mile High Stadium in Denver, Colorado.

13.9.5 Biomass Co-Firing Efficiency

P13.5 Determine the efficiency of a co-firing system utilizing biomass and coal with the following data: Roughly 18.57% of biomass was used by weight at a rate of 0.65 tonnes/h and coal at a rate of 3.5 tonnes/h.

The heating value of the biomass was 18 MJ/kg and that of coal was 31.3 MJ/kg. The output of the process was 17 MW. What was the ratio of biomass power input versus coal power (%)?

13.9.6 Conversion of Biomass into Small-Scale Electrical Power

P13.6 A small ranch consumes around 12,000 kWh of electricity each month. If a small-scale biomass-fueled power plant is established for this facility, how much biomass is required each 30-day month if the unit is to be operated at least 350 days of a year on a 24 h/day operation (tonnes/day and tonnes/year)? Assume an overall conversion efficiency of 45%. The biomass to use has a productivity of 30 wet tonnes/ha/yr (13.36 tons/acre/yr) with about 50% moisture. Relatively dry biomass at 10% moisture will be used for the conversion process with a heating value of 19 MJ/kg. How many hectares are required for this operation? Convert the units to the English system.

13.9.7 Alkali Index for Animal Manure

P13.7 Calculate the alkali index for animal manure using the data in Table 13.2. Comment on the values generated.

13.9.8 Base-to-Acid Ratio in Animal Manure Ash Sample

P13.8 Calculate the base-to-acid ratio for animal manure samples using the data in Table 13.2. Comment on the values generated.

13.9.9 Bed Agglomeration Index of Animal Manure Ash Sample

P13.9 Calculate the BAI for animal manure samples using the data in Table 13.2. Comment on the values generated.

13.9.10 Compressive Strength of Animal Manure Ash Pellet Sample

P13.10 Determine the compressive strength of animal manure ash pellets (in units of kPa) if the load measured just before crumbling is 526 N and the dimensions of the pellets are as follows: 2.54 cm in diameter by 1.65 cm in height. Note that 1 Pa = 1 N/m^2. Convert units in English system (psi).

13.9.11 Melting Materials During Combustion

P13.11 At high combustion temperature, some biomass components will turn into molten ash, the incombustible material following biomass combustion. When this material cools when exposed to lower temperature, it can stick to the furnace walls or components. What do you call this phenomenon?

a. fouling b. slagging c. slugging d. elutriation

13.9.12 Sticky Materials During Combustion

P13.12 At high combustion temperature, some ash components of biomass will melt and stick to internal surfaces of ducts and pipes that are cooled from the outside environment. The continued buildup of these formations will lead to blockage and ultimately clogging of these ducts and pipes. What do you call this phenomenon?

a. fouling b. slagging c. slugging d. elutriation

13.9.13 Slagging and Fouling Indices for Agricultural Biomass

P13.13 This agricultural biomass slagging and fouling index expresses the quantity of potassium oxides and sodium oxides per unit weight of the fuel energy in units of collective weights of this compound divided by the energy content of the biomass. Name this slagging and fouling index.

a. base-to-acid ratio b. agglomeration index c. alkali index d. R_f and R_s

13.9.14 Biomass Requirement for a Power Plant and Potential Revenue

P13.14 The 115-MW power plant facility, located in Sacul, Texas, uses wood as fuel. It is one of the largest biomass-fueled power plants in the country. It used a bubbling fluidized bed-boiler and a condensing steam turbine with an evaporative cooling tower. Determine the amount of biomass this plant requires each year, assuming a dedicated high-yielding hybrid poplar is used as fuel. Likewise, determine the potential savings in electrical power for this facility assuming power is sold at $0.12/kWh. Finally, how many hectares of land is required for this project?

13.9.15 Biomass Pellet Stoves

P13.15 A Canadian resident plans to use a pellet stove for household heating during the long cold winter. Estimate the amounts of pellets used per year, all associated costs, and report on savings compared with electrical heating or using propane or natural gas.

REFERENCES

Bailis, R., D. Ogle, N. MacCarthy and D. Still. 2007. The water boiling test (WBT). Publication by the Shell Foundation for the Household Energy and Health Program, WBT Version 3, January 2007, Shell Foundation, Shell Center, London.

Bryers, R. W. 1996. Fireside slagging, fouling, and high-temperature corrosion of heat-transfer surface due to impurities in steam-raising fuels. *Progress in Energy and Combustion Science* 22 (1): 29–120.

Capareda, S. C. 1995. Utilization of residue from agriculture and agro-industry. In: *Proceedings of the Study Meeting on Appropriate Energy Sources for Rural Areas held in Manila, Philippines, August 2–9, 1994* (S. C. Capareda, ed.). A Report of an APO Study Meeting Organized by the Asian Productivity Organization, Tokyo, Japan.

Energy Saver. 2022. Wood and pellet heating. USDOE Office of Energy Saver and Energy Efficiency & Renewable Energy, Washington, DC; Available at: https://www.energy.gov/energysaver/wood-and-pellet-heating#:~:text=In%20fact%2C%20pellet%20stoves%20are,and%2090%2C000%20Btu%20per%20hour. Accessed 29 September 2022.

Fernandez Llorente, M. J. and Carrasco Garcia, J. E. 2005. Comparing methods for predicting the sintering of biomass ash in combustion. *Fuel* 84: 1893–1900.

Hodges, A. 2011. Economic impact analysis of woody biomass energy development. In: *Wood to Energy: Using Southern Interface Fuels for Bioenergy* (C. Staudhammer, A. L. Hermansen-Baez, D. R. Carter and E. A. Macie, eds.). USDA Forest Service, Southern Research Station, Ashville, NC. General Technical Report GTR-132. pp. 67–77.

Hoffmann, D. and M. Weih. 2005. Limitations and improvement of the potential utilization of woody biomass for energy derived from short rotation woody crops in Sweden and Germany. *Biomass and Bioenergy* 28: 267–279.

IEA. 2007. Biomass for power generation and CHP. IEA Energy Technology Essentials. January 2007. Also available at: https://www.iea.org/techno/essentials3.pdf.

Jenkins, B. M., L. L. Baxter, T. R. Miles Jr. and T. R. Miles. 1998. Combustion properties of biomass. *Fuel Processing Technology* 54: 17–46.

Johnson, M., R. Edwards, V. Berrueta and O. Masera. 2010. New approaches to performance testing of improved cookstoves. *Environmental Science & Technology* 44 (1): 368–374.

Johnsson, F. 2007. Fluidized bed combustion for clean energy. Paper presented at the 2007 ECI Conference on the 12th International Conference on Fluidization-New Horizons in Fluidization Engineering held in Vancouver, Canada, May 13–17, 2007. Article No. 5 Vol. RP4. In: *Proceedings of the 12th International Conference of Fluidization - New Horizons in Fluidization Engineering Sponsored by ECI, Engineering Conferences International Symposium Series* (F. Berruti, X. Bi and T. Pugsley, eds.).

Levy, A. R., R. E. Barret, R. D. Giammar and H. R. Hazard. 1981. Coal combustion. Chapter 8. In: *Coal Handbook* (R. A. Meyers, ed.) Marcel Dekker, Inc., New York. pp. 359–431.

Maglinao, A. L. 2009. Instrumentation and evaluation of a pilot scale fluidized bed biomass gasification system. MS Thesis. Department of Biological and Agricultural Engineering, Texas A&M University, College Station, TX.

Maglinao, A. L. and S. C. Capareda. 2010. Predicting fouling and slagging behavior of dairy manure (DM) and cotton gin trash (CGT) during thermal conversion. *Transactions of the ASABE* 53 (3): 903–909.

Maglinao, A. Jr., S. C. Capareda, C. B. Parnell, Jr. and D. Carney. 2008. Demonstration and simulation of a fluidized bed gasification system for power generation. ASABE Paper No. 08-4935, ASABE, St. Joseph, MI.

Nishiyama, A., H. S. Himojima, A. Ishikawa, Y. Itaya, S. Kambara, H. Moritomi and S. Mori. 2007. Fuel and emissions properties of Stirling engine operated with wood powder. *Fuel* 86 (15): 2333–2342.

Nutalapati, D., R. Gupta, B. Moghtaderi and T. F. Wall. 2007. Assessing slagging and fouling during biomass combustion: A thermodynamic approach allowing for alkali/ash reactions. *Fuel Processing Technology* 88: 1044–1052.

Obernberger, I. and G. Thek. 2004. Techno-economic evaluation of selected decentralized CHP applications based on biomass combustion in IEA partner countries final report. International Energy Association (IEA) Task 32 Project Report performed in cooperation with the Institute for Resource Efficient and Sustainable Systems, Graz University of Technology, Graz, Austria.

Skrifvars, B. J., M. Hupa, R. Backman and M. Hiltunen. 1994. Sintering mechanisms of FBC ashes. *Fuel* 73 (2): 171–176.

Tortosia-Masia, A. A., F. Ahnert, H. Spliethoff, J. Loux and K. Hein. 2005. Slagging and fouling in biomass co-combustion. *Thermal Science* 9 (3): 85–98.

Vamvuka, D. and D. Zografos. 2004. Predicting the behavior of ash from agricultural wastes during combustion. *Fuel* 83 (14–15): 2051–2057.

Van Loo, S. and J. Koppejan (eds.). 2008. *The Handbook of Biomass Combustion and Co-Firing*. Earthscan, Sterling, VA.

Wark, K. 1983. *Thermodynamics*. 4th Ed. McGraw Hill Book Company, New York.

Wiltsee, G. 2000. Lessons learned from existing biomass power plants. NREL, USDOE Technical Report Under Su-contract No. AXE-8-18008. USDOE, Golden, CO.

Winegartner, E. C. (ed.). 1974. *Coal Fouling and Slagging Parameters*. Prepared by the ASME Research Committee on Corrosion and Deposits from Combustion Gases. The American Society of Mechanical Engineers, New York.

Zevenhoven-Onderwater, M., R. Backman, B. J. Skrifvars, M. Hupa, T. Lilienthal, C. Rosen, K. Sjostrom, K. Engvall and A. Hallgren. 2001. The ash chemistry in fluidized bed gasification of biomass fuels. Part II: Ash behavior prediction versus bench scale agglomeration tests. *Fuel* 80: 1503–1512.

ASSESS YOUR KNOWLEDGE

1. What is the main problem with combustion of high ash biomass?
   a. High moisture
   b. Low calorific value
   c. Low fixed carbon
   d. Slagging and fouling

2. This thermal conversion process utilizes the carbon and hydrogen of biomass to generate carbon dioxide, water, and heat.
   a. Liquefaction
   b. Combustion
   c. Gasification
   d. Pyrolysis

11. This is a slagging and fouling index, which is the ratio of iron oxide or ferric oxide to the sum of potassium oxide and sodium oxide.
    a. BAI
    b. Base-to-acid ratio
    c. Alkali index
    d. Slagging factor

12. The 115 MW biomass power plant in Sacul, Texas utilizes what type of thermal conversion reactor?
    a. Spreader stoker
    b. Gasifier
    c. Bubbling fluidized bed
    d. Fixed bed

3. This is a slagging and fouling index, which is the ratio of the sum of potassium oxide and sodium oxide to the energy content of the material being converted thermally.
   a. BAI
   b. AI
   c. Base-to-acid ratio
   d. Fouling factor

4. This is a slagging and fouling index, which is the ratio of the sum of the following inorganic compounds (ferric oxide, calcium oxide, magnesium oxide, potassium oxide, and sodium oxide) to the following inorganic compounds (silicon dioxide, titanium dioxide, and aluminum oxide).
   a. BAI
   b. AI
   c. Base-to-acid ratio
   d. Fouling factor

5. Determine the chemical formula for Jatropha shells. The elemental composition of this biomass by weight is as follows: C = 40.8%; H = 5.9%; O = 46.3; N = 2.2%; and S = 0.3%, with an ash content of 4.6%.
   a. $C_{3.4}H_{5.9}O_{2.89}N_{0.157}S_{0.009}$
   b. $C_{3.2}H_{5.6}O_{3.1}N_{0.04}S_{0.011}$
   c. $C_{40.8}H_{5.9}O_{46.3}N_{2.2}S_{0.3}$
   d. $C_{3.2}H_{5.6}O_{3.1}N_{0.04}S_{0.011}$

13. This factor used in the analysis of coal and is the ratio of the sum of all inorganic base compounds in the ash to the sum of acid compounds in the ash multiplied by the percentage of sulfur on dry coal.
   a. Fouling factor
   b. Alkali index
   c. Base-to-acid ratio
   d. Slagging factor

14. This factor used in the analysis of coal and is the ratio of all the inorganic base compounds in the ash to the sum of all acid compounds in the ash multiplied by the percentage of sodium oxide in the ash.
   a. Fouling factor
   b. Alkali index
   c. Base-to-acid ratio
   d. Slagging factor

15. Determine the air-to-fuel ratio (lb/lb) for the complete combustion of Jatropha shells. The elemental composition of this biomass by weight is as follows: C = 40.8%; H = 5.9%; O = 46.3; N = 2.2%; and S = 0.3%, with an ash content of 4.6%.
   a. 5.17
   b. 4.93
   c. 3.25
   d. 7.07

6. Determine the air-to-fuel ratio in units of (ft^3 air/lb fuel) for the complete combustion of Jatropha shells. The elemental composition of this biomass by weight is as follows: C = 40.8%; H = 5.9%; O = 46.3; N = 2.2%; and S = 0.3%, with an ash content of 4.6%. Assume the density of air to be 0.075 lbs/ft^3.
   a. 4.94
   b. 5.17
   c. 65.7
   d. 68.8

7. Determine the slagging and fouling indices for the following inorganic ash data: Al_2O_3 = 3.12; CaO = 27.41; Fe_2O_3 = 1.84; MgO = 10.90; MnO = 0.14; P_2O_5 = 4.98; K_2O = 5.28; SiO_2 = 32.46; Na_2O = 1.82; SO_3 = 6.12; TiO_2 = 0.22. The heating value of the biomass was 6,863 Btu/lb and the percentage of ash in the biomass was 18.62%. For this question, what is the descriptive fouling type for this biomass using the coal index R_f?
   a. Medium
   b. Severe
   c. High
   d. Low

16. Determine the volume of air in units of CFM for the complete combustion of Jatropha shells. The elemental composition of this biomass by weight is as follows: C = 40.8%; H = 5.9%; O = 46.3; N = 2.2%; and S = 0.3%, with an ash content of 4.6%. Assume the density of air to be 0.075 lbs/ft^3 and the biomass feed rate is 3 lbs/min.
   a. 4.94
   b. 19.72
   c. 65.9
   d. 197.2

17. Determine the amount of area required (in acres) to generate 1 MW of electrical power from the combustion of Jatropha shells. Assume the heating value of this biomass to be 8,275 Btu/lb and the productivity of land for Jatropha shells is about 5 tons/acre/yr. Furthermore, assume the overall conversion efficiency to electrical power of 50%. The plant will be operated for 350 days per year.
   a. 692
   b. 2,309
   c. 1,384
   d. 346

8. Determine the slagging and fouling indices for the following inorganic ash data: Al_2O_3 = 3.12; CaO = 27.41; Fe_2O_3 = 1.84; MgO = 10.90; MnO = 0.14; P_2O_5 = 4.98; K_2O = 5.28; SiO_2 = 32.46; Na_2O = 1.82; SO_3 = 6.12; TiO_2 = 0.22. The heating value of the biomass was 6,863 Btu/lb and the percentage of ash in the biomass was 18.62%. For this question, what is the value of bed agglomeration index for this biomass?
   a. 0.32
   b. 0.25
   c. 4.47
   d. 6.67

9. Determine the air-to-fuel ratio (lb/lb) for the complete combustion of rice straw. The elemental composition of this biomass by weight is as follows: C = 39.6%; H = 6%; O = 35.2%; N = 1.3%; and S = 0.192%, with an ash content of 17.7%.
   a. 5.17
   b. 6.2
   c. 3.25
   d. 4.2

18. Determine the slagging and fouling indices for the following inorganic ash data: Al_2O_3 = 3.12; CaO = 27.41; Fe_2O_3 = 1.84; MgO = 10.90; MnO = 0.14; P_2O_5 = 4.98; K_2O = 5.28; SiO_2 = 32.46; Na_2O = 1.82; SO_3 = 6.12; TiO_2 = 0.22. The heating value of the biomass was 6,863 Btu/lb and the percentage of ash in the biomass was 18.62%. For this question, what is the descriptive slagging type for this biomass using the coal index R_s?
   a. Severe
   b. Medium
   c. High
   d. Low

19. Determine the slagging and fouling indices for the following inorganic ash data: Al_2O_3 = 3.12; CaO = 27.41; Fe_2O_3 = 1.84; MgO = 10.90; MnO = 0.14; P_2O_5 = 4.98; K_2O = 5.28; SiO_2 = 32.46; Na_2O = 1.82; SO_3 = 6.12; TiO_2 = 0.22. The heating value of the biomass was 6,863 Btu/lb and the percentage of ash in the biomass was 18.62%. For this question, what is the value of the alkali index in units of lbs alkali/mmBtu for this biomass?
   a. 1.32
   b. 0.25
   c. 1.93
   d. 4.47

10. Determine the slagging and fouling indices for the following inorganic ash data: Al_2O_3 = 3.12; CaO = 27.41; Fe_2O_3 = 1.84; MgO = 10.90; MnO = 0.14; P_2O_5 = 4.98; K_2O = 5.28; SiO_2 = 32.46; Na_2O = 1.82; SO_3 = 6.12; TiO_2 = 0.22. The heating value of the biomass was 6,863 Btu/lb and the percentage of ash in the biomass was 18.62%. For this question, based on the base-to-acid ratio, what is the descriptive slagging and fouling potential for this biomass?
    a. Not to occur
    b. No idea
    c. It depends
    d. Certain to occur

20. Determine the air-to-fuel ratio in units of (ft^3 air/lb fuel) for the complete combustion of rice straw. The elemental composition of this biomass by weight is as follows: C = 39.6%; H = 6%; O = 35.2; N = 1.3%; and S = 0.192% with an ash content of 17.7%. Assume the density of air to be 0.075 lbs/ft^3.
    a. 4.94
    b. 5.17
    c. 82.7
    d. 68.8

Answers: 1d, 2b, 3b, 4c, 5a, 6c, 7b, 8b, 9b, 10d, 11a, 12c, 13d, 14a, 15b, 16d, 17a, 18a, 19c, 20c

ADDITIONAL LEARNING RESOURCES

- Access the Lecture Notes at the Instructor and Student Resources: https://routledgetextbooks.com/textbooks/instructor_downloads/

14 Biomass Sustainability Issues

LEARNING OBJECTIVES

Upon completion of this chapter, one should be able to

1. Describe the role of biomass for global sustainable development goals leading to a circular bioeconomy.
2. Describe the concept of circular bioeconomy and related issues such as carbon capture and sequestration (CCS), carbon intensity (CI), and carbon footprint (CF) work.
3. Formulate life cycle analyses (LCA) for the various biomass conversion processes and compare with other conventional processes.
4. Differentiate the various ways to analyze the energy expenditures from growing a biomass crop to generating a specific biofuel and enumerate all the processes involved for biofuel production.
5. Discuss the economics of various biomass energy conversion processes and compare economic analyses with other conversion processes.
6. Discuss other techniques for assessing the economic and environmental viability of various biomass conversion processes.

14.1 INTRODUCTION

Biomass resources have the important role of satisfying numerous sustainable development goals of many countries. These resources are already providing the world's food, feed, and shelter. In recent years, energy products in the form of biofuels, biomaterials, and high-value chemicals have emerged as additional outputs for biomass. Perhaps the most important role of biomass in the next decade is its potential as the climate changes (see Figure 14.1). It is a fact that the carbon dioxide emissions in the atmosphere are continually on the rise, leading to an increase in global temperatures and creating harsh environments. The focus on biomass use should prioritize the protection of the environment from impending catastrophic events.

A "circular economy" is now the popular goal for many countries to favor activities that prevent further degradation of the environment. It is defined as a systematic approach to economic development designed to benefit businesses, society, and the environment (Webster, 2015). A circular economy is regenerative or restorative in nature and aims to gradually ensure economic growth by careful use of finite resources as opposed to a linear economy where wastes are never recycled. With this concept comes numerous other related activities such as carbon capture and sequestration (CCS) and defining the carbon footprint (CF) of a process. This is

DOI: 10.1201/9781003294306-14

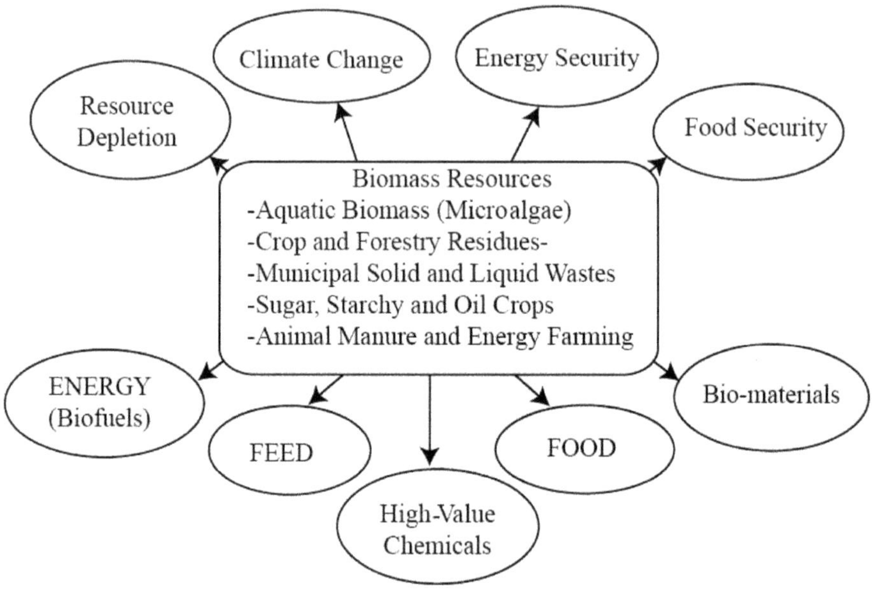

FIGURE 14.1 Role of biomass for a sustainable future.

where greenhouse gases (GHGs) released from renewable conversion processes are estimated to help protect human health and environment. When applied to the use of biomass, effective use of the bio-based resource is achieved by encouraging many different uses and developing a cycle of the natural systems where biomass is generated. A true circular economy avoids the use of non-renewable resources (e.g., fossil fuels) and preserves or enhances its use completely whenever possible. In the case of biomass, waste resources are ideally returned to the soil. This is where CCS processes come in. Waste biomass may be thermally converted to generate biochar that may be incorporated back into the soil. These will be part of the CF work for any biomass conversion process.

Life cycle analysis (LCA) is normally conducted when considering the conversion of any biomass into high-value materials, biofuels, and electrical power. LCA provides an idea of the contribution to sustainability, as well as the determination of net energy and GHG balance. There have been numerous attempts to determine the overall impact of a biomass refinery using the "cradle-to-grave" approach, or what some would call the "well-to-wheel" (WTW) approach. This chapter describes various ways of estimating the complete life cycle of biofuels. The new term is "cradle-to-cradle" for the circular bioeconomy.

The classic example of this analysis is outlined in a popular paper that disputes the positive energy balance for ethanol production (Pimentel, 2003). The paper concludes that it takes more energy to produce a liter of ethanol than is required to grow corn crop. Pimentel's paper suggests that the energy balance, economics, and environmental impacts of making ethanol from corn are negative. Specifically, the paper asserted that "about 29% more energy is used to produce a gallon of ethanol than the energy available from that same gallon of ethanol." This paper gave rise to serious debate about how these authors made their LCA, including what assumptions the authors made. The conclusions prompted the US Department of Agriculture (USDA) to come up with a new study to analyze the life cycle energy and cost for various conversion processes. The USDA analysis also included GHG emissions from biofuels. The inclusion of GHG analysis broadened the simple energy and mass balances approach to include environmental concerns (Delucchi, 2006). Different models that focus on specific parameter boundaries have been developed that estimate the complete LCA for different biofuels. Life cycles have been established to produce the most common biofuels as follows:

1. Corn to ethanol
2. Soybean to biodiesel
3. Switchgrass to ethanol
4. Wood to ethanol

The original LCA procedure was initiated at the US Department of Energy (USDOE) during the late 1990s (Spath et al., 1999; Mann and Spath, 1999, 2001) to compare the life cycle of conversion of coal and biomass to electricity. The USDOE report is a classic example of the LCA approach. The results are redrawn in Figure 14.2 for coal

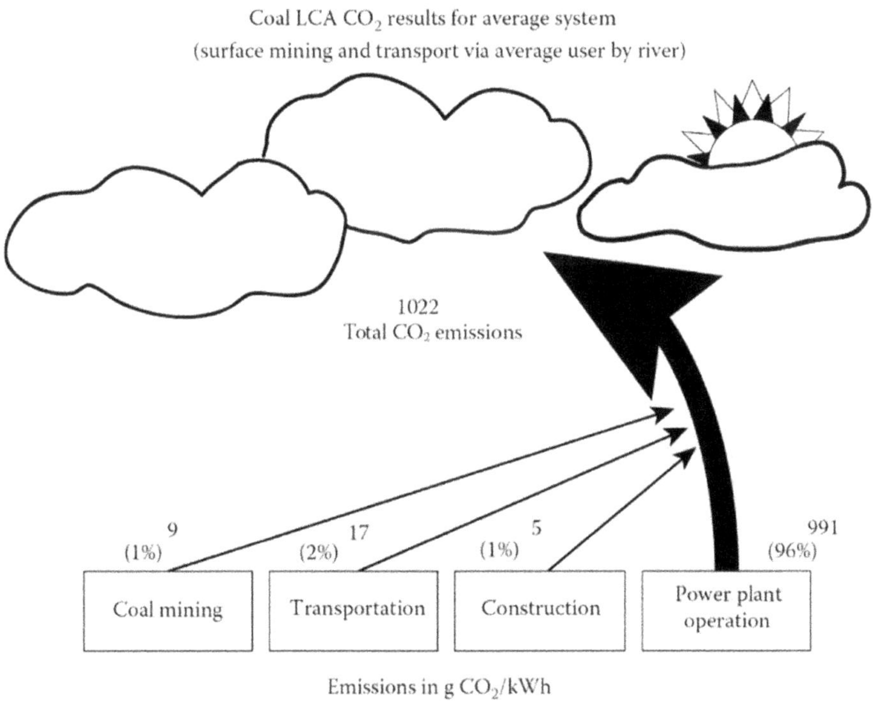

FIGURE 14.2 LCA for conversion of coal into electricity. (Modified from Mann, M. K. and P. L. Spath, The net CO_2 emissions and energy balances of biomass and coal-fired power systems. NREL, USDOE Project Report NREL/TP-430–23076 and NREL/TP-570–25119. National Renewable Energy Laboratory (NREL) of the USDOE, Golden, CO, 1999.)

as feedstock and Figure 14.3 for biomass materials as feedstock. The total CO_2 emission for coal conversion is over 1,020 g CO_2/kWh. By contrast, the net CO_2 emission for biomass materials is 46 g CO_2/kWh.

For biomass conversion into power, the USDOE analyses include activities from biomass feedstock production, transportation, and power plant construction and operation. The coal conversion analyses include coal mining, transportation, and power plant construction and operation. Contrary to many advocates of biomass conversion, there is net CO_2 balance if one includes every aspect of biomass production and conversion. The CO_2 balance is not zero; however, there is 95% carbon closure, which means that most of the CO_2 produced is used to grow the biomass. This is a fair estimate of LCA for biomass for power generation. There is now a range of software-based procedures to evaluate LCA of many biomass conversion processes. The procedures differ on the assumptions made during the conversion process. Thus, one must be familiar with all the aspects of biomass production and conversion to arrive at a fair estimate of energy and mass balance in biofuel production, including emissions, when the fuel is used.

Section 14.2 discusses the most common components of the LCA protocol. Although some of these components may have less impact on the overall LCA, they

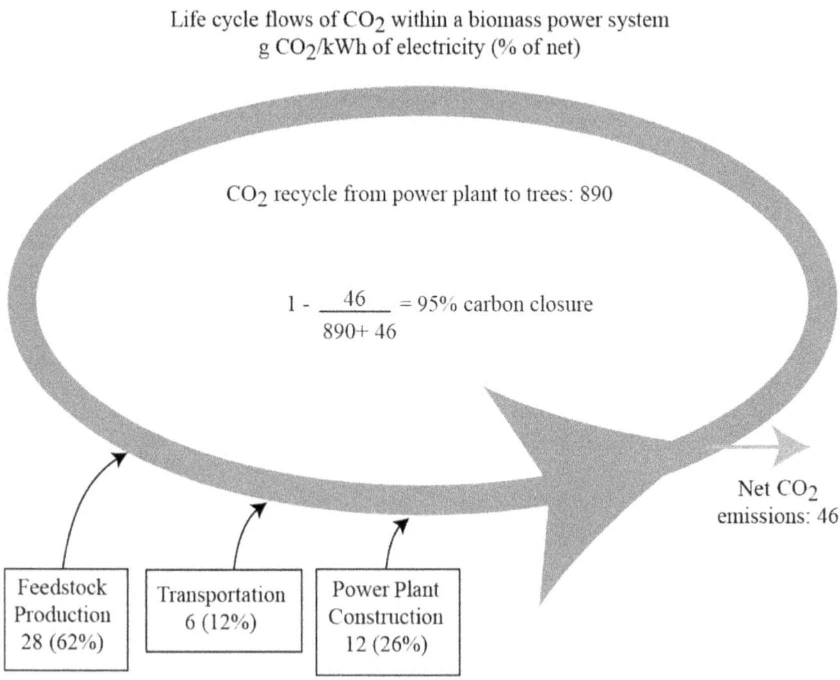

FIGURE 14.3 LCA for conversion of biomass materials into electricity. (Modified from Mann, M. K. and P. L. Spath, The net CO_2 emissions and energy balances of biomass and coal-fired power systems. NREL, USDOE Project Report NREL/TP-430–23076 and NREL/TP-570–25119. National Renewable Energy Laboratory (NREL) of the USDOE, Golden, CO, 1999.)

are still important for the overall trails for the conversion process. Perhaps the only free input is solar energy for crop growth, as well as water. In some locations where water is scarce, this input may also have some associated energy and cost.

14.2 WELL-TO-WHEEL (WTW) APPROACH

The most common components of typical LCAs are as follows:

1. Feedstock production—growing the crop to generate the biomass resource
2. Feedstock transport—transporting the biomass from land to point of use
3. Biofuel production—converting the biomass into fuel
4. Biofuel distribution and storage—delivering and storing biofuels
5. Dispensing biofuel—transferring biofuel to end users (e.g., a gas station or transfer station)
6. End use—using the fuel (e.g., engines, burners, or other combustion systems)

14.2.1 Feedstock Production

This component analyzes the energy (including exhaust emissions) required to grow the biomass. Energy analysis for feedstock production includes land preparation, fertilizer inputs, harvesting, and preprocessing of the biomass before it is brought to a refinery. Perhaps the most costly and significant inputs for biomass production are fertilizer and water usage. Other studies (e.g., Pimentel and Patzek, 2005) have a more extensive list of inputs for crop production. One must carefully identify assumptions made for the calculations and whether values used in the calculations are measured or assumed. A study (Pradhan et al., 2007) showed that differences in calculations vary with the assumed system boundaries, input amounts, and assumed energy equivalence, including the use or non-use of coproducts of the process.

14.2.2 Feedstock Transport

The energy and emissions associated with feedstock transport and logistics have become new delineated items in LCAs, primarily because widespread utilization of biomass requires complicated transport and logistics. Feedstock transport includes hauling crops from agricultural land to either a temporary storage site or directly to a refinery. Because of the complexities of biomass transport, numerous studies assumed a fixed or standard distance to the biorefinery. For example, a study by Ahmed et al. (1994) on soybean transport assumed an 80-km (50 mi) distance by truck. A similar study by Pimentel and Patzek (2005) assumed a total distance of 1,000 km (622 mi). However, despite this wide range of transport distance assumptions, the actual energy allocated to soybean transport as an example was quite small. Consequently, the differences in distance did not significantly affect the overall energy balance. Clearly, estimates of transport costs must be made. However, these costs account for a small contribution to overall energy input. Only large biofuel facilities where the hauling distance is longer than 80.5 km [50 mi] will experience significant transport energy, costs, and emissions.

14.2.3 Biofuel Production

The production of biofuels involves all the steps required to generate a product that qualifies as standard fuel for commercial sale. The various conversion processes for biofuels that were discussed in earlier chapters will each have their own unique energy, cost, and emissions level. For biodiesel, the important energy uses are in oil crushing, oil transport, transesterification process, and transport of biodiesel fuel for final distribution (Pradham et al., 2007).

For bioethanol production and conversion, whether a wet or dry milling process is used, the amount of energy used is typically around 70% (Shapouri et al., 2006), while transport is typically around 25%–27%. The remaining percentages are for corn transport and ethanol distribution. The major energy and costs for ethanol production and conversion come from the detailed factors and processes previously discussed. Electricity and natural gas for steam generation dominate the cost, followed by water usage and enzyme handling and processing. The other biomass conversion processes will use a similar approach in the estimate of energy, cost, and emissions generation. They should be applied in more detail as one becomes familiar with the ideal conversion process.

14.2.4 Biofuel Distribution and Storage

Biofuel distribution and storage may be a small energy and cost component of the LCA. However, it will always be included in the analysis. There are periods in the life cycle of a biofuel when these components may become significant. For example, distribution and storage costs are relevant when biodiesel production is hampered by high feedstock costs, when price competitiveness with diesel is an issue, and when there is a lack of market for the glycerin by-product. In this instance, the biofuel product, or coproduct (particularly glycerin), may have longer periods in storage, or it will encounter a longer transport distance for distribution (if a better cost-effective market is several hundreds of kilometers [or miles] away). In previous LCA studies cited (Shapouri et al., 2006), ethanol distribution and storage comprised only 2.5% of the energy allocation. When biodiesel was difficult to market in the United States, many entrepreneurs were able to capture the European market at economical returns, even when the distribution cost was rather high. Thus, logistics-oriented software is used to evaluate the economics of this type of transport and distribution.

14.2.5 Dispensing Biofuels

The energy, cost, and emissions concerning dispensing of biofuels before usage are also quite small, but the related parameters have always been included in newly developed models for analyzing life cycle. The primary reason for the inclusion of this component is that in some applications, the dispensing of biofuels could have a significant impact. For example, if a particular biofuel is not related to a commercially available transport fuel that has established dispensing stations (e.g., gasoline for bioethanol and diesel for biodiesel), the dispensing issues may be significant.

Consequently, these costs would affect the overall economics of use and the sale of the product. This is true for certain types of biofuels, for example, biofuels produced using fast pyrolysis processes. Mixing the bio-oil with either diesel or gasoline is not yet feasible. However, there is potential that this biofuel may follow the same distribution and dispensing route for crude oil. Thus, the dispensing system may be quite different, and consequently, may have to be developed. It is envisioned that bio-oil from fast pyrolysis, once successfully upgraded into commercial crude oil, may be mixed with crude oil for delivery to various refineries around the country. This concept will be a new paradigm in the way this biofuel can be handled and distributed, for example, in many oil fields around Texas and Oklahoma, whose sites are also agricultural crop production areas. As a result, bio-oil may be produced from above-ground biomass after harvest, converted, and upgraded on-site using mobile fast pyrolysis systems, then mixed with crude oil being pumped from those oil wells. The bio-oil dispensing route may then follow the complex pipelines originally allocated to crude oil from below the ground. A new system of quality monitoring and metering may be in place before this system could be adopted.

14.2.6 End Use of Biofuels

The last step in the life cycle of biofuel is end use. This portion of the LCA is also low in energy expenditures but important when significant activities abound. The use of biofuels is varied, so are the exhaust emissions when the biodiesel is finally combusted or used in engines, burners, or other conversion devices. In this stage, the biofuel will be compared with conventional fuels such as gasoline and diesel: comparative analyses may be made. Note the programs such as Greenhouse Gases Regulated Emissions and Energy use in Transportation (GREET) already have extensive data for ethanol and biodiesel usage, but not for newer biofuels such as bio-oil or biobutanol. In the future, LCA programs such as GREET should have those fuel calculations in place. For example, using 1 L of gasoline generates 3,012 g of CO_2-equivalent GHG emissions. Thus, replacing 1 L of gasoline with ethanol on a one-to-one energy basis requires 1.5 L of ethanol because ethanol has a lower heating value than gasoline. After adjusting for this energy equivalence (see Example 14.1 for calculating the CO_2 equivalent for the three major GHGs), replacing 1 L of gasoline with ethanol reduces GHG emissions by 38%, using a weighted industry average (i.e., 40% for dry mills and 36% for wet mills). This calculation applied the ASPEN-based process model using starch from corn (Shapouri et al., 2006).

Figure 14.4 shows the diagram of the life cycle for the analysis of corn ethanol beginning from corn cultivation to vehicle operations as described above. For GHG calculations, there are three main gases: carbon dioxide (CO_2), methane (CH_4), and nitrous oxide (N_2O). Each gas has a different 100-year global warming potential (GWP). The GWP is normally reported as CO_2 equivalents. The various GWPs for each compound are as follows: 1 for CO_2, 21 for CH_4, and 310 for N_2O (i.e., methane is 21 times more powerful than CO_2 and N_2O is 310 times more powerful than CO_2). These equivalents are recommended by the Intergovernmental Panel on Climate Change (IPCC) and called the Kyoto Protocol. IPCC is the leading international

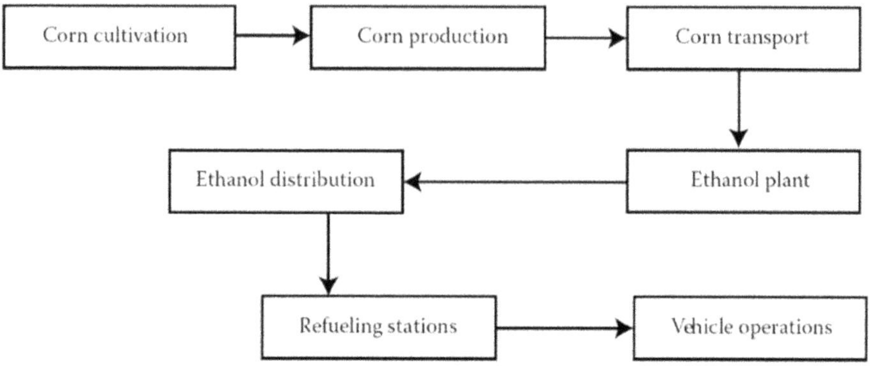

FIGURE 14.4 Diagram of LCA for corn ethanol.

body for the assessment of climate change. It was established by the United Nations Environmental Program and the World Meteorological Organization to provide the world with a clear scientific view of the current state of knowledge in climate change and its potential environmental and socioeconomic impacts.

According to the IPCC taskforce on National Greenhouse Gas Inventories, nitrous oxides will cause more global warming than either of those other two gases, for the same concentration. More specifically, the emission of 1 kg of methane has the same warming impact as 21 kg of CO_2. Thus, the GWP for methane is 21. Likewise, 1 kg of nitrous oxide has the same warming impact as 310 kg of CO_2. In equation form, this is shown in Equation 14.1.

$$CO_2 \text{ equivalent (tonnes)} = CO_2 \times 1.0 + CH_4 \times 21 + N_2O \times 310 \text{ tonnes} \quad (14.1)$$

Example 14.1: Greenhouse Gas Equivalency Calculations

Determine the equivalent CO_2 emissions from a conversion process whose gas emissions are as follows: 3 tonnes of CO_2, 2 tonnes of CH_4, and 0.05 tonnes of N_2O. Also, determine the percent contribution of each compound by weight of CO_2 equivalent.

Solution:

a. Equation 14.1 is used directly as follows:

$$CO_2E \text{ (tonnes)} = 3 \times 1.0 + 2 \times 21 + 0.05 \times 310 = 60.5 \text{ tonnes } CO_2$$

b. The percentage contribution of each compound in the overall GHG potential in CO_2 equivalent is calculated as follows:

$$CO_2 \text{ contribution (\%)} = \frac{3 \text{ tonnes}}{60.5 \text{ tonnes } CO_2} \times 100\% = 4.96\%$$

$$CH_4 \text{ contribution (\%)} = \frac{(2 \times 21) \text{ tonnes}}{60.5 \text{ tonnes } CO_2} \times 100\% = 69.4\%$$

$$\text{N}_2\text{O contribution (\%)} = \frac{(0.05 \times 310)\text{ tonnes}}{60.5\text{ tonnes CO}_2} \times 100\% = 25.62\%$$

The GHG calculations for actual energy-generating facilities, such as natural gas burners, are not as straightforward as those shown in Example 14.1. In most instances, the operation of a natural gas burner, for example, is a function of the percentage of utilization (effective operation in a year) or simply the utilization and the output rating as well as the fuel usage. In this case, the GHG estimate may be made by using Equation 14.2.

$$\text{GHG emissions} = \text{fuel usage} \times \text{emission factor} \tag{14.2}$$

The fuel usage is shown in Equation 14.3, where output rating and efficiency will depend on the manufacturer's published data.

$$\text{Fuel usage}\left(\frac{\text{m}^3}{\text{s}}\right) = \frac{\text{Output rating (MW)} \times \text{Utilization (decimal)}}{\text{Efficiency (decimal)} \times \text{HHV of fuel}\left(\frac{\text{MJ}}{\text{m}^3}\right)} \tag{14.3}$$

The emission factor varies with the type of conversion unit and must be acquired from the manufacturer. Usually, the manufacturer provides emissions in units of g/kWh or in units of g/m^3 of gas or g/L of fuel. For example, the emission factors for natural gas and diesel drivers based on fuel usage are summarized in Table 14.1. Example 14.2 provides an example for calculation.

Example 14.2: Gas Turbine Greenhouse Gas Equivalency Calculations

Determine the equivalent CO_2 emissions for a gas turbine whose emission factor is summarized in Table 4.1. The output rating of the unit is 26,070 kW, and the

TABLE 14.1
Emission Factors for Natural Gas and Diesel Drivers Based on Fuel Usage

Driver Type		CO_2	N_2O	CH_4	CO_{2e}
Reciprocating engine					
	2-cycle lean burn (g/m^3)	1,769	0.765	23.31	2,496
	4-cycle lean burn (g/m^3)	1,769	0.984	20.09	2,496
	4-cycle rich burn, turbo-charged (g/m^3)	1,769	0.533	3.69	2,012
Gas turbine (g/m^3)		1,769	0.077	0.138	1,796
Large diesel engine: >600 hp (kg/L)		2.744	0.0004	0.00014	2.871

Source: USEPA, *AP 42. Compilation of Air Pollutant Emission Factors,* 5th Ed. Vol. 3, Stationary Internal Combustion Engines. Technology Transfer Network Clearing House for Inventories and Emission Factors. Emission factors and Policy Application Center, Office of Air Quality Planning and Standards (OAQPS), Research Triangle Park, NC, 1995.

heating value of natural gas is assumed to be 37.2 MJ/m³ (1,000 Btu/ft³). Assume a very low thermal efficiency of 24%. The unit is to be operated 350 days in a year. Likewise, determine the percent contribution of each GHG by weight of CO_2 equivalent.

Solution:

a. The utilization is calculated first as follows:

$$\text{Utilization} = \frac{350}{365} = 0.9589$$

b. Equation 14.3 for fuel usage is then used to estimate fuel usage as follows:

$$\text{Fuel usage} = \frac{26.07\,\text{MW}}{0.24} \times \frac{\text{m}^3}{37.2\,\text{MJ}} \times \frac{\text{MJ}}{\text{MW}-\text{s}} \times 0.9589 = 2.8\frac{\text{m}^3}{\text{s}}$$

c. The GHG emissions will then be calculated using Equation 14.2 as follows:

$$CO_2\left(\frac{\text{tonnes}}{\text{yr}}\right) = 2.8\frac{\text{m}^3}{\text{s}} \times \frac{1769\,\text{g}}{\text{m}^3} \times \frac{\text{tonne}}{1\times10^6\,\text{g}} \times \frac{31.536\times10^6\,\text{s}}{\text{yr}} = 156{,}204\frac{\text{tonnes}}{\text{yr}}$$

$$N_2O\left(\frac{\text{tonnes}}{\text{yr}}\right) = 2.8\frac{\text{m}^3}{\text{s}} \times \frac{0.077\,\text{g}}{\text{m}^3} \times \frac{\text{tonne}}{1\times10^6\,\text{g}} \times \frac{31.536\times10^6\,\text{s}}{\text{yr}} = 6.8\frac{\text{tonnes}}{\text{yr}}$$

$$CH_4\left(\frac{\text{tonnes}}{\text{yr}}\right) = 2.8\frac{\text{m}^3}{\text{s}} \times \frac{0.138\,\text{g}}{\text{m}^3} \times \frac{\text{tonne}}{1\times10^6\,\text{g}} \times \frac{31.536\times10^6\,\text{s}}{\text{yr}} = 12.2\frac{\text{tonnes}}{\text{yr}}$$

d. The CO_2 equivalent is calculated as follows:

$$CO_2E\,(\text{tonnes}) = 156{,}204\times1.0 + 12.2\times21 + 6.8\times310 = 158{,}568.2\,\text{tonnes}$$

e. The percentage contribution of each compound in the overall GHG potential in CO_2 equivalent is calculated as follows:

$$CO_2\text{ contribution }(\%) = \frac{156{,}204\,\text{tonnes}}{158{,}568.2\,\text{tonnes }CO_2E} \times 100\% = 98.51\%$$

$$CH_4\text{ contribution }(\%) = \frac{(12.2\times21)\,\text{tonnes}}{158{,}568.2\,CO_2E} \times 100\% = 0.16\%$$

$$N_2O\text{ contribution }(\%) = \frac{(6.8\times310)\,\text{tonnes}}{158{,}568.2\,\text{tonnes }CO_2} \times 100\% = 1.33\%$$

14.2.7 GHG Emissions of Biomass Combustion

The CO_2, CH_4, and N_2O emissions are all produced during the combustion of wood. Nearly all fuel carbon or 99% in wood is converted to CO_2 during the combustion process (US Environmental Protection Agency [USEPA], 1995). This conversion is relatively independent of the firing configuration. CO_2 emitted from this source, according to USEPA AP 42, is generally not counted as part of GHG emissions because it is considered part of the short-term cycle of the biosphere. Formations of N_2O are minimized when combustion temperatures are kept high (>1,475°F) and excess air is kept to a minimum (<1%). Methane emissions are highest during periods of low-temperature combustion, or incomplete combustion such as during start-up or shut-down procedures for burners or boilers. Typically, conditions that enhance the formation of N_2O also enhance the emissions of CH_4 in the combustion of wood.

14.3 DISCUSSIONS OF SOFTWARE FOR LCA ANALYSIS

There are various biofuel models that may be used to evaluate the sustainability of use or production of certain biofuels. The most popular is the GREET Model. The Argonne National Laboratory (ANL) of the USDOE developed the GREET Model so that energy use and air emissions of various transportation fuels and vehicle technologies may be evaluated. As new biofuel conversion processes are developed, ANL would update the model to incorporate new changes and technical information. The transportation fuels that are included in the analysis are as follows: gasoline, diesel, compressed natural gas, liquefied petroleum gas, methanol, ethanol, biodiesel, electricity, and hydrogen. The GREET software is available at no cost to the public from the ANL website (ANL, 2021).

The GREET energy products model has incorporated new updates for bioenergy research. Enumerated below are newest energy products added in addition to previous pathways:

1. Ethanol pathways from corn starch, fiber, and cellulosic ethanol.
2. Biodiesel and renewable diesel from vegetable oil, tallow, and fatty acids.
3. Supply chain sustainability analysis fuel pathways: renewable gasoline, diesel, and hydrocarbon fuel pathways from lignocellulosic biomass and municipal wastewater sludge.
4. Sustainable aviation fuel.
5. Co-processing of bio-feedstocks in petroleum refineries.
6. Direct air capture and cryogenic carbon capture pathways for carbon dioxide for algae cultivation.
7. Feedstock production for biofuels.
8. Conventional waste management and waste-to-energy pathways.
9. Biopower.
10. Steam methane reforming for hydrogen production.

The stages in the GREET fuel-cycle model are shown schematically in Figure 14.5. The first stage of calculation relates to feedstock production, transportation, and storage. This will be followed by biofuel production, transportation, distribution, and storage. These two steps comprise the well-to-tank activities. The next sets of calculations involve biofuel utilization in vehicle operations. The vehicle operations include operations such as vehicle refueling, fuel combustion and conversion, fuel

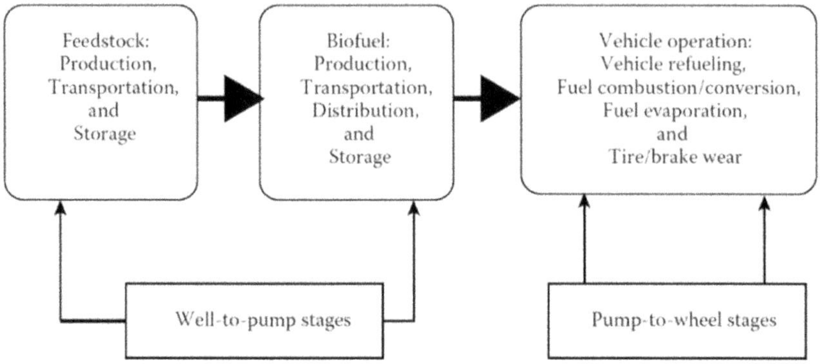

FIGURE 14.5 Stages covered in ANL GREET sustainability model.

evaporation, and tire and brake wear. These are called the tank-to-wheel activities. The combination is called the WTW analysis.

The GREET (2021) model was used to compare the GHG emissions and energy associated with the production and use of the biodiesel blend (B20) as well as the low-level blend of bioethanol (E10). The results are summarized in succeeding tables. The software is easy to use, especially with the graphical user interface (GREETGUI). The application-specific GUI accepts information from the user in the form of buttons, check boxes, and input text boxes. The GUI transfers this input to an underlying Microsoft Excel spreadsheet model (GREET). The program runs the model in the background and displays the results in the form of tables in another Excel output file.

The most recent version of the GREET Model (GREET 2021) was used to compare energy and GHG calculations using the biodiesel blend (B20) with diesel fuel and the ethanol blend (E10) with gasoline fuel. Passenger cars were used for both simulations using the CIDI (compression ignition-direct injection type) diesel engine for the biodiesel-blend fuel type and the SI (spark ignition) engine for the gasoline–ethanol blend. The biodiesel was assumed to come from 100% soybean biomass, and the ethanol was assumed to come from 100% corn biomass. Most GREET default options were used for the simulation.

For soybean farming, the energy production use was 16,560 Btu/bushel (640 MJ/tonne) and the extraction energy use was 3,992 Btu/lb of soy (9,265 MJ/tonne). For those not familiar with the English system of units, a bushel of soybeans is 60 lb (27.3 kg) but a bushel of shelled corn is only 56 lb (25.5 kg). The transesterification energy usage was 1,844 Btu/lb of fuel (4,280 MJ/tonne). The assumed corn energy production usage was 9,608 Btu/bushel (398 MJ/tonne). The assumption for ethanol production energy use for dry milling was 26,856 Btu/gal (7.5 MJ/L), and for wet milling, the assumed value was 47,409 Btu/gal (13.2 MJ/L). These values are very close to the values reported by Shapouri et al. (2006). The assumed CO_2 emission from domestic land use by corn farming was 447 g CO_2 per bushel of corn (17.6 g CO_2/kg). Comparative results of the simulation are summarized in Table 14.2 for the well-to-pump stage. The total energy consumed for the biodiesel route was less than the bioethanol route (110,753 vs. 278,090 J/MJ). However, the emissions of CO_2, CH_4, N_2O, and GHGs in general are significantly higher for the bioethanol route

TABLE 14.2
Comparative Results of Energy and GHG Emissions WTP for Two Biofuels Using GREET1 Software

Simulation Year 2000	Soybean-Based Biodiesel (B20) (J/MJ)	Low-Level Ethanol (E10) (J/MJ or kg/MJ)
Total energy (J/MJ)	110,753	278,090
Fossil fuels (J/MJ)	1,015,491	1,192,101
Coal (J/MJ)	16,544	19,654
Natural gas (J/MJ)	137,570	163,125
Petroleum (J/MJ)	861,303	1,009,321
Renewable (J/MJ)	91,963	82,015
Biomass (J/MJ)	89,290	78,896
VOC (kg/MJ)	0.0000099049	0.000028236
CO (kg/MJ)	0.000014355	0.000016361
NO_x (kg/MJ)	0.000028861	0.000034593
PM_{10} (kg/MJ)	0.0000019624	0.0000033251
$PM_{2.5}$ (kg/MJ)	0.0000015933	0.00000071501
N_2O (kg/MJ)	0.0000041349	0.0000026651
GHG-100 (kg/MJ)	0.0194	0.0226
VOC Urban (kg/MJ)	0.0000024744	0.000015198
CO Urban (kg/MJ)	0.000001868	0.0000030542
NO_x Urban (kg/MJ)	0.0000034545	0.000005296
PM_{10} Urban (kg/MJ)	0.00000048798	0.000000891
$PM_{2.5}$ Urban (kg/MJ)	0.00000037987	0.00000071501
SO_x Urban (kg/MJ)	0.0000039579	0.0000058095
CH_4 Urban (kg/MJ)	0.0000022377	0.00000290
CO_2 Urban (kg/MJ)	0.0043	0.0072
N_2O Urban (kg/MJ)	0.000000050534	0.000000087174

than for the biodiesel route. The overall GHG emission for ethanol is more than the biodiesel in its life cycle.

The GREET (2021) model estimated the emissions of the following major regulated pollutants: VOC, CO, NO_x, PM_{10}, $PM_{2.5}$, and So_x. The results are also summarized in Table 14.2 for total emissions or emissions in urban setting.

The above discussions on LCA analysis are rather linear. Future LCA will be modified to consider a circular economy, and instead of the WTW or cradle-to-grave approach, the new directions will demonstrate a cradle-to-cradle approach. This technique will be demonstrated in later sections of this chapter.

14.4 BIOFUEL ECONOMICS

The success of any biomass conversion system is dependent on its economic performance. In most economic analyses of biofuel conversion, the single most important factor is feedstock cost. For example, for biodiesel production, about 85% of the biodiesel cost comes from the cost of the feedstock (Piazza, 2008). This chapter reviews some of the economic parameters that affect the economic viability of major biofuel production processes as follows: biodiesel, bioethanol, biogas, pyrolysis, and gasification. In addition to the cost of feedstock, other limiting economic factors include the cost of chemical catalysts for biodiesel and pyrolysis of bio-oil, or the cost of enzymes for ethanol production. Perhaps the other factor that affects the viability of biofuel products is the relative price of crude oil in the commercial market; commercial gasoline and diesel costs are strongly dependent on those spot market prices as well.

14.4.1 Major Components of Biofuel Economic Analysis

The major cost components of the biofuel production process include feedstock costs, chemical costs for conversion, associated utilities, and management costs. Table 14.5 summarizes the relative cost of components (high, medium, or low) of major biofuels. The most cost-effective biofuels are those with the lowest overall cost when considering all vital economic parameters. Table 14.3 suggests that two biofuel production processes have the lowest overall costs: biogas and gasification. However, cost is only one factor in the sustainable production of biofuels.

14.4.2 Economics of Production Processes for Major Biofuels

14.4.2.1 Biodiesel Economics

The economics of making biodiesel from refined oil is well-understood. This understanding is based on many years of experience of refining oil from crops. In the United States, the majority of biodiesel comes from refined soybean oil. However, because of the rising cost of this refined oil versus the price of diesel fuel, there was a period when the cost of refined soybean oil was slightly higher than biodiesel in some stages of the US economy. Newer feedstock such as animal fats and oil, particularly chicken oil, emerged during the period of high soybean oil cost. Other emerging feedstocks include oil from microalgae biomass and waste oil from fast food chains.

Biodiesel operating costs in commercial systems include the following items:

1. Chemicals (catalysts, methanol, and some acids)
2. Manpower (plant manager and assistants)

TABLE 14.3
Relative Impact of Cost Components on Various Biofuels

Cost Component	Biodiesel	Bioethanol	Biogas	Gasification	Pyrolysis
Feedstock	High	High	Low	Low	Low
Chemicals	High	High	Low	Low	Medium
Labor/management	High	High	High	High	High
Taxes/insurance	Low	Low	Low	Low	Low
Capital costs	High	High	Low	Low	Medium
Storage facilities	High	Medium	Medium	Low	High

3. Utilities (electricity costs)
4. Maintenance (of process equipment, storage facilities, and others)
5. Taxes and Insurance (including the usual tax shelters and insurance)
6. Depreciation (for all process equipment)
7. Feedstock costs (vegetable or animal fats and oils)

An example of this operating cost analysis was reported by Willits (2007) with feedstock costs comprising about 85% of the overall cost. This is shown in Figure 14.6. The feedstock cost is followed by chemical costs around 11%, while all other costs are between 1% and 2%.

The conversion and processing costs were also reported with the highest cost component during conversion coming from chemicals, which is approximately \$0.22/gal (\$0.06/L). This is shown in Figure 14.7. This cost was followed by depreciation and manpower. Utilities, maintenance including taxes, and insurance are very small percentages of the overall conversion cost.

For the capital cost for the facility, the breakdown of the costs is distributed into eight major segments for a commercial biodiesel plant with the highest being the process equipment cost comprising 38% of the total cost and piping cost of about 20% of the capital cost. This is shown in Figure 14.8. Figure 14.9 shows the overall complete commercial biodiesel facility cost. The largest cost component of which includes the process equipment and building (65%). This is followed by tank farm (18%) and site work (10%). Utility and office building are less than 5% of the total capital cost (Willits, 2007).

The cost of commercial biodiesel plants varies according to the types of equipment used, and normally, would depend on whether a centrifuge is used for separation and filtering or if the plant employs a simple gravity filtration system. The design by Crown Iron Works (Waranica, 2007) is popular for the use of gravity separation systems.

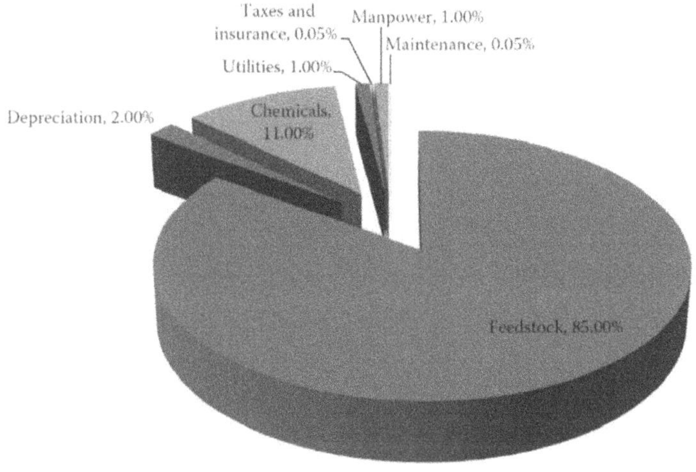

FIGURE 14.6 Breakdown of total operating cost for the production of biodiesel.

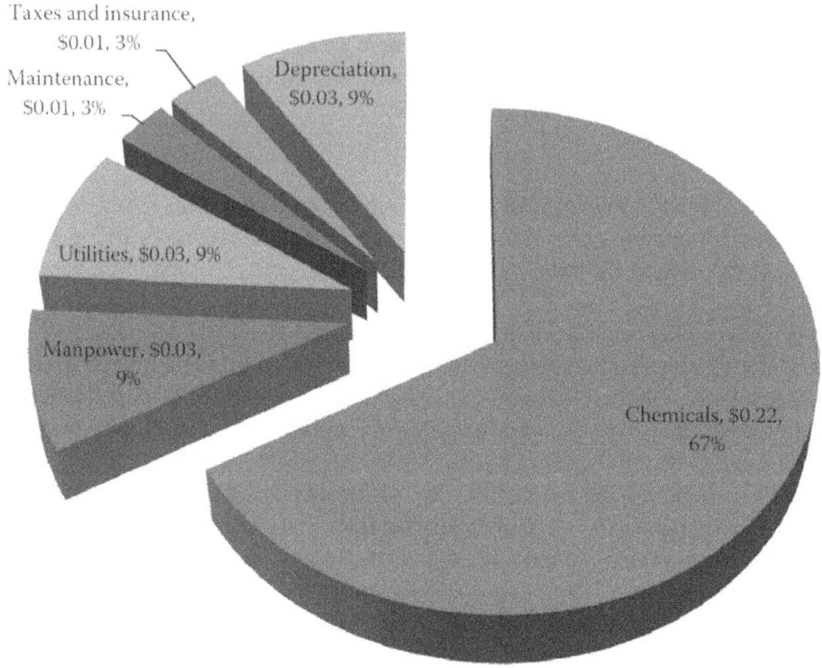

FIGURE 14.7 Breakdown of conversion and processing cost for making biodiesel.

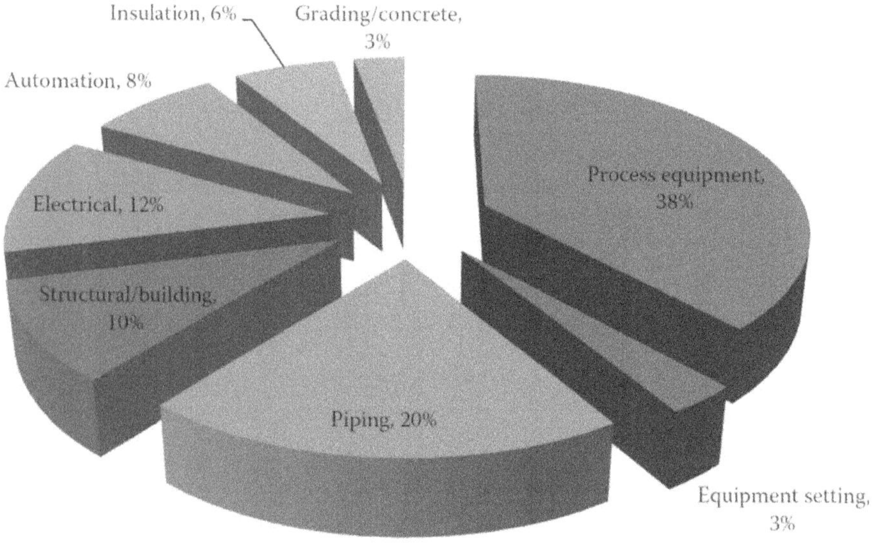

FIGURE 14.8 Breakdown of capital cost for commercial biodiesel plant.

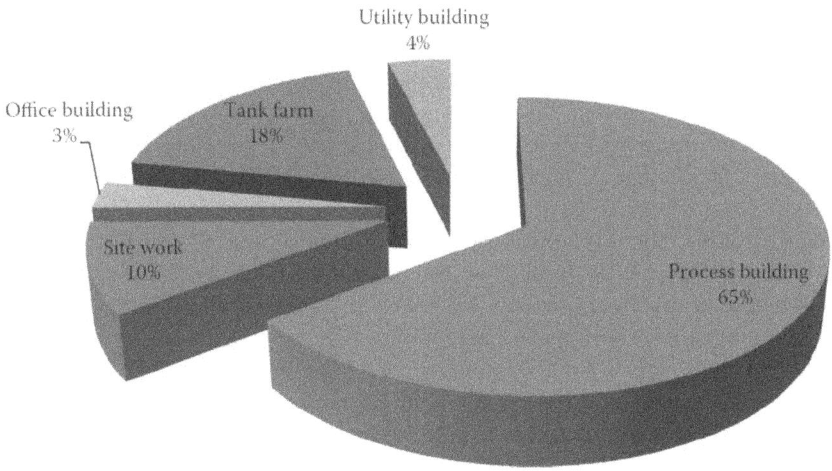

FIGURE 14.9 Capital cost composition in complete commercial biodiesel facility.

Designs from Desmet Ballestra (Willits, 2007) use mostly centrifuges for solids and glycerin separation.

14.4.2.2 Ethanol Economics

The economics for producing ethanol from corn (in the United States) and sugarcane (in Brazil) has been well-established. There was initial controversy about the supposedly negative net energy balance of corn-based ethanol. This controversy has been negated by the influx of numerous bioethanol plants (numbering more than 209 biorefineries) that have been built in the United States in the last decade. Bioethanol is the fastest growing renewable energy source in the world. The US production of ethanol in 2020 was around 65.5 billion liters [17.3 billion gallons] (USEIA, 2020b). The majority of US ethanol comes from corn.

One of the most comprehensive reports on the economics of ethanol is that published by the USDA in 2006 (Shapouri and Salassi, 2006). In this report, the ethanol production cost from Brazilian sugarcane is still the lowest, at $0.21/L ($0.81/gal), followed by ethanol from wet milling at $0.27/L ($1.03/gal) and ethanol from dry milling at $0.28/L ($1.05/gal) in the United States. The feedstock cost of the Brazilian ethanol comprises 37% of the total production cost and 63% from processing cost. The wet milling proportion is around 39% and 61%, respectively, while the dry milling feedstock and processing component is around 50:50. If ethanol is to be produced from sugarcane grown in the United States, the total production cost would be $0.63/L ($2.40/gal) and very close to that coming from sugar beets $0.62/L ($2.35/gal). Estimates of the capital expenditure costs to build facilities that would utilize sugarcane or sugar beets to produce ethanol in the United States would be expected to be higher than the capital costs for corn-based ethanol plants, because of higher feedstock preparation costs. A 20 MGY ethanol plant using sugarcane or sugar beets as a feedstock would be expected to have capital expenditure costs in the range of $2.10–$2.20 per gallon of annual capacity, compared with an estimate of $1.50 per gallon annual capacity for a corn-based facility. Cellulosic conversion of biomass into ethanol could reduce the overall costs of conversion and processing because the feedstock cost is relatively cheaper. Currently, the ethanol industry is on the positive economics side, despite the relatively higher cost of feedstock and the low cost of gasoline for transport fuel. However, with a federal government policy to stimulate ethanol production through the \$0.51 per gallon subsidy (Hurt et al., 2008), the economics becomes feasible in most corn-producing areas of the US Midwest. This ethanol subsidy is equivalent to about \$1.35/bushel of corn used, which is being paid for by this government program. Note that the energy value of ethanol by volume is less than that of gasoline. However, the octane rating of ethanol is close to 110, and thus, blending ethanol with low-octane gasoline will improve the octane rating of the mixed fuel blend. Ethanol will continue to be a big part of the transport fuel industry, especially as the industry has removed the use of the additive methyl tertiary butyl ester (MTBE) as a gasoline oxygenate because of its high toxicity. In addition, MTBE has been found to be harmful to groundwater. Ethanol becomes the only substitute for this additive. In fact, the justification for this subsidy is the environmental benefit of using ethanol, compared with the harmful effect of the MTBE oxygenates. MTBE was used as an additive for gasoline primarily to increase the octane rating.

The future direction of ethanol in the United States will be highly dependent on state and federal governmental policies as well as the transport fuel costs (compared with gasoline) and the corn markets. Corn production in the United States is still primarily for food, and only excess corn is used for biofuel production. Thus, it is believed that the cellulosic-based ethanol should take up the slack once the subsidy for corn-based ethanol is removed by the government.

14.4.2.3 Pyrolysis Economics

A study was funded by the USDOE to investigate the economics of converting cellulosic biomass (corn stover) using fast pyrolysis into bio-oil products; these products would be upgraded later into transportation fuels (Wright et al., 2010). The size of the facility envisioned for this project was 2,000 metric tonnes per day of corn stover feedstock. The bio-oil product would be upgraded by hydrotreatment and hydrocracking using hydrogen (produced by the pyrolysis) and additional hydrogen purchased from a commercial distributor. The hydrogen would be used to generate a mixture of naptha-range (gasoline-blend stock) chemicals and diesel-range products.

The study results indicated that petroleum fractions in the gasoline range from corn stover had a product value of \$0.82/L (\$3.09/gal) with on-site hydrogen production and \$0.56/L (\$2.11/gal) with purchased hydrogen. These values correspond to a \$0.21/L (\$0.83/gal) cost to produce the bio-oil. On this basis, the production cost would be \$1.73/L (\$6.55/gal) for the on-site hydrogen process and \$0.92/L (\$3.41/gal) for purchased hydrogen. These initial results suggest that pyrolysis-derived biofuels are competitive with other alternative fuels. However, the technology is still immature, which results in a higher level of uncertainty when applying the study's estimates and assumptions.

The capital cost for fast pyrolysis with integrated hydrogen production was estimated at \$287 million with a fuel yield of 134 million liters (25 million gallons) of gasoline and diesel-range products per year. The capital cost of the facility, if hydrogen is to be sourced commercially, is much lower at around $200 million, with a fuel yield of 220 ML (58 million gallons) of gasoline and diesel-range fuels. The initial assumed biomass cost for the above study was $50/short ton.

Note the difference between this study and that presented on algae pyrolysis, which has a lower bio-oil production cost of $0.61/L ($2.31/gal), based on a feedstock cost of only $30/dry ton. More research is required to improve the feasibility analysis and to validate the assumptions about a commercial facility.

14.4.2.4 Gasification Economics

Gasification technology to produce liquid hydrocarbon fuels has become more competitive, particularly with the use of improved Fischer–Tropsch (FT) technology (Lane, 2011). This technology is being investigated by ClearFuels Technology (Honolulu, HI). The target cost for transport biofuels is around $0.79/L ($3/gal). ClearFuels Technology uses a combination of high-efficiency biomass gasification followed by FT technology. The company claims to have developed a one-step rapid steam reforming process that converts all the biomass into syngas with no char, no liquid intermediates, no ash slagging or fouling problems, and low tar content. The USDOE funded ClearFuels Technology to build a 20 ton/day demonstration scale plant in Colorado and to develop a product demonstration unit that is based on FT technology and UOP's liquid fuel upgrading technology. UOP is a Honeywell Company based in Morristown, New Jersey.

Perhaps the most important benefit of gasification technology is waste utilization for heat generation and electrical power production. In many biorefineries, heat and electrical power are required. In an ethanol plant, a biomass gasification system would be an excellent addition to convert the biorefinery's residual waste into heat and electrical power. An economic study about the use of biomass gasification at ethanol plants was reported by Tifanny et al. (2009). In many ethanol plants, natural gas is used to produce process heat, especially in plants with greater than 380 million liters (100 million gallons) per year of capacity. Such a plant generates an abundant amount of dry distiller's grain (DDG) now commonly used to feed farm animals.

The above study resulted in favorable rates of return for investment in this biomass alternative, compared with using natural gas or coal for heat and electrical power. In the study, the mixture of corn stover and molasses syrup provided the highest rates of return. The factors favoring the use of this biomass through gasification is the relatively high cost of natural gas in larger plants, the low price for DDG, the relatively low price for ethanol, and higher corn prices. These factors contributed to recommending the use of all DDG by-products, including available corn stover in an ethanol facility, to offset the thermal and electrical energy requirements of the plant.

There are still numerous challenges that are being faced by gasification system developers and manufacturers. These challenges must be satisfactorily addressed before this technology and system may be adopted by commercial biorefineries. Supposedly, because of its high cost, gasification technology could be intimidating for investors. However, newer studies (Wu et al., 2002) showed that the capital cost for biomass gasification for power generation is still lower than that for coal. In an economic analysis of biomass gasification and power generation (BGPG) in China, it was revealed that the unit capital cost of the BGPG system is only 60%–70% of similar coal power plants, and the operational costs are much lower than those for conventional power plants. However, because of the relatively low efficiency of small-scale biomass power plants, the current BGPG technology will lose its economic feasibility when the capacity is less than 160 kW or the price of biomass is greater than \$24/ton (200 Yuan RMB/ton, 2,002 rates). The development of medium-scale BGPG plants with a capacity ranging from 1 to 5 MW is recommended. This range of power generation is suitable for dedicated agricultural industries, such as rice mills or timber mills. Electrical power is also required, but there is an abundance of biomass resources available on-site for power conversion (e.g., rice hulls or wood wastes).

Mobile biomass gasification systems are now being developed, such as the unit being commercialized at Texas A&M University (TAMU), which is based on a fluidized bed gasification system. A company (SDL Citadel Global, Dallas, TX) has licensed the TAMU gasification system to generate electrical power from segregated municipal solid wastes. This feedstock has a negative feedstock cost due to tipping fees and is an ideal feedstock for power generation in urban areas such as Dallas, Texas. The company built a 1 MW commercial facility in Dallas in 2015. The target capital cost for this facility should be within \$1–\$2 million/MW to be truly economical. These mobile systems can be brought to facilities where biomass is abundant, such that collection and transport of biomass does not add to operational costs. In the United States, dedicated facilities such as cotton gins are ideal for the application of this gasification system. A study was made to evaluate the benefits of on-site power production for cotton gins (Multer et al., 2010). The findings suggested that generating electricity for this facility utilizing the produced cotton gin trash could be an attractive venture in the future. A cotton gin that processes 20 bales per hour of cotton would need electrical power of at least 1 MW. Example 14.3 shows some typical power calculations for a cotton gin.

Example 14.3: Gasification Power Calculations for Cotton Gin

A 40 bales/h gin operates at 1,000 h per season. Each bale has around 500 lb of cotton gin trash (CGT) with a heating value of 7,000 Btu/lb. The electrical power requirement of this gin is around 50 kWh/bale. The CGT will be used to generate electrical power at an overall conversion efficiency of 15%. Do you think the trash generated by this gin is enough to generate the electrical power during the operation of the gin? If so, can an additional 200,000 Btu/bale of drying energy also be derived from the cotton gin trash? Discuss the results.

Solution:

a. The amount of trash generated is calculated as follows:

$$\text{Trash (tonnes)} = \frac{40\,\text{bales}}{\text{h}} \times \frac{1{,}000\,\text{h}}{\text{season}} \times \frac{500\,\text{lb CGT}}{1\,\text{bale}} \times \frac{\text{ton}}{2{,}000\,\text{lb}} = 10{,}000\ \text{tons}$$

b. If this trash is to be gasified and converted into power, the amount of power produced may be calculated as follows:

$$\text{Theoretical power}\left(\frac{\text{MMBtu}}{\text{h}}\right) = \frac{10{,}000\ \text{tons}}{1{,}000\ \text{h}} \times \frac{2{,}000\ \text{lb}}{1\ \text{ton}} \times \frac{7{,}000\ \text{Btu}}{\text{lb}} = 140\frac{\text{MMBtu}}{\text{h}}$$

$$\text{Actual power (MW)} = 140\frac{\text{MMBtu}}{\text{h}} \times \frac{1\ \text{MW}}{3.413\ \text{MM Btu/h}} \times 0.15 = 6.15\ \text{MW}$$

c. The electrical power consumption of the facility is calculated from the usage as follows:

$$\text{Required power (MW)} = \frac{50\ \text{kWh}}{\text{bale}} \times \frac{40{,}000\ \text{bales}}{1{,}000\ \text{h}} = 2\ \text{MW}$$

d. Thus, there is more power from the gin trash than the electrical power requirement of the facility.
e. The required drying power is calculated as follows:

$$\text{Required drying power}\left(\frac{\text{MBtu}}{\text{h}}\right) = \frac{200{,}000\ \text{Btu}}{\text{bale}} \times \frac{40{,}000\ \text{bales}}{1{,}000\ \text{h}} = 8\frac{\text{MMBtu}}{\text{h}}$$

$$\text{Required drying power (MW)} = \frac{8\ \text{MMBtu}}{\text{h}} \times \frac{\text{MW}}{3.413\ \text{MM Btu/h}} = 2.34\ \text{MW}$$

f. There is more power available that can be used to dry the cotton bales.

14.5 SUSTAINABILITY OF BIOFUELS PRODUCTION

The use of biomass for energy purposes conflicts with the use of biomass for feeding the world. Several research reports have been written about the food versus fuel nexus. The use of biomass as a fuel source affects the price of other commodities that also utilize this biomass. For example, the trend in corn prices in the United States has been greatly affected by the use of corn as the main feedstock for ethanol production. Sustainability has traditionally been defined as the capacity to endure. In the context of biomass conversion and use, for example, a "sustainable biofuel" was defined by the Biodiesel Sustainable Task Force as a product that "demonstrates environmental stewardship across its life cycle by maximizing net energy balance and minimizing the negative impacts on natural ecosystems, globally and locally; it adds value, creates wealth, and enhances the quality of life of the entire community and of stakeholders" (NBB, 2008). The USDOE simply defines a sustainable biofuel as one that is "economically competitive, conserves the natural resource base, and ensures social well-being" (USDOE and USDA, 2009). There are various ways of measuring the sustainability of biofuels, and some will be discussed later in this chapter.

14.5.1 Measuring Sustainability of Biofuels

Sustainability has three dimensions: social, economic, and environmental. What is lacking in these criteria is the technical aspect of the process that is directly related to the economic dimension. In this case, the primary basis for sustainability hinges on net energy balance. If the energy required to produce a certain biofuel is more than the energy required to generate the biomass, the biofuel will not be a sustainable energy resource. The first gauge of sustainability is the net energy balance. However, some authors use net energy ratio (NER) and net energy balance (NEB). The associated energy estimates to produce ethanol, for example, as reported by Shapouri et al. (2006) showed that the production system comprised corn production, transport, conversion, and distribution. The energy balance and energy ratios vary for dry or wet processes but with only slight differences. The weighted average energy required to produce ethanol from four various stages of biofuel production was around 20 MJ/L for both the wet milling and the dry milling processes, and without calculating the energy from coproducts. The corn production comprised 25.6% of the total energy expended, whereas the other subsystems are as follows: corn transport, 2.6%; conversion, 69%; and distribution, 2.1%. Detailed values for dry and wet milling processes are summarized in Table 14.5 with corresponding weighted average values (Table 14.4).

The NER was estimated at 1.06 as the average value and 1.192 for the net energy balance.

14.5.1.1 Net Energy Ratio (NER) Method

The NER is the amount of energy required to produce the biofuel divided by the energy content of the biofuel, as shown by Equation 14.4.

TABLE 14.4
Energy Associated with Ethanol Production from Corn without Coproducts Allocation

	Energy Use of Milling Process (MJ/L)		
Production Subsystem	**Dry**	**Wet**	**Weighted Average (MJ/L)**
Corn production	5.261	5.171	5.216
Corn transport	0.596	0.586	0.591
Ethanol conversion	13.133	14.591	13.862
Ethanol distribution	0.414	0.414	0.414
Total energy used	19.404	20.726	20.083
Net energy balance	1.871	0.513	1.192
Net energy ratio	1.10	1.02	1.06

Source: Shapouri, H., M. Wang and J. A. Duffield, Net energy balancing and fuel-cycle analysis. In: *Renewable-Based Technology Sustainability Assessment* (J. Dewulf and H. Van Langenhove, eds.). John Wiley & Sons Ltd., New York, 2006.

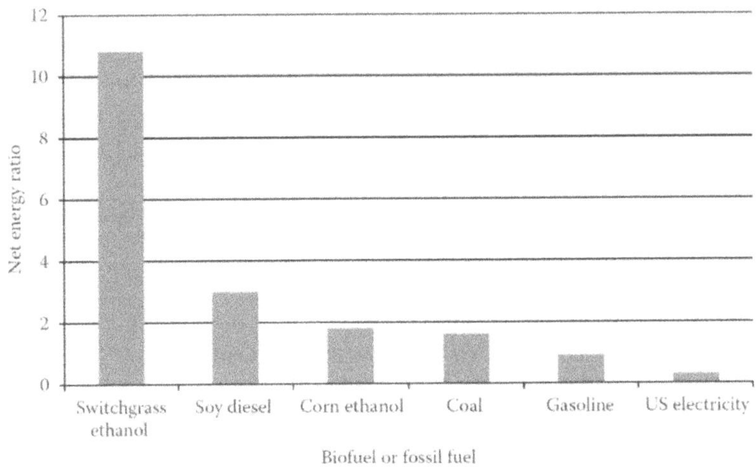

FIGURE 14.10 NERs of renewable and fossil fuel energy sources. (From Shapouri, H., M. Wang and J. A. Duffield, Net energy balancing and fuel-cycle analysis. In: *Renewable-Based Technology Sustainability Assessment* (J. Dewulf and H. Van Langenhove, eds.). John Wiley & Sons Ltd., New York, 2006.)

$$NER = \frac{\text{Energy content of biofuel (MJ)}}{\text{Energy required to produce the biofuel (MJ)}} \tag{14.4}$$

Example 14.4 shows a typical value of NER for biofuels. The NERs for various conversion processes varies, with switchgrass ethanol having the highest NER of more than 10 followed by soy diesel of around 3. Coal has a slightly lower NER than corn ethanol, whereas gasoline is close to 1. This is graphically shown in Figure 14.10.

Example 14.4: NER Calculations

Determine the NER to produce ethanol from corn. The following data was established: the heating value of ethanol (lower) was reported as 21.28 MJ/L and the reported total energy use to produce the corn feedstock was 20.08 MJ/L (Shapouri et al., 2003).

Solution:

a. The NER is calculated from the ratio of the given values as follows:

$$NER = \frac{21.28\ \text{MJ/L}}{20.08\ \text{MJ/L}} = 1.06$$

b. There is more energy in the ethanol product produced than the energy required for corn production, transport, conversion, and distribution. This value must be greater than 1.0 for the biofuel to be considered as having a positive net energy balance.
c. Note that the NEB value is low because the energy of coproducts is not included in the calculations.

14.5.1.2 Net Energy Balance (NEB) Method

The NEB is defined as the difference between the total energy required to produce the biofuel to the energy content of the biofuel as shown by Equation 14.5.

$$\text{NEB} = \text{biofuel HV (MJ)} - \text{energy required to produce the biofuel (MJ)} \tag{14.5}$$

Example 14.5 shows some typical values of NEB calculations for biofuels.

Example 14.5: Net Energy Balance Calculations

Determine the NEB to produce ethanol from corn using wet milling. The following data was established: the heating value of ethanol (lower) was reported as 21.28 MJ/L and the reported total energy use to produce the corn feedstock using wet milling process was 20.726 MJ/L (Shapouri et al., 2003).

Solution:

a. The NEB is simply calculated from the difference between these values as follows:

$$\text{NEB} = 21.28 - 20.726 = 0.554$$

b. There is more energy in a liter of ethanol than the energy required for corn production, transport, conversion, and distribution. This value must be greater than zero (or positive) for the biofuel to be considered as having a positive NEB.
c. Note that the NEB value is low because the energy of coproducts is not included in the calculations.

As discussed earlier, the energy balance calculations are dependent on the boundaries and assumptions used in the calculations. For example, if the energy content of the coproducts is included in the calculations, the associated energy conversion systems may have a reduced energy requirement for production. In this case, the energy of the coproducts is subtracted from the production process. This assumption lowers the overall total energy usage to around 16.027 MJ/L and broken down into 11.294 MJ/L for conversion, 4.251 MJ/L for production, 0.481 MJ/L for corn transport, and 0.414 MJ/L for distribution as weighted average values. Detailed values for both dry and wet milling processes are summarized in Table 14.5, including weighted average values. The NEB becomes 5.249 and the NER was improved to 1.330 (Shapouri et al., 2006).

Using the coproducts of biofuel production would significantly improve the NEB and the NER. Including corn ethanol coproducts, the NEB would improve from approximately 1.2 to an average of 5.3. Likewise, the NER would improve from 1.1 to 1.3. The NER is even higher when sugarcane is used for ethanol production: this would range from 8.3 to 10.2 (Shapouri et al., 2006).

TABLE 14.5
Energy Associated with Ethanol Production from Corn with Coproducts Allocation

	Energy Use of Milling Process (MJ/L)		
Production Subsystem	**Dry**	**Wet**	**Weighted Average (MJ/L)**
Corn production	4.314	4.188	4.251
Corn transport	0.489	0.474	0.481
Ethanol conversion	10.769	11.819	11.294
Ethanol distribution	0.414	0.414	0.414
Total energy used	15.572	16.482	16.027
Net energy balance	5.704	4.794	5.249
Net energy ratio	1.370	1.290	1.330

Source: Shapouri, H., M. Wang and J. A. Duffield, Net energy balancing and fuel-cycle analysis. In: *Renewable-Based Technology Sustainability Assessment* (J. Dewulf and H. Van Langenhove, eds.). John Wiley & Sons Ltd., New York, 2006.

14.6 CIRCULAR ECONOMY CONCEPTS AND ACTIVITIES

In recent years, circular economy, as applied to biomass use and conversion, has emerged as an economic and sustainability strategy (Yuan et al., 2008). Three elements must be present as follows (Korhonen et al., 2018):

a. closed system
b. use of pure biomass as renewable energy
c. overall systems analysis

In biomass conversion processes, waste generation is avoided to be considered a closed system. Biomass is considered a renewable energy resource. The overall strategy for conversion must be made by evaluating the overall systems from biomass production to product use. The American Society of Biological and Agricultural Engineers (ASABE) has developed a protocol for system identification and planning action to achieve circularity (ASABE TFACS, 2022). The protocol recommended the following eight steps:

1. **Review the System of Interest**—State the purpose, system scope and its perceived boundaries.
2. **Define the System**—Identify component boundaries, environment, flows into and from the system (including product, nutrients, energy, water, and losses). Identify upstream subsystems and flows and downstream subsystems(s) along value chain(s). Draw schematic diagram of the system, its boundaries, environment, components, important flows, and upstream and downstream subsystems (showing inputs and outputs).

3. **Problem Identification**–Identify problems that need to be addressed to make the system more circular. Focus on one or two resources, such as nitrogen (N), water, and energy. Refine system description and diagrams. Review the five circular bioeconomy systems (CBS) Principles adopted by ASABE and identify system losses, wastes inefficiencies, opportunities for adhering to the five CBS principles.

   The five CBS principles are as follows:
   a. Increase resource use efficiencies
   b. Design out waste and pollution
   c. Keep products and material in use
   d. Regenerate natural systems
   e. Provide economic benefits
4. **Objectives**—Identify 1–3 objectives that, if achieved, would enhance the circularity of your selected system. Describe your vision of the selected system after it is transformed and draw a modified diagram of your envisioned future system. Examples are as follows:
   i. Reduce dairy wastes loading to the environment by 75%
   ii. Achieve 50% supply of renewable heat and electrical power
   iii. Increase soil CCS
   iv. Reduce GHG emissions
   v. Increase profits by 10%
5. **Impacts**—Identify the positive and negative impacts that would be generated if the objective were achieved. For example:
   i. Focus on one or two resources such as water and electricity along with other energy requirements (e.g., natural gas and diesel)
   ii. Target priority objectives for transforming system or its components
   iii. Identify potential impacts of achieving objectives
6. **Barriers**—Identify potential barriers to achieving the objectives of increasing the circularity of the selected system.
7. **Actions**—Identify needed interventions and select specific actions.
8. **Prepare Short Report**—Enumerate objectives, show diagrams, objectives for increased circularity, potential impacts, anticipated timetable, barriers, needs, or opportunities.

The ASABE-proposed protocol attempts to convert a linear economy into circular economy. A linear economy is defined as the economic space where the value of products, materials, and resources is maintained in the economy for a short period of time with substantial generation of wastes. A circular economy, on the other hand, is defined as the economic space where the products, materials, and resources are maintained in the economy for as long as possible with minimal generation of wastes (Carus and Dammer, 2018).

The circular bioeconomy combines circular economy and bioeconomy concepts and aims for the cascading use of biomass from biological resources using a systems approach for economic development (ASABE TFACS, 2022). The process starts from recycling, reusing, and valorization of all waste generated during land management, use of energy and materials for crop production, processing and conversion, product distribution and use, and disposal of unused materials.

14.6.1 Benefits and Example of Creating Circular Bioeconomy Systems

The benefits of creating a circular bioeconomy system comes in many forms. Foremost is increased profitability. In many instances, a new biomass-based enterprise is created, and there is usually an increase in the use of renewable energy. All waste streams are recovered and reused for improved economic returns. The resource productivity of the feedstock is increased while reducing GHG emissions. Overall losses are reduced in significant amounts and recycling is enhanced. Figure 14.11 shows how a linear economy for milk production in a dairy farm is converted into a circular economy. In a linear economy, inputs include power and

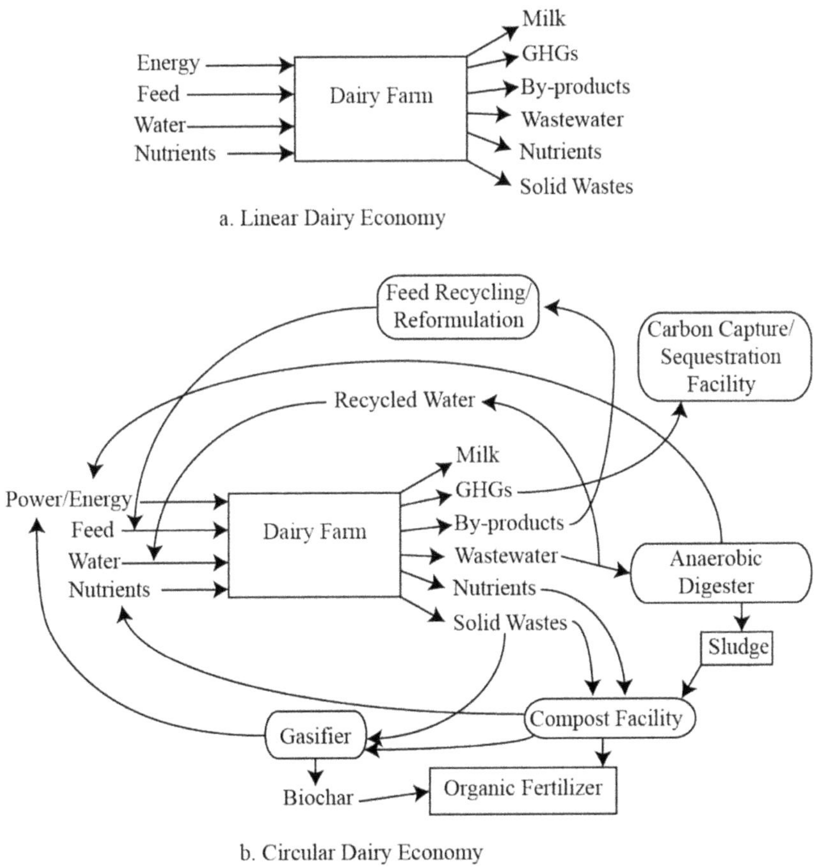

FIGURE 14.11 Linear economy for milk production in a dairy farm converted into a circular bioeconomy.

energy, feed, water, and nutrients and the outputs include milk, GHGs, by-products, wastewater, wasted nutrients, and solid manure wastes. To convert this system into a circular bioeconomy, GHGs may be captured and sequestered in a CCS facility. All organic by-products may be recycled and reformulated. An anaerobic digester (AD) may be installed to convert wastes into biogas, which may provide heat and power to the facility. Likewise, a gasifier may be installed to turn all separated solid wastes into additional heat, power, and energy. A compost facility may also be installed to handle all other wastes that in turn will be converted into fertilizer for the feed farm. Water is recycled for the facility, which may be used to flush manure and as irrigation water for the feed production farm. All nutrients are recycled whenever possible. The series of examples shows how much heat, power, and energy may be recovered for a 500-head dairy facility.

Example 14.6: Recoverable Power from Anaerobic Digestion Facility in a Dairy Farm

A 500-head hybrid dairy in Central Texas uses 723 kWh/yr/hd of electrical power (Capareda et al., 2010). The operator plans to install an anaerobic digester (AD) to begin addressing reduction in power for a circular economy for this facility. Assume the biogas production to be around 1.27 m^3/h/d of biogas with a heating value of 22.4 MJ/m^3. Determine the recovered energy (kWh) per year (365 days) and percent reduction in purchased power for this facility assuming an overall 35% conversion efficiency from biogas to electrical power in an internal combustion engine.

Solution:

a. The daily theoretical (P_t) and actual (P_a) power production is calculated as follows:

$$P_t\left(\frac{\text{MJ}}{\text{day}}\right) = 500\text{ hds} \times \frac{1.27\text{ m}^3}{\text{hd} - \text{day}} \times \frac{22.4\text{ MJ}}{\text{m}^3} = 14,224\frac{\text{MJ}}{\text{day}}$$

$$P_a(\text{kW}) = 14,224\frac{\text{MJ}}{\text{day}} \times \frac{1\text{ day}}{24\text{ h}} \times \frac{1\text{ h}}{3,600\text{ s}} \times \frac{1,000\text{ kJ}}{\text{MJ}} \times 0.35 = 164.6\text{ kW}$$

b. The total renewable energy produced by this AD facility over a year is calculated as follows:

$$\text{Energy}\left(\frac{\text{kW}}{\text{year}}\right) = 164.6\text{ kW} \times \frac{8,760\text{ h}}{\text{year}} = 1,441,896\frac{\text{kWh}}{\text{year}}$$

c. The total fossil-based energy utilized by this facility is calculated as follows:

$$\text{Energy}\left(\frac{\text{kWh}}{\text{year}}\right) = 500\text{ hds} \times \frac{723\text{ kWh}}{\text{hd} - \text{year}} = 361,500\frac{\text{kWh}}{\text{year}}$$

d. There is excess power indicating 100% reduction in electrical power requirements and excess power may be sold to the grid for additional revenue.

Example 14.7: Power and Biochar from Gasification Facility in Dairy Farm

A 500-head hybrid dairy facility in Central Texas plans to install an advanced mechanical solids separator after the morning and afternoon manure flush from its free stall. The manure will be dried (10% moisture content, MC) using part of the syngas (20% of power produced), and the remaining gaseous product will be used to generate electrical power. Estimate the potential power to be generated using an overall conversion efficiency of 15%. About 20% of the input dry manure will be converted into biochar as potential fertilizer. Likewise estimate the amount of biochar produced each year in units of metric tonnes (MT). The recoverable dry manure with 10% MC was found to be 2.3 kg/head/day. Its heating value was found to be 11.6 MJ/kg. The electrical power consumption for this facility is about 723 kWh/ yr/hd. If the biochar is sold for $100/MT, how much is the revenue from the biochar each year?

Solution:

a. The amount of dry manure generated each day is calculated as follows:

$$\text{Manure}\left(\frac{\text{MT}}{\text{Year}}\right) = 500\text{ hds} \times \frac{2.3\text{ kg}}{\text{hd} - \text{day}} \times \frac{365\text{ days}}{\text{year}} \times \frac{1\text{ MT}}{1{,}000\text{ kg}} = 420\frac{\text{MT}}{\text{yr}}$$

b. The potential power produced is calculated as follows assuming only 80% of the energy of dry manure.

$$P_t\left(\frac{\text{MJ}}{\text{day}}\right) = 500\text{ hds} \times \frac{2.3\text{ kg}}{\text{hd} - \text{day}} \times \frac{11.6\text{ MJ}}{\text{kg}} = 13{,}340\frac{\text{MJ}}{\text{day}}$$

$$P_a(\text{kW}) = 13{,}340\frac{\text{MJ}}{\text{day}} \times \frac{1\text{ day}}{24\text{ h}} \times \frac{1\text{ h}}{3{,}600\text{ s}} \times \frac{1{,}000\text{ kJ}}{\text{MJ}} \times 0.15 \times 0.80 = 18.5\text{ kW}$$

c. Hence, the amount of energy produced by the gasification system in a year is calculated as follows:

$$\text{Energy}\left(\frac{\text{kW}}{\text{year}}\right) = 18.5\text{ kW} \times \frac{8{,}760\text{ h}}{\text{year}} = 162{,}060\frac{\text{kWh}}{\text{year}}$$

d. The total fossil-based energy utilized by this facility is calculated as follows:

$$\text{Energy}\left(\frac{\text{kWh}}{\text{year}}\right) = 500\text{ hds} \times \frac{723\text{ kWh}}{\text{hd} - \text{year}} = 361{,}500\frac{\text{kWh}}{\text{year}}$$

e. This value is higher than the gasification product, and hence, the % reduction in fossil energy requirement is calculated as follows:

$$\text{Energy Reduction}\,(\%) = \frac{162{,}060\ \text{kWh yr}^{-1}}{361{,}500\ \text{kWh yr}^{-1}} \times 100\% = 44.8\%$$

f. About close to half of the electrical power requirement is provided by renewable energy resource.

g. The amount of biochar potentially produced is calculated as follows:

$$\text{Biochar}\left(\frac{\text{MT}}{\text{year}}\right) = 420\frac{\text{MT}}{\text{year}} \times 0.20 = 84\frac{\text{MT}}{\text{year}}$$

h. The revenue from the biochar each year is calculated as follows:

$$\text{Biochar Revenue}\left(\frac{\$}{\text{year}}\right) = 84\frac{\text{MT}}{\text{year}} \times \frac{\$100}{\text{MT}} = \frac{\$8.400}{\text{year}}$$

Example 14.8: Recovered Energy in Biochar, Bio-oil, and Syngas from Pyrolysis of Dairy Manure

A 500-head hybrid dairy in Central Texas plans to market the biochar from manure and sell the product as soil amendment. They plan to use a simple externally heated auger pyrolyzer system for this purpose. They wanted to optimize biochar yield by operating at a lower temperature (350°C) but will recover the syngas for drying wet manure and recover the bio-oil to be sold as alternate low-quality fuel for heating purposes. Biochar yield was maximized at 40% by weight with a heating value of 18.5 MJ/kg. The bio-oil yield was 15% by weight with a heating value of 15 MJ/kg. The remaining synthesis gas was used for drying wet manure. Assume recoverable dry manure yied is 5 kg/hd/day. The original heating value of the manure was found to be 11.6 MJ/kg. Determine the energy (MJ) and percentages (%) contained and recovered in each of the coproducts of this process.

Solution:

a. The amount of manure produced each day for the pyrolysis project is calculated as follows:

$$\text{Daily Manure Production (kg)} = 500\ \text{hds} \times \frac{5\ \text{kg}}{\text{hd} - \text{day}} = 2{,}500\frac{\text{kg}}{\text{day}}$$

b. The input energy from the dried manure is calculated as follows:

$$\text{Input Energy (MJ)} = 500\ \text{hd} \times \frac{5\ \text{kg}}{\text{hd} - \text{day}} \times \frac{11.6\ \text{MJ}}{\text{kg}} = 29{,}000\ \text{MJ}$$

c. The output energy contained in the biochar and its percentage is calculated as follows:

$$\text{Biochar Energy (MJ)} = \frac{2{,}500\ \text{kg}}{\text{day}} \times 0.4 \times \frac{18.5\ \text{MJ}}{\text{kg}} = 18{,}500\ \text{MJ}$$

$$\%\ \text{Energy in Biochar} = \frac{18{,}500\ \text{MJ}}{29{,}000\ \text{MJ}} \times 100\% = 63.8\%$$

d. The output energy contained in the bio-oil is calculated as follows:

$$\text{Biooil Energy (MJ)} = \frac{2{,}500\ \text{kg}}{\text{day}} \times 0.15 \times \frac{15\ \text{MJ}}{\text{kg}} = 5{,}625\ \text{MJ}$$

$$\%\ \text{Energy in Bio-oil} = \frac{5{,}625\ \text{MJ}}{29{,}000\ \text{MJ}} \times 100\% = 19.4\%$$

e. The syngas energy assuming the difference is estimated as follows:

$$\text{Syngas Energy (MJ)} = 29{,}000\ \text{MJ} - \left(18{,}500\ \text{MJ} + 5{,}625\ \text{MJ}\right) = 4{,}875\ \text{MJ}$$

$$\%\ \text{Energy in Syngas} = \frac{4{,}875\ \text{MJ}}{29{,}000\ \text{MJ}} \times 100\% = 16.8\%$$

14.6.2 Carbon Capture and Sequestration

Carbon dioxide (CO_2) capture and sequestration (CCS) is a set of technologies that can greatly reduce CO_2 emissions from new and existing energy conversion facilities such as bioethanol refineries that generate enormous amounts of exhaust gas, a primary contributor to global warming and climate change (USEIA, 2020a). This process usually involves three steps:

a. capture of CO_2 from biofuel production facilities or industrial processes
b. transport of the captured and compressed CO_2 (usually in pipelines)
c. underground injection and geologic sequestration

A bioethanol plant's CO_2 is relatively clean and pure, easy to compress, move, and store. This material could be sequestered way beneath the soil surface for a long period of time. Recent information (Schroeder, 2022) reports that 40%–45% of the CO_2 of the US market comes from ethanol production.

For bioethanol plants looking into on-site CCS, the early process involves drilling, sampling, and data collection (Geiver, 2022). Although not all ethanol plants are situated on potential on-site CCS activity, many plants are already integrating this process if their geographic footprint is ideal. The five basic steps are as follows:

a. **Obtain Permits:** Geologic research requires drilling permits.
b. **Prepare the Drill Site:** Pad preparation usually requires leveling a space of 121.95 m by 121.95 m [400 feet by 400 feet] space. The drilling rig needs a flat, workable surface. Pads usually take two weeks to construct.
c. **Drill the Hole**: Well drilling in the modern age is a multistep process. Each stage includes adding casing or using drilling mud and cement to ensure that the downhole effect does not impact existing fresh water sources.
d. **Gather Downhole Data**: After the shaft is drilled out and core samples of rock are collected, wireline trucks with sensors lower line into the well to gather more data.
e. **Close the Hole**: Following data collection, test hoes or wells are sealed temporarily while data is analyzed.

About 70% of the CO_2 produced in the United States goes to the food and beverage industry, while the other 30% is used for industrial purposes and for agriculture to metal trade. This is also one clear market for captured CO_2 in a biorefinery (Schroeder, 2022). Much of the proposition for sequestering carbon dioxide currently comes from federal tax credits from Section 45Q, which yields between $22 and $55 per sequestered metric tonne of CO_2 [$20–$50/tons]. Thus, biorefineries producing CO_2 have two options: merchant CO_2 and sequestered CO_2. Both activities will generate additional revenue from the government. Example 14.9 shows the theoretical maximum potential CO_2 produced in a bioethanol facility.

Example 14.9: Carbon Capture and Sequestration Example

A 3.785-million liter per year (1 MGY) bioethanol plant plans to sequester all the CO_2 produced by the facility and inject it into underground soil. It was found that the CCS technology was appropriate for the soil strata of the facility. Assume the ideal production rate from sugar to bioethanol of around 88 kg CO_2 per 180 kg of pure sugar. This process will also generate 92 kg of bioethanol. Assume the density of bioethanol to be 0.783 kg/L. Determine the total amount of CO_2 sequestered in units of metric tonnes (MT) per year for this facility and the potential maximum federal revenue.

Solution:

a. Based on the theoretical data given, the amount of CO_2 produced per liter of bioethanol is calculated as follows:

$$\frac{\text{kg}\,CO_2}{\text{kg Bioethanol}} = \frac{88\,\text{kg}\,CO_2}{92\,\text{kg bioethanol}} = \frac{0.9565\,\text{kg}\,CO_2}{\text{kg bioethanol}} = 0.9565\frac{\text{MT}\,CO_2}{\text{MT bioethanol}}$$

b. Next, the total amount of bioethanol produced by weight for this facility each year is calculated as follows:

$$\text{Bioethanol Production}\left(\frac{\text{MT}}{\text{yr}}\right) = \frac{3.785\times10^6\,\text{L}}{\text{yr}} \times \frac{0.783\,\text{kg}}{\text{L}} \times \frac{\text{MT}}{1{,}000\,\text{kg}} = 2963.655\frac{\text{MT bioethanol}}{\text{yr}}$$

c. The total amount of CO_2 produced each year on a weight basis is calculated as follows:

$$CO_2\ \text{Production}\left(\frac{\text{MT}}{\text{yr}}\right)=2{,}963.655\frac{\text{MT bioethanol}}{\text{yr}}\times\frac{0.9565\ \text{MT}\ CO_2}{\text{MT bioethanol}}=2834.74\frac{\text{MT}\ CO_2}{\text{yr}}$$

d. The maximum potential federal revenue for this process assuming the highest reported revenue potential of \$50/MT each year is calculated as follows:

$$\text{CCS Revenue}\left(\frac{\$}{\text{yr}}\right)=2834.74\frac{\text{MT}\ CO_2}{\text{yr}}\times\frac{\$50}{\text{MT}}=\$141{,}737/\text{yr}$$

14.6.3 Carbon Footprint (CF) of Facilities

CF research evaluates the total GHG emissions caused by a process or events. This terminology has now expanded to cover individuals, organizations, services, places, or products. The unit is expressed as carbon dioxide equivalent (CO_{2eq}). The four major contributors to CF values are the energy and power use; feed or food consumption, chilled water, cold water, hot water and natural gas use; transportation; and even office materials use. In biomass conversion, for example, to produce bioethanol from corn, the CF may mean the CO_2 emitted during land preparation, planting, harvesting, and conversion to the fuel, including transportation use in an internal combustion engine.

CF work may focus not only on energy but on other resources such as water. For example, research results show that US corn bioethanol has a smaller CF (541 L water/L ethanol) than Brazilian sugarcane bioethanol (1,115 L water/L bioethanol) (Mekonnen et al., 2018).

CF parameters may also be used in individual agricultural facilities such as animal farms. Table 14.6 summarizes some emission factors used for energy and utility usage. The process may be as simple as taking an inventory of all energy and utility use and referring to the table for conversion. Example 14.10 shows a very simple example for a dairy facility.

Example 14.10: Carbon Footprint on a Dairy Facility

Determine the overall carbon footprint (units of MT CO_{2eq}/yr) for a 500-head dairy facility in Central Texas. The estimated emissions for each dairy animal in terms of equivalent CO_{2e} emissions per head (Borhan et al., 2011) from ground-level sources was reported as 27.5 kg $CO2_{eq}$ hd/d resulting from collective methane, CO_2, and N_2O emissions. In addition, the dairy facility utilizes energy to operate the milking, housing, manure management, feeding, and water for a total annual energy usage of 723 kWh/hd/yr (Capareda, 2010). Natural gas usage was reported at 5 kWh/hd/d. Finally, diesel fuel usage was estimated at 155 L/hd/yr. The energy content of diesel was roughly 36.9 MJ/L.

Solution:

a. The annual gaseous emission is calculated as follows:

$$\text{GHG Emissions}\left(\frac{\text{MT CO}_{2e}}{\text{yr}}\right)=\frac{27.5\,\text{kg CO}_{2e}}{\text{hd}-\text{d}}\times 500\,\text{hds}\times\frac{365\,\text{d}}{\text{yr}}\times\frac{\text{MT}}{1{,}000\,\text{kg}}$$

$$=5{,}018.75\frac{\text{MT CO}_{2e}}{\text{yr}}$$

b. The equivalent CO_{2e} emissions from electrical power usage using Table 14.6 are calculated as follows:

$$\text{Equivalent Power Emissions}\left(\frac{\text{MT CO}_{2e}}{\text{yr}}\right)=723\frac{\text{kWh}}{\text{hd}-\text{yr}}\times 500\,\text{hds}\times\frac{0.424\,\text{kg CO}_{2e}}{\text{kWh}}$$

$$\times\frac{\text{MT}}{1{,}000\,\text{kg}}=153.3\,\frac{\text{MT CO}_{2e}}{\text{yr}}$$

c. The equivalent CO_{2e} emissions from the use of natural gas are calculated as follows:

$$\text{Equivalent Natural Gas Usage Emissions}\left(\frac{\text{MT CO}_{2e}}{\text{yr}}\right)=5\frac{\text{kWh}}{\text{hd}-\text{d}}\times\frac{365\,\text{d}}{\text{yr}}\times 500\,\text{hds}$$

$$\times\frac{0.181\,\text{kg CO}_{2e}}{\text{kWh}}\times\frac{\text{MT}}{1{,}000\,\text{kg}}=165.2\frac{\text{MT CO}_{2e}}{\text{yr}}$$

d. Diesel usage equivalent emissions are estimated as follows:

$$\text{Equivalent Diesel Usage Emissions}\left(\frac{\text{MT CO}_{2e}}{\text{yr}}\right)=155\frac{\text{L}}{\text{hd}-\text{yr}}\times 500\,\text{hds}\times\frac{36.9\,\text{MJ}}{\text{L}}\times\frac{1\,\text{kWh}}{3.6\,\text{MJ}}$$

$$\times\frac{0.253\,\text{kg CO}_{2e}}{\text{kWh}}\times\frac{\text{MT}}{1{,}000\,\text{kg}}=201\frac{\text{MT CO}_{2e}}{\text{yr}}$$

TABLE 14.6

Emission Factors Used for Energy and Utility Usage

Category	Unit Used	Conversion Factor	Source
Chilled Water	kWh or MJ	0.475 kg CO_{2e}/kWh	Nordin and Majid, 2013
Cold Water	m^3	0.18 kg CO_{2e}/m^3	HRWC, 2014
Electricity	1 kWh (3412 Btu)	0.424 kg CO_{2e}/kWh	USEIA, 2020a
Hot water	kWh or MJ	0.2265 kg CO_{2e}/kWh	USEIA, 2020a
Natural Gas	kWh or MJ	0.181 kg CO_{2e}/kWh	USEIA, 2020a
Diesel/Fuel Oil	kWh or MJ	0.253 kg CO2e/kWh	USEIA, 2020a

e. The overall CF for this facility is the sum of all equivalent emissions as follows:

$$\text{Total Facility Emissions}\left(\frac{\text{MT}\,\text{CO}_{2e}}{\text{yr}}\right)=(5{,}018.75+153.3+165.2+201)=5{,}538.25\frac{\text{MT}\,\text{CO}_{2e}}{\text{yr}}$$

f. More than 90% of these emissions are from ground-level greenhouse gas emissions (i.e., 5018.75/5538.25 = 0.9062).
g. Note that there may be other activities at the dairy farm that may contribute to additional emissions not accounted for in this example. The operator must develop a more comprehensive inventory of all possible greenhouse gas emissions.

14.6.4 Carbon Intensity

Carbon intensity (CI) is an emission intensity. It is defined as the emission rate of a given pollutant relative to the intensity of a specific activity. A common example is the number of grams of CO_2 released per MJ of energy produced. Electricity that is generated using fossil fuels is more carbon intensive as the process creates more CO_2 emissions than the renewable equivalent. Emission intensities are used to derive estimates of air pollutants such as GHGs based on the amount of fossil fuel combusted. Renewable biofuels are excellent choices when identifying a low CI fuel even for electricity production. Some biofuels produced from lignocellulosic biomass may have negative carbon intensities.

When maize stover is converted into bioethanol for vehicle use or biomass for electricity without CCS, the CI is reported at 20.3 and 21.2 g CO_{2e}/MJ (Gelfand et al., 2020). If switchgrass is used the same way, the CI is reduced to −5.2 and −47.5 gCO_{2e}/MJ, respectively. Table 14.7 summarizes some carbon dioxide emissions coefficients when estimating the CI of fuels and other biomass resources.

The carbon intensities of various biodiesel products were compared with that of fossil fuels and results are summarized in Table 14.8. Waste vegetable oil and waste animal fat had the lowest CI.

TABLE 14.7
Carbon Dioxide Emissions Coefficients by Fuels (USEIA, 2021 and Gelfand et al., 2020)

Fuel Type	g CO_{2e}/MJ	Fuel Type	g CO_{2e}/MJ
Propane	59.60	MSW	39.52
Diesel	70.27	Tire-Derived Fuel	81.49
Kerosene	69.37	Waste Oil	70.14
Coal	90.75	Miscanthus (power)	−260
Natural Gas	50.15	Bioenergy Sorghum	17
Gasoline	66.98	Corn (ethanol)	75

TABLE 14.8
Carbon Intensities of Various Biodiesel Types Compared with Fossil Fuel Carbon Intensity (Posada et al., 2012)

Fuel Type	g CO_{2e}/MJ	Biodiesel Type	g CO_{2e}/MJ
Fossil Fuel	85	Palm Oil	53
Rapeseed	47	Waste Vegetable Oil	10
Sunflower	35	Waste Animal Fat	10
Soybean	50	Jatropha	32

14.7 CONCLUSION

Biomass conversion is a relatively labor- and management-intensive undertaking. It is very similar to agricultural production. Large tracks of land may be required for wide-scale production, which requires similar agricultural inputs such as fertilizer, water, and abundant sunshine. The photosynthetic conversion of biomass is very low. The biomass resources are widely dispersed and contain high moisture, causing transport and logistics issues. There are harvest difficulties over large areas, and there will be landscape changes with environmental impact.

Feedstock cost is usually the most important economic component of biofuel production. Biomass resources with negative or zero cost will have an economic advantage. Examples include municipal solid wastes, municipal sewage, or animal manure. Using these feedstocks efficiently will also benefit the environment.

Markets for various biomass conversion products are on the rise locally and internationally. For example, when the market for biodiesel in the United States was strained because of the high cost of refined oil, the European market was targeted by various US producers. However, these US producers were met with different standards for biodiesel because the European market is canola-centric, whereas the US-based ASTM standard is described as neutral to any vegetable oil. It was shown that some biofuels are economically dependent on subsidies. In the United States, the growth of the biodiesel industry was originally driven by the feedstock subsidy in the initial years of its popularity. When this subsidy was removed, the positive economics of producing biodiesel became very difficult to meet. Some producers opted to shift to a lower-cost feedstock, such as chicken oil. Unfortunately, when this shift occurred, by the simple law of supply and demand, the price of chicken oil increased, thereby shifting the feedstock market to even cheaper oil, such as used oil from the fast food industry. Then, the cycle moved on to other feedstocks, such as algae.

The world biofuel industry will continue to face various economic issues concerning biofuel production. It was shown that the economic viability of most biofuels will be dependent on the value of other by-products of the process. The importance, for example, of distiller's grain for ethanol production has been discussed. The use and sale of biochar has become economically important.

Finally, because the world is still a fossil fuel-based economy, at present, the economics of biofuel production will always be analyzed against the price of crude oil. Making refinery-grade bio-oil from fast pyrolysis processes becomes highly dependent on the price of crude oil. Feedstock costs and crude oil prices are the most crucial factors affecting biofuel economy.

14.8 PROBLEMS AND DISCUSSION ISSUES

14.8.1 Greenhouse Gas Equivalency Calculations

P14.1 Natural gas is used in boilers and the amount of natural gas used was 1,000,000 m^3 in a year. Determine the equivalent CO_2 emissions from this system if the emissions are as follows (Canadian Association of Petroleum Producers, 2003):

- CO_2 = 1,891 g/m^3 fuel
- CH_4 = 0.0363 g/m^3 fuel
- N_2O = 0.0347 g/m^3 fuel

Likewise, determine the percent contribution of each compound by weight of CO_2 equivalent.

14.8.2 Gas Turbine Greenhouse Gas Equivalency Calculations

P14.2 Determine the equivalent CO_2 emissions for a gas turbine whose emission factors are as follows: CO_2 = 1,769 g/m^3, CH_4 = 0.138 g/m^3, and N_2O = 0.077 g/m^3. The output rating of the unit was 29.31 MW (100 MMBtu/h), and the heating value of natural gas is assumed to be 37.2 MJ/m^3 (1,000 Btu/ft^3). Assume the thermal efficiency to be 80% and the unit is to be operated 360 days in a year. Likewise, determine the percent contribution of each GHG by weight of CO_2 equivalent.

14.8.3 GREET Modeling Part 1

P14.3 Download the latest version of the GREET Model and run a scenario to estimate energy and GHG emissions from producing ethanol from wet milling of corn. The given parameters are as follows:

- Target year = 2020
- Report the units per 27.334 MJ of ethanol (i.e., 1 kg ethanol)
- Use GREET default assumption estimates
- The density of ethanol is 789 kg/m^3

Generate a table of results showing the following parameters:

a. Total CO_2 emissions from well to use (in units of kg)
b. CH_4 emissions (in units of g)
c. N_2O emissions (in units of g)
d. VOC (g)
e. CO (g)
f. NOx (g)

g. PM10 (g)
h. PM2.5 (g)
i. SO_x (g)
j. Biogenic carbon (BC) (mg)
k. Particle organic carbon (POC) (mg)
l. Well to use energy resources in units of MJ
m. Total water use (in units of m^3)

Determine yearly GHG emissions offset (CO_{2eq}) using ethanol blended with gasoline at an average blending rate of 10% and a yearly average gasoline fuel usage of 2,127 L (562 gallons) in the United States.

14.8.4 GREET Modeling Part 2

P14.4 Download the latest version of GREET Model and run a scenario to estimate the energy and GHG emissions from producing biodiesel from soybeans. The given parameters are as follows:

- Simulation year = 2020
- Report the units per 40 MJ of biodiesel (i.e., 1 kg biodiesel)
- Use GREET default assumption estimates
- The density of biodiesel is 880 kg/m^3

Generate a table of results showing the following parameters:

a. Total CO_2 emissions from well to use (in units of kg)
b. CH_4 emissions (in units of g)
c. N_2O emissions (in units of g)
d. VOC (g)
e. CO (g)
f. NO_x (g)
g. PM10 (g)
h. PM2.5 (g)
i. SOx (g)
j. Biogenic carbon (BC) (mg)
k. Particle organic carbon (POC) (mg)
l. Well to use energy resources in units of MJ
m. Total water use (in units of m^3)

Determine the daily GHG emissions offset (CO_{2eq}) using ethanol blended with diesel fuel at an average blending rate of 5% and with a daily diesel fuel consumption of 461.8 million L (122 million gallons) in the United States.

14.8.5 Infield Pyrolysis System for Immediate Biochar Field Application

P14.5 An infield biomass pyrolysis system is being developed to be attached to a combine/harvester. The goal is to immediately pyrolyze the biomass residues and return the biochar directly to the field. A small 3-row combine (76.2 cm/row) [30 inches/row] for harvesting sorghum plots was used for the process with an average harvest speed of 4.8 kph (kilometers per hour or.34 m/s) [3 mph]. The field efficiency is 65%. Simulate this system and determine the biomass waste harvest rate (tonnes/h) as well as biochar application rate (tonnes/ha) using sorghum crop. The sorghum biomass yield was estimated at 4.1 tonnes/ha [1.8 tons/acre] with an estimated biomass residue yield of 5.3 tonnes/ha [2.3 tons/acre] assuming 1:1.3 biomass to waste ratio. The biochar yield is 30% of the biomass waste input.

14.8.6 Wood Bioethanol Economics

P14.6 A short rotation woody crop is being planned for conversion into biofuels (particularly ethanol from woody biomass) in the State of Minnesota. About 80,972 ha [200,000 acres] of land will be devoted to this project. The estimated yield is 16.84 wet tonnes/ha/yr [7.5 wet tons/acre/yr]. The moisture content after harvest is roughly 46%. Determine the following:

a. The annual physical volume available for harvest on a dry basis in tonnes/ha/yr [tons/acre/yr].
b. The annual revenue for biomass producer if the selling price of woody crop is $88/dry tonne [$80/dry ton].
c. The amount of ethanol produced each year in million liters/year (or MGY, million gallons per year) if the ethanol yield is 229 L/dry tonne [55 gallons per dry ton].
d. The revenue from ethanol sale if the price of ethanol is $1.06/L [$4/gallon].
e. Comment on the revenue generated from the biomass and ethanol sale.

14.8.7 Pyrolysis Economics

P14.5 Determine the processing and upgrade cost for pyrolysis bio-oil so that it will be on a break-even cost with commercial crude oil that is sold for $90/barrel. The feedstock cost is $80/dry ton and the bio-oil yield is 50% by weight. The density of the bio-oil product is 1.1 kg/L [9.16 lbs/gal].

14.8.8 Gasification Economics

P14.6 A 5 MW biomass-fueled power plant is to be constructed in a locality using gasification technology. The project devoted 8,097 ha [20,000 acres] of land for this purpose. The heating value of biomass to be used is approximately 20 MJ/kg at around 10% moisture. The power plant is to be operated 350 days in a year. The overall biomass-to-electrical power conversion efficiency is 15%. What would be the appropriate biomass yield in "as received" basis (10% moisture) in units of tonnes/ha/yr such that this plant can have a sustainable supply of biomass? Assume that upon harvest, the biomass moisture content is 50%, what is the biomass yield in wet tonnes/ha/yr? How much biomass (at 10% moisture) is required each year in tonnes?

14.8.9 Net Energy Ratio (NER) Calculations

P14.7 Determine the NER to produce ethanol from corn. The following data was established: the heating value of ethanol (lower) was reported as 21 MJ/L, and the reported total energy use to produce the corn feedstock was 19.4 MJ/L (Shapouri et al., 2006).

14.8.10 Net Energy Balance (NEB) Calculations

P14.8 Determine the NEB to produce ethanol from corn using dry milling. The following data was established: the heating value of ethanol (lower) was reported as 21 MJ/L, and the reported total energy use to produce the corn feedstock using the dry milling process was 19.4 MJ/L (Shapouri et al., 2006).

14.8.11 Carbon Dioxide Equivalent Emissions per Unit of Energy

P14.11 Which among the utilities listed has the highest reported equivalent CO_2 emissions per kWh unit of energy used?
a. fuel oil b. natural gas c. hot water d. electricity

14.8.12 Best Biodiesel Type with Lowest Carbon Intensity

P14.12 Which among the biodiesel types listed below had the lowest reported carbon intensity?
a. waste oil b. palm oil c. soybean d. rapeseed

14.8.13 Capital Intensive Biofuel for Conversion Processes

P14.13 Which among the biofuel or conversion process enumerated below has the highest capital cost impact?
a. gasification b. pyrolysis c. biogas d. bioethanol

14.8.14 Converting a Cotton Gin Operation from Linear to Circular Economy

P14.14 The manager of a 90-bale per hour cotton gin facility located in the Panhandle of Texas plans to convert his facility from a simple linear economy to a circular economy to reduce GHG emissions. Enumerate and discuss all possible interventions to this facility to achieve near zero carbon economy. List all assumptions made.

14.8.15 Converting a Biodiesel Facility into Circular Bioeconomy

P14.15 The manager of a 11.355-million liters per year (3 MGY) biodiesel facility in Dayton, Texas, plans to reduce its carbon footprint during its yearly operation. It uses conventional fossil fuel (diesel and gasoline) for its truck/vehicle fleet and natural gas for heating and cooling systems. Develop a plan to significantly reduce its carbon footprint. List all assumptions made including references.

REFERENCES

Ahmed, I., J. Decker and D. Morris. 1994. How much energy does it take to make a gallon of soydiesel? Report prepared for the National Soydiesel Development Board (now National Biodiesel Board, NBB), Jefferson City, MO.

Argonne National Laboratory (ANL). 2021. The Greenhouse Gases, Regulated Emissions, and Energy Use in Transportation (GREET) Model. Energy Systems, Argonne National Laboratory of the US Department of Energy, Argonne, IL. Available at: https://greet.es.anl.gov/. Accessed 13 March 2021.

ASABE TFACS. 2022. ASABE's Circular Bioeconomy Systems (CBS) Initiative. Workshop presentation presented at the 2022 Annual International Meeting of the American Society of Agricultural and Biological Engineers held in Houston, Texas on 19 July, 2022. ASABE, St. Joseph, MI. Document created by the Task Force for Transitioning Food ad Agriculture to Circular Systems (TFACS). Circular Bioeconomy Systems (CBS) AN ASABE Initiative.

Canadian Association of Petroleum Producers. 2003. Calculating green house gas emissions guide. Global Climate Change Voluntary Challenge Guide Docs #55904, Calgary, Canada.

Capareda, S. C., S. Mukhtar, C. Engler and L. B. Goodrich. 2010. Energy usage survey of dairies in the Southwestern United States. *Applied Engineering in Agriculture* 26 (4): 66–675. American Society of Agricultural and Biological Engineers ISSN 0883–8542. ASABE, St Joseph, MI.

Carus, Michael and Lara Dammer. 2018. The "Circular Bioeconomy" Concepts, Opportunities and Limitations. Noa Institute Paper #9 on Bio-based Economy 2018-1. Nova Institute, Huerth, Germany.

Delucchi, M. A. 2006. *Life Cycle Analyses of Biofuels*. Institute of Transportation Studies, University of California, Davis, Davis, CA.

Geiver, Luke. 2022. Sequestering in place. Ethanol Producer Magazine, July 2022 Issue. Available at: https://ethanolproducer.com/., pp 24–29. Grand Forks, ND, 58203.

Hodges, A. W., T. J. Stevens and M. Rahmani. 2010. Economic Impacts of Expanded Woody Biomass Utilization on the Bioenergy and Forest Products Industries in Florida. Final report to Florida Department of Agriculture and Consumer Services - Division of Forestry, University of Florida, 36 pages, Revised February 23, 2010. Gainesville, FL.

Hurt, C., W. Tyner and O. Doering. 2008. *Economics of Ethanol*. Purdue University Extension Publication Series on Bioenergy: Fueling America Through Renewable Resources. Purdue University Cooperative Extension Service and Department of Agricultural Economics. Purdue University, West Lafayette, IN. Also available at: https://www.ces.purdue.edu/bioenergy.

Korhonen, Jouni, Antero Honkasalo and Jyri Seppala. 2018. Circular economy: The concept and its limitations. *Ecological Economics* 143: 37–46. Elsevier Science, Direct, UK.

Lane, J. 2011. The new economics of next-gen gasification. *Biofuels Digest* January 11, 2011 Issue.

Mann, M. K. and P. L. Spath. 1999. The net CO_2 emissions and energy balances of biomass and coal-fired power systems. NREL, USDOE Project Report NREL/TP-430-23076 and NREL/TP-570-25119. National Renewable Energy Laboratory (NREL) of the USDOE, Golden, CO.

Mann, M. K. and P. L. Spath. 2001. A life cycle assessment of biomass cofiring in a coal-fired power plant. *Clean Products and Processes* 3: 81–91.

Multer, C. L., C. B. Parnell Jr., R. O. McGee and S. C. Capareda. 2010. Benefits of onsite power production for cotton gins. In: *Proceedings of the 2010 Beltwide Cotton Conferences*. Cotton Foundation, Memphis, TN.

National Biodiesel Board (NBB). 2008. Sustainability Task Force Report presented at the National Biodiesel Conference held on Feb 3–6, 2008 in Orlando, FL. (E. Bockian Landsburg, Task Force Lead), NBB, Jefferson City, MO.

Nordin, A., Majid, M.A.A., 2013. Carbon dioxide emission analysis of chilled water production by using gas turbine exhaust heat. *International Journal of Environmental Science and Technology* 4 (6): 672–675. https://doi.org/10.7763/IJESD.2013.V4.436.

Piazza, M. 2008. Feedstock economics: A comparison of feedstocks. Paper presented at the 5th Practical Short Course on Biodiesel and Industrial Applications of Vegetable Oils, Texas A&M University, May 13–16, 2008 and Sponsored by the Food Protein Research and Development Center, Texas Engineering Experiment Station, College Station, TX.

Pimentel, D. 2003. Ethanol fuels: Energy balance, economics and environmental impacts are negative. *Natural Resources Research* 12 (2): 127–134.

Pimentel, D. and T. W. Patzek. 2005. Ethanol production using corn, switchgrass, and wood; Biodiesel production using soybean and sunflower. *Natural Resources Research* 14 (1): 65–76.

Posada, Francisco, Chris Malins and Ani Baral. 2012. Biodiesel carbon intensity, sustainability and effects on vehicles and emissions. The International Council on Clear Transportation. Briefing paper released January 2012, Washington, DC. Available at: https://theicct.org/wp-content/uploads/2021/06/ICCT_biodiesel-briefing_Jan12.pdf. Accessed September 6, 2022.

Pradhan, A., D. Shresta, J. Van Gerpen and J. Duffield. 2007. The energy balance of soybean oil biodiesel production: A review of past studies. *Transactions of the ASABE* 51 (1): 185–194.

Schroeder, Katie. 2022. The Ethanol Industry's Commercial Capture. Ethanol Producer Magazine, July 2022 Issue. Available at: https://ethanolproducer.com/., pp. 24–29. Grand Forks, ND, 58203.

Shapouri, H. and M. Salassi. 2006. The economic feasibility of ethanol production from sugar in the United States. Report published by the USDA Office of Energy Policy and New Uses (OEPNU) and the Office of Chief Economist (OCE), the US Department of Agriculture (USDA), Washington, DC.

Shapouri, H., J. A. Duffield and M. Wang. 2003. The energy balance of corn ethanol revisited. *Transactions of the ASABE* 46 (4): 959–968.

Shapouri, H., M. Wang and J. A. Duffield. 2006. Net energy balancing and fuel-cycle analysis. In: *Renewable-Based Technology Sustainability Assessment* (J. Dewulf and H. Van Langenhove, eds.). John Wiley & Sons Ltd., New York. pp. 73–86.

Spath, P. L., M. K. Mann and D. R. Kerr. 1999. Life cycle assessment of coal-fired power production. NREL, USDOE Project Report NREL/TP-570-25119. June 1999. National Renewable Energy Laboratory (NREL) of the USDOE, Golden, CO. Under Contract No. DE-AC-36-98-GO10337.

Tifanny, D. G., R. V. Morey and M. J. De Kam. 2009. Economics of biomass gasification/combustion at fuel ethanol plants. *Applied Engineering in Agriculture* 25 (3): 391–400.

USDOE and USDA. 2009. Sustainability of biofuels: Future research opportunities. Report from October 2008 Workshop. DOE/SC-0114, US Department of Energy Office of Science and US Department of Agriculture (https://genomicsgtl.energy.gov/biofuels/sustainability and https://www.ree.usda.gov).

USEIA. 2020a. How much carbon dioxide is produced when different fuels are burned? U.S. EIA, Washington, DC. Available at: https://www.eia.gov/tools/faqs/faq.php?id=73&t=11 (accessed 6 December 2020).

USEIA. 2020b. US fuel ethanol production capacity increased by 3% in 2019. Today in Energy (Issue 29 September 2020). US Energy Information Administration, Washington, DC. Available at: https://www.eia.gov/todayinenergy/detail.php?id=45316. Accessed 27 May, 2022.

USEIA. 2021. Carbon Dioxide Emissions Coefficients. S Energy Information Administration, 1000 Independence Avenue, SW, Washington DC. Available at: https://www.eia.gov/environment/emissions/co2_vol_mass.php. Accessed 2 August 2022. USEPA. 1995. *AP 42. Compilation of Air Pollutant Emission Factors*, 5th Ed. Vol. 3, Stationary Internal Combustion Engines. Technology Transfer Network Clearing House for Inventories and Emission Factors. Emission factors and Policy Application Center, Office of Air Quality Planning and Standards (OAQPS), Research Triangle Park, NC.

USEPA. 2009. ENERGY STAR and Other Climate Protection Partnerships. 2009 Annual Report. https://www.energystar.gov/ia/partners/annualreports/2009_achievements.pdf (accessed 19 March 2022).

Waranica, G. 2007. Equipment and technology for biodiesel production. Presented at the 4th Practical Short Course on Biodiesel and Industrial Applications of Vegetable Oils held on August 19–22, 2007 and organized by the Food Protein Research and Development Center, Texas Engineering Experiment Station, Texas A&M University, College Station, TX.

Webster, Ken. 2015. *The Circular Economy: A Wealth of Flows*. MacArthur Foundation Publishing, ISBN 0992778425. ISBN 0992778425. Cowes, UK.

Willits, J. 2007. Biodiesel plant design: Planning, site selection and critical parameters. Presented at the August 19–22, 2007, at the 4th Practical Short Course on Biodiesel and Industrial Applications of Vegetable Oils organized by the Food Protein Research and Development Center, Texas Engineering Experiment Station, Texas A&M University, College Station, TX.

Wright, M. M., J. A. Satrio, R. C. Brown, D. E. D. Daugaard and D. D. Shu. 2010. Techno-economic analysis of biomass fast pyrolysis to transportation fuels. USDOE Technical Project Report prepared under Task No. BB07-7510 and Report No. NREL/TP-6A-46586 under Contract No. DE-AC36-08GO28308. The National Renewable Energy Laboratory (NREL), Golden, CO.

Wu, C. Z., H. Huang, S. P. Zheng, and X. L. Yin. 2002. An economic analysis of biomass gasification and power generation in China. *Bioresource Technology* 83 (1): 65–70.

Yuan, Z., Bi, J., Moriguichi, Y., 2008. The circular economy: A new development strategy in China. *Journal of Industrial Ecology* 10: 4–8.

ASSESS YOUR KNOWLEDGE

1. Determine the % contribution of methane from the overall equivalent CO_2 emissions from a conversion process whose gas emissions are as follows: 10 tonnes of CO_2, 3 tonnes of CH_4, and 0.05 tonne of N_2O.
   a. 14.56
   b. 71.2
   c. 9.02
   d. 27.48

2. Determine the % contribution of N_2O from the overall equivalent CO_2 emissions from a conversion process whose gas emissions are as follows: 10 tonnes of CO_2, 3 tonnes of CH_4, and 0.05 tonne of N_2O.
   a. 14.56%
   b. 11.3%
   c. 9.02%
   d. 17.5%

11. A 400-head hybrid dairy in Central Texas uses 750 kWh/yr/hd of electrical power. The owner plans to install an anaerobic digester (AD) to begin addressing reduction in power for a circular economy for this facility. Assume the biogas production to be around 1.27 m^3/h/d of biogas with a heating value of 22.4 MJ/m^3. Determine the recovered energy (kWh) per year (365 days). Assume an overall 35% conversion efficiency from biogas to electrical power in an internal combustion engine.
    a. 1,153,692
    b. 750,652
    c. 657,023
    d. 361,123

12. A 400-head hybrid dairy in Central Texas uses 750 kWh/yr/hd of electrical power. The owner plans to install an anaerobic digester (AD) to begin addressing reduction in power for a circular economy for this facility. Assume the biogas production to be around 1.27 m^3/h/d of biogas with a heating value of 22.4 MJ/m^3. Determine the percent reduction in purchased power for this facility assuming an overall 35% conversion efficiency from biogas to electrical power in an internal combustion engine and for 365 days in a year.
    a. 57.6
    b. 100
    c. 65.7
    d. 36.1

3. Determine the equivalent CO_2 emissions from a conversion process whose gas emissions are as follows: 10 tonnes of CO_2, 3 tonnes of CH_4, and 0.05 tonne of N_2O.
   a. 88.50
   b. 67.42
   c. 9.02
   d. 27.48

13. A 400-head hybrid dairy facility in Central Texas plans to install an advanced mechanical solids separator after the morning and afternoon manure flush from its free stall. The manure will be dried (10% moisture content, MC) using part of the syngas (20% of power produced), and the remaining gaseous product (80%) will be used to generate electrical power. The potential power may be estimated using an overall conversion efficiency of 15%. About 20% of the input dry manure will be converted into biochar as potential fertilizer. Likewise, the amount of biochar produced each year may be estimated each year in units of metric tonnes (MT). The recoverable dry manure with 10% MC was found to be 2.3 kg/head/day. Its heating value was found to be 11.6 MJ/kg. The electrical power consumption for this facility is about 750 kWh/yr/hd. The biochar is sold for $100/MT. Determine the % reduction in energy usage from electrical power production.
    a. 100
    b. 20.5
    c. 43.3
    d. 88.4

4. Determine the % contribution of CO_2 from the overall equivalent CO_2 emissions from a conversion process whose gas emissions are as follows: 10 tonnes of CO_2, 3 tonnes of CH_4, and 0.05 tonne of N_2O.
   a. 41.56
   b. 31.3
   c. 11.3
   d. 27.48

14. A 400-head hybrid dairy in Central Texas plans to market the biochar from manure and sell the product as soil amendment. They plan to use a simple externally heated auger pyrolyzer system for this purpose. They wanted to optimize biochar yield by operating at lower temperature (350°C) but will recover the syngas for drying wet manure and recover the bio-oil to be sold as alternate low-quality fuel for heating purposes. Biochar yield was maximized at 40% by weight with a heating value of 18.5 MJ/kg. The bio-oil yield was 15% by weight with a heating value of 15 MJ/kg. The remaining synthesis gas was used for drying wet manure. Assume recoverable dry manure yied is 5 kg/hd/day. The original heating value of the manure was found to be 11.6 MJ/kg. Determine the energy contained in the recovered bio-oil (in units of MJ) from this pyrolysis process.
    a. 20,000
    b. 23,200
    c. 4,500
    d. 6380

5. A 400-head hybrid dairy in Central Texas uses 750 kWh/yr/hd of electrical power. The owner plans to install an anaerobic digester (AD) to begin addressing reduction in power for a circular economy for this facility. Assume the biogas production to be around 1.27 m^3/h/d of biogas with a heating value of 22.4 MJ/m^3. Determine the theoretical power (MJ/day) available from this facility based on the energy content of the manure.
   a. 57,125.5
   b. 65,751.1
   c. 11,379.2
   d. 14,224.0

15. An 11.355-million liter per year (3 MGY) bioethanol plant plans to sequester all the CO_2 produced by the facility and inject into underground soil. It was found that the CCS technology was appropriate for the soil strata of the facility. Assume actual production rate from sugar to bioethanol of around 80 kg CO_2 per 180 kg of pure sugar. This process will also generate 90 kg of bioethanol. Assume the density of bioethanol to be 0.750 kg/L. Determine the potential maximum federal credit revenue ($) for this process assuming revenue potential of $50/MT on a yearly basis.
   a. 757,000
   b. 296,300
   c. 378,500
   d. 147,000

6. A 400-head hybrid dairy facility in Central Texas plans to install an advanced mechanical solids separator after the morning and afternoon manure flush from its free stall. The manure will be dried (10% moisture content, MC) using part of the syngas (20% of power produced) and the remaining gaseous product will be used to generate electrical power. Estimate the potential actual power (kW) to be generated using an overall conversion efficiency of 15%. About 20% of the input dry manure will be converted into biochar as potential fertilizer. The recoverable dry manure with 10% MC was found to be 2.3 kg/head/day. Its heating value was found to be 11.6 MJ/kg. The electrical power consumption for this facility is about 750 kWh/yr/hd.
   a. 14.82
   b. 33.40
   c. 84.0
   d. 44.8

16. The overall carbon footprint (units of MT CO_{2eq}/yr) for a 400-head dairy facility in Central Texas will be estimated. The estimated emissions for each dairy animal in terms of equivalent CO_{2e} emissions per head (Borhan et al., 2011) from ground-level sources was reported as 27.5 kg CO_{2eq}/hd/d resulting from collective methane, CO_2, and N_2O emissions. In addition, the dairy facility utilizes energy to operate the milking, housing, manure management, feeding, and water for a total annual energy usage of 723 kWh/hd/yr (Capareda, 2010). Natural gas usage was reported at 5 kWh/hd/d. Finally, diesel fuel usage was estimated at 155 L/hd/yr. The energy content of diesel was roughly 36.9 MJ/L. Determine the equivalent CO_2 emissions from the diesel usage alone in units of MT CO_{2eq}/year.
   a. 510
   b. 652
   c. 201
   d. 161

7. A 400-head hybrid dairy facility in Central Texas plans to install an advanced mechanical solids separator after the morning and afternoon manure flush from its free stall. The manure will be dried (10% moisture content, MC) using part of the syngas (20% of power produced) and the remaining gaseous product (80%) will be used to generate electrical power. One may estimate the potential power to be generated using an overall conversion efficiency of 15%. About 20% of the input dry manure will be converted into biochar as potential fertilizer. The recoverable dry manure with 10% MC was found to be 2.3 kg/head/day. Its heating value was found to be 11.6 MJ/kg. The electrical power consumption for this facility is about 750 kWh/yr/hd. Determine the yearly electrical energy (kWh/year) generated by this facility through gasification assuming the power plant is operated 365 days in a year and 24 h/day.
   a. 148,255
   b. 129,823
   c. 84,560
   d. 44,898

17. The overall carbon footprint (units of MT CO_{2eq}/yr) for a 400-head dairy facility in Central Texas will be estimated. The estimated emissions for each dairy animal in terms of equivalent CO_{2e} emissions per head (Borhan et al., 2011) from ground-level sources was reported as 27.5 kg $CO2_{eq}$/hd/d resulting from collective methane, CO_2, and N_2O emissions. In addition, the dairy facility utilizes energy to operate the milking, housing, manure management, feeding, and water for a total annual energy usage of 723 kWh/hd/yr (Capareda, 2010). Natural gas usage was reported at 5 kWh/hd/d. Finally, diesel fuel usage was estimated at 155 L/hd/yr. The energy content of diesel was roughly 36.9 MJ/L. What percentage contribution to the overall CO_{2e}1 emissions comes from ground-level area sources?
   a. 90.53
   b. 65.52
   c. 20.10
   d. 30.5

8. An 11.355-million liter per year (3 MGY) bioethanol plant plans to sequester all the CO_2 produced by the facility and inject into underground soil. It was found that the CCS technology was appropriate for the soil strata of the facility. Assume actual production rate from sugar to bioethanol of around 80 kg CO_2 per 180 kg of pure sugar. This process will also generate 90 kg of bioethanol. Assume the density of bioethanol to be 0.750 kg/L. Determine the total amount of bioethanol produced each year by this facility in units of metric tonnes (MT).
   a. 7,570
   b. 2,963
   c. 1,475
   d. 8,516

18. The overall carbon footprint (units of MT CO_{2eq}/yr) for a 400-head dairy facility in Central Texas will be estimated. The estimated emissions for each dairy animal in terms of equivalent CO_{2e} emissions per head from ground-level sources was reported as 27.5 kg $CO2_{eq}$/hd/d resulting from collective methane, CO_2, and N_2O emissions. In addition, the dairy facility utilizes energy to operate the milking, housing, manure management, feeding, and water for a total annual energy usage of 750 kWh/hd/yr. Natural gas usage was reported at 5 kWh/hd/d. Finally, diesel fuel usage was estimated at 155 L/hd/yr. The energy content of diesel was roughly 36.9 MJ/L. Determine the equivalent CO_2 emissions from electrical power alone in units of MT CO_{2eq}/yr.
    a. 510.5
    b. 127.2
    c. 201.3
    d. 410.5

9. Determine the overall carbon footprint (units of MT CO_{2eq}/yr) for a 400-head dairy facility in Central Texas. The estimated emissions for each dairy animal in terms of equivalent CO_{2e} emissions per head from ground-level sources was reported as 27.5 kg $CO2_{eq}$/hd/d resulting from collective methane, CO_2, and N_2O emissions. In addition, the dairy facility utilizes energy to operate the milking, housing, manure management, feeding, and water for a total annual energy usage of 750 kWh/hd/yr. Natural gas usage was reported at 5 kWh/hd/d. Finally, diesel fuel usage was estimated at 155 L/hd/yr. The energy content of diesel was roughly 36.9 MJ/L.
   a. 4,435
   b. 1,652
   c. 2,010
   d. 5,108

19. An 11.355-million liter per year (3 MGY) bioethanol plant plans to sequester all the CO_2 produced by the facility and inject into underground soil. It was found that the CCS technology was appropriate for the soil strata of the facility. Assume actual production rate from sugar to bioethanol of around 80 kg CO_2 per 180 kg of pure sugar. This process will also generate 90 kg of bioethanol. Assume the density of bioethanol to be 0.750 kg/L. The sequestered CO_2 in units of metric tonnes (MT) per year for this facility is a potential maximum federal revenue. Determine the amount of CO_2 produced (in lbs) per gallon of bioethanol produced by this facility.
   a. 7.5
   b. 5.8
   c. 9.5
   d. 8.0

10. The overall carbon footprint (units of MT CO_{2eq}/yr) for a 400-head dairy facility in Central Texas will be estimated. The estimated emissions for each dairy animal in terms of equivalent CO_{2e} emissions per head from ground-level sources was reported as 27.5 kg/CO_{2eq}/hd/d resulting from collective methane, CO_2 and N_2O emissions. In addition, the dairy facility utilizes energy to operate the milking, housing, manure management, feeding, and water for a total annual energy usage of 750 kWh/hd/yr. Natural gas usage was reported at 5 kWh/hd/d. Finally, diesel fuel usage was estimated at 155 L/hd/yr. The energy content of diesel was roughly 36.9 MJ/L. Determine the equivalent CO_2 emissions from the use of natural gas alone in units of MT CO_{2eq}/year.
    a. 510
    b. 165
    c. 132
    d. 410

20. A 400-head hybrid dairy facility in Central Texas plans to install an advanced mechanical solids separator after the morning and afternoon manure flush from its free stall. The manure will be dried (10% moisture content, MC) using part of the syngas (20% of power produced) and the remaining gaseous product will be used to generate electrical power. For power production, the overall conversion efficiency is 15%. About 20% of the input dry manure will be converted into biochar as potential fertilizer. The amount of biochar produced each year in units of metric tonnes (MT) may also be estimated. The recoverable dry manure with 10% MC was found to be 2.3 kg/hd/d. Its heating value was found to be 11.6 MJ/kg. The electrical power consumption for this facility is about 750 kWh/yr/hd. If the biochar is sold for $100/MT, how much is the revenue from the biochar each year?
    a. $4,200
    b. $13,340
    c. $8,400
    d. $6,700

Answers: 1b, 2d, 3a, 4c, 5c, 6a, 7b, 8d, 9a, 10c, 11a, 12b, 13c, 14c, 15c, 16d, 17a, 18b, 19b, 20d

ADDITIONAL LEARNING RESOURCES

- Access the Lecture Notes at the Instructor and Student Resources: https://routledgetextbooks.com/textbooks/instructor_downloads/

Appendix A
Appropriate Units and Conversion

Throughout this textbook, units are frequently expressed in *Système International d'Unités* (SI) units, commonly referred to as metric system units. When appropriate, SI units are used with corresponding conversions into the English system. One should be familiar with converting from SI to English and vice versa. There are a few constants to remember when converting from one system of units to another. We will begin by enumerating several vital conversion constants from units of length to units of energy. The use of prefixes is also important because some unit numbers may be large for a given application and must be converted into its proper prefix. For example, in Chapter 1, we have reported energy resource values in exajoules (EJ), which is 1×10^{18} joules (J), because a single joule is such a relatively small unit of energy. Thus, the common prefixes are listed below:

- 1,000 (1×10^3) = kilo abbreviated as lowercase k
- 1,000,000 (1×10^6) = mega (M)
- 1,000,000,000 (1×10^9) = giga (G)
- 1,000,000,000,000 (1×10^{12}) = tera (T)
- 1,000,000,000,000,000 (1×10^{15}) = peta (P)
- 1,000,000,000,000,000,000 (1×10^{18}) = exa (E)

A.1 UNITS OF LENGTH

The most common conversion process performed from SI to English involves the units of length. In SI, the most ubiquitous length unit is meters (m) for small systems; however, for higher distances, the use of prefixes is also necessary. Thus, the most common constant to remember is listed as follows:

- 1 m = 3.28 feet (ft)
- 2.54 cm = 1 inch (in)
- 1.609 km = 1 mile (mi)

A.2 UNITS OF AREA

In the SI system, the most common area unit is the hectare (ha), which is equivalent to 2.47 acres. Perhaps this is the only conversion constant to remember in shifting one unit area to another. One can easily remember the order of magnitude of this area value by imagining that a hectare is an equilateral area having 100 m on each side. Thus, 1 ha is also equivalent to 10,000 m^2. These constants are summarized as follows:

- 1 ha = 2.47 acres
- 1 ha = 10,000 m^2

A.3 UNITS OF VOLUME

Units of volume are important when it comes to transporting fuels from a biorefinery to a commercial refinery. Most commercial refineries measure volume in terms of barrels, whereas those in the biorefinery business measure in terms of gallons. The difference between the two is about 42 (or 42 US gallons [gal] is 1 barrel). For some reason, the SI system of units does not have a clearly identified unit for volume on a larger scale, even though the unit of cubic meters is always used as well as liters (L). Thus, the constants to remember are summarized as follows:

- 1 barrel = 42 US gal
- 3.785 L = 1 gal
- 1,000 L = 1 m^3
- 1 ft^3 = 7.48 gal

A.4 UNITS OF DENSITY

Units of density are important in biomass energy conversions because biomass is such a low energy density material, coupled with a low bulk density feedstock. Density is defined as the unit of mass per unit of volume (kg/m^3) and is an important consideration when transporting biomass. One would want to have a biomass material with a very high density, such that transport costs are minimized. Density is a derived unit of mass and volume, and thus, there are no important constants to remember. A normal biomass density is around 15 kg/m^3, whereas a pelleted biomass will have a density of around 45 kg/m^3. These values are important in the transport of biomass and logistics.

A.5 UNITS OF MASS OR WEIGHT

It is easier to use the metric system for mass because the units of mass in the English system are in pound-mass (lb_m), or slugs, and the units of weight are in pound-force (lb_f). In English, lb_f is perhaps one of the most abused units of weight. Units of lb_f are sometimes confused with units of mass. In the SI system, it is quite easy to work with mass because the unit of mass is the kilogram, while the unit of weight is the newton, as defined in Newton's law of motion equation (see Equation A1.1) along with corresponding units. Thus, it is seen that the difference between a unit of mass and a unit of weight in the SI system is due to the acceleration of gravity. The third term in Equation A1.1 is a useful unity value to use in some conversion systems.

$$F = ma \quad N = kg \times \frac{m}{s^2} \quad 1 = \frac{kg - m}{N - s^2} \tag{A1.1}$$

The most useful constants to remember throughout this textbook are listed below. Most other SI-to-English system units of conversion of mass are these constants.

- 1 kg = 2.2 lb
- 1,000 kg = 1 metric ton (or simply 1 tonne)
- 1 ton = 2,000 lb
- 1,000 g = 1 kg

A.6 UNITS OF ENERGY

Energy is the ability to do work. It is also synonymous with the inherent property of a fuel and reported in units of joules. In this textbook, we refrain from using calories for energy unit since that term is associated with food intake, and not at all with biomass energy systems. For example, 1 L of milk would normally contain around 650 calories (cal); however, we will not report the energy contained in 1 liter of gasoline as 1,119,050 cal. One would simply report it as 47 MJ. One other important energy unit is kilowatt-hour (kWh), which is associated with electrical energy consumption in a household. A list of the most common constants to remember in converting one energy unit to another is as follows:

- 1,055 J = 1 Btu
- 1 kWh = 3.6 MJ = 3,412 Btu
- 1 J = 1 Newton-meter (Nm)
- 1 kcal = 4.1868 J
- 1 J = 1 kg × (m/s^2)

A.7 UNITS OF POWER

Power is the rate of doing work. Power is also the rate of using the energy contained in a fuel. The common SI unit of power is watts, which is equal to joules per second. The equivalent units in the English system are horsepower (hp) and Btu/h. Thus, the most important constants to remember are as follows:

- 746 W = 1 hp
- 1 hp = 33,000 ft-lb/min
- 1 W = 1 volt (V) × 1 ampere (A)

The reader should be aware that in most places outside of the United States, people regularly use the metric system, or the SI system of units, and thus, one would have to remember some common constants to convert one unit to another. It is necessary to remember that a mile is longer than a kilometer and that a hectare is much larger than an acre.

In summary, as a means of comparing biomass resources with conventional fuels, the relative energy levels of some conventional fuels are as follows:

- 1 barrel of crude oil (159 L or 42 gal) = 6,199 MJ (5,800,000 Btu)
- 3.785 L of gasoline (1 gal) = 131 MJ (124,000 Btu)
- 3.785 L of diesel (1 gal) = 147 MJ (139,000 Btu)
- 3.785 L of propane (1 gal) = 96 MJ (91,000 Btu)
- 28 L of natural gas (1 ft^3) = 1.1 MJ (1,026 Btu)
- 909 kg of coal (1 short ton or 2,000 lb) = 21,818 MJ (20,681,000 Btu)

Note that 1,000 kg (1 long ton or 1 tonne) of biomass will have an energy content of about 20,000 MJ, which is equivalent to over 3 barrels of crude oil, 578 L of gasoline, 515 L of diesel, 788 L of propane gas, and about one-half million liters of natural gas. However, when compared with coal, pine bark biomass has one-half the energy content, and thus requires twice as much material to equal the energy content of 1 tonne of coal.

Appendix B
Glossary of Terms in Biomass Conversion Processes

acetogenesis: The process through which acetate is produced by anaerobic bacteria from a variety of energy and carbon sources.

acidogenesis: Converting higher organics into simpler organics.

actuator: A mechanism that puts something into automatic action.

aerobic: Involving or requiring free oxygen.

air-gas mixture: A mixture of oxygen in air (together with nitrogen) and synthesis or any combustible gas.

alpha-amylase: An enzyme that acts specifically to accelerate the hydrolysis of starch to dextrins.

amylase: Any of several enzymes that accelerate conversion of starch to sugar.

anaerobic: Without air or in the absence of oxygen.

anaerobic digestion: A series of processes in which microorganisms break down biodegradable material in the absence of oxygen.

anhydrous: Free from water.

anhydrous alcohol (ethanol or methanol): Water-free alcohol, essentially 200 proof.

ash: The inorganic, noncombustible residue left after complete combustion of a material (at a minimum temperature of 700°C or 1,292°F [ASTM D3174]).

ash fusion test: A procedure used to determine the softening and melting behavior of ash. Ash of a material is formed into a cone and heated under prescribed conditions in a testing laboratory (ASTM D1857). The temperature points that are observed are the initial deformation temperature, softening temperature, hemispherical temperature, and fluid temperature. The higher the fusion temperatures of the ash of a material are, the lower the tendency of that material to slag in a combustion chamber.

auger: A conveying device used to move biomass from a source to a reactor or conversion system.

azeotrope: A constant boiling mixture whose vapor has the same composition as the liquid from which it came.

azeotropic distillation: Separation of the components of an azeotrope after another component has been added to change the characteristic of the azeotrope.

bedding: Materials used in many animal-feeding operations, whereby urine and animal excretions are deposited.

beer: The fermented mash.

bifuel system: A system that can operate on a combination of alternative and conventional fuel.

biobased product (biobased industrial product, bioproduct): Fuels, food, feed, chemicals, or industrial materials (e.g., pharmaceuticals, building materials, inks, coatings, or paper) commercially produced in whole or in part from biomass materials.

biochar: Charcoal created by pyrolysis of biomass.

biodiesel: Mono-alkyl esters of fatty acids derived from vegetable oils or animal fats when chemically reacted with an alcohol.

bioenergy: Energy derived from biomass. This includes biopower (see definition below), energy from biobased transportation fuels, and energy from biomass that is used for process or space heating.

biofuel: Fuel derived directly from living matter.

biogas: A gas containing methane and carbon dioxide produced from anaerobic digestion of animal manure.

biomass: All organic matter except fossil fuels. Specifically, organic materials that are plant or animal based, including but not limited to dedicated energy crops, agricultural crops and trees, food, feed and fiber crop residues, aquatic plants, forestry and wood residues, agricultural wastes, biobased segments of industrial and municipal wastes, processing by-products, and other non-fossil organic materials. 1 (see definitions below).

bio-oil: A product of pyrolysis of biomass. It is a dark brown, water-miscible liquid that contains oxygenated organic compounds.

biopower: Electricity or industrial process heat or steam generated from biomass or intermediate bioproducts.

bioproduct: See biobased product.

biorefinery: A facility that uses mechanical, thermal, chemical, and/or biochemical processes to convert biomass into value-added biobased products or key intermediates for the production of chemicals and other materials.

biosolids: Nutrient-rich organic material resulting from the treatment of wastewater and from anaerobic digestion of animal manures (see Section 3.61 for additional information on animal manures).

black liquor: Solution of lignin residue and the pulping chemicals used to extract lignin during the manufacture of paper.

briquettes: Biomass that is densified and molded into a distinguished three-dimensional form for ease of transport, storage, and use.

bushel: Volumetric measurement of grains and some agricultural product. It varies with grains such that a bushel of corn and milo weighs 56 lb, while this is also equivalent to 60 lb for soybean and wheat. A bushel of sunflower is only about 28 lb.

by-product: A secondary or incidental product that is derived from a manufacturing process or chemical reaction and is not the primary product or service being produced.

calorific: Pertaining to the amount of energy contained in a material usually measured in calories or equivalent energy units.

calorimeter: A device used to measure the energy content or heating value of combustible and organic material.

carbonization: A process similar in nature to pyrolysis, but which uses lower temperature and longer residence time in the reactor. The resulting products include approximately equal amounts of liquid, tar, and gas products.

catalyst: A substance or chemical that aids in the change in the chemical reaction without being consumed during the reaction.

catalytic: Involving a catalyst or acting as a catalyst.

cellulose: Linear polysaccharide consisting of several hundreds or thousands of linked D-glucose units.

cetane number (cetane rating): A measure of a fuel's ease of self-ignition. The higher the number, the shorter the ignition delay period and the better the fuel for a diesel engine.

char: A residue resulting from pyrolysis, carbonization, and gasification of biomass.

charcoal: Impure form of carbon, obtained as a residue when carbon-rich feedstock is partially burned or heated with limited access to air.

chemical oxygen demand (COD): A measure of the amount of oxygen required to oxidize the organic matter present in the sample.

closed-loop biomass: Primary biomass grown in a sustainable manner for the sole purpose of optimizing its value for bioenergy and bioproduct uses. This includes annual crops such as maize and wheat, and perennial crops such as trees, shrubs, and grasses such as switchgrass.

co-combustion: Combustion of biomass or combustible material together with a base fuel. The base fuel is usually a fossil fuel such as coal, natural gas, or diesel. Simultaneous combustion of two or more fuels in the same plant for energy production.

co-firing: The simultaneous use of two or more different fuels in the same combustion chamber of a power plant.

cogeneration system: The simultaneous production of both electricity and useful heat.

coke: The solid residue of impure carbon obtained from carbon-rich feedstocks after removal of volatile material by destructive distillation.

combustible: Able to catch fire or burn; flammable; or substance that ignites and burns readily.

combustion: Thermal conversion of a carbon-rich feedstock with an oxidant (excess air) to produce primarily heat energy, carbon dioxide, water, and ash.

coproduct: A secondary product with commercial value that is intentionally produced incidental to the manufacturing process or chemical reaction, and is not the primary product or service being produced.

corn stover: Biomass material from corn plant that consists of leaves and stalks. Corn stover makes up about half of the yield of corn plant biomass.

dedicated energy crops: Annual crops (such as maize) or perennial crops (such as trees and grasses) when grown specifically as feedstock for conversion to bioenergy and biobased products.

dehydrogenation: A chemical reaction that involves the removal of hydrogen.

denature: Make alcohol unfit to drink.

density, bulk: Mass per unit volume occupied by a large quantity of particulate material.

density, particle: Mass per unit volume of individual particles (including pores within the particle).

density, solid: Mass per unit volume of solid material excluding any interior pores.

devolatilization: A process (usually pyrolysis or gasification) whereby volatile materials are removed from carbon-rich feedstock.

dextrins: High molecular weight sugars; intermediates obtained in the conversion of starch to fermentable sugar.

dynamometer: A device used to measure internal combustion engine performance, such as brake horsepower, torque, and specific fuel consumption.

elemental analysis: Quantification of the basic compositional analysis of biomass and comprises measurement of carbon, hydrogen, oxygen, nitrogen, sulfur, and inorganic ash contents.

energy: The capacity to do mechanical work.

energy crops: Crop grown and harvested to be an energy feedstock.

energy ratio: Energy content of output (biomass or fuel) divided by the input energy.

enzyme: A complex substance that speeds up a reaction, usually in a given temperature range.

enzymatic hydrolysis: Hydrolysis through the action of enzymes (hydrolysis is defined below).

exothermic: A process that produces heat.

feedstock: A material that is further processed for conversion to bioenergy and/or biobased products.

fermentation: A process involving chemical changes in an organic substrate through the action of enzymes that are produced by microorganisms.

flash point: The temperature at which a combustible liquid ignites when flame is introduced.

fluid ash fusion temperature: The temperature at which the fused mass of ash has spread out in a nearly flat layer with a maximum height of 1.6 mm (0.063 in)—ASTM D1857.

fossil fuel: Fuels such as oil, natural gas, and coal formed in the ground over millions of years by chemical and physical changes in plant and animal residues under high temperature and pressure.

fouling: The accumulation of unwanted material on solid conveying surfaces to make the function inoperable or detrimental. In thermal conversion, these unwanted materials are the melted slag from inorganic ash components.

fuelwood: Wood material used for fuel and includes any wood or wood products that would comprise branch, stems, or trunks, and used primarily for heating or cooking.

gasification: Conversion of biomass into a combustible gas. Thermochemical conversion, typically occurring at 750°C–850°C (1,382°F–1,562°F) in an oxygen-deficient environment, of carbon-rich feedstocks into a gaseous fuel known as producer gas.

gelatinize: To make or become gelatinous or jellylike.

glucoamylase: An enzyme that acts specifically to convert, by hydrolysis, dextrins to glucose.

greenhouse gases: Gaseous constituents of the atmosphere, which trap the heat of the sun in the earth's atmosphere, producing what is commonly known as the greenhouse effect. Examples are water vapor, nitrous oxide, methane, carbon dioxide, hydrofluorocarbons (HFCs), perfluorocarbons (PFCs), and sulfur hexafluoride (SF6).

greenhouse heating: The heating effect due to trapping of the sun's radiant energy by greenhouse gases in the earth's atmosphere.

grindability: The ability of biomass to be reduced in size by physical or mechanical action.

hemispherical ash fusion temperature: The temperature at which the cone of ash has fused down to a hemispherical lump at which point the height is one-half of the width of the base.

higher heating value (or HHV): The full energy content of a fuel. It is the amount of heat produced when a fuel is fully combusted, all of the products of combustion are cooled to 25° (77°F), and the water vapor formed during combustion is condensed into liquid water.

hopper (or feed hopper): A funnel-shaped bin used to store biomass materials.

HPLC: High-performance liquid chromatography; sometimes also referred to as high-pressure liquid chromatography.

hydrolysis: A decomposition process that involves splitting chemical bonds of substances and the addition of the hydrogen cation and the hydroxyl anion of water or simply converting insoluble complex biomass into soluble compounds.

hydrophobic: No affinity or attraction with water.

initial deformation ash fusion temperature: The temperature at which the point of the ash cone begins to round (see ash fusion test).

knocking: An abnormal combustion of the air and fuel mixture outside of the flame front of the air and fuel mixture being combusted in the engine cylinder. This event results in an irregular knocking sound from which the name was derived.

landfill: A site for disposal of waste material. These materials are usually from domestic or commercial wastes, such as kitchen wastes, yard wastes, and other household-disposed items.

lignocellulose: Plant cell wall biomass composed primarily of cellulose, hemicelluloses, and lignin.

liquefaction: Thermochemical conversion process of carbon-rich feedstocks into liquid pyrolysis oil and coproducts. Liquefaction is usually conducted in an environment of moderate temperatures (300–400°C or 572–752°F) and pressures.

low-energy producer gas or low calorific value gas (LCV): Results from air-blown gasifiers in which the gas is diluted by nitrogen from the air to an energy density by volume about 1/5 that of natural gas (4–11 MJ/m^3).

lower heating value (LHV): The amount of heat produced when a fuel is fully combusted, all of the products of combustion are cooled to 25°C (77°F), and the water vapor formed during combustion is still in vapor form. It can be obtained by subtracting the latent heat of vaporization of water from the higher heating value.

mash: A form that enables conversion of starches to sugars by enzymes.

mesophillic bacteria: Methane-forming bacteria that function at temperatures near 35°C.

methanogenesis: Converting organic acid and hydrogen into methane.

miscible: Capable of being mixed at any ratio without separating into two phases.

moisture content: Amount of water contained in a material expressed on either a wet weight basis (wb) or a dry weight basis (db). The reported value must contain either the wb or db identification.

monosaccharide: A sugar that cannot be decomposed to simpler sugars through hydrolysis.

municipal solid waste (MSW): A waste stream consisting of postconsumer materials.

nitrogen purge: Nitrogen gas is used to evacuate a container with other gases such as air or oxygen to ensure that the container is oxygen-free.

nitrogen-fixing: Microorganisms that use nitrogen from the air and utilize this for their metabolic pathway and thereby make this nitrogen available for the plant uptake.

octane number: A measure of the antiknock properties of a spark ignition fuel.

oligomerize: A chemical process that converts simple compounds such as monomers into higher molecular weight compounds or polymers.

open-loop biomass: Biomass that can be used to produce energy and bioproducts even though it was not grown specifically for this purpose. Examples of open-loop biomass include agricultural livestock waste and residues from forest- and crop-harvesting operations.

oxidant: A substance that oxidizes another substance. In biomass thermal conversion, this substance refer to oxygen in air or pure oxygen that is used to convert carbonaceous compounds into its oxidized form.

phase separation: A result of water contamination; occurs when water and part of the alcohol bleed out of the alcohol-gasoline blend and settle to the bottom of the gas tank.

photosynthesis: A process used by plants, including algae, and some other organisms to convert light energy into chemical energy with the use of sunlight and water to produce biomass for energy and fuel.

plenum: A chamber that contains pressurized air or gas that is used to transfer heat energy for drying materials above the chamber or for fluidizing bed materials in fluidized bed systems.

polysaccharide: A carbohydrate that can be decomposed by hydrolysis to monosaccharides.

pressure: Force divided by area.

primary biomass: Biomass produced directly by photosynthesis and harvested or collected from the field or forest where it is grown. Examples are grains, perennial grasses and wood crops, crop residues, and residues from logging and forest operations.

producer gas: Product of biomass gasification consisting of carbon monoxide, hydrogen, carbon dioxide, methane, and nitrogen, which can be burned for heat or power or converted to liquid fuels and chemicals.

proof: Twice the percentage (by volume) of alcohol concentration; for example, 70% ethanol is 140 proof.

proximate analysis: Determination of moisture, volatile matter, and ash, and the calculation of fixed carbon in a material.

psychrophilic bacteria: Methane-forming bacteria that function at temperatures lower than 20°C.

pyrolysis: A chemical change brought about by the action of heat in the complete absence of an oxidant.

pyrolyze: The act of pyrolysis.

quad: 1 Quadrillion Btu (10^{15} Btu or 1.05×10^{15}kJ or GGJ). One quad is the energy equivalent of 172 million barrels ($27.35 \times 10^6 m^3$) of crude oil.

reactor: Vessel or equipment used in the conversion of feedstocks.

reflux: A liquid stream with a high alcohol concentration returned to the top of the distillation column.

renewable energy: Energy derived from a natural, managed, or cultivated resource that can be replaced as it is used. Examples are wind, solar, hydro, and biomass.

residue: Material that remains after a portion of biomass is removed by a process.

rumen: A chamber in the stomach of ruminant animals where microbial fermentation of ingested animal feed occurs. It is also known as paunch and is part of several other chambers for digestion of animal food.

ruminant: A mammal that digests plant-based food and converts it into useful energy. Important ruminants include cattle, goat, sheep, water buffalo, deer, camel, and antelope. Humans, dogs, and swine are non-ruminant and cannot process grass or other cellulosic and hemicellulose fibers.

saccharification: The breaking of complex carbohydrates (starch) into simple sugars (hydrolysis).

secondary biomass: Residues and by-product streams from food, feed, fiber, wood and materials processing plants (such as sawdust, black liquor, and cheese whey), and manures from concentrated animal-feeding operations.

slag: Ash that is or has been in a molten (or liquid) state.

slagging: An act of forming and accumulation of the slag or molten inorganic materials.

solubilize: To make a substance soluble or dissolvable.

sludge: Solids remaining after the digestion process of organics. It could also mean the solid materials remaining after wastewater treatment processes.

softening ash fusion temperature: Temperature at which the cone of ash has fused down to a spherical lump in which the height is equal to the width at the base (see ash fusion test).

solvent: Liquid substance that dissolves or disperses another substance.

soot: The fine black particles, chiefly composed of carbon, produced by incomplete combustion of carbon-rich feedstock.

stoichiometric: The calculation of the quantities of reactants and products in a chemical reaction. In a balanced chemical reaction, the amount of elements in the reactant must equal that in the product in a molar or mass basis.
sugar platform: A conversion approach that involves the breakdown of biomass into component sugars, which are then converted into products such as ethanol and other valuable fuels and chemicals.
syngas or synthesis gas: A synthetic gas composed of hydrogen and carbon monoxide, which is obtained by gasification of biomass and which can be transformed to fuels, products, power, and hydrogen.
tar: Dark brown or black bituminous (usually odorous) viscous liquid obtained by destructive distillation of carbon-rich feedstock.
terpenes: Large and diverse class of organic compounds found in various plants. These compounds have a very strong odor and may have some repellent properties with insects or other organisms.
tertiary biomass: Postconsumer residues and wastes, such as fats, greases, oils, construction and demolition wood debris, other waste wood from urban environments, as well as packaging wastes, municipal solid wastes and landfill gases.
thermochemical or syngas platform: A conversion approach that uses thermal processes, such as gasification and pyrolysis, to break down biomass into desired products or intermediate components, which are in turn converted into desired fuels and chemicals.
thermophilic bacteria: Methane formers that prefer temperatures near 40°C.
torrefaction: A mild thermal pretreatment of biomass under slightly elevated temperature from 200°C to 320°C without the presence of any oxidant.
total organic carbon (TOC): A means of measuring the organic matter content of biomass that is used for conversion when using anaerobic digestion processes.
transesterification: A chemical reaction of exchanging the alkoxy group of an ester by another alcohol. The products of a transesterification reaction are a new ester and a new alcohol. A catalyst is typically required for transesterification. An example is when an alcohol reacts with the triglycerides contained in vegetable oils and animal fats to produce biodiesel, with glycerin as a by-product.
ultimate analysis: See elemental analysis.
volatiles: Materials (such as vapors and gases) that vaporize from heated (e.g., pyrolysis) and non-heated (e.g., anaerobic digestion) biomass.
volatile solids: Materials available for conversion to biogas.
volatility: The state of readily vaporizing at relatively low temperatures.
wood fuel: See fuelwood.

Index

Note: **Bold** page numbers refer to tables; *italic* page numbers refer to figures and page numbers followed by "n" denote endnotes.

ablative pyrolysis 286
absorptivity, biomass **116,** 116–117
Accellerase 1500 194–197
Accellerase BG 194, 195
Accellerase XC 194–197
Accellerase XY 194, 195
acetogenesis 209, 211, **211,** 222
acetotrophic reaction 211
acid-insoluble lignin (AIL) percent 113, 114–115
acid-insoluble residue (AIR) percent 113–114
acidogenesis 209, 210, 222
acid-soluble lignin (ASL) 113
  percent 115–118
AED (aerodynamic equivalent diameter) 76–77
aerodynamic equivalent diameter (AED) 76–77
AFEX (ammonia fiber explosion) technique 187
A/F ratio *see* air-to-fuel (A/F) ratio
agricultural biomass 19, 84, 396, 405, 427, 428, 432, 433, 447
  slagging and fouling with 436–441
agricultural industry 373, 476
  crop residues 8–10
  future of 4–6
AIR *see* acid-insoluble residue (AIR)
air–fuel mixer 379
air-to-fuel (A/F) ratio 322, 338, 365
  calculations, in fluidized bed gasifier 342–343
  combustion 344–345
  during combustion 422
  defined 322
alkali index 438–440, **439**
alkaline hydrolysis 187
alligator grass *(Alternanthera philoxeroides)* 26
Alternanthera philoxeroides (alligator grass) 26
American Oil Chemists Society 136
American Society for Agricultural and Biological Engineering (ASABE) 91
American Society for Testing and Materials (ASTM) standards 40, 62, 75, 78–80, **80,** 81, 82, 91, 94
  BETA Lab 158, **158**
  characterization 134–135, 152–153
  D 93 (for flash point) 153
  D 445 154
  D1857 (Standard Test Method for Fusibility of Coal and Coke Ash) 438
  D 2500 185
  D 6584 188
  D 6751-07 152
  E 11–09 107
  E 711 86
  E 873-82 75
  E 1690–08 109
  E 1721-01 113
  E 1757-01 (Standard Practice for Preparation of Biomass for Compositional Analysis) 107
  E 1758-01 119
  proximate analysis 79
  summary of 122, **123**
  ultimate analysis 84, **84**
American Society of Agricultural Engineering (ASAE) 91
American Society of Biological and Agricultural Engineers (ASABE) 91, 257
  protocol 481–483
ammonia fiber expansion (AFEX) processes 103
ammonia fiber explosion (AFEX) technique 187
ammonia recycle percolation (ARP) technique 187
amylase 180, 183
anaerobic digester (AD) 210, 217, 221, 225, 227, 229, 235, 484
  circular bioeconomy systems 484–485
anaerobic digestion process 209–217, *214*
  acetogenesis 209, 211, **211,** 222
  acidogenesis 209, 210, 222
  advantages 223
  biomass/waste parameters important in 217–221
    biochemical oxygen demand 217–218
    chemical oxygen demand 218–219
    septic tanks 221, **221**
    total organic carbon 219
    total solids 261–262
    volatile suspended solids 261–262
  disadvantages 223
  hydrolysis 209–210
  methane production by 216–217
  methanogenesis 250, 252–257
  schematic representation *210,* 222, *222,* 249
  single-stage, conditions for *228,* 228–229
anaerobic filters 229–230
angle of friction 78

angle of repose 77–78
animal fat
  characteristics 136–137
  fatty acids in 137, **138**
  production of 135, **136**
animal manure, as source of biomass 17–19, **18,** 58, 216–217, 443
  combustion of 443
  production 230, **230**
ANL (Argonne National Laboratory) 467, *468*
aquatic plants 11, 12, 25–27
  for biomass energy systems development **26**
Aquatic Species Program 11
AR *see* auger reactor (AR)
Arachis hypogaea 148
area, units of 509–510
Argonne National Laboratory (ANL) 467, *468*
Arundo donax 25
ash
  coal 89
  composition of cotton gin waste and sorghum stalks 90, **90**
  of extractives-free biomass 114
  fusibility of **92–93**
  fusion temperature of 88
  measurement, proximate analyses and 80–83
  pellets
    compressive strength of 443–444
    dimensions of 442
    melting point 441–444
ASL *see* acid-soluble lignin (ASL)
Aspergillus awamori 173
Aspergillus niger 12, 42, 173
ASTM *see* American Society for Testing and Materials (ASTM) standards
atmospheric correction factor 192, **192**
auger-based pyrolyzer 48–50
auger reactor (AR) 283
  calibration curve for 284–285
  residence time in, estimation of 283

B100 39, 62
bagasse 249, 285, 445
BAI *see* bed agglomeration index (BAI)
barrel 28, 135
base-to-acid ratio for 440–441
batch atmospheric process 146
batch vacuum process 146
B20, blending terminology 39
  GREET model results for 468
Beckman Coulter Counter Multisizer 3 76
bed agglomeration index (BAI) 441
bed material
  average particle size of 358–360
  proper use of 370–371
benefit cost ratio (BCR) 162, 163, 168
benzene 178, 249, 251
BET analyzer *see* Brunauer, Emmet, and Teller (BET) analyzer
biochar 64–65, 295
  analysis 301–302, *302,* **303**
    results of 290, **290**
  applications of 47, 48
  circular bioeconomy systems 485–487
  production 64–65
  proximate analysis of 308, **308**
  yields 303–304, 307, 308, *308,* **309**
biochemical oxygen demand (BOD) 258–259
biodiesel 39–40, *40*; *see also* biofuels; diesel
  cloud point of 156
  conversion process 134
  economic parameters 162–164
    case study 167–168
  economics of making 470–473, *471–473*
  and engine performance 159–160
  flash point of 153–154
  industry, growth of 133, *134*
  kinematic viscosity of 154–156
  plants, design of 160–164
    Crown Iron Works 160, *161*
    hydrotreated vegetable oil (HVO) 161–162
    transesterification process 160
    Westfalia 161, *162*
  production 133–164, *134*
    oil crops and 13–17, **15, 15–16,** 23–24
    statistics 133, *134*
  samples, analysis of 158, **158**
  total and free glycerin in **157,** 157–158
  viscosity 139
  water and sediment 156–157
biodiesel blend (B20) 39, 62, 159, *159*
  GREET model results for 468
BioEnergy Testing and Analysis Laboratory (Beta Lab) 79, **79,** 87, 124
biofuels; *see also specific fuels*
  components of 379
  defined 378–379
  dispensing of 462–463
  distribution and storage 462
  economic 470–477
    components of analysis 470, **470**
  end use of 463–466
  models
    GREET/GREET1 model 467–469, *468,* **469**
  production of 462
  sustainability of 477–481
biogas
  applications/uses of 234–235
  composition of 215–216
  conversion process 224–225
  digesters (*see* digesters)
  energy content of 234
  production of 43–45, *44,* 209–235

anaerobic digestion process (*see* anaerobic digestion process)
microbes and 221–222, *222*
utilization 234–235
biological conversion process 41–45
biomass properties for 101–125
combining with thermal conversions 55–56
biomass
advantages 7
aquatic 11–12
carbohydrates 113, 119–122
combustion (*see* combustion)
components 103, **103**
compositional analysis of 73, *74,* 101–102, 249, **249**
conversion process (*see* conversion process, biomass)
cookstoves (*see* cookstoves)
disadvantages 7–8
as energy source 1–31
ethanol production 174, *174*
gasification 63, **63** (*see* gasification)
gasifier (*see* gasifier, application of)
laboratory analytical procedures 104, **105**
liquefaction process (*see* liquefaction, biomass)
methods 91, **92**
for analysis 91–93, **92–93**
moisture content in, calculation of 6–7
particle size distribution 255
pelleted
torrefied biomass *vs.* 258–259
physical properties of
angle of friction 78
angle of repose 77–78
bulk density 75–76, **252,** 252–253
heat capacity 78
particle density 78
particle size distribution 76–77
thermal conductivity 78
plants 1
pretreatment 123–124, *124,* 185–189
advantages/disadvantages of torrefaction as 265–266
chemical 186–188, *188*
combination of 189
enzymatic 188, 195–198
physical 185–186
properties
for biological conversion (*see* biological conversion process)
for thermal conversion (*see* thermal conversion process)
pyrolysis (*see* pyrolysis)
resources 8–27, 36 (*see also specific resources*)
storage issues 256
sweet sorghum 196, *196*
system flow alternatives for 74, *74*
thermal conductivity 78
thermal gravimetric analysis studies of 259–260
torrefaction (*see* torrefaction)
torrefied
moisture content of 255–256
pelleted biomass *vs.* 258–259
physical properties of 252–258
ultimate analyses 84–85, **85,** 87, **87**
usage in other countries 4, **5**
woody 76
BioMax Power System 335
bio-oil 48, 297, *297*; *see also* biofuels
analysis results 290
characterization processes 297–299
corn stover, analysis for 297–298, **298,** 302, **303**
corn stover properties 305, **305**
mixing 463
production of 62–63
product of LBL process 405
raw 297
simple distillation of 298, **299**
switchgrass 300–304, *301,* **302**
upgrade processes 299
yield of 280–281, 300, 306–308, *308,* **309**
bio-physico-chemical changes, torrefaction 247–249, *248,* **249**
biorefineries 28, 265, 510
sizes of 28
bleaching 145–146
bluestem 25
BOD *see* biochemical oxygen demand (BOD)
BOM process *see* Bureau of Mines (BOM) process
Botryococcus brunii 24
Boudouard reaction 392
breakeven point (BEP) 162
Brunauer, Emmet, and Teller (BET) analyzer 255
bubbling FBC system 428
bulk density 75–76, **252,** 252–253, 365, **365**
Bureau of Mines (BOM) process 402–404, *403*
efficiency calculations 403–404

calcium oxide 179
calibration curve 101, 106, 119, **119,** 284, 364
for Augers 284–285
calibration verification standards (CVS), recovery 119
calorific value 263–264, *264*
Camelina 21, 22
Candida acidothermophilum 176
Candida shehatae 176
carbohydrates 113, 119–122

carbon capture and sequestration (CCS) 457–458, 487–489
carbon content, of raw biomass 262–263
carbon conversion efficiency 52
carbon dioxide, production of 51
carbon footprint (CF) 489–491, **490**
carbon intensity (CI) 491, **491–492**
carbon-to-nitrogen (C/N) ratio 213
Carthamus tinctorius L 148
cassava 180, 181
cattails *(Typha latifolia)* 11
CCD process *see* classic cyclone design (CCD) process
cellulose 103–104
cellulosic biomass 57, 175, 199, 400, 474
  conversion, into ethanol 173, 184–185
CGT *see* cotton gin trash (CGT)
charcoal *see* biochar
chemical conversion process 39–41
chemical oxygen demand (COD) 218–219
China Lake Process 399–402, *400*
  schematic diagram *400*
  vs. fast pyrolysis 399–400
CHP plants *see* combined heat and power (CHP) plants
CIDI (compression ignition-direct injection type) 468
circular bioeconomy systems (CBS)
  anaerobic digester (AD) 484–485
  benefits 483–487
  biochar 485–487
  milk production 483, *483*
circulating FBC system 428
classic cyclone design (CCD) process 373
ClearFuels Technology 475
Clostridium ljungdahlii 199
Clostridium strain P11 56
Clostridium thermocellum 174
cloud point 156
C/N ratio *see* carbon-to-nitrogen (C/N) ratio
coal
  conversion into electricity, LCA and *459*, 459–460
  slagging/fouling indices 437–438
COD *see* chemical oxygen demand (COD)
co-firing biomass 433–434
cogeneration systems 60
ComAp Systems, LLC 64
combined heat and power (CHP) plants 433, 446
combustion 45, 54–55, 421–448
  A/F ratio during 422
  applications of 444–446
    combined heat and power plants 446
    cooking and heating 445
    electrical power generation 445–446
    mechanical power generation 445
  biomass cookstoves 428–432, *430, 432*
  defined 421
  direct 421
  efficiency of 426
  equations 421
  F/A ratio of 322–323, 325
  FBC system 426–428, *427*
  greenhouse gas emissions of 466–467
  heat of 88
  of methane 88
  MSW, air needed for 425–426
  spreader stoker system *424*, 424–426
  zone 324
Community Power Corporation (CPC) 335
compositional analysis, biomass
  methods for 106–122
  preparation of samples for 107–108
  problem-solving roadmap 101, *102*
condensibles, liquid portion of 249
constant boiling mixtures 174
continuous vacuum process, bleaching 146
controlled cooking test 430
control panel, component of bifuel system 379
convection oven-drying 107
conventional pyrolysis 280, 283
conversion process, biomass 36–65
  applications of 57–65
    biochar production 64–65
    bio-oil production 62–63
    electrical energy 58–59
    heat energy 58
    mechanical energy 60–62
    synthesis gas production/usage 63–64
  biological *41*, 41–45
  categories of 36, 39 (*see also specific categories*)
  chemical 39–41, *40*
  efficiency of 37
  hybrid 55–57
  overview of 37–39, *38*
  thermal 45–55
    combustion 45, 54–55
    gasification 45, 50–53
    pyrolysis 45, 47–50
    torrefaction 45, 46–47
    types of 45–46, *46*
cookstoves 428–432
  design of *430, 432*
  efficiency calculations 430
  pellet 432
  versions of 428–429
corn
  conversion into ethanol processes of 182–183
  production of 473
corn oil, production of 135, **136**
corn stover 6, 9–10, 84, 85, 106, 109, 116, **298, 303,** 374, 375
  bio-oil properties 305, **305**

pyrolysis of 304–306
cotton gin trash (CGT) 300, **439,** 439–440
ash pellets 441–444
compressive strength of **442,** *443,* 443–444
base-to-acid ratio for 440–441
chemical formula for 323
gasification power calculations for 476–477
COVID pandemic 3
crops
extracting oils from 139–141
hemicellulose content in 104
herbaceous 24–25
hydrocarbon 23–24
oil 13–17, **15–16,** 23–24
residue, as source of biomass 8–10, **9**
sugar 12–13, **13, 14,** 23, **23**
yields (*see* crop yields)
crop yields
estimation, for agricultural crops **9**
oil crops **15, 15–16, 23**
sugar and starchy crops **13, 14,** 23, **23**
crossdraft gasifier 328–329
Crown Iron Works 471
biodiesel plant design 160, *161*
crude oils 131, 141
refining processes 141–150
bleaching 145–146
criteria for 148
degumming 142–143
deodorization 146–148
dewaxing 144–145
neutralization 143–144
CVS *see* calibration verification standards (CVS)
cyclones
design procedure 373
less efficient 373
pressure drop in 377
primary/low-efficiency 373–376
secondary/high-efficiency 376
cylindrical above-ground digesters 227

D 93 (Standard Test Method for Flash Point by Pensky–Martens Closed Cup Tester) 153
D 445 (Standard Test Method for Kinematic Viscosity of Transparent and Opaque Liquids [and Calculation of Dynamic Viscosity]) 154
D 1217 (Standard Test Method for Density and Relative Density [Specific Gravity] of Liquids by Bingham Pycnometer) 155
D1857 (Standard Test Method for Fusibility of Coal and Coke Ash) 438
D 2500 (Standard Test Method for Cloud Point of Petroleum Products) 185
D2709 (Standard Test Method for Water and Sediment in Middle Distillate Fuels by Centrifuge) 157
D 6584 (Standard Test Method for Determination of Free and Total Glycerin in B100 Biodiesel Methyl Ester by Gas Chromatography) 188
D 6751-07 (Standard Specification for Biodiesel Fuel Blend Stock [B100] for Middle Distillate Fuels) 152
DDG *see* distillers' grain (DDG); dry distiller's grain (DDG)
deforestation 421
degree of carbonization 264
degumming 142–143
dehydration, ethanol 178–180
density
bulk 75–76, **252,** 252–253, 365, **365**
of ethanol 175
oil 139
particles 75
units of 510
deodorization 146–148
destructive distillation *see* pyrolysis
Desulfomarculum 221
Desulfovibrio 221
dewaxing 144–145
calculations 145
diesel; *see also* biodiesel
properties of 139
digesters, biogas
cylindrical above-ground 227
design 230–233
dome-type 225–226
Indian gobar gas plant 226
minimum size 232
second-generation 227–229
Taiwanese Rectangular Digester Design 226–227
dimethyl ether, production of 394
direct liquefaction process 402–407
advantages of 410
Bureau of Mines 402–404
disadvantages of 411
Laurence Berkeley Laboratory (LBL) process 404–407
Pittsburgh Energy Research Center 402–404
disaccharides 180; *see also* sugars
distillers' grain (DDG) 182, 183
dome-type digesters 225–226
downdraft gasifier 328, *328, 329,* 336, *336*
drop-in fuels 400
dry distiller's grain (DDG) 475
dry milling process 182, *182*
duckweed *(Lemma)* 11
durability, measurement of 257–258
dynamic viscosity, calculation 155

E 711 (Standard Test Method for Gross Calorific Value of Refuse- Derived Fuel by Bomb Calorimeter) 86
E 873-82 (Standard Test Method for Bulk Density of Densified Particulate Biomass Fuels) 75
E 1721-01 (Standard Test Method for Determination of Acid- Insoluble Residue in Biomass) 113
edible fats, fatty acid composition of 137, **138–139**
edible tallow, production of 135, **136**
E10, ethanol fuel 62
  GREET model results for 468
efficiency, of biomass conversion 37
Eichornia crassipes (water hyacinth) 11, 25, 26
electrical energy 58–59
electric power
  combining with heat 60
  conversion of biomass into 435–436
  generation
    biogas and 234
    biomass combustion process for 421, *423,* 445–446
    case studies, in United States 434–436, **435**
    FBG and 378–380
    gasifiers and 335–336
  units of 511–512
elephant grass 25
endothermic process, defined 392
energy
  balance in pyrolysis process 278–280
  biomass as source of 1–31
  consumption, in United States 1–4, **3, 4,** 10
  densification 246–247
    advantages 257
  electrical 58–59
  heat 58
  production, energy associated with 478, **478**
  ratios 136
  renewable 3
  sources of 2
  units of 511
  wood 20
energy farming 8, 20–27
  fast-growing trees 21–22
  herbaceous crops 24–25
  hydrocarbon crops 23–24
  oil crops 23–24
  sugar/starchy crops 23
Engine Power Test Code—Spark Ignition and Compression Ignition—Net Power Rating 159
engines, biodiesel fuel, testing of 159–160
Enterobacter 222
equivalence ratio 365
  defined 322–323
equivalent spherical diameter (ESD) 76, 77
Escherichia coli 176, 222
ESD *see* equivalent spherical diameter (ESD)
ethanol 42, 62
  cellulosic
    commercial plants 198
    studies of 192–198
  combustion of 179–180
  conversion of cellulosic biomass into 173, 184–185
  dehydration methods 178–180
    multieffect distillation 179
    solvent extraction 178–179
    tertiary azeotropic distillation 178
    vacuum distillation 178
  density of 175
  economics for producing 473–474
  extraction procedure 109–110
  fermentation **42,** 42–43
  production
    enzymes 197, *197*
    overview 173–175
    pathways 174, *174*
    pilot-scale *189,* 189–192, *190*
    raw materials used in 42
    from starchy crops 180–184
    from sugars 173
    from sweet sorghum 176, *177*
  sorghum grain conversion into 183
  water and 174, 175, 178–180
  yields of 12–13, **13**
ethanol blend (E10), GREET model results for 468
Eucalyptus grandis, bulk density of 249, 252, **252,** 253
Eucalyptus saligna 249
Euphorbia 23
Euphorbiaceae species 24
eutectic point analysis 88–91
extractables, as biomass content 104
extractives-free biomass
  ash content of 113–114
  ODW of 112
  percent ASL calculation basis on 116
  percent total solids of 110–111

F/A ratio *see* fuel-to-air (F/A) ratio
fast/flash pyrolysis 280, 284–285
  applications of products from 295–297
fast-growing trees 21–22; *see also specific types*
  criteria for energy purpose 21–22
fats
  animal (*see* animal fat)
  edible **138–139**
  fatty acids in 137, **138**
fatty acid

categories 137
composition of 136, 137–139, **138–139**
free 143–144
list of 137, **138**
FBC system *see* fluidized bed combustion (FBC) system
FBG *see* fluidized bed gasifier (FBG)
FC *see* fixed carbon (FC)
feedstock production, as component of LCA 461
feedstock transport, as component of LCA 461
fermentation 176
ethanol **42,** 42–43
sweet sorghum juice 193, *193, 194*
of synthesis gas 56
FFAs *see* free fatty acids (FFAs)
Fischer-Tropsch (FT) technology 337, 396–399, *397,* 475
5-day BOD (BOD_5) 258
fixed-bed gasifier 328–329
turndown ratio 338–339
fixed carbon (FC), measurement, proximate analyses 79, 82–83
flash point, of biodiesel 153–154
defined 153
flash vacuum pyrolysis 285–286
fluid catalytic cracking (FCC) process 409
fluidization 356
cold tests 371–372
fluidized bed combustion (FBC) system 423, 426–428, *427*
advantage of 428
bubbling 428
circulating 428
zones in 428, *429*
fluidized bed gasifier (FBG) 329–333, *330, 331,* **331**
advantages of 330, 340, 355
A/F ratio calculations in 342–343, *343*
design of
particle size of bed material 358–360
dimensions of, designing 372–373
and electrical power generation 378–380
gas distribution systems for 356, 358, *358*
operational protocol 367–370
pilot-scale 359, **359**
quality of 356, *357*
synthesis gas and 378
terminal velocity 362
turndown ratio 338–339
use of 333
volumetric flow rates 363–364
0.3048, operation of 364–372
fluidized bed reactor 282, 356
fluidizing media in 357
residence time in 282
fouling 54, 58, 73, 158, 428, 445–447
with agricultural biomass 436–441
factor 88–91, *89, 90*
fourier transform infrared (FTIR) 297, 304
free fatty acids (FFAs) 141, 143–144
removal of, deodorization and 146
freeze-drying 107
FTIR (Fourier transform infrared) 297
FT technology *see* Fischer-Tropsch (FT) technology
fuel cells 234
fuels; *see also* biodiesel; ethanol
B100 description 39
fuel-to-air (F/A) ratio 322–323
of biomass combustion 325
fuelwood, as source of biomass 10–11, **11,** 22
for combustion system 54–55
fusion temperature, of ash 88

gas chromatograph (GC) 137, 157
with mass spectrometer 297
operating conditions for 157, **157**
gasification 45, 50–53, 321–345, *342, 343*
biomass 63, **63**
chemistry of 323–326
economics 475–477
effect of steam in 63
experiments, biomass properties used in 366, **367**
overview 321–323
power calculations for CGT 476–477
vs. pyrolysis 322
studies *339,* 339–345
temperature 342
gasifiers 50–51, *51, 52,* 63
applications of 333–337
electrical power production 335–336
heat production 334, *334*
mechanical power generation 336–337
crossdraft 328–329
development of 333
downdraft 328, *328, 329,* 336, *336*
fixed-bed 328–329
fluidized bed 329–333, *330,* 331, **331,** 370, **371**
advantages of 330, 340, 355
A/F ratio calculations in 342–343, *343*
use of 333
operation of 0.3048 fluidized bed 364–372
preparation 368
stove 334
superficial velocity in 332–333
throughput 338
turndown ratio 337–339
types of 326–333
updraft 326–328, *327*
zones of 323–324, *324*
gas power valve 379
gas-to-liquid (GTL) process 390

gas train 379
GC *see* gas chromatograph (GC)
General Electric (GE) 59
GHG *see* greenhouse gas (GHG)
global warming potential (GWP) 463
glycerin, in biodiesel 157–158
gobar gas plant, Indian 226
grass
  alligator 26
  species for biofuel production 25
  switchgrass 25
greenhouse gas (GHG) 459
  balance 458
  calculations 463
  emissions of biomass combustion 466–467
  equivalency calculations 464–465
  gas turbine 465–466
Greenhouse Gases, Regulated Emissions, and Energy Use in Transportation (GREET) model 463, 467–469, **469**
  stages of 467, *468*
GREET model *see* Greenhouse Gases, Regulated Emissions, and Energy Use in Transportation (GREET) model
grindability 254
GTL (gas-to-liquid) process 390
GWP (global warming potential) 463

H/C ratio *see* hydrogen to carbon (H/C) ratio
HDO *see* hydrodeoxygenation (HDO)
hearth zone 324, 326
heat
  capacity 78
  combination with power 60
  of combustion 88
  energy 58
  generation, biogas and 234
  thermal sensible, production of 334, *334*
heating rate, pyrolysis processes 282–286
heating value (HV) 139, 230, **231,** 279, 302, **303,** 307, **307**
  analysis 86–91
    heat of combustion 88
    stoichiometry of reactions 88
  moisture or ash 287–288
  synthesis gas 290–291, **291,** 292–293, 391
Helianthus annus 148
hemicellulose 103, 104, 184, 249
herbaceous crops 24–25
HHV (high heating value) 79, 86, 87
high-efficiency cyclones 376
high heating value (HHV) 79, 86, 87
high-performance liquid chromatography (HPLC) 125, 192–193
  sugar analysis 119, **119**
high-rate biogas digesters 227–229
  anaerobic filters 229–230
high-temperature FT process (HTFT) 396–398
HPLC *see* high-performance liquid chromatography (HPLC)
HRT *see* hydraulic retention time (HRT)
HTL *see* hydrothermal liquefaction (HTL)
HTU process *see* hydrothermal upgrading (HTU) process
hybrid conversion process 55–57
hybrid poplar 21
hybrid sorghum 23, 183
hydraulic retention time (HRT) 215
  laboratory batch digestion system 213, *213*
hydrocarbon crops 23–24
  examples of 24
hydrodeoxygenation (HDO) 409
hydrogen content, biomass 263
hydrogenotropic reaction 211
hydrogen to carbon (H/C) ratio 261, 264
hydrolysate 113
hydrolysis 209–210
  alkaline 187
  strong acid 187
  weak acid 186–187
hydrolytic process 409–410
hydrophobicity 246, 256–257
hydrothermal liquefaction (HTL) 406, *408,* 408–409
hydrothermal upgrading (HTU) process 409–410
  conversion efficiency for 410
hydrothermolysis 186
hydrotreated vegetable oil (HVO) 161–162, 165

Imperial Chemical Industries (ICI) 393
Indian gobar gas plant 226
indirect liquefaction process 391–402
  advantages of 411
  China Lake Process 399–402, *400*
  disadvantages of 411
  FT technology 396–399, *397*
  methanol production from synthesis gas 392–394, *394*
  mobil process 394–396
inert gas 47, 75, 141, 246, 277, 282, 286, 370, 447
Intergovernmental Panel on Climate Change (IPCC) 463
internal rate of return (IRR) 162, 163, 168
International Rice Research Institute (IRRI) 428
invertase 180
ionic liquids 186, 188
IPCC (Intergovernmental Panel on Climate Change) 463
IPCC taskforce on National Greenhouse Gas Inventories 464

IRRI (International Rice Research Institute) 428
IRRI-IPA-QALAN 429

Jatropha seed 139, 140, **141,** 164
John Deere test engine 159, *159*
sample specifications for 190, **191**

ketones 56, 249, 250, 301, 398
kinematic viscosity, of biodiesel 154–156
calculation 155–156
kitchen performance test 430
Kjeldahl method 111
KOH solutions 143, 144
Kyoto Protocol 463

laboratory, analysis, items in 105–106
laboratory analytical procedures (LAPs) 101, 104
determination of extractives 109–110
list of 104, **105**
percent solids determination 108–109
preparation of samples for compositional analysis *102,* 106, 107–108
LabVIEW software 341
landfill gas 3, 5, 59, 378, 379
LAPs *see* laboratory analytical procedures (LAPs)
Laurence Berkeley Laboratory (LBL) process 404–407, *405*
bio-oil product of 405
temperature for 404–405
LBL process *see* Laurence Berkeley Laboratory (LBL) process
LCA *see* life cycle analysis (LCA)
LCV *see* low calorific value (LCV)
lecithin 142
Lemma (duckweed) 11
length, units of 509
Leucaena leucocephala 21–22
life cycle analysis (LCA) 458
biofuels
dispensing of 462–463
distribution and storage 462
end use of 463–466
production of 462
for conversion of coal into electricity *459,* 459–460, *460*
for corn ethanol 463, *464*
feedstock production 461
feedstock transport 461
software/programs for 467–469
GREET/GREET1 model 467–469, *468,* **469**
well-to-wheel approach 461–467
lignin 103, 104, 118, 134–135, 184, 248–249
solubilization 187
lipids 136, 161, 222, 249, 250
liquefaction, biomass
direct 402–407
advantages of 410
Bureau of Mines 402–404
disadvantages of 411
Laurence Berkeley Laboratory (LBL) process 404–407
Pittsburgh Energy Research Center 402–404
HTU process 409–410
hydrothermal 406, *408,* 408–409
indirect 391–402
advantages of 411
China Lake Process 399–402, *400*
disadvantages of 411
FT technology 396–399, *397*
methanol production from synthesis gas 392–394, *394*
mobil process 394–396
overview 390
liquid biofuels *see* bio-oil
liquid condensibles 250–251
liquid hot water treatment 186
LLC (Roscoe, IL) 64
longer chain fatty acids 137
Lo Trau Rice Hull Stove 429
low calorific value (LCV) 321
gas, composition of 341, **341**
low-efficiency cyclones 373–376, *374*
low-temperature FT process (LTFT) 396, 398
lyophilization 107

Malvern Mastersizer 76
manure, as source of biomass 17–19, **18,** 58, 216–217, 443
combustion of 443
production 230, **230**
mass balances, microalgae 288–290
mass, units of 510–511
MC *see* moisture content (MC)
mechanical energy 60–62
mechanical power
generation
as biomass combustion application 445
gasifier and 336–337
medium-chain fatty acids 137
methane
combustion of 88
conversion of acetic acid into 211
production of 43–45
yield, woody biomass types 230, **231**
Methanobacterium formicium 221
Methanobacterium propionicium 221
methanogenesis 250, 252–257
methanol
production, from synthesis gas 392, 394, *394*
yields 392
methanol-to-gasoline (MTG) process 395

methyl tertiary butyl ester (MTBE) 474
microalgae
energy balances 290–295
mass balances 288–290
pyrolysis of 287–295
microbes 42
and biogas reduction 221–222, *222*
microwave pyrolyzers 286
miscanthus 25
MixAlco™ process 55, 187
Mobil Corporation 394, 395
mobil process 394–396
moisture content (MC) 82, 83, 86–88, 93
calculation of 6–7
proximate analyses and 79, 80
of torrefied biomass 255–256
molecular sieves 179
monosaccharides 175; *see also* sugars
Mossgas/PetroSA plant 397, 398
MSW *see* municipal solid waste (MSW)
MTBE (methyl tertiary butyl ester) 474
multieffect distillation 179
municipal sludge 6
municipal solid waste (MSW) 3, 5–6, 8, 19–20, 221, **221,** 423, 425–426

Nannochloropsisoculata
compositional analysis of 101
total composition of 122, **122**
napier grass 25, 184
NASS (National Agricultural Statistics Service) 135
National Agricultural Statistics Service (NASS) 8, 135
National Renewable Energy Laboratory (NREL) 101, 104, 106, 111, 113, 116, 122, 186
laboratory analytical procedures (*see* laboratory analytical procedures (LAPs))
net energy balance (NEB) 480–481, **481**
net energy ratio (NER) 478–479, *479*
net present value (NPV) 162, 163, 168
neutralization 143–144
New Hope Power Partnership (power plant) 335
nitrogen, content of 246, 265

O/C ratio *see* oxygen to carbon (O/C) ratio
ODW *see* oven dry weight (ODW)
oil crops, as source of biomass 13–17, **15, 15–16,** 23–24
examples of 24
oils
crude 131, 141
density 139
extraction processes 139–141
fatty acid composition of 162, **164**
production, in United States **160,** 160–161
properties 139, **140**
raw 136
refining processes 141–150
bleaching 145–146
criteria for 148
degumming 142–143
deodorization 146–148
dewaxing 144–145
neutralization 143–144
stages 146, *147*
soybean 13–15
transesterification process 40, 151–152
vegetable 135–137
yield, calculation 140, **141**
Oklahoma State University (OSU) 56–57, *57*
1D1D cyclone 373
1D2D cyclone 374, 375, **375**
1D3D secondary cyclone 376, **376**
organic acids 44, 56, 187, 209–211, 222, 224, 296, 298
organosolv 186, 187
oven dry weight (ODW) 109
of extractives-free sample 112
percent protein of extractives-free 111–112
overall product yield 295, *295,* 305, *305*
oxidation region 324
oxidative delignification 186, 187
oxygen to carbon (O/C) ratio 261, 264
ozone 187, 188

Pacific Northwest National Laboratory (PNNL) 406, 408
Panicum virgatum (switchgrass) 25
particles
density of 75
size 76–77, **77**
particle size distribution (PSD) 73, 76–77, 255
payback period (PBP) 162, 163, 168
PCA *see* plate count agar (PCA)
pelletability *257,* 257–258
Pellet Fuels Institute (PFI) 432
pellet stoves 432
Penicillium funiculosum 194
Pensky–Martens Flash Point Analyzer 153
pentane 178
percent acid-insoluble lignin (%AIL) 113, 114–115
percent acid-insoluble residue (%AIR) 113
determination of 113–114
percent acid-soluble lignin (%ASL) 115–118
calculation, on basis of extractives-free biomass 116
percent extractives, determination of 109–110
percent total solids
determination of 108–109
of extractives-free biomass 110–111

PERC process *see* Pittsburgh Energy Research Center (PERC) process
PetroSA/Mossgas plant 397, 398
PFI Standards Program 432
photosynthesis 1
pilot-scale
  fluidized bed gasifier (FBG) 359, **359**
  production of ethanol *189,* 189–192, *190*
Pittsburgh Energy Research Center (PERC) process 402–404, *403*
plants
  aquatic 11, 12, 25–27, **26**
  biomass 1
  commercial cellulosic ethanol 198
  commercial plants 182
  for energy production 21
  hydrocarbon 23–24
plate count agar (PCA) 192
plenum 50
polysaccharides 175; *see also* sugars
polyunsaturated fatty acids 137
Pongamia 164
power *see* electric power
pressure drop, cyclones 377
pretreatment, biomass 123–124, **124,** 185–189
  advantages/disadvantages of torrefaction as 265–266
  chemical 186–188
  combination of 189
  enzymatic 188, 195–197
  physical 185–186
producer gas *see* synthesis gas
protein
  analysis of 111–112
    percent AIL and 115
  as biomass content 104
proximate analyses 79–83
  and moisture content measurement 80
  and VCM measurement 80–81
PSD *see* particle size distribution (PSD)
Public Utilities Regulatory Policies Act of 1978 (PURPA) 60
PURPA *see* Public Utilities Regulatory Policies Act of 1978 (PURPA)
pycnometer 75
pyrolysis 45, 47–50, 277–310
  ablative 286
  of corn stover 304–306
  cotton gin trash 300
  economics 474–475
  energy balance in 278–280
  fast/flash 280, 284–285
    China Lake Process *vs.* 399–400
  flash vacuum 285–286
  gasification *vs.* 322
  of microalgae **287,** 287–295
    energy balances 290–295
    mass balances 288–290
  overview 277–282
  reactor, components of 277–278
  schematic representation of 278, *278*
  slow/conventional 280, 283
  sorghum biomass 307–309
  switchgrass 300–304, *301,* **302**

raw biomass 255–257
  carbon content 262–263
  hydrogen content 263
  Van Krevelen diagram 261–262
raw oil 40, 134, 136
RBD *see* refined, bleached, and deodorized (RBD)
refined, bleached, and deodorized (RBD) 40, 141, 146, 148, 151
refining processes, oils 141–150
  bleaching 145–146
  criteria for 148
  degumming 142–143
  deodorization 146–148
  dewaxing 144–145
  neutralization 143–144
renewable energy 1, 3, 7, 8, 89, 473, 481, 483, 484, 486
Renewable Fuels Standard (RFS) 133
residence time
  auger reactor, estimation of 283
  estimation of 332
  in fluidized bed reactor 282
reynolds number (Re) 360
RFS *see* Renewable Fuels Standard (RFS)

SAA technique *see* soaking in aqueous ammonia (SAA) technique
saccharifying enzyme 180, 191
Saccharomyces cerevisiae 12, 42, 173, 176, 192, 199
Saccharomyces uvarum 176
SAE *see* Society for Automotive Engineers (SAE)
SAE J 1349 standard (Engine Power Test Code—Spark Ignition and Compression Ignition—Net Power Rating) 159
Sasol plant 397, 398
saturated fatty acids 137
scanning electron microscope (SEM) 255
screw conveyor 364, *365*
  capacity of 366–367
second-generation digesters 227–229
second law of efficiency 37, 61
sediment, as biodiesel parameter 156–157
seeds, oil extraction from 139–141
SEM *see* scanning electron microscope (SEM)
septic tanks 221, **221**
Shell Middle Distillate Synthesis (Shell SMDS) 398

short-chain fatty acids 162
SI *see* Systems International (SI)
sieve screen, sizes of 76, **77**
slagging/fouling indices
  with agricultural biomass 436–441
  alkali index 438–440, **439**
  base-to-acid ratio 440–441
  bed agglomeration index 441
  for coal 437–438
  factor 88–91, *89, 90*
slow/conventional pyrolysis 280, 283
slugging 356
SME (soybean methyl ester) 136
SMR (steam-methane reforming) 392
soaking in aqueous ammonia (SAA) technique 187
solid products, from torrefaction process 250
solids retention time (SRT) 213, *213*
solvent extraction 141, 178–179, 309
solvolysis 186
sorghum biomass pyrolysis **307,** 307–309
sorghum grain, conversion into ethanol 183–184
sorghum stalks, chemical formula for **321,** 321–322
soybean methyl ester (SME) 136, 158, 159
soybean oil 13–15
  production of 135, **136**
sphericity 255
  defined 358–359
spreader stoker system 424–426
  schematic representation *424*
SRS *see* sugar recovery standards (SRS)
SRT *see* solids retention time (SRT)
standard atmosphere and pressure (STP) 290
Standard Specification for Biodiesel Fuel Blend Stock [B100] for Middle Distillate Fuels 188
standard temperature and pressure (STP) 52, 58
Standard Test Method for Bulk Density of Densified Particulate Biomass Fuels 75
Standard Test Method for Cloud Point of Petroleum Products 156
Standard Test Method for Density and Relative Density [Specific Gravity] of Liquids by Bingham Pycnometer 155
Standard Test Method for Determination of Acid-Insoluble Residue in Biomass 113
Standard Test Method for Determination of Carbohydrates in Biomass by High-Performance Liquid Chromatography 119
Standard Test Method for Determination of Free and Total Glycerin in B100 Biodiesel Methyl Ester by Gas Chromatography 157
Standard Test Method for Flash Point by Pensky–Martens Closed Cup Tester 153–154
Standard Test Method for Fusibility of Coal and Coke Ash 438
Standard Test Method for Gross Calorific Value of Refuse- Derived Fuel by Bomb Calorimeter 86
Standard Test Method for Water and Sediment in Middle Distillate Fuels by Centrifuge 157
starchy crops; *see also* sugars
  cultivation of 181
  ethanol production from 180–184
  as source of biomass 12–14, **13, 14,** 23, **23**
starved air 425
steam-methane reforming (SMR) 392
steep 183
stoichiometry 88, 421
storage, of biomass 256
STP *see* standard atmosphere and pressure (STP)
strong acid hydrolysis 187
sucrose 42, 175, 180, 192
sugar beets 176, 473
sugarcane 12, 23, 25, 39, 42, 104, 176, 184, 198, 199, 249, 285, 335, 445, 473, 480, 489
sugar recovery standards (SRS) 119
sugars
  advantages of 176
  analysis, HPLC system for 119, **119**
  classification 175
  conversion into ethanol 42–43
  crops (*see also specific crops*)
    as source of biomass 12–14, **13,** 13–17, **14,** 23, **23,** 23–24
  disadvantages of 176
  ethanol production from 173
  formula for 175–176
  percentage of, calculation 120, **121**
sulfur 265
sulfuric acid 115, 183
superficial velocity, calculation of 332–333
surface area 255
sustainability, biofuels
  measurement 477–481
    NEB method 480–481, **481**
    NER method 478–479, *479*
sweet sorghum 176, 196, *196*
  biomass 196, *196*
  ethanol production from 176–178, *177*
  juice 192, 193, *193*
switchgrass *(Panicum virgatum)* 25
switchgrass pyrolysis 300–304, *301,* **302**
Syngenta 393
synthesis gas 47, 50, 277, 295, 321
  direct use, and heat/steam production 378
  fermentation 56
  heating value of 391

liquid fuel production from 337
methanol production from 392, 394, *394*
production of 56, 63–64
usage of 63–64, 333
yield 300
system flow alternatives 74, *74*
Systems International (SI) 27
and heat capacity 78

Taiwanese Rectangular Digester Design 226–227
TAMU *see* Texas A&M University (TAMU)
tank-to-wheel (TTW) activities 468
tar 50
TDR *see* turndown ratio (TDR)
TDS (total dissolved solids) 220
temperature
gasification 342
for LBL process 404–405
terminal velocity, FBG 362
tertiary azeotropic distillation 178
Texas A&M University (TAMU) 476
THCs *see* total hydrocarbons (THCs)
thermal conductivity 78
thermal conversion process 45–55, 280
and biological conversions 56–57
biomass properties for 73–94
heating value analysis 86–91
proximate analyses 79–83
ultimate analysis 83–85
combustion (*see* combustion)
gasification 45, 50–53
pyrolysis 45, 47–50
solid/liquid/gaseous products for, yields of 280, **280**
torrefaction 45, 46–47
types of 45–46, *46*
thermal gravimetric analysis, biomass 259–260
thermodynamics efficiency 37, 61
thermophilic organisms 253
throughput, gasifiers 338
TOC (total organic carbon) 219
topping cycle 234
torrefaction 45, 46–47, 246–267, *247*
advantages of 47
as pretreatment process 265–266
bio-physico-chemical changes 247–249, *248*, **249**
chemical composition of 261–265
calorific value 263–264, *264*
carbon content 262–263
degree of carbonization 264
H/C ratio 264–265
hydrogen content 263
nitrogen 265
sulfur 265
disadvantages of
as pretreatment process 266
products 249–251, *250*
energy densification 246–247
gaseous 251
liquid condensibles 250–251
solid 250
total dissolved solids (TDS) 220
total hydrocarbons (THCs) 300
total organic carbon (TOC) 219
total solids (TS) 219–221
total suspended solids (TSS) 220
transesterification process 40, *151*, 151–152, 160
trees; *see also specific types*
fast-growing 21–22
Trichoderma reesi 42, 173, 194, 199
triglyceride 136, *137*
TS *see* total solids (TS)
TTW activities *see* tank-to-wheel (TTW) activities
turndown ratio (TDR) 337–339, 355, 372, 380
fixed-bed gasifier 338–339
2D2D cyclone 373
Typha latifolia (cattails) 11

UASB *see* upflow anaerobic sludge blanket (UASB)
ultimate analysis 83–85, 421
of common biomass 84, **85**
data for selected solid and biomass fuels **87**
United States
agricultural industry in 4–6
commercial cellulosic ethanol plants 198
energy consumption 1–4, 10
power generation case studies in 434–436, **435**
vegetable oil and animal fats production 135, **136**
University of Minnesota (UM) design 286
unsaturated fatty acids 137
updraft gasifiers 326–328, *327*
upflow anaerobic sludge blanket (UASB) 229, 230, 235
USDA (US Department of Agriculture) 459
US Department of Agriculture (USDA) 459
US Department of Energy (USDOE) 11, 25, 27, 73, 78, 101, 106, 286, 310, 445, 446
biomass conversion into power 460
reports 459
US Department of Labor 154
US Department of Transportation 154
US Environmental Protection Agency (USEPA) 19, 432, 466–467
UV-VIS spectroscopy 115, 116, 125

vacuum distillation 178
Vanguard Refractory Grade Aggregate 370
Van Krevelen diagram 264–265
raw biomass 261–262

vapor recompression distillation 178
VCM *see* volatile combustible matter (VCM); volatile organic matter content (VCM)
vegetable oil; *see also specific types*
  characteristics 136–137
  fatty acids in 137, **138**
  production of 135, **136**
  refining calculations **148,** 148–150, *149*
viscosity 139, **140**
  kinematic 154–156
voidage 361, **361**
volatile combustible matter (VCM) 79, 80–81, *81,* 82, 83, 88, 93
volatile organic matter content (VCM) 250
volatile suspended solids (VSS) 261–262
volumetric flow rates, FBG 363–364
volume, units of 510
VSS (volatile suspended solids) 261–262

wastes
  agriculture and 4–6
  compositional analysis of 249, **249**
  parameters, in anaerobic digestion process 217–221
wastewater treatment plants 221
water
  as biodiesel parameter 156–157
  and ethanol 174, 175, 178–180
  extraction 109
water-boiling test 430, 431
water hyacinth *(Eichornia crassipes)* 11, 25, 26
weak acid hydrolysis 187
well-to-tank (WTT) activities 467
well-to-wheel approach, LCA 458, 461–467
Westfalia biodiesel plant design 161, *162*
wet milling process 182, *182*
wheat grass 25
wildland–urban interface (WUI) areas 434
Wiley Mill 106, 107
wood energy 20
woody biomass 76
  for power generation 434–435, **435**
  types, methane yield of 230, **231**
WTT (well-to-tank) activities 467
WUI (wildland-urban interface) areas 434

0.3048 fluidized bed gasifier
  operation of 364–372
ZSM-5, zeolite catalyst 394–395, 398, 412
Zymomonas mobilis 176

For Product Safety Concerns and Information please contact our EU representative GPSR@taylorandfrancis.com
Taylor & Francis Verlag GmbH, Kaufingerstraße 24, 80331 München, Germany

www.ingramcontent.com/pod-product-compliance
Ingram Content Group UK Ltd.
Pitfield, Milton Keynes, MK11 3LW, UK
UKHW022212190926
475702UK00003B/315
* 9 7 8 1 0 3 2 2 7 8 3 3 9 *